THE EVOLUTION
OF THE IGNEOUS ROCKS

Fiftieth Anniversary Perspectives

H. S. YODER, JR., EDITOR

1979

PRINCETON UNIVERSITY PRESS

PRINCETON, NEW JERSEY

Dedicated to the memory of

NORMAN L. BOWEN

CONTENTS

PREFACE

The book *The Evolution of the Igneous Rocks* by N. L. Bowen taught us that the principles of physical chemistry can be applied successfully to petrological processes. The major petrogenetic questions discussed by Bowen in his lectures at Princeton University in the spring of 1927 persist today. Only the answers appear to change with time. The purpose of the present volume is to provide a new view of those questions in the light of almost fifty years of accumulated observations using the principles so clearly set out by Bowen.

Each of these questions is herein reexamined by a student recognized for his dedicated study of the specific problem, retaining the underlying philosophy of Bowen's text published in 1928. That philosophy, in brief, consists of several parts. The first, and probably most important, involves recognition of a set of observations in the field that appear to be related. Second, simplification of those relations into a set of experiments analogous, in terms of composition, temperature, pressure, and other relevant parameters, to those believed to have existed in nature. Third, execution of those experiments in an unambiguous manner that can be reproduced by others. Fourth, interpretation of the results and application of the principles to specific natural occurences. Fifth, reexamination of the field relations and testing in the field the principles deduced from experiment. The process is reiterated until a satisfactory interpretation results that will account for the field observations.

Bowen's philosophy has stood the test of time. It is the wish of the authors of these chapters, paralleling in this volume those of Bowen's book, to commemorate that philosophy. The reader will discover that the observations of the past fifty years have resulted in new interpretations, some distinctly at variance with those of Bowen. It is believed, however, that adherence to Bowen's philosophy will eventually lead to the most satisfactory interpretation of the "facts" as they are perceived by the coming generations of students.

H. S. Yoder, Jr.

Geophysical Laboratory
Carnegie Institution of Washington
Washington, D.C.

THE EVOLUTION
OF THE IGNEOUS ROCKS

Fiftieth Anniversary Perspectives

Chapter 1

THE PROBLEM OF THE DIVERSITY
OF IGNEOUS ROCKS

G. MALCOLM BROWN

Department of Geological Sciences, University of Durham, Durham, England

The science of igneous petrogenesis is a study of the origin and evolution of rocks that have formed through the generation of magmas by melting processes, and the cooling of those magmas from a liquid, or liquid and crystalline state, into a glassy to crystalline state. Field and textural observations do not constitute sufficient evidence for defining a rock as igneous in origin (Barth, 1962, p. 51) but when allied with experimental studies on silicate melts similar in composition to the rock in question, the origin of the rock can be verified. Experimental studies, primarily by N. L. Bowen, formed the only firm basis from which the science of igneous petrogenesis could develop. If there were no diversity of igneous rocks, there would be no science of igneous petrogenesis. Large-volume flows of apparently homogeneous basalt are problematical, but only when viewed in relation to the more common type of diverse flow. The "problem" of the diversity of igneous rocks is more a matter of "problems" that will not be solved until all the complex processes of igneous petrogenesis have been understood. Bowen was as aware of that, in 1928, as one is now, and selected the topic of diversity for the opening chapter of his book. He stressed the importance of igneous rock associations and the need to explain differences in mineral assemblages and rock compositions, both between and within those associations, according to definite physicochemical processes. One should first consider, briefly, what type of diversity led Bowen to make the investigations that were described in the rest of his book, and then view the subject in the light of contemporary knowledge.

© 1979 by Princeton University Press
The Evolution of the Igneous Rocks: Fiftieth Anniversary Perspectives
0-691-08223-5/79/0003-12$00.60/0 (cloth)
0-691-08224-3/79/0003-12$00.60/0 (paperback)
For copying information, see copyright page

Bowen endorsed the concept of "petrographic provinces" (Judd, 1886), but pointed to the importance of time as well as place in such groupings and introduced "rock associations" as a preferable name. Such an association of rocks, sharing certain properties and connected by a "community of origin," was said to have been derived from a single original magma. He favored a basaltic magma in the parental role but did not think it was requisite in all cases. The question of whether more than one type of parental magma was involved did not tempt Bowen to provide an opinion. A division into alkaline and subalkaline types he thought useful, yet he did not quibble with R. A. Daly's view that there were no essential differences between the basaltic magmas of the various associations. At that time, Bowen was much more concerned with diversity produced within, rather than between rock associations. He stressed the continuous nature of variations in composition within a rock association, such that the division of such a series into specific members could only be arbitrary. Then followed his major discussion on the processes likely to be responsible for diversity within series, i.e., differentiation processes. He discounted compositional gradients in the liquid phase (temperature or gravity controlled), except insofar as pressure gradients could affect the concentration of volatile components. His support was entirely for processes involving separation of distinct phases either by gaseous transfer, liquid immiscibility, or crystallization. He went on to demonstrate the dominant role of fractional crystallization (associated sometimes with the additional effects of assimilation).

Petrologists would agree that there have been few major changes in emphasis since Bowen's exposition, regarding the general concept of igneous rock associations and the main processes responsible for variation within the associations. They would emphasize that their main concern is still with the diversity of what are now called *primary*, *parental*, and *derivative* magmas and their rock products, but particularly with the definition of what constitutes a primary magma. Since 1928 the subject has been expanded enormously through access to information on ocean-floor igneous rocks, and the theory of global plate tectonics has required a reappraisal of what are now called *petrogenic provinces*. The vast amounts of experimental data on multicomponent systems, referred to in subsequent chapters, have led to an expansion of knowledge about the deep regions of primary magma generation in the Earth's upper mantle, and about the processes of crystal fractionation, liquid immiscibility, and the role of volatile components in further promoting diversity in the patterns of magmatic evolution. Further advances relate to the application of thermodynamics to an understanding of the liquid state, rates of crystallization and nucleation, solid-liquid-gas equilibrium relations, and the effects of temperature and pressure on mineral equilibria. Extensive studies on the distribution of chemical elements and certain isotopes, and on element partitioning between crystalline and liquid phases, have been accompanied by consideration of the relative importance of fractional crystallization, fractional melting, and equilibrium partial melting in the derivation of particular natural silicate liquids (e.g., Gast, 1968; Shaw, 1970).

In order to recognize diversity in any population it is necessary first to define the criteria for distinctions and the limits of tolerance beyond which uniformity becomes diversity. Igneous rocks as a whole are uniform according to the criteria and broad limitations defined in the opening sentence of this chapter, and thus diverge from metamorphic and sedimentary rocks. The introduction of sets of criteria such as liquidus temperature, silica content, or locality within a specified area of the Earth's crust would lead to the recognition of diversity likely to have petrogenic significance, whereas criteria such as rock grainsize or precise age of crystallization would have less significance. As more is learned about field associations and the physicochemical processes governing the evolution of magmatic rocks, so one is better able to discard irrelevant distinguishing criteria and introduce more significant ones. Early geologists recognized a significance in the diverse chemical compositions of basalt, trachyte, and rhyolite. Later, chemical diversity even among basalts was recognized, initially only as tholeiitic and alkalic basalt magma types and chiefly according to SiO_2 and $Na_2O + K_2O$ contents (Bailey et al., 1924; Kennedy, 1933) but subsequently, according to quartz-orthopyroxene-olivine-nepheline normative ratios, as quartz tholeiite, tholeiite, olivine tholeiite, olivine basalt, and alkalic basalt (Yoder and Tilley, 1962). A further refinement in basalt classification has developed from the use of certain trace element and isotopic data. In the early days of ocean-floor exploration, the few dredged basalt samples from ocean-ridge regions were classed as uniform, olivine tholeiites rich in alumina (Engel, Engel, and Havens, 1965). They were called oceanic (or abyssal) tholeiites, but subsequent studies have shown that oceanic tholeiites are of two types (e.g., Schilling, Anderson, and Vogt, 1976), those from the ridges being more depleted in K, light rare-earth elements, and radiogenic Sr and Pb isotopes than those from shallow oceanic platforms such as the Azores, Hawaii, Iceland, and the Galapagos islands. Since Schilling (1975) has discovered chemical gradations between the two types along the axis of the Mid-Atlantic Ridge, one must recognize diversity even among oceanic (abyssal) tholeiites. The story of andesite is similar in that as more sophisticated criteria were applied, such rocks were subdivided into hawaiite of alkalic, icelandite of tholeiitic, and orogenic andesite of calc-alkalic affinities. It was then customary to assume that island-arc andesites were all of calc-alkalic type and similar in composition and source, but along the axis of the Lesser Antilles arc, three chemically distinct types of andesite have been erupted over the past 3 million years, which show appreciable diversity in, for example, Ni, Cr, and K contents (Brown *et al.*, 1977). The basalts of the Moon provide another example of how increasing diversity was recognized as more samples were collected and more sophisticated methods were applied during the mission program (summary in Taylor, 1975). Similar developments have taken place in the recognition of diversity among acidic and ultramafic rock suites.

Lest one falls into the trap of giving almost every rock a different name according to its major and trace element composition and locality, limits need to be drawn and the assumption made that in time, those limits will be redefined. One can only apply limits according to the *present* state of knowledge on the processes involved in igneous-rock evolution.

If the Earth has been formed by the homogeneous accretion of solar-nebula matter, or was homogenized by total melting of a heterogeneous accretion of such matter, then the starting mixture would have been uniform and probably close to a chondritic composition. Diversity began with the formation of a core and mantle. The two earliest Earth melts would thus have been metallic and ultramafic, so the first differentiation process was probably due to the immiscibility of metal-sulphide and silicate liquids. One still does not know how and when the Earth's primitive crust was formed, or what its composition was. It could perhaps have been andesitic (e.g., by sequential accretion to give a veneer of low-temperature, volatile-rich condensates; Larimer and Anders, 1967), ultramafic (by uniform crystallization from the surface downward), or a complex assemblage such as the anorthosites, norites, troctolites, and gabbros of the Moon's crust. On the Moon, the major process of fractional crystallization began shortly after formation, as evidenced by crystallization ages of 4.5×10^9 years for dunite and troctolite fragments (Papanastassiou and Wasserburg, 1976). On Earth, the only unmetamorphosed basic or ultrabasic igneous rocks that one can examine in any detail are not part of a primitive crust but are younger rocks related to the melting of a source in the Earth's upper mantle. Although some can be categorized as the products of "primary" magmas, no such magmas can be viewed as being chronologically "primitive." The younger the magma, the less likely it is to have been derived from mantle material unmodified by earlier melting events.

The starting composition, in a consideration of diversity from a common parentage, is best referred to as a *primary magma*. The source of primary magmas is universally accepted as the upper mantle, but because the source is not likely to be homogeneous in composition and because the pressure-temperature conditions of magma generation will vary, a range in the composition of primary magmas is to be expected. If the magma is truly primary, its composition will reflect the conditions of crystal-liquid equilibrium at the stage when the melt was initially removed from the residual crystalline phases at source. Source depths will vary according to the tectonic environment, as shown by various studies summarized by Yoder (1976, Ch. 3). These suggest that the source depths for most primary magmas are probably about 100 ± 50 km. It would seem unlikely that such magmas (e.g., basaltic) could reach high levels in the crust without becoming modified in composition by lower-pressure crystal fractionation. If this has occurred, then the basalt compositions will not reflect the conditions of primary generation. Such non-primary basalt magmas would, however, possess compositions such that on cooling they can give rise to derivative magmas, and in that sense they may be called *parental magmas*. Although, in the stricter sense, such basalts are derivative magmas, their role as parents to a wide range of lower-temperature derivative magmas within the crustal environment is far more significant.

If extensive olivine fractionation has occurred prior to basaltic magma eruption, then the primary magmas would have been picritic rather than basaltic in composition (O'Hara, 1968). Hence the only candidates for primary-magma products within the crust would be rocks such as komatiites (Viljoen and Viljoen, 1969). This view is not shared by Ringwood (1975), who advances arguments, chiefly based on nickel and chromium distribution in basalts, against extensive olivine fractionation during the ascent of basaltic magmas within the mantle. An extensive discussion of this critical problem is provided by Yoder (1976, Ch. 8), mainly by reference to melting studies on eclogites, two olivine tholeiites, and a nepheline-normative olivine basanite. In no case was olivine produced in the high-pressure mineral assemblages. This and other experimental evidence is advanced to favor the hypothesis that the compositions of basaltic magmas have been modified during ascent from their mantle source. Hence they are derivatives of the primary magmas. Yoder excludes from this category those alkalic basalts that have transported mantle nodules to the surface; such rapid ascent would imply the inability to fractionate smaller olivine crystals and hence imply a closer approach to a primary composition for the basalt host. He does not include reaction between nodules and the host basalts which could, during ascent to lower *P-T* conditions, have modified the primary basalt compositions slightly in certain cases.

The compositions of basalt magmas are such that they possess high liquidus temperatures relative to most other crustal magmas, and hence cooling and differentiation will give rise to a range of lower-temperature derivative magmas (e.g., Tilley, Yoder, and Schairer, 1963, 1965). The experimental evidence will be discussed in subsequent chapters. Field and analytical data indicate spatial and chemical relationships that support the experimental data, such as the basaltic chilled margins (and basaltic pore material in the basal layers) of strongly fractionated intrusions, and the relative abundance of those chemical elements, in basalts, that are depleted by crystal-liquid fractionation.

There is compositional diversity among basalt magmas, much of the variation being gradational. The distinctions are often subtle, and are more easily recognized by reference to the composition of the derivative magmas of each apparently basaltic parent. Consideration of such "igneous suites" (Harker, 1909), "magma series" (Bailey et al., 1924), or "rock associations" (Bowen, 1928) often shows that the derivative magmatic offspring possess chemical characteristics inherited from the parent magma (called *consanguinity* by Iddings, 1892). A subtle compositional difference between two basalts may imply a difference in the composition of the mantle source material (including variation in volatiles content), varying amounts of partial melting, variation in the degree of equilibrium versus fractional melting, or a different history of fractionation and reaction during ascent from the mantle source. A combination of several of these effects seems more likely than a single process in the generation of basalt-magma diversity. Experimental studies help in distinguishing between liquid-solid equilibria at varying pressures, but more information on the partitioning of trace elements at varying pressures would also be of value.

Tholeiitic and alkalic basalt magmas are generally recognized as major parental types. It is still difficult to decide whether specific olivine tholeiite, (alkali) olivine basalt, quartz tholeiite, and high-alumina basalt magmas are to be viewed as having a parental rather than derivative status, but olivine tholeiite in oceanic-ridge regions and the type of high-alumina tholeiitic basalt found in several island-arc regions appear to qualify. Ultrapotassic and nephelinitic magmas are difficult to attribute to a derivative origin in regions where alkalic basalts are rare or absent. Special conditions that gave rise to high thermal activity in localized belts during the Precambrian are advocated to explain the genesis of Archaean komatiite magmas (e.g., Nesbitt and Sun, 1976) and mid-Proterozoic, labradoritic anorthosite magmas (Bridgewater and Windley, 1973). This type of association of magma types with chronology, applicable also to the abundance of tholeiites and scarcity of basanites among Precambrian volcanics, needs consideration in petrogenic modeling. It may well indicate long-term changes in the thermal regimes and compositions at the mantle sources.

On a global scale, it is no longer logical to draw a sharp distinction between oceanic and continental magma provinces unless related to a particular period in the configuration of lithospheric plates. Subcontinental mantle will become suboceanic mantle through the effects of sea-floor spreading. The abyssal oceanic-ridge tholeiites, however, and the alkalic basalts of ocean-floor platforms and oceanic islands, belong to a definable petrogenic province where the environment is established at a particular period of tectonic evolution. Ultrapotassic and nephelinitic magmas are most abundantly associated with kimberlites and carbonatites where rifting of thick continental lithosphere has disturbed an environment in which continental stability has persisted for a long time (e.g., East Africa), but oceanic islands may contain potassic rocks (Tristan da Cunha) and nephelinitic rocks (Hawaii). Some continental tholeiites have probably suffered from contamination by sialic crust, but those generated during ancient lithospheric rifting need not have been affected, so the observed spectrum of compositions, in terms of dispersed elements and $^{87}Sr/^{86}Sr$, is to be expected. Alkalic basalts may occur in a similar continental environment (such as the Scottish Tertiary Province). The basalts of island arcs may range from nepheline-normative basanites to quartz-normative tholeiites within a short period of arc evolution. Arcs such as the Lesser Antilles, formed along a north-south zone between plates of oceanic lithosphere to the west and east, are not to be viewed as the products of a continental environment.

The remaining major igneous rock assemblages are the ultramafic bodies of orogenic belts and the (dominantly) granitic plutons of continental shields and folded geosynclinal belts. The ultramafic assemblages (like those occurring as xenoliths in lavas) are, probably, mostly derived directly from the upper mantle. Orogeny may cause emplacement of fragments of either subcontinental or suboceanic mantle, as "fertile" lherzolites with compositions that could yield basaltic partial melts, or as "barren" harzburgites, depleted in elements that enter early partial melts, respectively (Nicolas and Jackson, 1972). Some ultramafic material may represent mafic cumulates of the ocean-crust ophiolite suite (Jackson, 1971). The granitic melts of orogenic regions are mostly attributable to the melting or partial melting of sialic crust of variable composition and a complex history of elemental and isotopic depletions. Mixing of granitic and more basic melts has also occurred under these conditions.

Most magmas will, on cooling, become *differentiated* (Becker, 1897). Only very rapid flow from the source will inhibit crystal-liquid, liquid-liquid or gas-liquid differentiation. After crystallization begins, as pointed out by Bowen (1928), the liquid composition has changed, and this is still viewed as the most important of the magmatic differentiation processes.

Several mechanisms may then operate to produce *crystal sorting*. The liquid may be squeezed from the crystal interstices (*filter-pressing*) but this is unlikely to occur in non-orogenic environments. In certain circumstances where laminar flow of magma occurs within a conduit, the crystals could be concentrated in the axial region of higher velocity flow. Such *flowage differentiation* (Bhattacharji and Smith, 1964) would be most likely during dike or sill emplacement. The most widespread type of crystal sorting will be through *gravity crystal sorting*. Floating of low density crystals has been described (Sahama, 1960) but most crystal phases have been concentrated by sinking. The classic method of observing the effects of gravity settling of crystals is to examine basic layered intrusions (Wager and Brown, 1968). If the crystals are prevented from complete reaction with the liquid, the residual liquids will show gradual changes in composition as the process continues. This process of *fractional crystallization* has been demonstrated particularly in the case of the Skaergaard layered intrusion (Wager and Deer, 1939; Wager and Brown, 1968). The natural evidence provides strong support for the fractionation process in general, as described by Bowen (1928). The successive residual liquids are equivalent to the "derivative magmas" of igneous petrogenesis.

In the case of the Skaergaard, Bushveld, Palisades sill, and other differentiated tholeiitic intrusions, however, the successive derivative liquids show the strong iron enrichment proposed for basaltic liquids by Fenner (1929). Bowen (1928, p. 110) disagreed strongly with Fenner's views on the subject, but contrasted trends of silica and iron enrichment in calc-alkalic and tholeiitic basalt differentiation, respectively, are now recognized. The importance of variable oxygen fugacities in controlling the trends (Osborn, 1959), although debatable in detail, is a particularly valuable concept.

Intrusive bodies and surface volcanoes are not necessarily separate phenomena, and this point was stressed by reference to the Rhum intrusion (Brown, 1956). It was proposed that a layered intrusion could lie beneath an active volcano and that fractional crystallization in the subsurface magma chamber would give rise to compositional changes in the erupted lavas. Without such a "subtraction reservoir," erupted lavas should reflect their mantle source and would all be basaltic, modified only by possible volatile loss and contamination. The recognition of layered gabbroic xenoliths in Hawaiian lavas (e.g., Jackson, 1968) has lent support to the hypothesis and conversely, the interpretation of the lower oceanic crust as consisting of layered gabbroic and ultrabasic rocks, produced in localized magma reservoirs above which oceanic basalts were erupted (Cann, 1970), has evolved from this concept. The crystal fractions, first referred to as accumulations (Bowen, 1928, p. 166), are nowadays referred to as *igneous cumulates* (Wager, Brown, and Wadsworth, 1960).

The other processes likely to give rise to magma differentiation, referred to by Bowen, are liquid immiscibility and gaseous transfer. Liquid immiscibility has gained support from the abundant evidence of sulphide-silicate melt immiscibility; from the proposed separation of iron-rich and silica-rich melts at the latest stage of crystallization of lunar basalts (Roedder and Weiblen, 1971) and of the Skaergaard intrusion (McBirney, 1975); and from silicate-liquid immiscibility with sodium-carbonate liquids (Koster van Groos and Wyllie, 1966). But the process is still believed to be minimal in importance for explaining the greater part of igneous-rock diversity. The separation of a volatile fluid-phase could result in the transfer of the more mobile elements either within magma bodies or to the wall rocks.

A few natural examples of mixing of contrasted magmas have been documented (e.g., Larsen, Irving, and Gonyer, 1938; Wager *et al.*, 1965), and a review of the problems of mixing and immiscibility of basaltic and rhyolitic magmas is given by Yoder (1973). If the mixed components can be partly identified, such as the relict phenocrysts of porphyritic felsite and porphyritic ferrodiorite in the Skye marscoite hybrid (Wager *et al.*, 1965), then the process could be contrasted with differentiation by use of the term "integration." The mechanism of assimilation, involving reaction between magma and invaded wall-rock material, was described by Bowen (1928) and later defined more specifically by Shand (1947, p. 69). Such reaction is expected to accompany most igneous events although, in most cases, the other processes would then operate in the differentiation of the modified magma. Only fractional crystallization could account for the regular, gradational variations observed so frequently in suites of parental plus derivative magmatic products.

References

Bailey, E. B., Clough, C. T., Wright, W. B., Richey, J. E., and G. V. Wilson, Tertiary and Post-Tertiary geology of Mull, Loch Aline and Oban, *Mem. Geol. Surv. Scot.*, 1924.

Barth, T. F. W., *Theoretical Petrology*, 2d ed., John Wiley and Sons, New York and London, 416 pp., 1962.

Becker, G. F., Some queries on rock differentiation, *Am. J. Sci.*, 3, 21–40, 1897.

Bhattacharji, S., and C. H. Smith, Flowage differentiation, *Science*, 145. 150–153, 1964.

Bowen, N. L., *The Evolution of the Igneous Rocks*, Princeton University Press, Princeton, 334 pp., 1928.

Bridgewater, D., and B. F. Windley, Anorthosites, post-orogenic granites, acid volcanic rocks and crustal development in the North Atlantic Shield during the mid-Proterozoic, *Spec. Publ. Geol. Soc. S. Africa, 3*, 307–317, 1973.

Brown, G. M., The layered ultrabasic rocks of Rhum, Inner Hebrides, *Phil. Trans. Roy. Soc. Lond.*, ser. B, *240*, 1–53, 1956.

Brown, G. M., Holland, J. G., Sigurdsson, H., Tomblin, J. F., and R. J. Arculus, Geochemistry of the Lesser Antilles volcanic island arc, *Geochim. Cosmochim. Acta, 41*, 785–801, 1977.

Cann, J. R., New model for the structure of the ocean crust, *Nature, 226*, 928–930, 1970.

Engel, A. E. J., Engel, C. G., and R. G. Havens, Chemical characteristics of oceanic basalts and the upper mantle, *Bull. Geol. Soc. Am. 76*, 719–734, 1965.

Fenner, C. N., The crystallization of basalts, *Am. J. Sci., 18*, 225–253, 1929.

Gast, P. W., Trace element fractionation and the origin of tholeiitic and alkaline magma types, *Geochim. Cosmochim. Acta, 32*, 1057–1086, 1968.

Harker, A., *The Natural History of Igneous Rocks*, Methuen, London, 384 pp., 1909.

Iddings, J. P., The origin of igneous rocks, *Bull. Phil. Soc. Washington, 12*, 128–130, 1892.

Jackson, E. D., The character of the lower crust and upper mantle beneath the Hawaiian Islands, *Proc. 23rd Int. Geol. Congr. 1*, 131–150, 1968.

———, The origin of ultramafic rocks by cumulus processes, *Fortschr. Mineral., 48*, 128–174, 1971.

Judd, J. W. On the gabbros, dolerites and basalts, of Tertiary age, in Scotland and Ireland, *Q. J. Geol. Soc. Lond., 42*, 49–97, 1886.

Kennedy, W. Q., Trends of differentiation in basaltic magmas, *Am. J. Sci., 25*, 239–256, 1933.

Koster van Groos, A. F., and P. J. Wyllie, Liquid immiscibility in the system Na_2O–Al_2O_3–SiO_2–CO_2 at pressures to 1 kilobar, *Am. J. Sci., 264*, 234–255, 1966.

Larimer, J. W., and E. Anders, Chemical fractionations in meteorites—II: Abundance patterns and their interpretation, *Geochim. Cosmochim. Acta, 31*, 1239–1270, 1967.

Larsen, E. S., Irving, J., and F. A. Gonyer, Petrologic results of a study of the minerals from the Tertiary volcanic rocks of the San Juan Region, Colorado, *Am. Mineral., 23*, 227–257, 417–429, 1938.

McBirney, A. R., Differentiation of the Skaergaard intrusion, *Nature, 253*, 691–694, 1975.

Nesbitt, R. W., and S.-S. Sun, Geochemistry of Archaean spinifex-textured peridotites and magnesian and low-magnesian tholeiites, *Earth Planet. Sci. Letters, 31*, 433–453, 1976.

Nicolas, A., and E. D. Jackson, Répartition en deux provinces des péridotites des chaînes alpines logeant la Méditerranée: implications géotechniques, *Bull. suisse Min. Petrog., 52/3*, 479–495, 1972.

O'Hara, M. J., The bearing of phase equilibria studies in synthetic and natural systems on the origin and evolution of basic and ultrabasic rocks, *Earth Sci. Reviews*, *4*, 69–133, 1968.

Osborn, E. F., Role of oxygen pressure in the crystallization and differentiation of basaltic magma, *Am. J. Sci.*, *257*, 609–647, 1959.

Papanastassiou, D. A., and G. J. Wasserburg, Early lunar differentiates and lunar initial $^{87}Sr/^{86}Sr$, in *Seventh Lunar Sci. Conf. Abstracts*, Part 2, Lunar Science Institute compilation, Texas, pp. 665–667, 1976.

Ringwood, A. E., *Composition and petrology of the Earth's mantle*, McGraw-Hill, New York, 618 pp., 1975.

Roedder, E., and P. W. Weiblen, Petrology of silicate melt inclusions, Apollo 11 and Apollo 12 and terrestrial equivalents, in *Proc. Second Lunar Sci. Conf.*, vol. 1, A. A. Levinson, ed., The MIT Press, Cambridge Mass., pp. 507–528, 1971.

Sahama, Th. G., Kalsilite in the lavas of Mt. Nyiragongo (Belgian Congo), *J. Petrol.*, *1*, 146–171, 1960.

Schilling, J.-G., Azores mantle blob: rare-earth evidence, *Earth Planet. Sci. Letters*, *25*, 103–115, 1975.

Schilling, J.-G., Anderson, R. N., and P. Vogt, Rare earth, Fe and Ti variations along the Galapagos spreading centre, and their relationship to the Galapagos mantle plume, *Nature*, *261*, 108–113, 1976.

Shand, S. J., *Eruptive Rocks*, 3d ed., Wiley, New York, 488 pp., 1947.

Shaw, D. M., Trace element fractionation during anatexis, *Geochim. Cosmochim. Acta*, *34*, 237–243, 1970.

Taylor, S. R., *Lunar Science: A Post-Apollo View*, Pergamon, New York, 372 pp., 1975.

Tilley, C. E., Yoder Jr., H. S., and J. F. Schairer, Melting relations of basalts, *Ann. Rept. Geophys. Lab.* (*Carnegie Institution of Washington Yearbook*, *62*), 77–84, 1963.

Tilley, C. E., Yoder Jr., H. S., and J. F. Schairer, Melting relations of volcanic tholeiite and alkali rock series, *Ann. Rept. Geophys. Lab.* (*Carnegie Institution of Washington Yearbook*, *64*), 69–82, 1965.

Viljoen, M. J., and R. P. Viljoen, Evidence for the existence of a mobile extrusive peridotitic magma from the Komati formation of the Onverwacht Group, *Spec. Publ. Geol. Soc. S. Africa*, *2*, 87–112, 1969.

Wager, L. R., and G. M. Brown, *Layered Igneous Rocks*, Oliver and Boyd, Edinburgh, 588 pp., 1968.

Wager, L. R., Brown, G. M., and W. J. Wadsworth, Types of igneous cumulates, *J. Petrol.*, *1*, 73–85, 1960.

Wager, L. R., and W. A. Deer, Geological investigations in East Greenland: pt. III, The petrology of the Skaergaard intrusion, Kangerdlugssuaq, East Greenland, *Medd. Grønland*, *105*, no.4, 1–352, 1939.

Wager, L. R., Vincent, E. A., Brown, G. M., and J. D. Bell, Marscoite and related rocks of the Western Red Hills complex, Isle of Skye, *Phil. Trans. Roy. Soc. Lond.*, ser. A, *257*, 273–307, 1965.

Yoder Jr., H. S., Contemporaneous basaltic and rhyolitic magmas, *Am. Mineral.*, *58*, 153–171, 1973.

———, *Generation of Basaltic Magma*, National Academy of Sciences, Washington, D.C., 265 pp., 1976.

Yoder Jr., H. S., and C. E. Tilley, Origin of basaltic magmas: an experimental study of natural and synthetic rock systems, *J. Petrol.*, *3*, 342–532, 1962.

Chapter 2

SILICATE LIQUID IMMISCIBILITY
IN MAGMAS

EDWIN ROEDDER,

U.S. Geological Survey, Reston, Virginia

INTRODUCTION

The history of science is littered with the shells of cast-off concepts, each of which seemed eminently suitable to hold to perfection all aspects of the then-known body of facts. One such concept is that silicate liquid immiscibility in magmas is a mechanism to form various rock types. It was proposed by Zirkel and Rosenbusch early in the development of petrology to explain the juxtaposition of rocks having quite disparate compositions, usually without intermediate types. Rocks such as basalt and rhyolite, and various pairs of dike rocks, were thus generally and conveniently assumed to have formed simply by the "splitting" of a formerly homogeneous magma into two immiscible magmas of contrasting composition. These would be expected to separate like oil and water, due to density differences.

Early studies at the Geophysical Laboratory of the Carnegie Institution of Washington verified that the systems $CaO-SiO_2$, $MgO-SiO_2$ and $FeO-SiO_2$ did indeed have extensive fields of immiscibility. Greig (1927) showed, however, that these fields were eliminated by the addition of even small amounts of alkalies or alumina. Inasmuch as the resulting ternary immiscibility fields were similar, Greig used a pseudoternary plot of $(CaO + MgO + FeO)-SiO_2-(Na_2O + K_2O + Al_2O_3)$ to show the relationship of immiscibility to the composition of igneous rocks. No igneous

© 1979 by Princeton University Press
The Evolution of the Igneous Rocks: Fiftieth Anniversary Perspectives
0-691-08223-5/79/0015-43$02.15/0 (cloth)
0-691-08224-3/79/0015-43$02.15/0 (paperback)
For copying information, see copyright page

rock composition even approached the limits of the immiscibility field; even those rocks having the lowest total $(Na_2O + K_2O + Al_2O_3)$ contained twice as much as could be found in any synthetic mixture showing immiscibility (~ 10 vs. 5 wt. %). Furthermore, synthetic immiscibility took place only at excessively high temperatures ($\sim 1700°C$). Thus, experimental verification of immiscibility was found only in geologically unlikely compositions, at geologically unreasonable temperatures. Bowen (1928) then showed that if immiscibility had indeed occurred in natural magmas, the expected lines of evidence for it had not been found in the rocks. In addition, Bowen demonstrated that crystallization differentiation could produce both continuous and discontinuous compositional variations in igneous rocks; he described ample field and textural evidence to support this view. Silicate immisicibility as a potential petrological process was thus put to rest.

In 1951, Roedder, however, reported stable silicate immiscibility between much more geologically possible compositions at geologically reasonable temperatures in the system $K_2O-FeO-Al_2O_3-SiO_2$. No new evidence was presented at that time for immiscibility in natural rocks. In the last decade, however, immiscibility has again been proposed as the origin for some crystalline igneous rocks generally consisting of globular syenitic or granitic masses in a more mafic matrix (e.g., Philpotts and Hodgson, 1968; Ferguson and Currie, 1971, 1972; Brooks and Gélinas, 1975), and Roedder and Weiblen (1970) found actual menisci between glasses in the interstitial melts in the Apollo 11 lunar mare basalts. Similar glass-glass menisci were later found in most of the mare basalts and in a variety of terrestrial basalts (Roedder and Weiblen, 1971; De, 1974). These findings provide new evidence for possible silicate immiscibility that was not available in Bowen's time, but some have suggested that the pendulum has perhaps swung too far (Philpotts, 1977).

This chapter includes a discussion of the validity of several types of laboratory and field evidence for and against immiscibility, some examples from the literature, and its present status as a significant petrological process. The literature on immiscibility is replete with non sequiturs and erroneous interpretations; these will not be aired here.

GENERAL BEHAVIOR OF SYSTEMS INVOLVING IMMISCIBILITY

STABLE EQUILIBRIUM IMMISCIBILITY

Bowen (1928) illustrated stable immiscibility with a simple hypothetical binary system that unfortunately eliminated many features of the more complex systems that more closely approach natural magmas. Perhaps the most important of these features is the very gross difference in behavior on cooling for various bulk compositions. Figure 2-1 illustrates some of the possible modes for a simple ternary system A–B–C in which immiscibility in the system A–B is eliminated by the addition of C. Changes in the phase assemblage on cooling, scaled from the diagram, are shown for six compositions. Note that both the sequence of formation and relative amounts of melts L_1 (along M–Q; rich in A), L_2 (along N–Q; rich in B), and crystals of A, can vary widely, within a single system. Other compositions and diagrams with other topologies can yield additional types of behavior. Thus in some organic systems immiscibility decreases and is finally eliminated with decreasing temperature. Fenner (1948) made a major point of this in defending the petrological significance of immiscibility, but no silicate system shows such a lower consolute point.

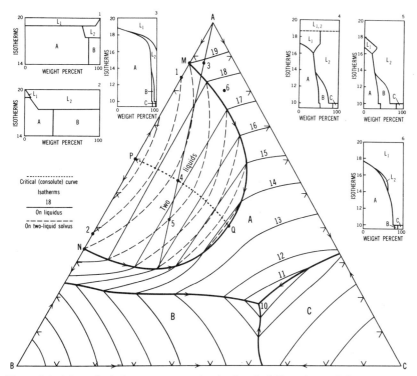

Figure 2-1. A hypothetical ternary system showing immiscibility. The small diagrams show the quantitative phase relationships for six marked starting compositions from liquidus to solidus, obtained by applying the lever rule to the ternary diagram.

The essential feature of immiscibility is simply that at some temperature a given melt can no longer exist stably, and splits into two melts. The melt can arrive at this state by simple cooling (points 1, 2, 4, and 5), or by cooling and crystallization (points 3 and 6). The splitting is, of course, a function of the composition of the melt itself, and is completely independent of the presence, absence, or the amount, of crystals. The immiscibility produces a unique partitioning of all components between the two. In silicate systems the partition coefficients for silica and the divalent oxides are very disparate (e.g., A and B, respectively, in Fig. 2-1), but those for alkalies and alumina are more nearly unity.

Compositions in the area PQM will first form a small amount of a conjugate melt in the area PQN, and conversely. Just as in any emulsion, the melt will now consist of globules of the new (dispersed) melt embedded in a host (continuous) melt. The tielines connecting the conjugate melts are given on Fig. 2-1 only for melts in equilibrium with crystals of A (i.e., along MQ and NQ), but all points on the surface of the solvus have such tielines, which will in general be similar in orientation to those on the ruled surface MNQ. With further cooling, at equilibrium, each of the two melts will exchange constituents with the other. The amounts moving in the two directions across the interface (i.e., the meniscus) between the two melts will generally be different, so the relative amounts of the two melts will also change. These amounts can be found by applying the simple lever rule, as has been done in Fig. 2-1.

Compositions along the line PQ show a different behavior on cooling. On contacting the solvus, they will split into approximately equal amounts of two melts. These two will have *almost identical* compositions at first, and on cooling will gradually diverge from PQ along curving paths (not shown) on the solvus. These paired melts will contact the liquidus surface simultaneously, one along MQ and the other along NQ, at points on a tieline on that ruled surface. Thus they *must* have identical liquidus temperatures. Crystallization of A will then result in the bulk composition of the two-melt mixture moving directly away from A, with continuous change in the compositions and amounts of the two melts along MQ and NQ as required by the tielines, until only one melt remains. After this occurs, further crystallization will be unaffected by the previous immiscible state except in the displacement of the isotherms by the field of immiscibility.[1]

---

[1] This displacement of the isotherms, and the extreme effectiveness of Al_2O_3 in reducing the size of the field of immiscibility in the system $CaO–SiO_2$, was found to be of considerable industrial importance thirty years ago in the steel industry. The service life of silica brick refractories (essentially just SiO_2 plus a few percent CaO) was found to be significantly decreased by each 0.1% Al_2O_3 present as impurity.

METASTABLE IMMISCIBILITY

Greig (1927) presented data on the system $BaO-SiO_2$ that revealed a novel sigmoidal curvature for the liquidus of SiO_2—concave upward near SiO_2, nearly flat from 5 to 25 wt. % BaO, then falling off rapidly beyond 25% BaO (i.e., concave downward). He showed, on the basis of the early work of Roozeboom, that such a curvature is indicative of a field of immiscibility either above (stable) or below (metastable) the liquidus. Later work by Seward et al. (1968) proved that the field of immiscibility is *below* the liquidus in this system, as was suggested also by the work of Ol'shanskii (1951) and of Tewhey and Hess (1974).

Metastable subliquidus immiscibility is widely exploited in the glass industry, and its possible presence in magmatic systems is of more than passing interest to geologists. If such fields of immiscibility occur beneath flat or sigmoidal liquidi, as in the systems albite–fayalite, leucite–forsterite–SiO_2, and orthoclase–diopside, the very likely depression of the liquidus to this temperature range by other components present in rocks may result in the formation of stable immiscibility. The additives may, of course, affect the immiscibility field also, so the results are neither implicit nor predictable. Irvine (1975) has pointed out, in addition, that the effects of nonideality of the melts related to immiscibility should extend well beyond the actual limits of the immiscibility itself, stable or metastable, with the effect that liquidus temperature contours and phase boundaries are shifted away from the immiscibility gap.

RATE PROCESSES

After globules of a second melt have nucleated, all changes in composition of the two melts require diffusion across the interface and through both melts. The growth of a globule of one melt in another is identical with the growth of a crystal of a solid solution from a melt, with one important and major difference—the ease with which equilibrium may be achieved in the two cases. Diffusion in melts will probably be many orders of magnitude faster than in crystals (Hofmann, 1975). It is thus feasible to have a magma crystallizing plagioclase under fractionating, nonequilibrium conditions (i.e., yielding strongly zoned crystals) at the same time that the melt is splitting under essentially equilibrium conditions.[2] Under some

---

[2] The rates of diffusion within the two melts may also differ greatly, so one might envisage a case wherein low-silica melt globules retain a uniform composition, but are at equilibrium only with respect to the surface layers of a much more viscous and compositionally zoned high-silica melt. Such behavior has been observed in synthetic glasses (Baylor and Brown, 1976).

circumstances it appears possible that local disequilibrium from local fractionation can produce a melt that splits while the main body of the melt does not (Rutherford, Hess, and Daniel, 1974).

Because the two melts may be very different in density, they will tend to separate in a gravitational field, as long as the presence of crystals does not prevent it. The rate of such separation will be a direct function of the size and effective density difference of the dispersed globules, and an inverse function of the effective viscosity of the continuous host phase. The relative movement will, of course, be independent of any bulk movements due to convection or vesiculation, for example. Crystals will affect both density and viscosity. Although any crystal present in one of a pair of conjugate melts must, by definition, be in equilibrium with both, the *amount* of crystals formed in each phase on cooling may be very different, and, if the tieline extended intersects the crystal composition point, *all* such crystals may form in only one of the two melts. Also, if the globules do not stay in perfect chemical exchange with the host, crystallization can result in their being no longer in equilibrium with each other.

Fractionation will occur if the globules sink (or rise) far enough to preclude further equilibration with the main host melt. The two resultant magmas will then behave as new compositions. On cooling the six compositions shown on Fig. 2-1 after such a separation, the L_2 melts will immediately leave the line NQ and show no further immiscibility; the L_1 melt in each case will continue to separate new globules of L_2 and crystals of A. If the crystals are solid solutions, equilibrium will require continuous exchange of materials not only between the two melts but also with the crystals. Thus, the fractionation caused by crystal separation may be modified by melt separation.

Nucleation and growth of crystals in the two melts are also rate processes, and it is important to remember that these rates may be many orders of magnitude slower for silica-rich melts than for the conjugate melts rich in divalent (RO) oxides.

THEORETICAL BASIS OF IMMISCIBILITY

EQUILIBRIUM PROCESSES

No difference exists thermodynamically between the process of separation from a melt, of a crystal or of a globule of a second melt. In both, the melt becomes saturated with respect to the new phase and precipitates it. Ever since immiscibility was found to be of value in glass processing it has prompted considerable study by glass technologists (e.g., see Charles, 1973). Particularly in the last decade or so it has become increasingly apparent that immiscibility, rather than being exceptional, is probably an inherent feature of silicate glasses including even such systems as Na_2O-SiO_2. The glass technologist frequently deals with glasses far below the liquidus, and hence most of the extensive literature pertains to metastable immiscibility in extremely viscous systems. In nature, one presumably deals with stable immiscibility at or above the liquidus, in more fluid systems, but the processes are exactly parallel.

Immiscibility is usually explained on the basis of structural models of the liquid state. Unfortunately, these models are generally established with data from studies of glasses formed on quenching that may not maintain the melt structure (Boon and Fyfe, 1972). Many of the models are necessarily qualitative or empirical, because they involve the properties of hypothetical molecular species in solution, and no consensus has been reached. All involve the breakup of the chainlike polymerized structure of a melt consisting of tetrahedral Si (the "network former," with substitutional Al and interstitial alkalies) by the addition of other constituents (the "modifying oxides") until suddenly a new, modifier-rich melt forms (Hess, 1971). The entropy and enthalpy terms of the free energy change of this mixing process can presumably be nearly equal and opposite, so the occurrence, extent, and temperature of immiscibility can be extremely sensitive to composition.[3]

According to Levin (1967), the ionic field strength of the modifying cation and the number of oxygens per modifier cation are the most important parameters in immiscibility, but the results of Muan (1955) in the system $FeO-Fe_2O_3-SiO_2$ seem to contradict these ideas. The deviations from ideal mixing due to melt structure that result in immiscibility become more complex in multicomponent systems (Nakagawa and Izumitani, 1972) and in some systems may yield *three* coexisting immiscible melts (Haller *et al.*, 1970).

Numerous attempts have been made to predict immiscibility in multicomponent systems, e.g., Barron (1972), but these are at best only as reliable as the available thermodynamic data, and the free energy differences are very small (Hess, 1975). On the basis of calculated Gibbs free energy, Currie (1972) showed that the system leucite–fayalite–SiO_2

---

[3] As an example of this sensitivity, there are large changes in the immiscibility in several systems between liquid water and various organic liquids on substitution of D_2O for H_2O. Thus, there is a closed-loop immiscibility field that is 78.5°C wide in the system β–picoline–D_2O, but complete miscibility in the equivalent system with H_2O (Andon and Cox, 1952; Cox, 1952).

should have immiscibility (as was found), but that the equivalent system with sodium should not. Naslund (1976), however, has shown experimentally that the sodium system also has immiscibility, but at slightly higher f_{O_2} (10^{-9} vs. 10^{-12} atm).

RATE PROCESSES

The processes of nucleation and growth of small globules of one melt within another are sufficiently similar to those of crystallization that the classical theories of crystal nucleation and growth are applicable. Inasmuch as the kinetics of nucleation, the distribution of the nuclei, and the resultant morphologies of the two phases all are of great concern to the glass technologist (e.g., see Doremus, 1973), many studies have been made of the pertinent systems. Conventional nucleation theory fails near the critical point (Goldburg and Huang, 1975), and in particular, it has been found that the mechanism followed for the exsolution of a glass in the metastable region is quite different from that for a glass inside the spinodal region where exsolution is spontaneous (Cahn, 1968). Although there may be little possibility that these kinetic phenomena will be of concern in geological examples having lower viscosity, one kinetic aspect may be important. This aspect is the possibility of metastable immiscibility preceding and aiding nucleation and growth of crystalline phases, as is apparently common in ceramic systems (Zdaniewski, 1975).

NATURAL OCCURRENCE OF IMMISCIBILITY

LITERATURE REVIEW

Numerous recent papers suggest that immiscibility has been involved in the origin of certain specific rocks. From this literature, summarized in Table 2-1, it is apparent that such occurrences generally consist of coexisting felsic and mafic rocks, and include both over- and undersaturated types.[4] Two examples from Table 2-1 will be described briefly here, as they illustrate several features common to many of the occurrences. The first is late-stage immiscibility during crystallization of the lunar basalts, and the second is pre-eruption immiscibility in Archean tholeiites.

---

[4] Many other reports of similar assemblages do not invoke immiscibility, and hence are not listed in Table 2-1. Although the degree of saturation with silica is petrologically very important, many of the available analyses are ambiguous in this respect because of introduction or replacement of preexisting minerals with carbonate, analcite, or both.

LUNAR BASALTS

OCCURRENCE AND PETROGRAPHY   Immiscibility between late residual Fe-rich basaltic melt and a potassic granite melt was first reported from high-Fe, high-Ti mare basalts returned by Apollo 1 (Roedder and Weiblen, 1970). The basalts from the different Apollo sites vary significantly in composition (Weiblen and Roedder, 1976), but all show evidence of late stage immiscibility. The residual melt ($\sim 5\%$), enriched in Si, Fe, and K, split to form two melts, one high in iron, and the other high in silica. These are found as two coexisting, intermixed glasses, one brownish, with a high index of refraction, the other colorless with a very low index of refraction (Fig. 2-2). The contacts are as sharp as can be resolved by the optical microscope (<1 μm), and the electron microprobe (<2 μm). Most occurrences consist of spheres of high-Fe glass embedded in high-Si glass filling interstices between late crystals, although the reverse is not unusual. All larger masses of high-Fe glass have devitrified to a mass of ~ 1 μm crystals (mainly pyroxene and ilmenite?) but some still retain the sharp menisci against the high-Si glass (Fig. 2-3). Splitting has occurred with as much as 11.6 vol. % melt remaining (see Fig. 2-3; Roedder and Weiblen, 1972a). Because the high-Fe melt contains $>75\%$ normative pyroxene, and crystallizes very easily, it is preserved only in special situations, out of contact with pyroxene (e.g., embedded in high-Si glass or plagioclase, Figs. 2-2, 2-3, and 2-4). It is also found as much larger inclusions in late ilmenite (Roedder and Weiblen, 1971, their Fig. 23). The high-Si glass frequently occurs as a sharply defined zone of melt inclusions in the outer parts of late pyroxene and plagioclase crystals, particularly where these are adjacent to mesostasis.

From the distribution and compositions of inclusions, Roedder and Weiblen suggest that during crystallization the interstitial melt became greatly enriched in Fe and K until its gross composition was essentially that of the high-Fe melt, at which time small globules of high-Si melt separated and were trapped, presumably by adhering to the growing crystals due to preferential wetting, to form inclusion zones marking the onset of immiscibility. All high-Fe melt that was in contact with pyroxene crystallized to form a late high-Fe zone (or pyroxferroite), leaving only the high-Si melt as a glassy residue in most areas of mesostasis.

COMPOSITION   Electron microprobe analyses of the two glasses (see Table 2-1 for references), although they vary from one site or sample to another, generally correspond in composition to titaniferous ferropyroxenite and quartz-rich potash granite (Table 2-2). The differences in

Table 2-1. Summary of recent literature in which silicate-silicate liquid immiscibility has been proposed for natural rock. (All papers present analyses except as noted.)

Reference	Locality	Composition		Size or range of globules	Notes
		Host	Globules		
Anderson, 1971	Hat Creek, Calif. high-alumina olivine tholeiite	Ferrodacite	Phosphatic ferrobasalt	15 μm	See also Anderson and Gottfried, 1971.
Brooks & Gélinas, 1975	Various Archean and modern rocks	Various	Various		No analyses—see also Gélinas and Brooks, 1976.
Butler, 1973	Davie County, North Carolina Piedmont	Pl-microcline-qz-hnbl	Augite-hornblende	20 mm	
Carman et al., 1975	Rattlesnake Mtn. Sill, Big Bend, Tex.	Analcime-monzonite	Plagioclase syenite	5-20 mm	
Currie, 1975	Ice River alkaline complex, British Columbia	Ijolitic (feldspar-free)	Syenitic		Rock types only analyzed; ocellar lamprophyres occur also.
De, 1974	Deccan traps, India	"High silica" glass	"High iron" glass	2-10 μm	No analyses given.
" "	Skaergaard complex, East Greenland	Alkali-Al-rich Sandwich Horizon rock	Ferrodiorites in Upper Zone		Rock types only; analyses from literature.
Dence et al., 1974	West Clearwater Lake impact crater, Quebec	Rhyolitic glass	Clays ± calcite & qz; widely variable	10-500 μm	
Drever, 1960	Igdlorssuit, West Greenland	Picrite	Dolerite ± natrolite, mt	1-3 mm	
Ferguson & Currie, 1971	Callander Bay dikes, Ontario	Monchiquitic & camptonitic	Leucocratic—much carbonate or K-spar	0.5-25 mm	
Ferguson & Currie, 1972	Barberton Mtn. Land, Transvaal	Basaltic komatiite lavas	Pl, carbonate, qz mainly	1-100 mm	

Reference	Location			Size	Remarks
Furnes, 1973	Pillow lavas, Solund, Norway	Chlorite-epidote	Leucoxene mainly	0.05-3 mm	No analyses given.
Gélinas & Brooks, 1976	Various Archean & modern rocks	Various	Various		1473 analyses plotted.
Gélinas, Brooks, & Trzcienski, 1976	Abitibi metavolcanic belt (Archean), Quebec	Tholeiitic rhyodacitic to ultramafic lavas	Low-K rhyolite	5-50 mm	In part pillow lavas, all iron-rich.
Holgate, 1954	Various	Various basalts	Quartzose xenoliths		Special type of immiscibility involving "transfusion across semipermeable membrane."
Kogarko, Ryabchikov, and Sørenson, 1974	Ilimaussaq, Greenland	Iron-rich melteigite	Calcite-analcite zeolite		
"	"	Lujavrite	Analcime-rich		
McBirney, 1975	Skaergaard, East Greenland	Upper Zone C	Granophyre		Analyses given as graphical plot only; see McBirney & Nakamura, 1974
McBirney & Nakamura, 1974	"	"	"	1-10 µm	
Mason & Melson, 1970	Disko, Greenland (iron-bearing basalts)	Granitic(?) residual melt	FeO-rich(?)	1-6 µm	No analyses given.
Mutanen, 1976	Komatiite provinces in Finland	Peridotitic komatiite	Albitite		Compositions given only for mafic rocks.
Olsen & Jarosewich, 1970	Weekeroo Station meteorite	Pyroxene-rich	Feldspar-rich		Bulk analysis only.
Paakkola, 1971	Central Finnish Lapland	Ultrabasic lavas	Spilitic lavas		Separation occurred at depth.

Abbreviations used: pl—plagioclase; qz—quartz; hnbl—hornblende; mt—magnetite; ±—with or without.

Table 2-1. (*Continued*)

Reference	Locality	Composition — Host	Composition — Globules	Size or range of globules	Notes
Philpotts, 1971	Sill just north of the Island of Montreal, Quebec	Teschenite	Nepheline syenite to quartz syenite	2-5 mm	
Philpotts, 1972	Ste. Dorothée sill, Monteregian petrological province, Quebec	Fourchite	Nepheline syenite	5-50 mm	New analyses, not same as Philpotts & Hodgson, 1968.
Philpotts, 1976	Monteregian petrological province, Quebec	Fourchite to quartz diorite	Nepheline syenite to granite	Various	
Philpotts, A. R., (personal communication)	Triassic diabase in Connecticut	High index of refraction (high-Fe?)	Low index of refraction (high-Si?)	$\sim 10\ \mu m$	Manuscript in preparation
Philpotts & Hodgson, 1968	Ste. Dorothée sill, Monteregian petrological province, Quebec	Fourchite	Analcite syenite	5-50 mm	
Popov, 1972	Lamprophyres, N. Tien-Shan, U.S.S.R.	Alkaline gabbrodiorite	Microsyenite	2-10 mm	Globules rarely 40 mm
Quick & Albee, 1976	Lunar sample 12013	Mafic (SiO_2-48; Al_2O_3-14; FeO-13; K_2O-0.05)	Felsic (SiO_2-74; Al_2O_3-12; FeO-1; K_2O-7)	~ 1 cm	Two intermixed "possibly unrelated" melts, with partitioning as in immiscible melts.
Roedder & Weiblen, 1970	Apollo 11 mare basalts	Ferropyroxenite, in part glassy	Potassic granite glass	$\leq 15\ \mu m$	Analyses for 33 high-Si and 6 high-Fe melts.
Roedder & Weiblen, 1971	Apollo 11 & 12 mare basalts	"	"	$\leq 20\ \mu m$	Analyses for 17 high-Si and 6 high-Fe melts.

Reference	Rock type	Phase 1	Phase 2	Size	Comments
Roedder & Weiblen, 1972a	Apollo 12, 14 and 15 mare, KREEP, and feldspathic basalts	Ferropyroxenite, in part glassy	Potassic granite glass	≤20 μm	Analyses for 38 high-Si and 17 high-Fe melts.
Roedder & Weiblen, 1972b	Luna 16 mare basalts	Ferropyroxenite, devitrified	Potassic granite glass	≤10 μm	Analyses for 19 high-Si and 3 high-Fe melts.
Roedder & Weiblen, 1975	Apollo 17 mare basalts	Ferropyroxenite, devitrified	Potassic granite glass	≤20 μm	Analyses for 32 high-Si melts (Apollo 17) and 1 (Apollo 12), all in ilmenite.
Scofield & Noble, 1977	Calc-alkalic rhyodacite volcanic center, Julcani, Peru	Silicic	Mafic	≤75 μm	Occurs in melt inclusions in phenocrysts.
Switzer, 1975	Apollo 15 basalt	Potash granite glass	Ferropyroxenite glass	100 μm	Two types of high-Fe glass analyzed.
Upton, 1965	South Greenland	Camptonite	Analcite-calcite-zeolites	1-10 cm	
Vogel and Wilband, 1978	Composite dike, South Carolina	Lamprophyre	Granite		
Weiblen & Roedder, 1973	Apollo 15 mare basalts	Ferropyroxenite, generally devitrified	Potassic granite glass	≤20 μm	Analyses for 18 high-Si and 8 high-Fe melts.
Weibe, 1977	Nain anorthosite, Labrador	Granite	Ferrodiorite	~100 cm	Both are late differentiates.
Yeats et al., 1973	Cretaceous deep sea basalts, Pacific	Tholeiitic basalts	Enriched in Si, Al, & Na; depleted in Fe & Mg (red globules)	100-600 μm	Opaque globules also present, depleted in Si, Al, & Na, & enriched in Fe & Mg.

Abbreviations used: pl—plagioclase; qz—quartz; hnbl—hornblende; mt—magnetite; ±—with or without.

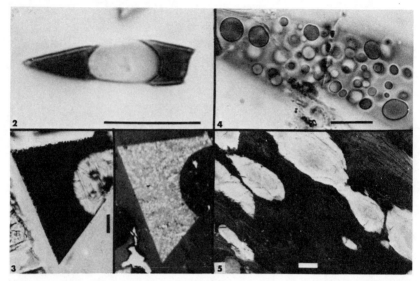

Figure 2-2. Primary inclusion of immiscible high-Si (light) and high-Fe (dark) melts, now glasses, in plagioclase from lunar sample 12057. Bar = 10 μm. Analyses given by Roedder and Weiblen (1971).

Figure 2-3. Interstitial inclusion consisting of a globule of immiscible high-Si melt in high-Fe melt between plagioclase crystals in lunar sample 14310. High-Fe melt is now a microcrystalline mass, opaque in transmitted light (left) and bright in reflected light (right). Bar = 10 μm. Analyses given by Roedder and Weiblen (1972a).

Figure 2-4. Immiscible silicate melts in basalt collected at 1020°C from the 1965 Makaopuhi lava lake, Hawaii. Dark globules are high-Fe glass, embedded in high-Si glass, between two plagioclase crystals. Bar = 10 μm. Roedder and Weiblen (1971).

Figure 2-5. Varioles of low-K rhyolite composition in tholeiitic basalt, Abitibi metavolcanic belt (Archean), south of Rouyn, Noranda, Quebec, Canada. Bar = 1 cm. Sample courtesy C. Brooks.

the composition of the high-Si melts between the various samples are small, but those for the high-Fe melts are larger (Roedder and Weiblen, 1972a). The high-Fe melt can be somewhat under- or oversaturated. Thus, sample 14310 has 14% normative Fe-rich olivine in the high-Fe melt and 46% normative quartz in the high-Si melt.

The alkali contents of the high-Si melts are surprisingly high and variable. Thus, magmas starting with as little as 0.028% K_2O (e.g., sample 15555) can end with a residual high-Si melt containing 6.95% K_2O. As the plagioclase also contains some K_2O, the amount of high-Si melt must be very small. When the values of K_2O/Na_2O for high-Si glasses are compared with their host rocks, (Roedder and Weiblen, 1972a,b; Weiblen

Table 2-2. Averages of CIPW norms for high-Si (156 analyses) and high-Fe (38 analyses) glasses in lunar samples (references in Table 2-1)

	High-Si		High-Fe*	
Q	42.8		1.1*	
C	1.2		—	
or	39.3 ⎫		1.5 ⎫	
ab	3.4 ⎬ 50.3		1.3 ⎬ 13.0	
an	7.6 ⎭		10.2 ⎭	
wo	0.1 ⎫		16.8 ⎫	
en	0.5 ⎬ 4.2		6.2 ⎬ 74.5**	
fs	3.6 ⎭		51.5 ⎭	
il	1.3		9.2	
ru	0.1		—	
cm	—		0.1	
ap	0.1		2.1	
	100.0		100.0	

* Some are slightly oversaturated and others slightly undersaturated.
** The composition of the pyroxene would be $Wo_{22.6}En_{8.3}Fs_{69.1}$.

and Roedder, 1973), a perplexing, almost regular but inverse relation is found. This inverse relation might result from a combination of the amount and stage of crystallization of plagioclase (removing Na_2O), versus the total volume of crystallization (enriching K_2O).

Fragments of granitic composition have been found in all the lunar soils and in many of the breccias, including, particularly, sample 12013. Although many of these fragments appear to be the high-Si melt from various types of basalts (Hess *et al.*, 1975, p. 900), there are some compositional differences, particularly in Na:K:Ca and Mg/Fe (e.g., Roedder and Weiblen, 1972a, p. 269; Quick and Albee, 1976).

The lunar basalts are deficient in both alkalies and alumina, relative to most terrestrial rocks. Several terrestrial basalts that are far from exotic in bulk composition, however, show similar late stage immiscibility (Fig. 2-4).

COMPARISON WITH SYSTEM $K_2O-FeO-Al_2O_3-SiO_2$ A preliminary diagram for the ternary system leucite–fayalite–SiO_2, a plane in the system $K_2O-FeO-Al_2O_3-SiO_2$, is shown in Fig. 2-6. It reveals a field of low-temperature immiscibility, in compositions relatively high in alkalies and alumina, that is separate[5] from the high-temperature immiscibility field

[5] Visser and Koster van Groos (1976a, b) have reported a much larger, single field of immiscibility, but Roedder (1977) suggests that they were observing metastable subliquidus immiscibility formed on quench.

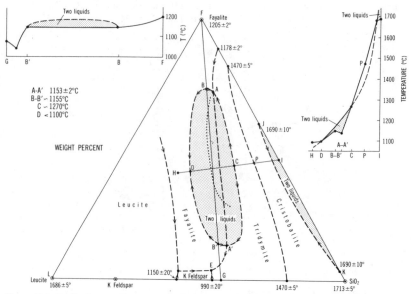

Figure 2-6. Preliminary diagram of the system leucite–fayalite–SiO$_2$ (adapted from Roedder, 1951), showing fields of immiscibility (shaded) at high temperature (along J–K), and at low temperature (A–B–D–B'–A'–C–A). All compositions run in 1-atm. N$_2$ in metallic iron. Dotted line is 1180°C isotherm on upper surface of two-liquid solvus (Watson, 1976). The inset figures are T-X sections along the lines G–F and H–I.

along the FeO–SiO$_2$ sideline. The upper surface of the stable low-temperature solvus slopes down from a maximum at point C to a minimum at D, on the basis of quench runs on a few homogeneous glass compositions (Roedder, 1951), and on some runs involving approaches to equilibrium from both within and without the solvus (Watson, 1976). The probability of a connecting field of subliquidus metastable immiscibility is self-evident from the geometry of a T-X section (H–I, Fig. 2-6). The fayalite field has a very flat liquidus, and metastable subliquidus immiscibility under it can be observed on slow quenching (Roedder, 1959, p. 282). The low-temperature immiscibility field $ABDB'A'CA$ in Fig. 2-6 represents a section through an immiscibility phase volume in the system K$_2$O–FeO–Al$_2$O$_3$–SiO$_2$. As seen in Fig. 2-7, this quaternary volume is lens shaped and lies astride the plane representing a K$_2$O/Al$_2$O$_3$ mole ratio of 1:1. It is important to note also that although this volume comes within $\sim 2\%$ K$_2$O or Al$_2$O$_3$ of the two adjacent ternary systems

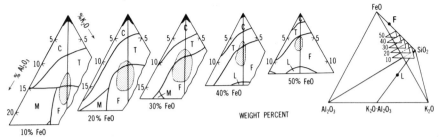

Figure 2-7. The high-silica part of the system K_2O–FeO–Al_2O_3–SiO_2, showing sections at various FeO contents. All compositions run in 1-atm. N_2 in metallic iron. The high-temperature field of immiscibility along the FeO–SiO_2 sideline is in black; the phase volume of low-temperature immiscibility astride the 1:1 K_2O:Al_2O_3 plane is shaded. Primary phase fields are shown for cristobalite (C), tridymite (T), mullite (M), fayalite (F), and leucite (L). Figure 2-6 is the ternary section L–F–SiO_2 in this system. Preliminary diagram drawn on the basis of partial or complete quench runs on $\sim$250 compositions (Roedder, 1953a, b). A more recent compilation, based on all available quench data (Roedder, 1978), differs mainly in that the internal field of immiscibility does not reach 50% FeO, but closes off between 40% and 50% FeO.

FeO–Al_2O_3–SiO_2 and K_2O–FeO–SiO_2, its presence is not evident in either diagram (Schairer and Yagi, 1952; Roedder, 1952).

When these data are compared with the lunar glasses, there is a close analogy (Fig. 2-8A) in both tielines connecting conjugate liquids and the boundaries of the immiscibility field. The close analogy is probably fortuitous, as the lunar glasses contain as much as 18% of constituents other than those four oxides. Furthermore, the mole ratio alkalies/alumina in the lunar high-Si melt is $\sim$0.66, rather than the 1.0 of Fig. 2-6.[6] (The ratio for the high-Fe melt is even lower, but has large analytical uncertainties.) The immiscibility volume in the system K_2O–FeO–Al_2O_3–SiO_2 is asymmetrical about the 1:1 plane (Fig. 2-7) with the larger part on the Al-rich side, extending to $K_2O/Al_2O_3 = 0.4$.

ARCHEAN VARIOLITIC THOLEIITES

OCCURRENCE AND PETROGRAPHY   Gélinas, Brooks and Trzcienski (1976) have described a series of slightly metamorphosed Archean tholeiitic lavas

___

[6] Some individual conjugate liquids in Fig. 2-6 may deviate from a 1:1 ratio (Watson, 1976; Naslund, 1976; Watson and Naslund, 1977).

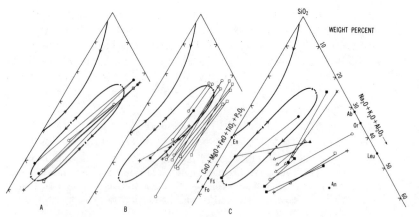

Figure 2-8. Pseudo-ternary diagram showing field of low-temperature immiscibility in the system leucite–fayalite–SiO$_2$, adapted from Weiblen and Roedder (1973; see also Fig. 2-6) and tielines for various conjugate melts and rock pairs. All compositions recalculated on the basis of plotted oxides only.

A: Solid circles—coexisting glasses in lunar basalts from Apollo 11, 12, and 15; crosses—coexisting glasses in lunar basalt 14310; open circles—synthetic Apollo 11 sample after equilibration at 1045°C; squares—Apollo 15 grain. The last item is from Switzer, (1975); all other data from Roedder and Weiblen (1970, 1971, 1972a) and Weiblen and Roedder (1973). Other lunar samples are similar but have been omitted for clarity.

B: Solid circles—synthetic Skaergaard sample after equilibration 5–10°C above its liquidus (McBirney and Nakamura, 1974); open circles—varioles and matrices of Archean variolites (Gélinas, Brooks, and Trzcienski, 1976); crosses—ocelli and matrices of Barberton Mountain Land (Transvaal) basaltic komatiites (Ferguson and Currie, 1972); squares—rhyolite-basalt pairs that did *not* show immiscibility in experiments at 950° and 1200°C (Yoder, 1973).

C: Solid circles—phosphatic ferrobasalt globules in ferrodacite residuum, Hat Creek high-alumina tholeiites (Anderson, 1971); open circles—phonolite globules in basanite (*not* considered to represent immiscibility by Mackenzie and White, 1970); crosses—leucocratic ocelli in Callander Bay monchiquite (Ferguson and Currie, 1971); open squares—nepheline syenite ocelli in Ste. Dorothée fourchite sill (Philpotts, 1972); solid squares—other Monteregian ocelli pairs (Philpotts, 1976); solid triangles—augite-hornblende orbicules in granodiorite (from orbicular crystallization?; Butler, 1973); open triangles—globules in tholeiitic deep sea basalts (Yeats *et al.*, 1973); × —composite dike, Winnsboro, S. C. (Vogel and Wilband, 1976).

from near Rouyn-Noranda (Abitibi metavolcanic belt, Quebec, Canada) that contain large numbers of spheroidal masses called varioles, of strongly contrasting color (Fig. 2-5). The varioles, which have diameters of 5–50 mm, have sharp contacts with the matrix, and are found in both massive and pillowed flows. Their occurrence is stratigraphically limited to form definite marker horizons traceable for nearly 100 km, according to the work of Wilson, and of Dimroth, Boivin, Goulet, and Larouche, quoted by Gélinas, Brooks, and Trzcienski (1976). Although regionally continuous, they may be discontinuous along strike on a local scale.

The details of the field occurrence, in particular such features as the coalescence of flow-deformed varioles along the center of some flows and in the axial regions of lava tubes, their distribution in pillows, and their deformation where in contact, all seem to point to the formation of the varioles *prior* to extrusion. Both the varioles and the matrix consist of a fine-grained assemblage of secondary minerals formed from the primary assemblage during the metamorphism of this belt (prehnite-pumpellyite to lower greenschist facies). Both contain coarser dendritic, acicular, skeletal mafic crystals (now pseudomorphs) that may cut across the interface. These skeletal mafic minerals have swallow-tail terminations and frequently are hollow, ornamented chains, arranged in fan-shaped groups. They are thought to have been quench crystals of olivine and some pyroxene, originally present in a glass that is now devitrified. The matrix also shows perlitic fractures with axiolitic acicular crystals. The contacts between the varioles and the matrix are optically and chemically sharp, to $<10 \ \mu$m.

COMPOSITION   The variolites (i.e., whole rocks) and the matrix for the varioles are relatively Fe enriched and have extremely low K (generally <100 ppm). The matrix ranges from a Fe-rich rhyodacite through basalt to an ultramafic composition with SiO_2 as low as 39%. The varioles have a low-K rhyolite composition ($\leq 0.02\%$ K_2O). On the Greig diagram (Fig. 2-8B), the tielines between variole and matrix are essentially parallel to those in the system K_2O–FeO–Al_2O_3–SiO_2, but the compositions of the matrix are not coincident with the edge of the immiscibility field in this system; such differences are predictable, as these rocks contain only sodium and no potassium, so the immiscibility field should be different.

Gélinas, Brooks, and Trzcienski (1976) also report lavas that contain small varioles having compositions essentially the same as the matrix, as are commonly described in the literature. They assume that these originated through crystallization, but it is tempting to suggest that they might represent immiscibility with only a small compositional difference, near to a conjugate liquid line (e.g., *P–Q*, Fig. 2-1).

EVIDENCE SUGGESTIVE OF IMMISCIBILITY

EVIDENCE FROM GENERAL FIELD RELATIONS

Bowen (1928) pointed out that the association or alternation of two compositionally different rocks does not establish immiscibility as the process whereby they originated; other processes, in particular crystal fractionation, appeared to him to provide a much better explanation. Furthermore, he stated that as liquid immiscibility must take place over a range of temperatures, there should be many examples in nature in which the "obvious and unfailing" evidence for it would have been quenched in and preserved as two glasses.

Most field occurrences believed to represent immiscibility provide only suggestive evidence of immiscibility, because most rocks are crystalline. In addition, immiscibility can and does yield residual melts high in silica and alkalies, similar to those expected from crystal fractionation. So the association or alternation of felsic and mafic rock types that has been considered favorable evidence for immiscibility is ambiguous at best. Furthermore, Yoder (1973) has shown that the well-known bimodal distribution can be the result of other processes.

EXPECTED BEHAVIOR IN MAGMA CHAMBER   What phenomena can be expected when immiscibility occurs in a magma chamber? Field evidence from the layered intrusions provides at least a partial analogy. Although the rheological properties of the crystal-melt mixtures involved there are complex and not completely known, descent of denser masses of melt plus crystals (density currents) have apparently caused deep scouring of the crystal mush at the sides and bottom of the chamber (Wager and Deer, 1939).

Immiscibility will yield a density contrast of ~ 0.4 g/cm^3 (McBirney and Nakamura, 1974) or even as much as 0.87 g/cm^3 (Philpotts, 1972), the same as that between many crystals and melt. Separation of globules of melt should be more rapid and effective, however, for several reasons. *First*, the geometry of immiscibility fields is such that it is possible to have very gross changes with a very small drop in temperature (e.g., Fig. 2-6). Similar drastic changes can indeed occur in crystallizing systems (e.g., Roedder, 1959, Figs. 48, 51, 54, and 55), but these require the removal of large amounts of latent heat, whereas immiscibility would probably involve relatively little latent heat (Charles, 1973). *Second*, a melt can split into roughly equal volumes of two melts, rather than merely form a few globules. Physical separation would then be a very different process. *Third*, as pointed out by Bowen (1928), globules of a given phase will coalesce if they touch, thus increasing the average size of globule and the speed of gravitational separation. There also can be considerable movement due to surface tension forces during the coalescence (Delitsyn, Melent'yev, and Delitsyna, 1974a), hastening equilibration. Haller (1965) has developed the mathematical theory of coalescence, based on studies of borosilicate glasses. (Fat globules on the surface of a hot thin soup can provide an excellent two-dimensional display of this surprisingly vigorous turbulence.)

Globules of a dispersed phase might even move through the interstices of a crystal mush, as long as the density difference can overcome the surface tension tending to keep them spherical, and as long as they do not wet the crystal surfaces (i.e., the contact angles are 180°). If the dispersed liquid phase wets a given crystal surface in preference to the continuous host phase, these adhering globules are no longer free agents to sink or swim. The writer has found evidence for such preferential wetting of olivine and pyroxene by heavy, liquid-iron globules in meteorites (e.g., Murray) and synthetic melts, and of plagioclase by light globules of potash granite composition in the lunar mare basalts (Roedder and Weiblen, 1971, 1972a).

If the dispersed phase is felsic, it should rise in a basaltic liquid. Because volatiles such as H_2O (and hence CO_2) tend to partition into this felsic phase, they will also be concentrated upward, introducing considerable ambiguity, however, as it is well known that another kind of "immiscibility"—boiling or vesiculation—also results in sweeping such volatiles upward.

If immiscibility occurs before eruption, once again the bulk of the possible evidence for it may be eliminated. Bowen (1928) stated that continued separation of the conjugate melt would have to occur during the bulk of the crystallization of a magma, and hence evidence for it should be plainly visible. Whereas this would be true for a composition such as he chose on his hypothetical binary diagram (his Fig. 1), a binary composition that intersects the solvus on the opposite side (as is also probable in many multicomponent natural systems), *can* leave the immiscibility field and crystallize completely independently, and reveal no evidence of its former state as one of a pair of immiscible melts.

If the separation of immiscible melts is incomplete for any reason, intermediate rocks would result, and subsequent crystallization could eliminate the evidence of immiscibility. It is also possible that the evidence of immiscibility now found in the rocks represents merely the last bit of continued separation, after a previous subcrustal, nearly complete separation of the immiscible melts.

EVIDENCE ON AN OUTCROP SCALE

Bowen (1928) applied a temporary coup de grâce to immiscibility by stating that although it should yield globules of one melt in another in nature, such globules "are utterly lacking." [Although the terms orbicular, ocellar, spherulitic, variolitic, and globular may each have specific definitions (e.g., Phillips, 1973) the usage in the literature is so overlapping and imprecise that the single term globules will be used here.] Bowen showed that the few examples of globules in the literature at that time did not meet some of the theoretically necessary consequences of immiscibility. More field data, however, have been accumulated since then. This evidence for immiscibility consists basically of four features, discussed in the following sections.

DISTORTED GLOBULES   Nonspherical globules might be caused by shear during flow, and the larger the globule, the more likely that its shape will be nonspherical. This distortion is most obvious in former lava tubes or in the centers of lava sheets (e.g., Gélinas, Brooks, and Trzcienski, 1976). It is also apparent in the globules found toward the edges of dikes (Philpotts and Hodgson, 1968). Philpotts (1977) points out that because of internal pressures from surface-tension forces, small globules should deform larger ones at the point of contact. This phenomenon is indeed noticeable in globules a few micrometers in size, but as the pressures in millimeter globules will be three orders of magnitude smaller, the effect may not provide a useful criterion for immiscibility.

Late stage felsic melts have been present in and around many mafic igneous rocks. A necessary, but insufficient, proof of immiscibility requires that the surrounding rock also was a melt, with or without crystals. Thus, in a sill or flow, lower density globules may have flattened tops, except where there has been sufficient coalescence of material against the overlying crystal mush to form small diapir-like embayments. Philpotts (1972) has used such diapirs and the distribution and shape of the globules to calculate the density contrast and the interfacial tension between the two melts, as well as the viscosity of the rising globules.

COALESCENCE OF GLOBULES   All stages of coalescence will be visible if quenching has been effective. As felsic globules will likely have a lower solidus than a mafic host, their coalescence is more probable in those parts that are expected to solidify last, such as in the cores of lava tubes (Gélinas, Brooks, and Trzcienski, 1976). The end product would be the accumulation of larger tabular bodies of light melt against the then top of the chamber.

GAS VESICLES WITHIN GLOBULES   Several investigators (e.g., Ferguson and Currie, 1971) have reported spherical gas vesicles in the centers or tops of felsic globules, and others (e.g., Philpotts and Hodgson, 1968) have found spherical masses of carbonate (amygdules) in a similar position. Such gas vesicles can be explained by the lower temperature for bulk crystallization of the felsic melt. After the mafic host is a rigid mass, shrinkage of the felsic melt on crystallization would produce a bubble, as occurs in silicate melt inclusions trapped in rigid crystals. This bubble may be filled with carbonate or quartz at some later time to form an amygdule.

SIZE VARIATION OF GLOBULES   At contacts of a dike, pillow, or flow, the globules may become larger away from the contact, because the slower quench gives more time for coalescence. Globules may even be absent in the quenched zone. A corollary is that compositional differences between globules and host also tend to increase away from such contacts (Gélinas, Brooks, and Trzcienski, 1976), presumably for the same reason. Philpotts suggests, however, that globule nucleation rates varied with distance from the contact in the Ste. Dorothée sill (1972, his Fig. 4), and that this was the controlling factor in that particular occurrence.

EVIDENCE FROM LABORATORY STUDIES OF NATURAL ROCKS

DATA FROM MICROSCOPY   The best evidence for immiscibility is the presence of a meniscus[7] between different glasses. The lunar basalts do show actual blebs of one glass in another of contrasting composition, as do some terrestrial samples (e.g., Fig. 24 in Roedder and Weiblen, 1971). Some contain major amounts of glass (e.g., Carstens, 1963, p. 34), but in most samples one or both melts are now partly or completely crystalline or devitrified. Crystallization can completely mask the former menisci, but nucleation and crystallization features can still provide important evidence.

Nucleation Features   Nucleation may occur at the margins of an immiscible globule, not only because of the crystals in the surrounding mush, but because *any* interface can aid nucleation. It is also possible that in viscous melts, crystal growth from the rim inward would be too slow and nucleation might occur throughout the mass first, but later devitrification or metamorphism would yield similar results.

---

[7] Meniscus is used here to refer to a phase boundary, and does not include gradational interfaces, however abrupt, due to inadequate mixing.

Spherulitic Crystallization   There has been much discussion of the problems of recognition of spherulitic crystallization, as though it eliminates the possibility of immiscibility; actually it can occur in any melt, immiscible or not, and both processes yield spherical masses, i.e., "globules." There should generally be no necessary connection between the physical centers of globules of melt and the nucleation centers for spherulitic growth. Also, there is no reason for globules formed by immiscibility to have a quench crystal or microphenocryst nucleus in the center, unless this crystal itself acted as a nucleus for the formation of the original globule, and thereafter stayed in the center. Spherulitic growth from this nucleus could then cause confusion in interpretation.

Spherulites are frequently polymineralic, but seldom have compositions identical with the original medium (e.g., Kesler and Weiblen, 1968), so they may mimic immiscibility. If the globules cut across the flow banding without disturbing it, (Grieg, 1928, p. 385), they are almost certainly spherulites. Several other types of evidence for spherulitic growth, such as radial fibrous crystal habit in the globules, are given by Greig.

Primary Crystallization, i.e., at or Near the Liquidus   The nature of any primary crystals is very important in evaluating the possibility of immiscibility, as they, or at least their outer surface, should be in equilibrium with both melts. The primary origin can be established only on the basis of texture, but if the crystals are thin, skeletal, hollow, or chainlike (e.g., Gélinas, Brooks, and Trzcienski, 1976), and particularly if they cross the interface between liquids, there is little doubt.[8] Drever (1960) reported a thin veneer of magnetite crystals outlining the globule interfaces (possibly adhering there to minimize surface energy?). The occurrence of many crystals throughout the globule (e.g., Philpotts, 1972), is evidence for the presence of a melt, but does not establish its origin. Similarly, stratified fillings (e.g., amphibole at bottom, nepheline and feldspar in middle, and carbonate at top; Philpotts and Hodgson, 1968) show the sequence of crystallization but not the origin of the filling.

DATA FROM COMPOSITION

Composition of Individual Phases   Although Bowen (1928) stressed the necessity of identical liquidus temperature and composition for the crystals forming in two immiscible melts, and this is true in theory, the

---

[8] Note, however, that, if the "globules" are from spherulitic growth, phenocrysts previously present can also appear to cut across the interface. Also, the morphology of coeval crystals of a given phase in the different melts may be very different.

practical application of this test is difficult at best. The total amounts of phases forming from the two are generally very different, introducing problems of physical reequilibration of solid solutions with melt. The first few tiny crystals of a given phase should be the same in both, but where little of this phase forms, solid diffusion with later layers could eliminate the early core, whereas in the conjugate melt, the large crystals of this early composition would be expected to persist as cores. Similarly, the outside rims of a given solid solution would only be identical if the two melts have maintained equilibrium during cooling. Equilibrium is certainly not expected to be maintained, because, as Bowen so clearly demonstrated, fractionation is the rule, not the exception. So the test of identical composition is applicable only where a few crystals are present in glass (e.g., Gélinas, Brooks, and Trzcienski, 1976), and even here, if these are "quench crystals" that grew rapidly, the possibility of original disequilibrium is great.

Bulk Composition Although it is difficult to obtain valid data from crystalline assemblages, uniformity of composition of globules, and of host, is a necessary but insufficient proof of immiscibility. Later deuteric or metamorphic alteration must be evaluated. Thus, late carbonate filling some gas vesicles is easy to recognize, but disseminated carbonate or analcite may be primary or secondary. Also, during hydration there may be gross changes in Fe^{2+}/Fe^{3+}. If there is a range in compositions it must be systematic for both globule and host, because tielines connecting these two melts can rotate as temperature changes. Alleged conjugate melts from several recent reports are plotted on Fig. 2-8. Some of these differ so much from the system K_2O–FeO–Al_2O_3–SiO_2 that a comparison of the tielines may be misleading. Note, however, that because anorthite ("An" on Fig. 2-8C) plots to the lower right of the diagram, crystallization of basic plagioclase will drive the residual liquid toward the field of immiscibility. The tielines for melts rich in alkalies plus alumina are more nearly horizontal on Fig. 2-8C (see also Ferguson and Currie, 1971), indicating much smaller differences in the amounts of SiO_2-network former in the two melts ($\sim 12\%$ compared with $> 30\%$ for the lunar samples of Fig. 2-8A). Some claim that the siliceous melt formed is never peralkaline, but Naslund (1976) has found such compositions.

Several features of bulk rock composition have been used as evidence of immiscibility (Philpotts, 1976, 1977). Thus, immiscibility is suggested if the ocellar rocks in a suite fall in the region of immiscibility, but the nonocellar rocks fall outside, or, if the deep seated rocks have a compositional gap but the near surface rocks have a continuum, with ocellar rocks in this gap.

Gradients at Interface At a meniscus between glasses from quenched conjugate melts, all composition gradients should be infinite (step-like). Electron microprobe traces over the boundary should be as sharp as the instrumental resolution, unless the boundary has been blurred by devitrification, alteration, or annealing during quench. If the apparent "menisci" are actually from the mixing of two normally miscible magmas, or from late stage melts filling vesicles, both the gradients at the interfaces and the gross difference in composition of globules and host should decrease under slower cooling (e.g., toward the center of a flow), the exact reverse of that found by Gélinas, Brooks, and Trzcienski (1976).

Evidence from Experimental Studies

NATURE OF LABORATORY EVIDENCE FOR IMMISCIBILITY The most convincing laboratory evidence of immiscibility is the formation of menisci between melts (now glasses) from an originally homogeneous melt. Because the surface tension at an interface between silicate melts may be low, and the viscosity high, these menisci need not be spherical. In experimental runs, globule size increases with time at constant temperature. Nucleation and growth of immiscible globules can and frequently does occur on slow quench, so it is important to avoid this source of ambiguity. Just as crystals formed on quench are generally minute or skeletal, immiscibility on quench generally yields only minute globules (<2 μm), and the size will not vary with run duration. Very commonly such globules will be unevenly distributed, being most prevalent in the slower cooled center of the charge.

In addition to the presence of menisci, the two glasses must have uniform properties such as color and index of refraction, or better still, uniform compositions, as determined by electron microprobe. Mafic melts will frequently devitrify during quench; this may change the color drastically but usually has only minor effects on the index of refraction or composition.

Apparent menisci between glasses are not adequate in themselves, because they may result merely from sluggish diffusion between two viscous melts, e.g., in the laboratory heating of mixtures of basaltic and rhyolitic glasses (Yoder, 1973). The obvious solution is to prove that the splitting is *reversible*, or that it occurs with various starting materials (e.g., Watson, 1976). Another important proof is the internal consistency of the data from adjacent compositions or temperatures, and, or course, their validity in terms of the phase rule. The various sources of ambiguity such as compositional gradients in the original charge, surface oxidation of iron, and reaction with the container, will generally yield glasses with variable compositions and inconsistent phase boundary data.

COMPOSITION   The partitioning of major elements between the two melts may be determined experimentally, and insofar as this partitioning is that found in nature, it provides strong supporting evidence for immiscibility. Numerous attempts have been made to simulate natural immiscibility with synthetic systems covering a wide range of compositions (Ferguson and Currie, 1971, 1972; Hess *et al.*, 1975; Khitarov *et al.*, 1973; McBirney and Nakamura, 1974; Markov *et al.*, 1972; Massion and Koster van Groos, 1973; Milyukov, 1970; Naslund, 1976; Philpotts, 1971; Philpotts and Hodgson, 1968; Pugin, 1975; Roedder and Weiblen, 1970; Rutherford *et al.*, 1974; Suleimenov *et al.*, 1973; Visser and Koster van Groos, 1976a,b; and Watson, 1976). Immiscibility was found in all these studies, but many had inadequate control of variables such as total FeO, P_{O_2}, and H_2O, so interpretation of the results is uncertain. Most were in the range 1000°-1200°C, but the immiscibility range was frequently narrow (e.g., 20°, Ferguson and Currie, 1971) and hence could easily be missed. The resulting experimental tielines are generally subparallel (e.g., see Fig. 2-8), but the extent of immiscibility differs widely. This difference probably results from the fact that Fig. 2-8 is a grossly oversimplified representation of the multicomponent natural systems.[9] What is needed are additional systematic studies of simple systems to gain a clearer understanding of the effects of each constituent, and additional detailed studies in which the entire phase assemblage is followed from the liquidus, through immiscibility, to the solidus (e.g., Rutherford *et al.*, 1974; Hess *et al.*, 1975).

Although the distribution of major elements during immiscibility yields a granitic or syenitic alkali-aluminosilicate melt (Irvine, 1976) resembling that from crystal fractionation, there are important differences in the partitioning of some minor elements. Thus, Roedder and Weiblen (1971) found P to be strongly partitioned into the lunar high-Fe melt, the opposite of that found in terrestrial crystal fractionation, where P is concentrated into the granitic residuum (Anderson and Greenland, 1969). The partitioning of P between immiscible melts has since been experimentally verified and measured by Ryerson and Hess (1975) and by Watson (1976). Watson (1976) also determined the partitioning and the effect on immiscibility in $K_2O-FeO-Al_2O_3-SiO_2$ for thirteen other elements. Additional criteria for immiscibility stem from other studies of isotopes and trace-element partitioning (Avdonin, 1975; Lovering and

---

[9] For example, a pyroxene-rich cumulate would plot with the high-Fe melts on Fig. 2-8

Wark, 1975; McCarthy and Hasty, 1976; Ryerson and Hess, 1975; and Gélinas, Brooks, and Trzcienski, 1976), and still other studies of this type are in progress. Extrapolation from these studies to natural systems requires evaluation, particularly of the effects of P_{H_2O}, P_{total}, and the fugacity of oxygen. Thus, Nakamura (1974) showed that immiscibility in the system leucite–fayalite–SiO_2 is greatly reduced or eliminated at 15 kbar, but Naslund (1976) showed that immiscibility increased greatly with oxygen fugacity, and that a similar field forms and expands as oxygen increases in the analogous system with Na. In other compositions, Markov et al. (1972) found immiscibility in hydrous alkalic ultrabasic melts up to 18 kbar, and Khitarov et al. (1973) showed that immiscibility in olivine tholeiitic melts increased with pressure up to 14 kbar. The role of P_{H_2O} in these studies is uncertain, because it is usually not the only variable. Philpotts (1971) reported that dry mixtures of teschenite and its nepheline-analcite ocelli showed no immiscibility, but they did exhibit immiscibility under 5000 psi H_2O. It is generally assumed, however, that water plays a role similar to alkalies in reducing immiscibility. Naslund and Watson (1977) have found that pressures up to 5 kbar, with or without CO_2, increase the size of the immiscibility volume in the system K_2O–FeO–Al_2O_3–SiO_2.

It is quite obvious that position on the Greig plot (Fig. 2-8) in itself is not an adequate criterion. Irvine (1975) uses a novel plot based on cationic norms that helps show the relation between immiscibility and plagioclase-mafic cotectics, but no plot is satisfactory for multicomponent rock systems. The many compositional parameters, and their interactions, make the understanding of immiscibility "seem very remote, but attainable" (Hess, 1971).

ALTERNATIVE EXPLANATIONS OF FEATURES ATTRIBUTED TO IMMISCIBILITY

Much of the evidence for immiscibility in nature is ambiguous, and so some of the features discussed above have been attributed to the following alternative processes (see also discussions and references in Carstens, 1963; Phillips, 1973; Ferguson and Currie, 1971; Gelinas, Brooks, and Trzcienski, 1976; Philpotts, 1976, 1977).

SPHERULITIC CRYSTALLIZATION   Fibrous crystallization, generally in a glass, radiating out from a common point of nucleation, produces spherical structures called spherulites. Greig (1928), showed that globular structures at Agate Point, Ontario, had radial textures that cut across presumed flow banding in the enclosing rhyolite. Both the radial texture and the crosscutting relations strongly suggest spherulitic crystallization after emplacement of a glassy flow, rather than immiscibility. Both immiscible melt globules and spherulites can show a flattened contact with contiguous neighbors; the globules by failure to break the film of host melt and mingle, the spherulites by leaving a layer of material rejected by the advancing crystallization fronts.

ORBICULAR CRYSTALLIZATION   Many occurrences of orbicular rocks are described in the literature, often with radial structure, concentric mineral bands, or both, and sometimes with country rock xenoliths as cores (e.g., Bryhni and Dons, 1975). Although immiscibility has been considered, it is highly unlikely, and there is no consensus as to the mechanism(s) by which these various features form. They presumably represent sequential crystallization from a fluid medium. Some have formed by the recrystallization of xenoliths. Holgate (1954) noted that during the reaction of quartzose xenoliths with mafic magmas in Scotland, a corona of pyroxene forms, and quartzo-feldspathic melts form within this corona. He attributed this to a "transfusion process' through a "semipermeable membrane" (the pyrozene corona), "due to the intervention of a condition of liquid immiscibility," and extended this concept to the origin of basalts and rhyolites in the crust. Although his invocation of immiscibility in the Skaergaard rocks was perceptive (McBirney and Nakamura, 1974), the interpretation of the data, and particularly of the physical mechanism involved, were seriously questioned by Roedder (1956); see also Holgate, (1956).

FILLING OF VESICLES WITH LATE MELTS   One of the most frequent suggestions for the origin of felsic globules and amoeboid "pegmatitic segregations" in most mafic rocks is that gas vesicles formed, and then late stage melts, enriched in alkali aluminosilicates by normal crystal fractionation, filled these vesicles to form "segregation vesicles" (Smith, 1967). The major but unstated problem with this origin is the disposition of the gas whose expansion against the magma was responsible for the vesicle in the first place. It is difficult to understand how the pressure in the vesicle could drop so far that the cavity could be refilled with another fluid, particularly one with an even higher concentration of volatiles. Smith (1967) has explained this paradox by assuming that after the vesicles formed, external pressure increased because of flow of the lava into deeper water, thus forcing the residual melt into the vesicles, even to the point of eliminating the gas phase completely. He does not explain, however, how a lava could continue to flow when it has crystallized to the point that the vesicles within it will not collapse when the external pressure is doubled. The exact reverse situation provides a possible explanation for some of these features. After the surrounding host has become rigid, shrinkage on crystallization in the core of the mass could result in redistribution of the remaining residual melt.

Considerable attention has been given to the behavior of adjacent gas vesicles that happen to contact. If the host melt is of low viscosity, they will first form a cuspate border and then quickly coalesce to a single sphere. Contrary to several published statements, the surface-tension forces acting on two contacting globules of immiscible fluid are identical in nature whether they be gas or melt. The force at a gas-melt interface will certainly be greater than at a melt-melt interface, but these differences may be minor compared with those of another independent variable, the viscosity of the host.

Bowen (1928) stated that there should be a continuum in the sizes of globules from immiscibility, and hence it is sometimes erroneously assumed that if there is no continuum, immiscibility has not occurred. Both gas and melt globules nucleate and grow via essentially identical processes of diffusion (unless from spinodal decomposition). Slow cooling can yield a single stage of nucleation and uniformly sized blebs of melt (or gas), as is commonly done in various manufacturing processes. Fast cooling will form nuclei at various stages, and hence a continuum of globule sizes, but this is true for both gas and immiscible melts.

FORMATION OF AMYGDULES   The formation of amygdules of obviously secondary minerals such as chalcedony is similar to the preceding mechanism and differs merely in that it takes place later in the cooling history. Ambiguity arises only when the filling consists of minerals such as zeolites, analcite, and carbonates that may be phases from a late-stage, possibly immiscible, volatile-rich fluid.

One very important, but frequently ignored, feature is the uniformity of filling composition, because globules formed by immiscibility should be essentially uniform in composition. Thus, "globules" that range from carbonate in one to zeolite in another (or even to empty vesicles) are almost certainly amygdules. Globules from immiscibility could, of course, occur in the same outcrop with amygdules.

MAGMA MIXING   Magma mixing has been invoked in lieu of immiscibility to explain the association of basic and silicic glasses at many volcanoes. Some volcanoes alternate eruptions of the two compositions, but more commonly there is a switch in a single eruption, as at Hekla, from early rhyolitic to later andesitic, with a mixed rock at the time of the switch, consisting of stringers and blebs. No adequate mechanism has been suggested, however, for dispersing the one melt into another of a different density, with which it is presumed to be miscible, in the form of isolated millimeter- to centimeter-sized blebs that have *sharp* contacts.[10]

Tuff flows may also cause ambiguity with respect to their origin. In these, obvious scoriaceous or pumiceous fragments of the two compositions are frequently mixed on a hand-specimen scale. If such tuff flows were subsequently welded, and particularly if mildly metamorphosed, the similarity to rocks formed by immiscibility might be striking (R. L. Smith, personal communication).

An additional field problem in using magma mixing to explain variolitic lavas (e.g., Gélinas, Brooks, and Trzcienski, 1976) is that in some localities the magma pairs are always found associated with each other, and never as individual flows with one or the other composition. It is difficult to imagine a plumbing system within the volcanic vent that would always tap *both* supplies, and do this with sufficient uniformity to provide 100-km-long mappable formations with an emulsion-like mixture (Wilson, 1941; Dimroth *et al.*, 1973).

OTHER TYPES OF IMMISCIBILITY IN MAGMAS

Immiscibility has occurred in most magmas, if the term is used in its more general sense for two or more coexisting noncrystalline polycomponent solutions, at equilibrium, differing in properties and generally in composition (Roedder and Coombs, 1967, p. 419). More limited definitions can result in semantic problems and misunderstanding. Thus, a fluid phase that has separated from some magmas is a hydrosaline melt with ≤ 70 wt. % NaCl and density ≥ 1.4 g/ml (Roedder and Coombs, 1967; Roedder, 1971). Most investigators would term this immiscibility, yet there may well be a continuum in both composition and density between such hydrosaline melts and low-density steam (i.e., vesiculation). Immiscibility has also been documented in a series of systems of silicates plus chlorides or fluorides, mainly by Russian workers (e.g., Kogarko *et al.*, 1974; Delitsyn *et al.*, 1974b).

---

[10] E.g., Gélinas, Brooks, and Trzcienski (1976) show natural gradients of < 10 μm, yet Yoder (1973) developed gradients ≤ 300 μm wide in only one hour in some of his basalt-rhyolite experiments. Similarly, the gradients in untreated samples from Breiddalur, Iceland, were found to be ≤ 5 μm (Yoder, 1973, p. 164), yet this "emulsion" flow is 20 feet thick (Walker, 1963), and hence cooled very slowly. The intermixing is intimate with individual blebs as small as 20 μm (Blake *et al.*, 1965), so Yoder's experiment showing homogenization of these glasses in one hour at 1200° could have been above the solvus. This solvus would, therefore, have to be subliquidus for these compositions, but not necessarily metastable, since crystals are present in both glassy phases of the rock (Yoder, 1973).

Another type of immiscibility, particularly in basaltic magmas, involves the separation of an immiscible sulfide-rich melt. This melt, which may carry important amounts of copper and nickel, separates early in the cooling history (Kullerud and Yoder, 1965; Skinner and Peck, 1969; Skinner and Barton, 1973; MacLean and Shimazaki, 1976) and may form above the olivine liquidus in some of these rocks (Roedder and Weiblen, 1970; Roedder, 1976).

Immiscibility has been suggested for the large phosphatic iron deposits such as Kiruna, Sweden (Asklund, 1949), and more recently, Kodal, Norway (Bergstøl, 1972), as well as the Khibiny apatite-nepheline ores in the USSR (Kogarko *et al.*, 1974), and even for chromite ores (Pavlov *et al.*, 1975). Although phosphorous is enriched in the high-iron melt in silicate immiscibility (up to 6%; Anderson, 1971), and many of the high-iron silicate melts found in nature do yield considerable oxide phase on crystallization, these melts have more SiO_2 than FeO. Naslund (1976) found, however, that the high-Fe melt in the system $FeO-Fe_2O_3-KAlSi_3O_8-SiO_2$ at high P_{O_2} contained as much as 80% $FeO + Fe_2O_3$. Philpotts (1967) has shown that mixtures of apatite, magnetite, and silicate (diorite) can produce *three* immiscible melts (phosphate-rich, phosphate-oxide, and silicate), but a wider range of possible compositions needs to be studied. Natural immiscibility between phosphatic and high-silica melts has been observed from the combustion of organic-rich sediments (Bentor and Kastner, 1976).

In CO_2-bearing systems, there is unambiguous field and laboratory evidence of immiscibility between carbonate-rich melts and various silicate melts, particularly those that are alkalic. The homogenization of such carbonate-silicate melt pairs has been actually observed in inclusions in apatite crystals from ijolite (Rankin and Le Bas, 1974). Silicate immiscibility frequently involves very appreciable amounts of carbonate, especially in the more alkali-rich phases (Table 2-1), and carbonate may be essential to such immiscibility. Several simple carbonate-bearing systems have been explored, particularly by Wyllie and coworkers (e.g., Koster van Groos and Wyllie, 1973), but the natural systems may well include significant Na_2O, K_2O, CaO, MgO, Al_2O_3, SiO_2, P_2O_5, CO_2, and H_2O, as well as lesser but possibly important FeO, TiO_2, Nb_2O_5, F, Cl, and rare earths. In fact, *three* fluid phases have been proposed by Ferguson and Currie (1971) for the Callander Bay dikes—a kaersutite-olivine-lamprophyre melt, a carbonate-rich melt, and a feldspar-zeolite-rich syenite melt.

Although none of these is exactly parallel to silicate-silicate immiscibility, they can provide insight that might be helpful. The physical behavior, the types of phase diagrams, and the nature of the evidence for them are all essentially the same, and for some, there may well be a continuum with immiscible silicate compositions.

CONCLUSIONS—PETROLOGICAL SIGNIFICANCE OF SILICATE IMMISCIBILITY

In recent years experimental laboratory proof of liquid immiscibility has been found with geologically reasonable compositions and temperatures in a variety of silicate systems, usually yielding a felsic, alkali-aluminosilicate melt and a mafic melt rich in Fe, Mg, Ca, and Ti. Although relatively small changes in composition can initiate or eliminate immiscibility, and vary the partition coefficients, immiscibility has been experimentally verified in such a wide range of rock compositions that it might be a general feature of many systems.[11] As many of the simpler oxide systems that do not show stable immiscibility have flat or sigmoid liquidi, they may have metastable subliquidus immiscibility. Adding other constituents may lower the liquidus to the point that it contacts the solvus but these same constituents can also decrease the extent and temperature of immiscibility, so the net effect is difficult to predict. Only the top of the solvus is exposed generally, and hence the temperature range for stable immiscibility may be small. The possible significance of metastable immiscibility in petrology, the limits under which silicate immiscibility occurs in compositions simulating natural rocks, and the possible connection with the immiscible separation of melts of carbonate-, oxide-, sulfide-, or phosphate-rich composition have yet to be investigated.

Field and laboratory evidence is adequate to prove that immiscibility occurred during the formation of a relatively few natural rocks, and to suggest it in others. The process is such that preservation of unambiguous evidence can be expected to be relatively exceptional, even if the process were common. It is also unfortunate that the most extensive occurrence of variolitic rocks believed to be from immiscibility (Gélinas, Brooks, and Trzcienski, 1976) is in metamorphosed Precambrian rocks, and glassy basalts with such blebs have not been found. Therefore one must extrapolate from the meager data base available to the assumption that immiscibility was probably much more common than just the few proven examples. The compositional range for which there is reasonable evidence for immiscibility is striking, and is far wider than several published

---

[11] Burnham (1975, p. 116) states that this evidence stems from either "disequilibrium or the action of a volatile-rich phase," because he believes that immiscibility in these compositions is precluded on theoretical grounds.

statements and predictions would indicate. This range includes lavas from low-K ultrabasic and basaltic komatiites (low-Al) to high-K nepheline basalts, high-Al olivine tholeiites, normal and high-Fe tholeiites, and lamprophyres, to various high-Ti lunar mare, "KREEP," and feldspathic basalts with a sevenfold range in K/Na ratio. An additional wide range of volcanic rocks may also have undergone immiscibility (Gélinas and Brooks, 1976).

Although the range of rock types is large, the most important parameter, the composition of the residual melt before splitting, is sometimes rather obscure (Roedder and Weiblen, 1977; Weiblen and Roedder, 1976). High or very high FeO/MgO seems to be a common but not universal feature. The composition of the felsic melt also varies widely, although most commonly rhyolitic, syenitic, or nepheline-syenitic melts are found.

Most significant, perhaps, is the great range in silica for the two melts. Because the mafic melt is generally undersaturated,[12] and many of the conjugate felsic ones are oversaturated, crossing the silica-saturation barrier is possible as a result of immiscibility. In the system leucite–fayalite–SiO_2 (Roedder, 1951), melts with normative silica are in equilibrium with iron-rich melts with considerable normative (and modal) leucite and fayalite. Also, some of the lunar immiscible potash granite melts with 43% normative quartz have olivine-normative conjugate melts (Table 2-2). Philpotts (1976) has explored the petrological significance of this difference in silica content in various terrestrial occurrences.

If immiscibility occurs early in the crystallization history of a magma, as seems most likely in alkalic systems (Philpotts, 1976), its effects will be more significant, but physical separation and crystallization both tend to eliminate the evidence. Some of the best proof of immiscibility in nature has consisted of minute globules of one glass embedded in another. In most examples this has happened at such a late stage that it in itself would have very little significance for petrology. But such evidence is not restricted to the last trace of liquid; one lunar basalt (sample 14310) split while there was ~12% melt (Roedder and Weiblen, 1971), and a terrestrial basalt split when there was at least 44% melt (Roedder and Weiblen, 1972a).

Most evidence for immiscibility comes from surface or near-surface rocks containing both fractions. The possibility exists, but it remains to be established, that any major field occurrence of *separate* bodies of rocks of similar composition, such as the associations lamprophyre–syenite, camptonite–nepheline syenite, basalt–rhyolite, and particularly the gabbro–granophyre association of the layered intrusions, is from

---

[12] Anfilogov (1975) claims that at least one of two immiscible melts *must* be undersaturated, but there is much experimental evidence to the contrary.

subcrustal immiscibility. Perhaps the best evidence for immiscibility in such associations is in the experimental work of McBirney and Nakamura (1974) on the Skaergaard rocks. They also have met the important test of essentially identical liquidus temperatures for the two melts (p. 349), but their conclusion has been challenged by Goles (1975) on the basis of trace element studies.

The gross composition of the felsic liquid in much immiscibility is granitic, and very similar to granite derived by crystal fractionation. The partitioning of minor constituents in the two processes is different, however, and future studies of trace element and isotopic distributions may well indicate that at least some terrestrial granites were formed by immiscibility, although most are probably from the reworking of crustal materials or crystal fractionation. Because some of the partition coefficients between crystal and liquid are far from unity (Schnetzler and Philpotts, 1970), there may be problems in interpretation of such data. Still less easy to prove, but perhaps even more likely, is the possibility of formation of a primeval granitic crust on the earth (or moon) by immiscibility; the present rocks showing evidence of immiscibility might merely represent the sweating out of the last remnants from the lithosphere.

The history and evolution of the concept of immiscibility in petrology illustrates well the workings of science. The pendulum has swung widely, from early uncritical acceptance, to later total disparagement, and now back to respectability based on direct observation and demonstration. It is perhaps appropriate to suggest that, had the present body of data been available to Bowen in 1928, he would have written a very different chapter.

Acknowledgments   The writer is indebted to many persons for discussions and particularly for many helpful reviews of the manuscript. As it would be impossible to rank these numerous contributions, the individuals are listed alphabetically: L. Gélinas, T. N. Irvine, A. R. McBirney, H. R. Naslund, A. R. Philpotts, M. J. Rutherford, K. J. Schulz, R. L. Smith, J. H. Stout, E. B. Watson, P. W. Weiblen, and H. S. Yoder, Jr. It is also fitting in this context for the writer to express his appreciation and indebtedness for the stimulation and encouragment he received from N. L. Bowen at the Geophysical Laboratory thirty years ago.

References

Anderson, A. T., Alkali-rich, SiO_2-deficient glasses in high-alumina olivine tholeiite, Hat Creek Valley, California, *Amer. J. Sci.*, *271*, 293–303, 1971.
Anderson, A. T., and D. Gottfried, Contrasting behavior of P, Ti, and Nb in a

differentiated high-alumina olivine–tholeiite and a calc-alkaline andesite suite: *Geol. Soc. Amer. Bull.*, *82*, 1929–1942, 1971.

Anderson, A. T., and L. P. Greenland, Phosphorus fractionation diagrams as a quantitative indicator of crystallization differentiation of basaltic liquids, *Geochim. Cosmo. Acta*, *33*, 493–505, 1969.

Andon, R. J. L., and J. D. Cox, Phase relationships in the pyridine series. Part I. The miscibility of some pyridine homologues with water, *J. Chem. Soc. (London)*, *1952*, 4601–4606, 1952.

Anfilogov, V. N., Liquation in magmas and petrographic criteria for its recognition, *Geokhimiya, 1975*, *7*, 1035–1042, 1975 (in Russian; translated in *Geochem. Internat. 12*, 4, 54–61).

Asklund, B., The differentiation problem of the apatite iron ores, *Geol. Fören. Stockh. Förhand.*, *71*, 127–176, 1949 (in Swedish).

Avdonin, V. V., Role of the process of liquation in ore genesis, *Akad. Nauk S.S.S.R. Doklady*, *224*, 909–911, 1975 (in Russian).

Barron, L. M., Thermodynamic multicomponent silicate equilibrium phase calculations, *Am. Mineral.*, *57*, 809–823, 1972.

Baylor, Richard, Jr., and J. J. Brown, Jr., Phase separation of glasses in the system $SrO-B_2O_3-SiO_2$, *Amer. Ceram. Soc. Jour.*, *59*, 131–136, 1976.

Bentor, Y. K., and M. Kastner, Combustion metamorphism in Southern California, *Science*, *193*, 486–487, 1976.

Bergstøl, S., The jacupirangite at Kodal, Vestfold, Norway, *Mineral. Deposita*, *7*, 233–246, 1972.

Blake, D. H., R. W. D. Elwell, I. L. Gibson, R. R. Skelhorn, and G. P. L. Walker, Some relationships resulting from the intimate association of acid and basic magmas, *Quart. Jour. Geol. Soc. Lond.*, *121*, 31–49, 1965.

Boon, J. A., and W. S. Fyfe, The coordination number of ferrous ions in silicate glasses, *Chem. Geol*, *10*, 287–298, 1972.

Bowen, N. L., *The Evolution of the Igneous Rocks*, Princeton University Press, Princeton, N.J., 332 p., 1928.

Brooks, Christopher, and L. Gélinas, Immiscibility and ancient and modern volcanism, *Carnegie Institution of Washington Year Book 74*, 240–247, 1975.

Bryhni, Inge, and J. A. Dons, Orbicular lamprophyre from Vestby, southeast Norway, *Lithos*, *8*, 113–122, 1975.

Burnham, C. W., Thermodynamics of melting in experimental silicate-volatile systems, *Fortschr. Miner.*, *52*, Spec. Issue: IMA-Papers 9th Meeting, Berlin-Regensburg 1974, 101–118, 1975.

Butler, J. R., Orbicular rocks from Davie County, North Carolina piedmont, *Southeastern Geol.*, *15*, 127–139, 1973.

Cahn, J. W., Spinoidal decomposition, *Metallur. Soc. of A.I.M.E., Trans.*, *242*, 166–180, 1968.

Carman, M. F., Jr., M. Cameron, B. Gunn, K. L. Cameron, and J. C. Butler, Petrology of Rattlesnake Mountain sill, Big Bend National Park, Texas, *Geol. Soc. Amer. Bull.*, *86*, 177–193, 1975.

Carstens, H., On the variolitic structure, *Norges Geol. Undersokelse Arb.*, *223*, 26–42, 1963.

Charles, R. J., Immiscibility and its role in glass processing, *Amer. Ceram. Soc. Bull.*, *52*, 673–680, 686, 1973.

Cox, J. D., Phase relationships in the pyridine series. Part II. The miscibility of some pyridine homologues with deuterium oxide, *J. Chem. Soc. (London)*, *1952*, 4606–4608, 1952.

Currie, K. L., A criterion for predicting liquid immiscibility in silicate melts, *Nature, Phys. Sci.*, *240*, 66–68, 1972.

———, The geology and petrology of the Ice River alkaline complex, British Columbia, *Geol. Survey of Canada Bull.*, *245*, 59–65, 1975.

De, Aniruddna, Silicate liquid immiscibility in the Deccan traps and its petrogenetic significance, *Geol. Soc. Amer. Bull.*, *85*, 471–474, 1974.

Delitsyn, L. M., B. N. Melent'yev, and L. V. Delitsyna, Liquation in melts, its origin, evolution, and stabilization, *Akad. Nauk. S.S.S.R, Doklady*, *219*, 190–192, 1974a (in Russian; translated in *Doklady Acad. Sci. U.S.S.R*, *219*, 143–145, 1975).

Delitsyn, L. M., B. N. Melent'yev, and L. V. Delitsyna, The system acmite–nepheline–villiaumite and the differentiation of alkalic magma. *Akad. Nauk S.S.S.R., Doklady*, *214*, 1, 186–189, 1974b (in Russian; translated in *Doklady Acad. Sci. U.S.S.R.*, *214*, 194–196).

Dence, M. R., W. von Engelhardt, A. G. Plant, and L. S. Walter, Indications of fluid immiscibility in glass from West Clearwater Lake impact crater, Quebec, Canada, *Contr. Mineral. and Petrol.*, *46*, 81–97, 1974.

Dimroth, E., P. Boivin, N. Goulet, and M. Larouche, Tectonic and volcanological studies in the Rouyn-Noranda area, *Open-File Manuscript, Qué. Dept. Nat. Resources*, 60 pp., 1973 [as quoted by Gélinas, *et al.*, (1976)].

Doremus, R. H., Phase Separation, Chapter 4 in *Glass Science*, John Wiley and Sons, New York, 1973.

Drever, H. I., Immiscibility in the picritic intrusion at Igdlorssuit, W. Greenland, *Report of 21st Session, Internat. Geol. Cong.*, Part XIII, 47–58, 1960.

Fenner, C. N., Immiscibility of igneous magmas, *Amer. J. Sci.*, *246*, 465–502, 1948.

Ferguson, John, and K. L. Currie, Evidence of liquid immiscibility in alkaline ultra-basic dikes at Callander Bay, Ontario, *J. Petrol.*, *12*(3), 561–585, 1971.

Ferguson, John, and K. L. Currie, Silicate immiscibility in the ancient "basalts" of the Barberton Mountain Land, Transvaal, *Nature, Phys. Sci.*, *235*, 86–89, 1972.

Furnes, Harald, Variolitic structure in Ordovician pillow lava and its possible significance as an environmental indicator, *Geology*, *1*, 27–30, 1973.

Gélinas, L., and C. Brooks, Ancient and modern volcanics and proposed miscibility in silicate systems, *Reports Dépt. de Génie Minéral, Ecole Polytech. de Montréal, EP76-R-5*, C-1 to C-32, 1976.

Gélinas, L., C. Brooks, and W. E. Trzcienski, Jr., Archean variolites—quenched immiscible liquids, *Canad. J. Earth Sci.*, *13*, 210–230, 1976.

Goldburg, W. I., and J. S. Huang, Phase separation experiments near the critical point, *in* Fluctuations, Instabilities, and Phase Transitions, *Proc. of NATO Advanced Study Inst., Geilo, Norway, April, 1975*, Tormod Riste, ed., New York, Plenum Press, 87–106, 1975.

Goles, G. G., Comparisons between dikes of the Kangerdlugssuag region, east Greenland, and rocks of the Skaergaard intrusion (abst.), *Amer. Geophys. Union Trans.*, *EOS*, *56*, 12, 1070, 1975.

Greig, J. W., Immiscibility in silicate melts, *Am. J. Sci.*, *13*, 1–44, 133–154, 1927.

———, On the evidence which has been presented for liquid silicate immiscibility in the laboratory and in the rocks of Agate Point, Ontario, *Am. J. Sci.*, *265*, 373–402, 1928.

Haller, W., Rearrangement kinetics of the liquid-liquid immiscible microphases in alkali borosilicate melts, *J. Chem. Phys.*, *42*, 686–693, 1965.

Haller, W., D. H. Blackburn, F. E. Wagstaff, and R. J. Charles, Metastable immiscibility surface in the system $Na_2O–B_2O_3–SiO_2$, *J. Amer. Ceram. Soc.*, *53*, 34–39, 1970.

Hess, P. C., Polymer model of silicate melts, *Geochim. Cosmo. Acta*, *35*, 289–306, 1971.

———, $PbO–SiO_2$ melts: structure and thermodynamics of mixing, *Geochim. Cosmo. Acta*, *39*, 671–687, 1975.

Hess, P. C., M. J. Rutherford, R. N. Guillemette, F. J. Ryerson, and H. A. Tuchfeld, Residual products of fractional crystallization of lunar magmas: An experimental study, *Proc. Sixth Lunar Sci. Conf.*, *Geochim. Cosmo. Acta Suppl. 6*, Vol. 1, 895–909, 1975.

Hofmann, A. W., Diffusion of Ca and Sr in a basalt system, *Carnegie Institution of Washington Year Book 74*, for 1974–1975, 183–189, 1975.

Holgate, Norman, The role of liquid immiscibility in igneous petrogenesis, *J. Geology*, *62*, 439–480, 1954.

———, The role of liquid immiscibility in igneous petrogenesis: a reply, *J. Geology*, *64*, 89–93, 1956.

Irvine, T. N., The silica immiscibility effect in magmas, *Carnegie Institution of Washington Year Book 74*, for 1974–1975, 484–492, 1975.

———, Metastable liquid immiscibility and $MgO–FeO–SiO_2$ fractionation patterns in the system $Mg_2SiO_4–Fe_2SiO_4–CaAl_2Si_2O_8–KAlSi_3O_8–SiO_2$, *Carnegie Institution of Washington Year Book 75*, for 1975–1976, 597–611, 1976.

Kesler, S. E., and P. W. Weiblen, Distribution of elements in a spherulitic andesite, *Am. Mineral.*, *53*, 2025–2035, 1968.

Khitarov, N. I., V. A. Pugin, I. A. Soldatov, and I. D. Schevaleevsky, On the immiscibility in olivine tholeiite (experimental data), *Geokhimiya*, *1973* (12), 1763–1771, 1973 [in Russian; translated in *Geochem. Internat.*, *10*, (6), 1293–1299 (pub. 1974)].

Kogarko, L. N., I. D. Ryabchikov, and H. Sørenson, Liquid fractionation, in *The Alkaline Rocks*, H. Sørensen, ed., John Wiley and Sons, New York, pp. 488–500, 1974.

Koster van Groos, A. F., and P. J. Wyllie, Liquid immiscibility in the join $NaAlSi_3O_8–CaAl_2Si_2O_8–Na_2CO_3–H_2O$, *Am. J. Sci.*, *273*, 465–487, 1973.

Kullerud, G., and H. S. Yoder, Jr., Sulfide-silicate relations, *Carnegie Institution of Washington Year Book 64*, for 1964–1965, 192–193, 1965.

Levin, E. M., Structural interpretation of immiscibility in oxide systems: IV,

Occurrence, extent, and temperature of the monotectic, *J. Amer. Ceram. Soc.*, *50*, 29–38, 1967.

Lovering, J. F., and D. A. Wark, The lunar crust—chemically defined rock groups and their potassium-uranium fractionation, *Proc. Sixth Lunar Sci. Conf.*, *Geochim. Cosmo. Acta Suppl. 6*, Vol. 2, 1203–1207, 1975.

McBirney, A. R., Differentiation of the Skaergaard intrusion, *Nature*, *253*, 691–694, 1975.

McBirney, A. R., and Yasuo Nakamura, Immiscibility in late-stage magmas of the Skaergaard intrusion, *Carnegie Institution of Washington Year Book 73*, for 1973–1974, 348–352, 1974.

McCarthy, T. S., and R. A. Hasty, Trace element distribution patterns and their relationship to the crystallization of granitic melts, *Geochim. Cosmo. Acta*, *40*, 1351–1358, 1976.

MacKenzie, D. E., and A. J. R. White, Phonolite globules in basanite from Kiandra, Australia, *Lithos*, *3*, 309-317, 1970.

MacLean, W. H., and H. Shimazaki, The partition of Co, Ni, Cu, and Zn between sulfide and silicate liquids, *Econ. Geol.*, *71*, 1049–1057, 1976.

Markov, V. K., V. V. Nasedkin, and Yu. N. Ryabinin, Liquation in ultramafic alkalic magma at high pressures, *Akad. Nauk S.S.S.R., Doklady*, *207*, 428–429, 1972 (in Russian; translated in *Doklady Acad. Sci. U.S.S.R.*, *207*, 154–155, 1973).

Mason, B., and W. G. Melson, Comparison of lunar rocks with basalts and stony meteorites, *Proc. Apollo 11 Lunar Sci. Conf.*, *Geochim. Cosmo. Acta Suppl. 1*, Vol. 1, 661–671, 1970.

Massion, P. J., and A. F. Koster van Groos, Liquid immiscibility in silicates, *Nature, Phys. Sci.*, *245*, 60–63, 1973.

Milyukov, Y. M., Immiscibility in magma and artificial silicate melts and glasses, *Akad. Nauk S.S.S.R., Izvestiya, ser. geol.*, *1970*, no. 6, 126–131, 1970 (in Russian; translated in *Internat. Geol. Rev.*, *13*, 1247–1250, 1971).

Muan A., Phase equilibria in the system $FeO-Fe_2O_3-SiO_2$, *AIME Trans.*, *203*, 965–976, 1955.

Mutanen, Tapani, Komatiites and komatiite provinces in Finland, *Geologi*, *28*, 49–56, 1976 (in English).

Nakagawa, K., and T. Izumitani, Effect of a third component upon the immiscibility of binary glass, *Physics and Chem. of Glasses*, *13*, 85–90, 1972.

Nakamura, Yasuo, The system $Fe_2SiO_4-KAlSi_2O_6-SiO_2$ at 15 kbar, *Carnegie Institution of Washington Year Book 73*, for 1973–1974, 352–354, 1974.

Naslund, H. R., Liquid immiscibility in the system $KAlSi_3O_8-NaAlSi_3O_8-FeO-Fe_2O_3-SiO_2$ and its application to natural magmas, *Carnegie Institution of Washington Year Book 75*, for 1975–1976, 592–597, 1976.

Naslund, H. R., and E. B. Watson, The effect of pressure on liquid immiscibility in the system $K_2O-FeO-Al_2O_3-SiO_2-CO_2$ (abst.), *Geol. Soc. America Abstracts with Programs*, *9* (7), 1110–1111, 1977.

Olsen, Edward, and Eugene Jarosewich, The chemical composition of the silicate inclusions in the Weekeroo Station iron meteorite, *Earth Plan. Sci. Letters*, *8*, 261–266, 1970.

Ol'shanskii, Ya. I., Equilibrium of two immiscible liquids in silicate systems of alkali earth metals, *Akad. Nauk S.S.S.R., Doklady*, 76, (1), 93–96, 1951 (in Russian).

Paakkola, Juhani, The volcanic complex and associated manganiferous iron formation of the Porkonen-Pahtaraara area in Finnish Lapland, *Commis. Géol. de Finlande, Bull.*, No. 247, 1–83, 1971.

Pavlov, N. V., I. I. Grigorieva, and A. I. Tsepin, Chromite nodules as indicators of liquation in magmatic melt, *Acad. Nauk S.S.S.R., Izvestiya, Geol. Series, 1975*, no. 11, 29–45, 1975 (in Russian).

Phillips, W. J., Interpretation of crystalline spheroidal structures in igneous rocks, *Lithos*, 6, 235–244, 1973.

Philpotts, A. R., Origin of certain iron titanium oxide and apatite rocks, *Econ. Geol.*, 62, 303–315, 1967.

———, Immiscibility between feldspathic and gabbroic magmas, *Nature, Phys. Sci.*, 229, 107–109, 1971.

———, Density, surface tension and viscosity of the immiscible phase in a basic, alkaline magma, *Lithos*, 5, 1–18, 1972.

———, Silicate liquid immiscibility: its probable extent and petrogenetic significance, *Am. J. Sci.*, 276, 1147–1177, 1976.

———, Archean variolites—quenched immiscible liquids: discussion, *Can. J. Earth Sci.*, 14, 139–144, 1977.

Philpotts, A. R., and C. J. Hodgson, Rôle of liquid immiscibility in alkaline rock genesis, *23rd Internat. Geol. Cong.*, 2, 175–188, 1968.

Popov, V. S., Globular texture of lamprophyres, *Vses. Mineral. Obshch., Zapiski*, 51, 370–379, 1972 (in Russian).

Pugin, V. A., Conversion of olivine tholeiites to alkali magmas, *Geokhimiya, 1975*, no. 8, 1216–1222, 1975 (in Russian; translated in *Geochem. Internat.*, 12, 4, 196–201).

Quick, J. E., and A. L. Albee, 12013 revisited—two clastladen melts (abst.), in *Lunar Science VII*, 712–714, Houston, Texas, Lunar Science Institute, 1976.

Rankin, A. H., and M. J. Le Bas, Liquid immiscibility between silicate and carbonate melts in naturally occurring ijolite magma, *Nature*, 250, 206–209, 1974.

Roedder, Edwin, Low temperature liquid immiscibility in the system $K_2O-FeO-Al_2O_3-SiO_2$, *Am. Mineral.*, 36, 282–286, 1951.

———, A reconnaissance of liquidus relations in the system $K_2O \cdot 2SiO_2-FeO-SiO_2$, *Am. J. Sci. Bowen Volume*, 435–456, 1952.

———, Liquid immiscibility in the system $K_2O-FeO-Al_2O_3-SiO_2$ (abst.), *Geol. Soc. Amer. Bull.*, 64, 1466, 1953a.

———, High-silica portion of the system $K_2O-FeO-Al_2O_3-SiO_2$ (abst.), *Geol. Soc. Amer. Bull.*, 64, 1554, 1953b.

———, The role of liquid immiscibility in igneous petrogenesis: a discussion, *J. Geology*, 64, 84–88, 1956.

———, Silicate melt systems, in *Physics and Chemistry of the Earth*, Vol. 3, Pergamon Press, London, pp. 224–297, 1959.

———, Fluid inclusion studies on the porphyry-type ore deposits at Bingham, Utah, Butte, Montana, and Climax, Colorado, *Econ. Geol.*, 66, 98–120, 1971.

————, Petrologic data from experimental studies on crystallized silicate melt and other inclusions in lunar and Hawaiian olivine, *Am. Mineral.*, *61*, 684–690, 1976.

————, Liquid immiscibility in K_2O–FeO–Al_2O_3–SiO_2, a discussion, *Nature*, *267*, 558–559, 1977.

————, Silicate liquid immiscibility in magmas and in the system K_2O–FeO–Al_2O_3–SiO, an example of serendipity, *Geochim. Cosmo. Acta*, *42*, 1597–1617, 1978.

Roedder, Edwin, and D. S. Coombs, Immiscibility in granitic melts, indicated by fluid inclusions in ejected granitic blocks from Ascension Island, *J. Petrol.*, *8* (3), 417-451, 1967.

Roedder, Edwin, and P. W. Weiblen, Lunar petrology of silicate melt inclusions, Apollo 11 rocks, *Proc. Apollo 11 Lunar Sci. Conf., Geochim. Cosmo. Acta Suppl. 1*, Vol. 1, 801–837, 1970.

Roedder, Edwin, and P. W. Weiblen, Petrology of silicate melt inclusions, Apollo 11 and Apollo 12 and terrestrial equivalents, *Proc. Second Lunar Sci. Conf., Geochim. Cosmo. Acta, Suppl. 2*, Vol. 1, 507–528, 1971.

Roedder, Edwin, and P. W. Weiblen, Petrographic features and petrologic significance of melt inclusions in Apollo 14 and 15 rocks, *Proc. Third Lunar Sci. Conf., Geochim. Cosmo. Acta, Suppl. 3*, Vol. 1, 251–279, 1972a.

Roedder, Edwin, and P. W. Weiblen, Silicate melt inclusions and glasses in lunar soil fragments from the Luna 16 core sample, *Earth Plan. Sci. Letters*, *13*, 272–285, 1972b.

Roedder, Edwin, and P. W. Weiblen, Anomalous low-K silicate melt inclusions in ilmenite from Apollo 17 basalts, *Proc. Sixth Lunar Sci. Conf., Geochim. Cosmo. Acta Suppl. 6*, Vol. 1, 147–164, 1975.

Roedder, Edwin, and P. W. Weiblen, Compositional variation in late-stage differentiates in mare lavas, as indicated by silicate melt inclusions, *Proc. Eighth Lunar Sci. Conf., Geochim. Cosmo. Acta, Suppl. 8*, Vol. 2, 1767–1783, 1977.

Rutherford, M. J., P. C. Hess, and G. H. Daniel, Experimental liquid line of descent and liquid immiscibility for basalt 70017, *Proc. Fifth Lunar Sci. Conf., Geochim. Cosmo. Acta, Suppl. 5*, Vol. 1, 569–583, 1974.

Ryerson, F. J., and P. C. Hess, The partitioning of trace elements between immiscible silicate melts (abst.), *Amer. Geophys. Union Trans.*, *56*, 470, 1975.

Schairer, J. F., and Kenzo Yagi, The system FeO–Al_2O_3–SiO_2, *Amer. J. Sci. Bowen Volume*, *250A*, 471–512, 1952.

Schnetzler, C. C., and J. A. Philpotts, Partition coefficients of rare-earth elements between igneous matrix material and rock-forming mineral phenocrysts—II, *Geochim. Cosmo. Acta*, *34*, 331–340, 1970.

Scofield, Nancy, and D. C. Noble, Two-phase glass inclusions in phenocrysts in rhyodacite represent metastable liquid immiscibility in undercooled silicic melt (abst.), *Geol. Soc. America Abstracts with Programs*, *9* (7), 1165–1166, 1977.

Seward, T. P., III, D. R. Uhlmann, and David Turnbull, Phase separation in the system BaO–SiO_2, *J. Amer. Ceram. Soc.*, *51*, 278–285, 1968.

Skinner, B. J., and P. B. Barton, Jr., Genesis of mineral deposits, *Ann. Rev. Earth and Plan. Sci.*, *1*, 183–211, 1973.

Skinner, B. J., and D. L. Peck, An immiscible sulfide melt from Hawaii, in *Magmatic*

Ore Deposits, H. D. B. Wilson, ed., Econ. Geol. Monograph 4, pp. 310–322, 1969.

Smith, R. E., Segregation vesicles in basaltic lava, *Am. J. Sci.*, *265*, 696–713, 1967.

Suleimenov, S. T., M. Sh. Sharafiev, T. A. Abduvaliev, T. D. Nurbekov, and I. I. Sorokina, Microphase separation in glass containing certain crystallization initiators, in *Phase Separation Phenomena in Glass*, E. A. Porai-Koshits, ed., translated by Consultants Bureau, 208 pp., 1973.

Switzer, G. S., Composition of three glass phases present in an Apollo 15 basalt fragment, *Mineral. Sci. Invest., 1972–1973, Smithsonian Contrib. Earth Sci.*, *14*, 25–30, 1975.

Tewhey, J. C., and P. C. Hess, Two-phase region of $SrO-CaO-SiO_2$ melts (abst.), *Amer. Geophys. Union Trans.*, *55*, 483, 1974.

Upton, B. G. J., The petrology of a camptonite sill in South Greenland, *Grønlands Geol. Undersøgelse, Bull. 50*, 1–19 (*Meddel. om Grønland, 169* (11), 1–19), 1965.

Visser, Wiekert, and A. F. Koster van Groos, Liquid immiscibility in the system $Na_2O-K_2O-FeO-Al_2O_3-SiO_2$ (abst.), *Amer. Geophys. Union Trans.*, *57*, 340, 1976a.

Visser, Wiekert, and A. F. Koster van Groos, Liquid immiscibility in $K_2O-FeO-Al_2O_3-SiO_2$, *Nature*, *264*, 426–427, 1976b.

Vogel, T. A., and J. T. Wilband, Coexisting acidic and basic melts: geochemistry of a composite dike, *J. Geology*, *86*, 353–371, 1978.

Wager, L. R., and W. A. Deer, Geological investigations in East Greenland. Part III. The petrology of the Skaergaard intrusion, Kangerdlugssuaq, East Greenland, *Meddel. om Gronland, 105* (4), 335 pp. (re-issue, 1962), 1939.

Walker, G. P. L., The Breiddalur central volcano, eastern Iceland, *Quart. Jour. Geol. Soc. London*, *119*, 29–63, 1963.

Watson, E. B., Two-liquid partition coefficients: experimental data and geochemical implications, *Contrib. Mineral. Petrol.*, *56*, 119–134, 1976.

Watson, E. B., and H. R. Naslund, The effect of pressure on liquid immiscibility in the system $K_2O-FeO-Al_2O_3-SiO_2-CO_2$, *Carnegie Institution Washington Year Book 76*, for 1976–1977, 410–414, 1977.

Weiblen, P. W., and Edwin Roedder, Petrology of melt inclusions in Apollo samples 15598 and 62295, and of clasts in 67915 and several lunar soils, *Proc. Fourth Lunar Sci. Conf., Geochim. Cosmo. Acta, Suppl. 4*, Vol. 1, 681–703, 1973.

Weiblen, P. W., and Edwin Roedder, Compositional interrelationships of mare basalts from bulk chemical and melt inclusion studies, *Proc. Seventh Lunar Sci. Conf., Geochim. Cosmo. Acta Suppl. 7*, Vol. 2, 1449–1466, 1976.

Wiebe, R. A., Parental magma and liquid line of descent for a massive anorthosite pluton in the Nain anorthosite massif, Labrador (abst.), *Geol. Soc. America Abstracts with Programs*, *9* (7), 1226, 1977.

Wilson, M. E., Noranda District, Quebec, *Geol. Surv. Can. Mem.*, *229*, 148p., 1941.

Yeats, R. W., W. C. Forbes, K. F. Scheidegger, G. R. Heart, and T. H. Van Andel, Core from Cretaceous basalts, Central Equatorial Pacific, Leg 16. Deep sea drilling project. *Geol. Soc. Amer. Bull.*, *84*, 871–882, 1973.

Yoder, H. S., Jr., Contemporaneous basaltic and rhyolitic magmas, *Am. Mineral.*, *58*, 153–171, 1973.

Zdaniewski, W., DTA and x-ray analysis study of nucleation and crystallization of $MgO-Al_2O_3-SiO_2$ glasses containing ZrO_2, TiO_2, and CeO_2, *J. Amer. Ceram. Soc.*, *58*, 161–169, 1975.

Chapter 3

FRACTIONAL CRYSTALLIZATION
AND PARTIAL FUSION*

D. C. PRESNALL

Hawaii Institute of Geophysics, University of Hawaii, Honolulu, Hawaii
(Permanent address: Department of Geosciences, The University of Texas at Dallas,
P.O. Box 688, Richardson, Texas)

INTRODUCTION

In order to develop a comprehensive theory for the evolution of the
igneous rocks, a principal requirement is to explain the chemical diversity
of different rock types and the often continuous nature of chemical
changes between these types. At the same time, the frequency of occurrence
of all rock types is not the same, and it is equally important to explain the
existence of large volumes of chemically similar magma. Two very
important processes controlling the chemistry of igneous rocks are partial
fusion at the source region and fractional crystallization as the magmas
move upward to the earth's surface. Because phase relationships are
strongly dependent on pressure, the depth of liquid separation from the
parent material and the depth range over which fractional crystallization
occurs are also important. Two other processes, mixing of magmas and
assimilation (see Chapters 7 and 10), will not be dealt with here, but they
probably occur to some degree in almost every magmatic process and
may be important in explaining the origin of andesites (Anderson, 1976)
and granitic batholiths (Presnall and Bateman, 1973).

Before discussing the generation and evolution of magmas, it is necessary to define two terms, primary magma and parental magma. These
terms have sometimes been used interchangeably, but this practice will

© 1979 by Princeton University Press
The Evolution of the Igneous Rocks: Fiftieth Anniversary Perspectives
0-691-08223-5/79/0059-17$00.85/0 (cloth)
0-691-08224-3/79/0059-17$00.85/0 (paperback)
For copying information, see copyright page

not be followed here. *Primary magma* will refer to a magma as it exists immediately after separation from its source region. *Parental magma* will refer to a magma that has produced other magma compositions by fractional crystallization. Both primary and parental magmas may be completely liquid or may consist of a mixture of liquid, crystals, and gas. By these definitions, the parental magma of a gabbroic layered intrusion emplaced high in the earth's crust (assuming a single filling of the magma chamber) would be the magma as it exists just after emplacement and prior to cooling and fractional crystallization. This parental magma might have experienced fractional crystallization as it moved from its site of origin in the mantle to its site of final emplacement, and therefore would not be primary. Also, a primary magma might rise to the surface rapidly enough to prevent fractional crystallization. It would therefore not be parental, but would still be primary.[1]

Bowen (1928) considered fractional crystallization of basaltic parental magma to be the most important mechanism for producing the chemical diversity of common igneous rocks. Partly because plate tectonic concepts have forced a complete reevaluation of previous theories of petrogenesis, the development of concepts based on partial fusion has accelerated, and the number of primary and parental magma types is at present uncertain and controversial. There have been proposals that partial fusion is important in controlling the compositions of various types of basalt (Kuno, 1959; D. H. Green and Ringwood, 1967; Kushiro, 1968), andesite (T. H. Green and Ringwood, 1968; Yoder, 1969; Dickinson, 1970), granite (Tuttle and Bowen, 1958; Winkler, 1974), and rhyolite (Yoder, 1973).

In this chapter, attention will be directed first to evidence bearing on the question of whether or not fractional crystallization of basaltic parental magma is a sufficiently important petrologic process that it is primarily responsible for producing the great volumes of felsic intrusive rocks in granitic batholiths of orogenic regions. There appears to be essentially universal agreement that fractional crystallization is important in modifying magma compositions, but controversy persists over the importance of this process in producing granitic material from a basaltic parental magma. Following this discussion of fractional crystallization will be a section devoted to an expansion of some theoretical aspects of

---

* Hawaii Institute of Geophysics Contribution No. 781

[1] For another discussion of the usage of these terms, see Carmichael, Turner, and Verhoogen (1974, pp. 44–46).

partial fusion published earlier (Presnall, 1969), with an emphasis on comparisons between types of fusion processes likely to occur in nature and theoretical relationships deduced from simple phase diagrams.

FRACTIONAL CRYSTALLIZATION

Bowen (1928, 1941) distinguished two types of crystallization. During *equilibrium crystallization*, crystals formed on removal of heat continually react and reequilibrate with the liquid. During *fractional crystallization*, the crystals are immediately isolated from the system as soon as they are formed and are thereby prevented from further reaction with the liquid.

At the heart of Bowen's scheme of petrogenesis was the idea that andesites and the continuous array of rocks to rhyolite and granite are produced by fractional crystallization of basaltic parental magma. Basaltic, andesitic, and granitic rocks are typically associated together in orogenic zones, but field evidence supporting the proposition that they are co-magmatic and related by fractional crystallization is generally not definitive.

Layered gabbroic intrusions are the best documented and clearest examples of extreme fractional crystallization of basalt, but the differentiation trend toward iron-enrichment observed in these intrusions does not correspond to the typical association of basalt, andesite, and granite observed in orogenic regions along continental margins. Because these layered intrusions sometimes appear to have produced small amounts of felsic differentiates after the main iron-enrichment trend, it is instructive to compare the composition of the late differentiates from a well-documented layered gabbroic intrusion with the composition of a typical granitic batholith; for if Bowen's scheme of magmatic evolution is to apply to the majority of igneous rocks, it must apply to large granitic batholiths. For this comparison, the Skaergaard intrusion and the Sierra Nevada batholith are chosen.

It is commonly accepted that the middle stages of differentiation in layered gabbroic intrusions such as the Skaergaard intrusion do not correspond to the middle stages of the sequence basalt–andesite–dacite–rhyolite. It is not commonly emphasized, however, that the compositions of the late-stage crystallization products from the Skaergaard intrusion also do not correspond to late-stage crystallization products (granites) in typical large granitic batholiths. To show this lack of correspondence, a comparison will be made between the late-stage crystallization products from the Skaergaard intrusion and a typical Sierra Nevada granite.

Because some might consider that the entire batholith represents a late-stage crystallization residue from a basaltic parental magma, comparisons will also be made with Sierran quartz monzonites and granodiorites, which are more typical of the batholith as a whole.

Rocks formed during the last stages of crystallization of the Skaergaard magma are the melanogranophyres, followed by the Sydtoppen transitional granophyre and the Tinden acid granophyre (Wager and Brown, 1967). Initial ^{87}Sr-^{86}Sr ratios for the Sydtoppen and Tinden granophyres are much higher than those for the melanogranophyres and the other rocks of the main layered sequence; and it appears, based on this evidence, that the melanogranophyres are the last rocks to crystallize that can be attributed solely to fractional crystallization (Hamilton, 1963; W. P. Leeman, personal communication, 1976). In Table 3-1, the composition of the melanogranophyres (columns 1-3) is compared with that of a granite (called alaskite by Bateman *et al.*, 1963) from the Sierra Nevada batholith (column 7), and it is seen that the differences in SiO_2, TiO_2, total iron oxide, CaO, and K_2O are quite large. These differences also show in the norms. Further, the melanogranophyres contain ferrohedenbergite and iron-rich olivine, minerals usually not found in granitic rocks of the orogenic zones.

In their Figure 102, Wager and Brown (1967) plotted the normative compositions of the melanogranophyres in the "granite system," $NaAlSi_3O_8$–$KAlSi_3O_8$–SiO_2. Traditionally, this diagram has been used for plotting norms of granitic rocks [Bowen (1954, p. 11) and Tuttle and Bowen (1958, pp. 127–128) originally used it for plotting rocks containing 80 percent or more normative ab + or + q], and by plotting the compositions of the granophyres in this system, there is the implication that they are similar to granites. In Table 3-1 it is seen that only 55 to 62 percent of the normative constituents of these rocks consist of ab + or + q (columns 1-3), as compared with 94 percent for the Sierran granite (column 7). Thus, although differentiation of the Skaergaard magma seems to have produced a small amount of a rock called melanogranophyre, it would be a distortion to assert that this rock approximates the composition of a granite.

If the compositions are examined from a broader perspective to include more typical Sierran rocks such as quartz monzonite and granodiorite, it can be seen in Table 3-1 (columns 1-3) that the melanogranophyres from the Skaergaard intrusion contain only 65 to 70 percent of normative ab + an + or + q, whereas Presnall and Bateman (1973, p. 3185) noted that 98 percent of all the analyses of typical felsic rocks from the central Sierra Nevada batholith contain more than 79 percent of these constituents. The differences between the melanogranophyres and a typical

Table 3-1. Comparison of Skaergaard late-stage differentiates with Sierra Nevada rocks

	1	2	3	4	5	6	7
SiO_2	57.30	58.81	60.23	64.86	69.60	74.11	75.4
TiO_2	1.42	1.26	1.18	0.57	0.42	0.18	0.10
Al_2O_3	11.13	12.02	11.19	16.12	14.89	13.7	13.3
Fe_2O_3	5.15	5.77	5.52	1.90	1.07	0.60	0.3
FeO	11.75	9.38	9.11	2.52	1.99	0.88	0.74
MnO	0.28	0.21	0.24	0.09	0.07	0.05	0.08
MgO	0.81	0.72	0.51	1.55	0.91	0.32	0.12
CaO	4.83	5.03	5.11	3.80	2.70	1.29	0.48
Na_2O	3.51	3.87	3.92	3.44	3.18	3.44	4.1
K_2O	1.83	2.25	1.94	4.03	4.45	4.92	4.5
P_2O_5	0.35	0.71	0.27	0.23	0.12	0.06	0.01
H_2O^+	1.27	0.21	0.80	0.51	0.31	0.18 ⎫	0.46
H_2O^-	0.15	0.19	0.10	0.06	0.08	0.12 ⎭	
Total	99.78	100.43	100.12	99.68	99.79	99.88	99.6
q	14.95	15.43	17.74	18.90	22.52	31.92	32.58
or	10.82	13.30	11.47	23.91	26.13	28.94	26.69
ab	29.70	32.74	33.17	28.82	26.72	28.85	34.58
an	9.21	8.79	7.21	16.68	12.51	6.40	2.50
di	10.88	10.05	14.34	0.36	—	—	—
hy	11.82	7.33	4.43	5.91	4.53	1.72	1.36
mt	7.47	8.37	8.00	2.78	1.57	0.89	0.46
il	2.70	2.39	2.24	1.07	0.78	0.32	—
ap	0.84	1.68	0.64	0.62	0.29	—	—
c	—	—	—	—	0.31	0.51	0.71
ab + or + q	55.47	61.47	62.38	71.63	75.37	89.71	93.85
ab + or + an + q	64.68	70.26	69.59	88.31	87.88	96.11	96.35

Identification of columns

1. 5264, melanogranophyre of Skaergaard Upper Border Group γ (Wager and Brown, 1967, Table 9).
2. 3047, melanogranophyre of Skaergaard Upper Border Group γ (Wager and Brown, 1967, Table 9).
3. 4332, melanogranophyre of Skaergaard Upper Border Group γ (Wager and Brown, 1967, Table 9).
4. Granodiorite of McMurry Meadows, east-central Sierra Nevada (Bateman et al., 1963, Table 3, analysis 7).
5. Tungsten Hills quartz monzonite, east-central Sierra Nevada (Bateman et al., 1963, Table 3, analysis 17).
6. Quartz monzonite of Cathedral Peak type, east-central Sierra Nevada (Bateman et al., 1963, Table 3, analysis 25).
7. Alaskite of Cathedral Peak type, east-central Sierra Nevada (Bateman et al., 1963, Table 3, analysis 26).

granodiorite and two quartz monzonites from the Sierra Nevada batholith are shown in Table 3-1 (columns 4-6), and it is clear that fractional crystallization of the Skaergaard magma did not produce even the less felsic rock compositions represented in the Sierra Nevada batholith.

Data from other layered gabbroic intrusions also provide no evidence for the production of granitic magma by fractional crystallization from basalt, although these data are not as well documented as for the Skaergaard intrusion. In the case of the Bushveld complex, Wager and Brown (1967, p. 403) state that "The acid rocks . . . provide a bewildering array of granites, granophyres, and felsites which belong to various phases of igneous activity. Unfortunately they occur together at the level of the roof-zone of the main Bushveld basic intrusion, and include pre-Bushveld lavas, post-Bushveld granites, and possibly rheomorphic equivalents of the early, Rooiberg felsite lavas." Strontium isotope data on the Bushveld complex are not conclusive with regard to the question of a common origin for the granitic and mafic rocks, but Davies et al. (1970) believe that a separate origin is probable.

An obvious way to interpret all of these field relationships is to conclude, with Wager and Deer (1939), that the granitic rocks found in orogenic zones are not produced by fractional crystallization from a basaltic parent. Based on studies of thin sections of basalts, Fenner (1929) had reached a similar conclusion much earlier and believed that fractional crystallization of basalt yields an iron-rich residual liquid.

It is well known that a large amount of experimental data tends to support the fractional crystallization model for the production of granitic magma from a basaltic parent (for example, see Bowen, 1914, 1915; Andersen, 1915; Bowen and Schairer, 1938; Schairer and Bowen, 1938, 1947; Schairer, 1954, 1957; Schairer and Yoder, 1960), but all of these data are for simplified systems that do not exactly duplicate natural magmas. Moreover, not all simplified systems support Bowen's theory. The system $MgO-FeO-SiO_2$ (Bowen and Schairer, 1935) shows a clear trend of enrichment of residual liquids in iron; and in addition, certain fractional crystallization paths in this system show initial crystallization of Mg-rich olivine, then an absence of olivine, and finally crystallization again of more Fe-rich olivine, a sequence of events later observed in the Skaergaard intrusion by Wager and Deer (1939). In order to resolve the question of whether fractional crystallization leads to an iron-rich or alkali-and silica-rich residual liquid, it would be necessary to study a system of sufficient complexity that the interaction between these two trends could be examined. The simplest system showing this interaction would contain no fewer than eight components, for in addition to FeO and Fe_2O_3, it would be necessary to include the alkalies Na_2O and K_2O plus other oxides (SiO_2, CaO, MgO, Al_2O_3) necessary to form the most common igneous minerals. No system of this complexity has yet been studied and one is left with the dilemma that the existing data on simpler systems support both Fenner and Bowen.

Kennedy (1955) suggested a way out of the dilemma by proposing that the Bowen trend was the result of fractional crystallization of an oxidized basaltic magma and the Fenner trend was produced by fractional crystallization of a reduced magma. Based on the system MgO–FeO–Fe_2O_3–SiO_2, Osborn (1959) restated Kennedy's idea and showed that for certain starting compositions, iron enrichment was produced during fractional crystallization in a system closed to oxygen, whereas silica enrichment was produced by fractional crystallization at constant oxygen fugacity. He showed that the initial oxidation state of the magma exercises no control on the later stages of fractional crystallization. It was later pointed out by Presnall (1966) that iron enrichment could be suppressed also by buffering the oxygen fugacity of the magma with a gas of constant composition (decreasing oxygen fugacity with temperature). Osborn (1959, pp. 643–644) suggested that for basalts crystallizing in orogenic zones, water-bearing sediments could act as a source of oxygen for buffering the oxygen fugacity of the magma, but no field test has ever been developed for determining whether fractional crystallization of basalts in orogenic zones takes place with oxygen fugacity held constant or buffered by a gas of constant composition. More importantly, the most complex iron-bearing systems investigated so far (Presnall, 1966; Roeder and Osborn, 1966) contain no Na_2O or K_2O, and thus are still not complex enough to evaluate completely the interaction between iron enrichment and alkali-silica enrichment. This problem could be solved by crystallization experiments on natural rock compositions, but published results are difficult to interpret in terms of fractional crystallization for one or more of the following reasons: (1) the experiments describe equilibrium rather than fractional crystallization, (2) oxygen fugacity is not controlled, (3) the amount of iron lost to the capsule is unknown, and (4) equilibrium is not demonstrated. It is a remarkable fact that more than half a century of experimental studies has failed to prove conclusively whether or not fractional crystallization of basaltic magma is capable of yielding a granitic residual liquid at any pressure.

Even if it is shown by future experimental studies that granitic residual liquids can be so produced, these end products of fractional crystallization would occur in very small amounts (Morse, 1976), and the existence of large batholiths such as the Sierra Nevada would remain unexplained.

It might be argued that an escape from the volume argument would be to follow Hamilton and Myers (1967) and assume that batholiths are thin tabular bodies that extend only to very shallow depths. In this case, the volume of granitic rocks would be small and one could follow Bowen (1948, pp. 87–88) and postulate that the middle and lower parts of the thickened crust beneath the batholith are gabbroic. If the granite were produced from such a lower crust by fractional crystallization, any water that was present would be concentrated in the granitic liquids moving upward; and the residual material left behind would be nearly anhydrous, a condition that would result in the formation of eclogite or garnet granulite (Ringwood and D. H. Green, 1966, pp. 412–414; Ito and Kennedy, 1971; D. H. Green and Ringwood, 1972) with a P-wave velocity and density much higher than is observed for the lower crust beneath the Sierra Nevada batholith (Bateman and Eaton, 1967). Thus, the middle and lower crust beneath the Sierra Nevada cannot have a basaltic composition resulting from fractional crystallization that produced felsic rocks emplaced at a higher level, and the volume argument remains as an unsurmounted obstacle to the fractional crystallization model.

PARTIAL FUSION

Presnall (1969, p. 1179) distinguished two types of fusion. During *equilibrium fusion*, the liquid produced on heating continually reacts and re-equilibrates with the crystalline residue. During *fractional fusion*, the liquid is immediately isolated from the system as soon as it is formed and is thereby prevented from further reaction with the crystalline residue. Fractional fusion can be thought of as an infinite number of infinitely small equilibrium fusion events. *Partial fusion* is defined as fusion of some portion less than the whole and thus could refer to equilibrium fusion, fractional fusion, or some process intermediate between these two extremes. Geometrical methods for deducing crystal and liquid paths during equilibrium and fractional fusion were developed in detail by Presnall (1969), and for the following discussion a general knowledge of these methods will be assumed (see also Roeder, 1974; Morse, 1976).

Bowen (1928) emphasized that during fractional crystallization a compositionally continuous series of magmas is produced. Fractional fusion is fundamentally different in that the liquid path is discontinuous, and large volumes of compositionally distinct magmas can be produced with no intermediate magma types. Presnall (1969) illustrated these differences on ternary phase diagrams, and it might be supposed that more complex phase relationships in multicomponent rock systems together with deviations from perfect fractional fusion could blur these differences so that the two processes could not be distinguished in nature. This possibility will now be examined. In a ternary system, large volumes of chemically homogeneous melt can be produced during fractional fusion if melting occurs at either a peritectic or eutectic invariant point. For example, in Fig. 3-1, composition v lies within the solidus field of pyroxene + olivine

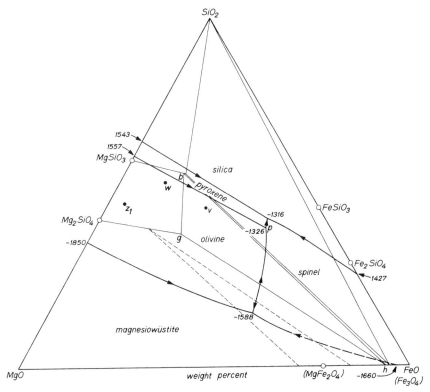

Figure 3-1. The system MgO–iron oxide–SiO$_2$ at the oxygen fugacity produced by equilibrium decomposition of CO$_2$, redrawn from Presnall (1969, Fig. 7) and interpolated from diagrams and data by Muan and Osborn (1956), Phillips, Sōmiya, and Muan (1961), and Speidel and Osborn (1967). All iron oxide is calculated as FeO and compounds in parentheses have been projected from the oxygen apex of the triangle Mg–Fe–O. Heavy lines (dashed where inferred) with arrows indicating decreasing temperature are liquidus boundary lines. Light solid lines (dashed where inferred) show compositions of phases in equilibrium at solidus temperatures. Temperatures of invariant points are indicated in degrees C. Further explanations of this diagram are given by Presnall (1966, pp. 755–760; 1969, pp. 1187–1189).

+ spinel (triangle b–g–h), and fractional fusion of any amount of composition v up to 28 percent of the total mass yields a melt that does not change in composition from the peritectic p. This percentage is determined by applying the lever rule to line r–v–p (Fig. 3-2). In nature, however, the number of components is likely to be larger than the number of phases and it cannot be expected that fusion will take place at an invariant point. The natural situation can be approximated in Fig. 3-2 by considering fractional fusion of composition w. In this case, the crystalline assemblage just below the solidus temperature consists of pyroxene and olivine. The first melt is not produced at an invariant point but is defined by the intersection f_1 of the boundary line i–p with the tangent (not shown) to the solidus fractionation line at w. As fractional fusion proceeds, the composition of the melt changes continuously along the boundary line toward i and reaches f_2 just as the composition of the crystalline residue reaches d. Point f_2 is defined by the tangent to the solidus fractionation line at d. Thus, when fractional fusion occurs with the liquid composition moving along a univariant liquidus boundary line, highly variable magma compositions can be produced.

Also it is common for large amounts of liquid to be produced along a univariant line with a very restricted range of compositions. Consider, for example, composition z_1, which, just below its solidus temperature, consists of about 70 percent olivine m and 30 percent pyroxene n. Such a mixture could be thought of as a simplified mantle composition. Frac-

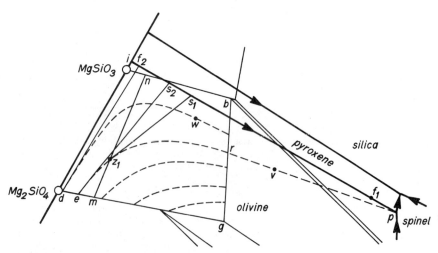

Figure 3-2. An enlarged portion of Fig. 3-1, redrawn from Presnall (1969, Fig. 8). Dashed lines are solidus fractionation lines.

tional fusion of z_1 would produce an initial melt at s_1, determined by the tangent to the solidus fractionation line at z_1; and as the crystal path moves along the solidus fractionation line from z_1 to e, the liquid path moves from s_1 to s_2. Just as the crystal path arrives at e, the amount of liquid produced is 33 percent, given by a lever (not shown) from e through the bulk composition z_1, to the boundary line i–p (Morse, 1976). Thus, fractional fusion of any proportion up to 33 percent of the total mass could occur, yet the range of possible liquids would vary only between s_1 and s_2. The effect is very similar to the production of a melt at an invariant point, and it is apparent that increasing the number of components (and thereby increasing the variance) does not alter the conclusion that large volumes of compositionally uniform magma can be generated during fractional fusion. When the crystal path reaches e, the residual crystals consist only of olivine, so further heating would produce no more liquids on the boundary line i–p. Instead, there would be a temperature hiatus during which no liquids would be produced; and then at very high temperatures, a second series of liquids along the olivine join would be generated. No liquids between s_1–s_2 and the olivine join would be produced.

In the above discussion, perfect fractional fusion has been assumed, and although Yoder (1976, p. 176) has suggested that fractional fusion would be expected under stress conditions, many situations involving deviations from this type of process may exist. Deviations from perfect fractional fusion would imply the production of finite amounts of melt prior to separation of this melt from the source, a process equivalent to a stepwise sequence of equilibrium fusion events. Also, there may not be complete separation of melt from residual crystals at each step.

Consider first a single equilibrium fusion event with complete separation of melt from crystals. In Fig. 3-3, the first part of the crystal path followed during equilibrium fusion of z_1 is the curved line z_1–z_2–a_1, defined by dropping tangents to solidus fractionation lines from z_1, and the corresponding part of the liquid path is s_1–s_3. The amount of liquid is 33 percent (when the lever a_1–z_1–s_3 is restored to true scale). This percentage is seemingly the same as the amount of liquid produced between s_1 and s_2 during fractional fusion as discussed above, but it will be noted that the levers used in the two cases (a_1–z_1–s_3 and e through z_1 to the olivine-pyroxene boundary line) are not quite identical.

The geometry of stepwise equilibrium fusion is illustrated in Fig. 3-3, and in order to separate construction lines, the amount of liquid generated at each step has been kept fairly large. For a liquid produced in the first step at s_4, the bulk composition of the crystalline residue would lie at z_2 on the crystal path for equilibrium fusion of z_1. The lever z_2–z_1–s_4 would

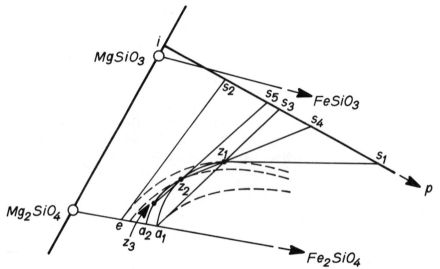

Figure 3-3. Distorted diagram to show stepwise equilibrium fusion of composition z_1. Labeled points are the same as in Fig. 3-2, and for the undistorted relationships, Fig. 3-2 should be used.

give the proportion of liquid separated from the source region for this step. After complete removal of liquid s_4, z_2 may be considered as a new starting composition for a second equilibrium fusion step. The composition of the crystalline residue does not continue along the line z_1–z_2–a_1 but moves out along the line z_2–z_3–a_2, the crystal path for equilibrium fusion of z_2. If the liquid separated at the second fusion step lies at s_5, the crystalline residue will lie at z_3, and the proportion of z_2 fused is given by the lever z_3–z_2–s_5. The *total* amount of fusion after these two fusion steps is given by a lever (not drawn) from z_3 through the original starting composition z_1 to the boundary line i–p.

For the third equilibrium fusion step (not shown), it will be assumed that the amount of melting is great enough that the crystalline residue reaches the olivine join and moves a short distance along it toward Mg_2SiO_4. If the liquid separated at this step is in equilibrium only with olivine, the composition of the liquid must lie in the primary phase volume of olivine between line i–p and the olivine join. For the next and all succeeding fusion steps, the composition of the residue lies on the olivine join, which means that the composition of the extracted liquids will also lie on this join. Only the first of the liquids in equilibrium with pure olivine lies off the olivine join, and it is important to note that this relationship applies no matter how many equilibrium fusion steps are postulated.

As noted above, the amount of melt produced along the pyroxene-olivine boundary line during perfect fractional fusion is nearly the same (33 percent) as for a single equilibrium fusion event producing a liquid at s_3 (Figs. 3-2 and 3-3). It can be seen in Fig. 3-3 that during stepwise equilibrium fusion, the composition of the crystalline residue will arrive at the olivine join always between e and a_1, regardless of the number of fusion steps. Thus, the total amount of liquid produced along the olivine-pyroxene boundary line is also approximately 33 percent.

An interesting feature of stepwise equilibrium fusion is that the compositional range of liquids produced along the olivine-pyroxene boundary line is even more restricted than during perfect fractional fusion. For a large number of small steps, the compositional range would be only slightly reduced; whereas for the large steps illustrated in Fig. 3-3, the compositional range is more strongly reduced (s_4–s_5 vs. s_1–s_2).

Thus, equilibrium fusion in small steps with complete separation of melt from crystals at each step gives results very similar to those produced by perfect fractional fusion; that is, a large volume of melt (33 percent of the starting composition at z_1) is produced with a very restricted range of composition on the boundary line i–p, and another large volume of melt is produced along the olivine join. Two differences are (1) a small amount of one liquid is produced with a composition intermediate between the two main types, and (2) the range of liquid compositions produced along i–p is even more restricted during stepwise equilibrium fusion than during perfect fractional fusion.

Now consider equilibrium fusion in small steps but with incomplete separation of melt from crystals at each step. For composition z_1 (Fig. 3-3), the first fusion step would produce a liquid at s_4 and a crystalline residue at z_2 as before. If some of the liquid at s_4 is left behind with the crystalline residue, the bulk composition remaining at the source region would not be at z_2 as before, but would lie on the straight line between z_2 and z_1. The first several fusion steps would produce melts along the boundary line i–p between s_1 and s_2 as the composition of the residue (in this case a mixture of crystals and liquid) moves in steps toward the olivine join. Just as in stepwise equilibrium fusion with complete separation of melt from crystals, the range of liquid compositions would be more restricted than during perfect fractional fusion. After the last liquid on the boundary line i–p is produced, continued heating would yield transitional liquids between i–p and the olivine join. The liquids extracted would never quite reach the olivine join because the bulk composition being fused (olivine plus liquid left behind from the previous step) would always lie slightly off this join. Even though the remaining liquids extracted would technically be transitional between the boundary line i–p and the olivine join, most would lie so close to the olivine join that if a volume frequency curve were prepared for the liquids, a pronounced minimum would exist in the compositional regional between s_1–s_2 and the olivine join. Of course, the temperatures required to produce melts near the olivine join are extremely high, which illustrates the point that there may often be only enough thermal energy to produce a single primary magma type (e.g., s_1–s_2) from a given source region.

From this discussion, it is concluded that large volumes of compositionally uniform magmas can be produced in natural fusion processes when these processes deviate from perfect fractional fusion at invariant points. Specifically, the variance can be greater than zero, fusion can take place as a series of stepwise equilibrium fusion events, and part of the liquid produced at each fusion step can be left behind in the source region. It should be remembered, however, that fusion of certain starting compositions (like w in Fig. 3-2) does not produce large volumes of chemically homogeneous magma. Also, if some of the residual crystals are retained with the liquid when it separates from the source region, chemically diverse magmas could be produced due to variations in the amount of retained crystals for successive fusion steps. The occurrence of large volumes of chemically homogeneous basalts in many regions (Presnall, 1969) suggests, at least for these regions, that compositions like w are not appropriate for the mantle and variable retention of residual crystals is not important.

DISCUSSION

Even though fractional crystallization of basaltic magma is believed to be an unlikely mechanism for the generation of large granitic batholiths in orogenic zones, the process of fractional crystallization is believed to be very important to petrogenesis in general, and to the origin of granitic batholiths in particular. Detailed arguments have been presented elsewhere (Presnall and Bateman, 1973) in support of the proposition that the felsic rock types in the Sierra Nevada batholith were produced by fractional crystallization not of a basaltic parent but of a parental magma similar in composition to andesite. Thus, the problem of generating large volumes of granitic magma disappears if another parental magma in addition to basalt exists. A combination of partial fusion and fractional crystallization processes involving more than one parental magma type may be expected to produce large volumes of certain parental magma types superimposed by a range of magma compositions of lesser volume produced by fractional crystallization of the parental types. Except for a difference in the number of parental magma types, such a scheme is essentially the same as that of Bowen. The task at hand is to develop criteria for determining with greater certainty which magma types are primary, which are parental, and which are derivative.

Acknowledgments It will be recognized that many of the thoughts expressed here, however imperfect, were inspired by the remarkable genius of N. L. Bowen. I wish to thank M. O. Garcia, T. N. Irvine, S. A. Morse, and H. S. Yoder, Jr. for thoughtful reviews of the manuscript. Financial support was provided by the Earth Sciences Section, National Science Foundation, NSF Grant DES74-22571.

References

Andersen, O., The system anorthite–forsterite–silica, *Am. J. Sci.*, *39* (Fourth Ser.), 407–454, 1915.

Anderson, A. T., Magma mixing: petrological process and volcanological tool, *J. Volcanol. Geotherm. Res.*, *1*, 3–33, 1976.

Bateman, P. C., L. D. Clark, N. K. Huber, J. G. Moore, and C. D. Rinehart, The Sierra Nevada batholith—a synthesis of recent work across the central part, *U.S. Geol. Survey Prof. Paper 414-D*, D1-D46, 1963.

Bateman, P. C., and J. P. Eaton, Sierra Nevada batholith, *Science*, *158*, 1407–1417, 1967.

Bowen, N. L., The ternary system: diopside–forsterite–silica, *Am. J. Sci.*, *38* (Fourth Ser.), 207–264, 1914.

———, The crystallization of haplobasaltic, haplodioritic, and related magmas, *Am. J. Sci.*, *40* (Fourth Ser.), 161–185, 1915.

———, *The Evolution of the Igneous Rocks*, Princeton University Press, Princeton, 332 pp., 1928.

———, Certain singular points on crystallization curves of solid solutions, *Natl. Acad. Sci. Proc.*, *27*, 301–309, 1941.

———, The granite problem and the method of multiple prejudices, in *Origin of Granite*, Geol. Soc. Am. Mem. 28, 79–90, 1948.

———, Experiment as an aid to the understanding of the natural world, *Proc. Acad. Nat. Sci. Philadelphia*, *106*, 1–12, 1954.

Bowen, N. L., and J. F. Schairer, The system $MgO-FeO-SiO_2$, *Am. J. Sci.*, *29* (Fifth Ser.), 151–217, 1935.

Bowen, N. L., and J. F. Schairer, Crystallization equilibrium in nepheline–albite–silica mixtures with fayalite, *J. Geol.*, *46*, 397–411, 1938.

Carmichael, I. S. E., F. J. Turner, and J. Verhoogen, *Igneous Petrology*, McGraw-Hill Book Co., New York, 739 pp., 1974.

Davies, R. D., H. L. Allsopp, A. J. Erlank, and W. I. Manton, Sr-isotopic studies on various layered mafic intrusions in southern Africa, in *Symposium on the Bushveld igneous complex and other layered intrusions, Geol. Soc. S. Africa, Spec. Pub. 1*, 576–593, 1970.

Dickinson, W. R., Relations of andesites, granites, and derivative sandstones to arc-trench tectonics, *Rev. Geoph. Space Phys., 8*, 813–860, 1970.

Fenner, C. N., The crystallization of basalts, *Am. J. Sci., 18* (Fifth Ser.), 225–253, 1929.

Green, D. H., and A. E. Ringwood, The genesis of basaltic magmas, *Contrib. Mineral. Petrol., 15*, 103–190, 1967.

Green, D. H., and A. E. Ringwood, A comparison of recent experimental data on the gabbro-garnet granulite-eclogite transition, *J. Geol., 80*, 277–288, 1972.

Green, T. H., and A. E. Ringwood, Genesis of the calcalkaline igneous rock suite, *Contrib. Mineral. Petrol., 18*, 105–162, 1968.

Hamilton, E. I., The isotopic composition of strontium in the Skaergaard intrusion, East Greenland, *J. Petrol., 4*, 383–391, 1963.

Hamilton, W., and W. B. Myers, The nature of batholiths, *U.S. Geol. Survey Prof. Paper 554-C*, C1-C30, 1967.

Ito, K. and G. C. Kennedy, An experimental study of the basalt-garnet granulite-eclogite transition, in *The structure and physical properties of the earth's crust, Am. Geoph. Union Geoph. Mon. 14*, 303–314, 1971.

Kennedy, G. C., Some aspects of the role of water in rock melts, in *Crust of the Earth—a symposium, Geol. Soc. Am. Spec. Paper 62*, 489–503, 1955.

Kuno, H., Origin of the Cenozoic petrographic province of Japan and surrounding areas, *Bull. Volcanol., 20* (Second Ser.), 37–76, 1959.

Kushiro, I., Compositions of magmas formed by partial zone melting of the earth's upper mantle, *J. Geoph. Res., 73*, 619–634, 1968.

Morse, S. A., The lever rule with fractional crystallization and fusion, *Am. J. Sci., 276*, 330–346, 1976.

Muan, A., and E. F. Osborn, Phase equilibria at liquidus temperatures in the system $MgO-FeO-Fe_2O_3-SiO_2$, *Am. Ceramic Soc. J., 39*, 121–140, 1956.

Osborn, E. F., Role of oxygen pressure in the crystallization and differentiation of basaltic magma, *Am. J. Sci., 257*, 609–647, 1959.

Phillips, B., S. Sōmiya, and A. Muan, Melting relations of magnesium oxide–iron oxide mixtures in air, *Am. Ceramic Soc. J., 44*, 167–169, 1961.

Presnall, D. C., The join forsterite–diopside–iron oxide and its bearing on the crystallization of basaltic and ultramafic magmas, *Am. J. Sci., 264*, 753–809, 1966.

———, The geometrical analysis of partial fusion, *Am. J. Sci., 267*, 1178–1194, 1969.

Presnall, D. C., and P. C. Bateman, Fusion relations in the system $NaAlSi_3O_8-CaAl_2Si_2O_8-KAlSi_3O_8-SiO_2-H_2O$ and generation of granitic magmas in the Sierra Nevada batholith, *Geol. Soc. Am. Bull., 84*, 3181–3202, 1973.

Ringwood, A. E., and D. H. Green, An experimental investigation of the gabbro-eclogite transformation and some geophysical implications, *Tectonophysics, 3*, 383–427, 1966.

Roeder, P. L., Paths of crystallization and fusion in systems showing ternary solid solution, *Am. J. Sci.*, *274*, 48–60, 1974.

Roeder, P. L., and E. F. Osborn, Experimental data for the system MgO–FeO–Fe_2O_3–$CaAl_2Si_2O_8$–SiO_2 and their petrologic implications, *Am. J. Sci.*, *264*, 428–480, 1966.

Schairer, J. F., The system K_2O–MgO–Al_2O_3–SiO_2:I, Results of quenching experiments on four joins in the tetrahedron cordierite–forsterite–leucite–silica and on the join cordierite–mullite–potash feldspar, *Am. Ceramic Soc. J.*, *37*, 501–533, 1954.

———, Melting relations of the common rock-forming oxides, *Am. Ceramic Soc. J.*, *40*, 215–235, 1957.

Schairer, J. F., and N. L. Bowen, The system, leucite–diopside–silica, *Am. J. Sci.*, *35A*, 289–309, 1938.

Schairer, J. F., and N. L. Bowen, The system anorthite–leucite–silica, *Bull. Soc. Geol. Finlande*, *20*, 67–87, 1947.

Schairer, J. F., and H. S. Yoder, The system albite–forsterite–silica, in *Carnegie Institution Washington Year Book 59*, 69–70, 1960.

Speidel, D. H., and E. F. Osborn, Element distribution among coexisting phases in the system MgO–FeO–Fe_2O_3–SiO_2 as a function of temperature and oxygen fugacity, *Am. Mineral.*, *52*, 1139–1152, 1967.

Tuttle, O. F., and N. L. Bowen, Origin of granite in the light of experimental studies in the system $NaAlSi_3O_8$–$KAlSi_3O_8$–SiO_2–H_2O, *Geol. Soc. Am. Mem. 74*, 153p., 1958.

Wager, L. R., and G. M. Brown, *Layered Igneous Rocks*, W. H. Freeman and Co., San Francisco, 588 pp., 1967.

Wager, L. R., and W. A. Deer, Geological investigations in East Greenland. Pt. 3. The petrology of the Skaergaard intrusion, Kangerdlugssuag, East Greenland, *Medd. om Grønland*, *105*, no. 4, 352 pp., 1939.

Winkler, H. G. F., *Petrogenesis of Metamorphic Rocks*, third edition, Springer-Verlag, New York, 320 pp., 1974.

Yoder, H. S., Jr., Calcalkaline andesites: experimental data bearing on the origin of their assumed characteristics, in *Proceedings of the andesite conference*, *Oregon Dept. Geol. Mineral. Ind. Bull. 65*, 77–89, 1969.

———, Contemporaneous basaltic and rhyolitic magmas, *Am. Mineral.*, *58*, 151–173, 1973.

———, *Generation of Basaltic Magma*, Natl. Acad. Sci., Washington, D.C., 265 pp., 1976.

Chapter 4

CRYSTALLIZATION IN SILICATE SYSTEMS

ARNULF MUAN

The Pennsylvania State University, University Park, Pennsylvania

INTRODUCTION

One of the most complex examples of crystallization in oxide and silicate systems is that of magmas. While the thermodynamic basis for understanding liquid-solid equilibria has been available since Gibbs' pioneering work a hundred years ago (1876, 1878), a quantitative treatment of the crystallization of magmas is complicated by the many composition parameters and external parameters involved, as well as by lack of attainment of equilibrium.

Inasmuch as the fundamental principles are the same for complex poly-component magmas as for compositionally simpler binary, ternary, or quaternary systems, the studies of key systems of the latter type have provided much insight into the crystallization processes of magmas. Much light has been shed on these processes by considering crystallization in the simple systems under two idealized conditions, perfect equilibrium and perfect fractionation. These two conditions represent extremes between which actual processes may take place. The present chapter deals with crystallization relations in these simpler systems as a basis for understanding crystallization sequences in the more complicated magma systems. In order to simplify relations further, the total pressure will be assumed to be constant (at 1 atm) in all considerations of crystallization processes presented in this chapter.

A very large amount of phase-equilibrium data for silicate systems has become available during the past few decades. Many of the systems studied by the early investigators, notably those of the Geophysical Laboratory of the Carnegie Institution of Washington, serve as excellent examples to demonstrate the principles involved in the crystallization of igneous magmas. It is a tribute to these men, both for their insight in choosing systems and for the care and high quality of their experimental work, that the same systems, for which the phase relations were determined more than fifty years ago, to a large extent are still used as examples. Main additions to the examples presented in the analogous chapter of Bowen's (1928) book *The Evolution of the Igneous Rocks* are iron oxide-containing systems, which play a very important role for an understanding of magma crystallization but had not been extensively studied experimentally at the time of the appearance of Bowen's book.

The examples presented in the following discussion are limited in number and were chosen to include common rock-forming minerals while still being simple enough to demonstrate the principles clearly.

BINARY SYSTEMS

Examples of binary systems presented in the following are arranged in order of increasing complexity based on the presence or absence of solid solubility among the phases and their tendency for compound formation. In addition, one example is given in which the gas phase plays an important role.

SYSTEMS WITHOUT APPRECIABLE SOLID SOLUTION AMONG THE PHASES

Eutectic Systems without Compound Formation A temperature-composition diagram ("t-x diagram") for the system $CaAl_2Si_2O_8$–$CaMgSi_2O_6$ at a constant total pressure of 1 atm is shown in Fig. 4-1 (Bowen, 1915; Osborn, 1942).[1] For a mixture X, crystals of anorthite start to separate out at the liquidus temperature (1328°C), and as heat is withdrawn, increasing amounts of anorthite crystallize as the liquid composition changes from a to E. The liquid thus becomes increasingly rich in the $CaMgSi_2O_6$ component until at the eutectic temperature, 1270°C, the liquid has also become saturated with diopside, at which point this phase co-precipitates with anorthite, and the liquid is consumed.

[1] For the sake of simplicity, and as an approximation, it will be assumed here that the end members are pure crystalline phases without any appreciable solid solubility.

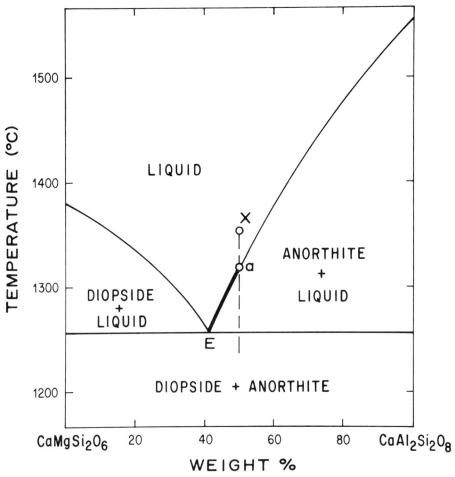

Figure 4-1. Phase relations in the system $CaMgSi_2O_6$–$CaAl_2Si_2O_8$ (Bowen, 1915; Osborn, 1942), illustrating paths of crystallization for a selected mixture.

If, during the crystallization process, anorthite crystals are withdrawn from the liquid, for instance by crystal settling, the course of liquid change during crystallization would not be different from the one described above.

The sequence of phase changes during melting of mixtures in the system shown in Fig. 4-1 is the reverse of that taking place during crystallization. For example, heating a crystalline mixture of composition X (50% anorthite, 50% diopside) produces a liquid of eutectic composition at 1270°C, with temperature remaining constant until diopside has completely dissolved. With further supply of heat to the system, the amount of anorthite gradually decreases as the composition of the liquid changes from E to a. At the latter temperature the last trace of anorthite crystals is consumed.

Binary Systems with Compound Formation The system $MgO-SiO_2$ (Bowen and Andersen, 1914) illustrates characteristic crystallization sequences involving congruently as well as incongruently melting compounds.

Consider first the crystallization of mixtures in the field of forsterite (Mg_2SiO_4), as shown in Fig. 4-2. This phase melts congruently, and the composition point Mg_2SiO_4 divides the diagram into two distinctly different parts as far as paths of crystallization are concerned. A mixture X located to the SiO_2 side of the Mg_2SiO_4 composition point will start precipitating forsterite at point a, and with decreasing temperature the liquid composition will change along the path from a to P. In contrast to this behavior, mixtures containing less silica than that of the congruently melting compound (Mg_2SiO_4) will change their liquid composition in the direction of the periclase–forsterite eutectic point, as shown for mixture Y in Fig. 4-2. There is no way in which crystallization of mixture Y, either under equilibrium or fractionation conditions, can yield products containing either metasilicate or silica as a phase. The Mg_2SiO_4 composition point may, therefore, be considered a "thermal divide" that cannot be crossed during the crystallization process.

Consider next the crystallization sequences of mixtures in the vicinity of the metasilicate composition, $MgSiO_3$, as illustrated in Fig. 4-3. During equilibrium crystallization of a mixture X of composition between $MgSiO_3$ and Mg_2SiO_4, increasing amounts of forsterite (Mg_2SiO_4) separate out, and the liquid composition changes from a to P. At the latter point, protoenstatite[2] becomes stable relative to the phase assemblage forsterite plus liquid, and the liquid phase is consumed by the reaction forsterite + liquid → protoenstatite at the peritectic temperature ($1557°C$). The final product of crystallization of mixture X at this temperature is thus protoenstatite and forsterite.

In a mixture of composition $MgSiO_3$ the sequence of phase changes during equilibrium cooling is identical to that described above except that in this case liquid and forsterite are consumed simultaneously at the

[2] Because of the complexities of the pyroxene structure and the difficulty of retaining the high-temperature modifications by quenching, there is still some uncertainty regarding the structural characteristics of this phase in the temperature region covered in Fig. 4-3 (Smith, 1959).

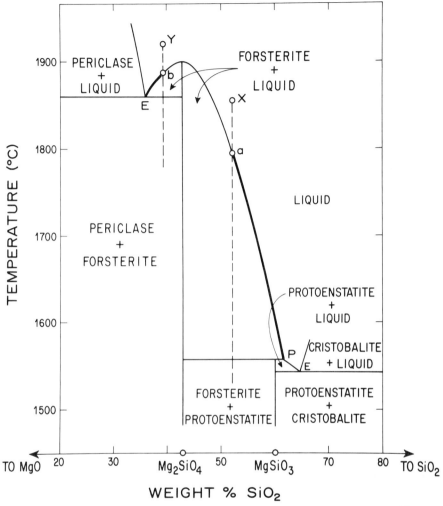

Figure 4-2. Phase relations in the system MgO–SiO$_2$ (Bowen and Andersen, 1914), illustrating paths of crystallization of selected mixtures in the vicinity of the orthosilicate (Mg$_2$SiO$_4$) composition.

peritectic temperature. Hence, in this case the crystallization yields a single phase as end product.

For a mixture of composition Y between MgSiO$_3$ and P (Fig. 4-3), the latter being the composition of the liquid at the peritectic point, the

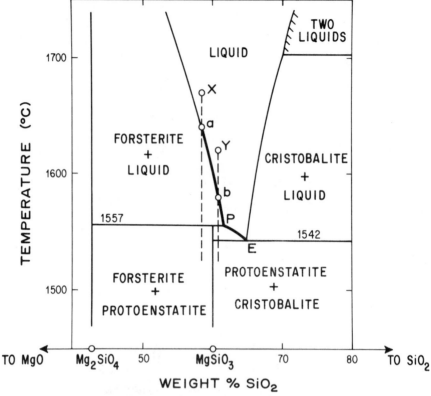

Figure 4-3. Phase relations in the system MgO–SiO₂ (Bowen and Andersen, 1914), illustrating paths of crystallization for selected mixtures in the vicinity of the
metasilicate (MgSiO₃) composition.

sequence of phase changes during equilibrium crystallization is similar to
those described above, except that in this case forsterite is completely
consumed at the peritectic temperature, while some liquid remains. With
increasing amounts of protoenstatite crystallizing as temperature decreases further, the liquid composition changes from P to E. At the latter
composition the liquid is saturated with silica (cristobalite), which co-
precipitates with protoenstatite, leaving a two-phase assemblage of these
crystalline phases as end products of the crystallization.

If equilibrium is not attained during cooling, the forsterite crystals may
not react with the liquid to form protoenstatite at the peritectic point as
described above. Instead, the liquid having reached composition P at the

peritectic temperature may crystallize to protoenstatite as temperature decreases from that of the peritectic to that of the eutectic (E), and to a mixture of protoenstatite and silica (cristobalite) at the eutectic, while the forsterite that formed at temperatures between the liquidus and the peritectic temperature fails to react completely with the liquid at the peritectic temperature or below and, hence, persists to temperatures below the solidus. This failure of reaction could happen, for instance, by formation of a layer of protoenstatite on the forsterite crystals, thus inhibiting the latter from further reacting with the liquid as temperature decreases below the peritectic. Rapid crystallization may, therefore, yield a product consisting of all three crystalline phases, protoenstatite, silica (cristobalite), and forsterite, even though the latter two phases are not stable together. A similar result is obtained if the forsterite crystals having formed in the temperature range between the liquidus and the peritectic temperature are removed from the liquid, for instance by crystal settling. In general, if there is relative movement between crystals (forsterite) and liquid, there will be an excess of forsterite crystals in some parts of the system and an excess of liquid in other parts. The former parts would crystallize to forsterite and protoenstatite, the latter parts to protoenstatite and silica (cristobalite). Such fractional crystallization, therefore, not only may yield differences in the relative amounts of the crystalline phases formed, it may also yield different phase assemblages. With extreme degree of fractionation, even a mixture having a composition close to, but on the silica side of, Mg_2SiO_4 may thus give rise to the formation of some liquid having the composition of the protoenstatite–silica (cristobalite) eutectic and crystallize to a mixture of these two crystalline phases.

The path of melting of a mixture of composition between $MgSiO_3$ and Mg_2SiO_4 is not the reverse of the path of fractional crystallization described above. A crystalline mixture of protoenstatite ($MgSiO_3$) and forsterite (Mg_2SiO_4) is stable relative to any liquid-containing phase assemblage until the peritectic temperature is reached. If, however, the liquid thus formed at the peritectic temperature is subsequently separated from the crystals and cooled below the eutectic temperature (1542°C), a crystalline aggregate (cristobalite plus clinoenstatite) will form which then upon heating yields some liquid of eutectic composition E at 1542°. Thus, a process of *repeated* fusion and crystallization, as opposed to a *simple* fusion process, may produce results similar to those obtained in a single course of fractional crystallization.

If a system has more than one incongruently melting compound, the number of possible fractional crystallization sequences is correspondingly multiplied, compared with those just described for the system $MgO-SiO_2$.

The principles are the same, however, and, hence, the considerations presented above can be extended to systems involving several compounds.

BINARY SYSTEMS WITH SOLID SOLUTION AMONG THE PHASES

A suitable and well-known example of a binary system with complete solid solution is $NaAlSi_3O_8$–$CaAl_2Si_2O_8$ (Bowen, 1913). Crystallization under equilibrium conditions in this system is shown in Fig. 4-4, illustrating changes in composition of liquid (a–a') and crystalline (b–b') phases in the temperature range between liquidus (a) and solidus (b') for a mixture of composition X. It is seen that both liquid and crystalline phases change their compositions continuously as temperature decreases and crystallization proceeds. This implies diffusion of ions between and within the coexisting phases. As this process is relatively slow, particularly in the crystalline phase, equilibrium is attained only if cooling takes place very slowly. In most cases, however, there is insufficient time for equilibrium to be attained at each temperature, resulting in zoned feldspar crystals, the interior of the crystals being more calcium-rich than the outer zone. This zoning of the crystals will remain after the liquid phase has disappeared unless the composition gradients are later removed by solid-state diffusion at subsolidus temperatures.

For the purpose of understanding the process of fractional crystallization, $NaAlSi_3O_8$–$CaAl_2Si_2O_8$ may be thought of as a system involving an infinite number of incongruently melting compounds for which the compositions of the crystalline and the liquid phase both change in infinitesimally small steps with decreasing temperature. Hence, an infinite multiplication of fractional crystallization possibilities exists in systems of this type. During fractional crystallization of mixtures in this system (see Fig. 4-5), the liquid composition changes along the liquidus curve, from a toward a', and the solid composition along the solidus curve, from b toward a'. If the fractionation is perfect (i.e., there is *no* reaction between crystalline and liquid phases), the last trace of liquid will have the composition (a') of the end-member component with the lowest melting point (in this case $NaAlSi_3O_8$).

As with the crystallization-melting relations in the system MgO–SiO_2 for the composition range involving incongruent melting, the path of melting in the system $NaAlSi_3O_8$–$CaAl_2Si_2O_8$ is not the reverse of the path of fractional crystallization. For instance, a solid solution of composition X (50% $NaAlSi_3O_8$–50% $CaAl_2Si_2O_8$) upon heating cannot develop a liquid phase below the solidus temperature for this mixture, $\sim 1260°C$. Only by *repeated* cycles of fusion and crystallization will a

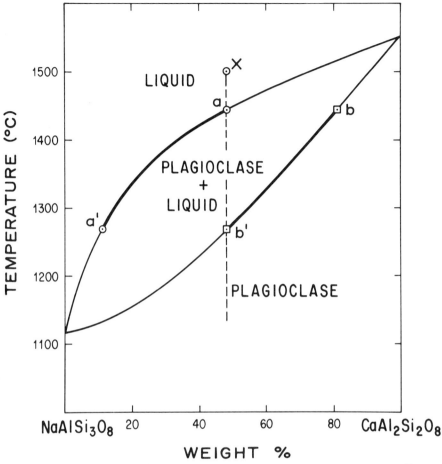

Figure 4-4. Phase relations in the system $NaAlSi_3O_8$–$CaAl_2Si_2O_8$ (Bowen, 1913), illustrating paths of equilibrium crystallization for a selected mixture.

melting process produce results similar to those obtained during fractional crystallization.

The diagrams displaying essentially no mutual solid solubility between the end members (e.g., $CaAl_2Si_2O_8$–$CaMgSi_2O_6$, Fig. 4-1) and complete mutual solid solubility (e.g., $NaAlSi_3O_8$–$CaAl_2Si_2O_8$, Fig. 4-4) are extremes. In most cases there is limited mutual solid solubility between the end members. These cases need not be considered here, as the principles involved are identical and the results obtained are similar to those

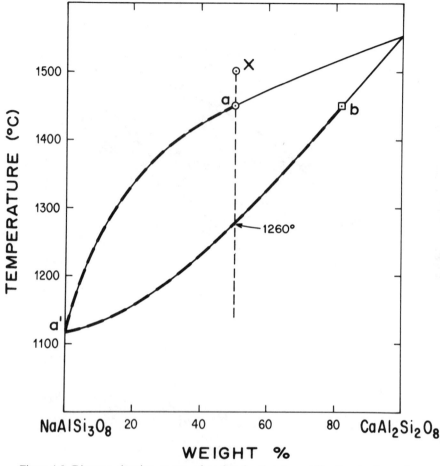

Figure 4-5. Diagram showing curves of perfect fractional crystallization in the system $NaAlSi_3O_8$–$CaAl_2Si_2O_8$ (Bowen, 1913).

described above. An example, described in detail by Bowen (1912), is the system $NaAlSiO_4$–$CaAl_2Si_2O_8$.

BINARY SYSTEMS IN WHICH THE GAS PHASE PLAYS AN IMPORTANT ROLE

One of the major developments in igneous petrology since the appearance of the original edition of Bowen's *The Evolution of the Igneous Rocks* has been the improved understanding of crystallization sequences in systems involving relatively unstable transition-metal oxides, where the equilibrium relations are strongly dependent on the oxygen fugacity of the system. The petrologically most important among such components is iron oxide. Iron occurs in three states of oxidation, Fe^0, Fe^{2+}, and Fe^{3+}, and the relative concentrations of these species in silicate phases, at a given temperature, depend on the oxygen fugacity of the system.

In order to illustrate the relations as clearly as possible, it is instructive to consider the simplest possible case, the binary system Fe–O. The part of this system of particular interest in the present context is shown in Fig. 4-6 (Darken and Gurry, 1945, 1946; Muan, 1955, 1958), comprising the composition range from FeO to Fe_2O_3. This diagram shows the stability fields of the various condensed phases (i.e., liquid and crystalline phases) as a function of temperature and Fe_2O_3/FeO at a constant total pressure of 1 atm. Whereas the diagram in this respect is similar to those commonly used to represent phase relations in binary systems ("t-x diagrams") (see, for instance, Figs. 4-1 to 4-5), an additional feature has been introduced in Fig. 4-6 in order to show the interdependence between the compositions of the condensed phases and the oxygen fugacity of the system, viz., the family of oxygen isobars superimposed on the diagram as light dash-dot curves. These curves traverse areas where one condensed phase is present (e.g., the area of liquid) diagonally, i.e., temperature and oxygen fugacity may be varied simultaneously. In areas where two condensed phases are present, on the other hand (e.g., magnetite + liquid), the oxygen fugacity is fixed if temperature is chosen, and vice versa. The oxygen isobaric curves, therefore, are horizontal in such areas.

Consider the path of crystallization of a mixture X (Fig. 4-7) as it is cooled under equilibrium conditions from a temperature above the liquidus to a temperature slightly below the solidus. If it is first assumed that the total composition (i.e., Fe_2O_3/FeO) of the mixture remains constant (a condition closely approximated in a closed system) as temperature is decreased, the liquid composition follows the path X–a–b, while the coexisting wüstite phase changes its composition along the path a'–b', with disappearance of the liquid phase at the temperature of the latter point. These relations are similar to those discussed in previous paragraphs of this chapter, where it was tacitly assumed in all derivations of crystallization paths that the total composition remains constant. In the system Fe–O, however, the total composition ($= Fe_2O_3/FeO$) will normally not remain constant. For instance, if the equilibrium cooling of mixture X is carried out at the constant oxygen fugacity prevailing at the liquidus temperature, Fe_2O_3/FeO of the mixture will increase as temperature decreases in accordance with the exothermic nature of the oxidation of Fe^{2+}

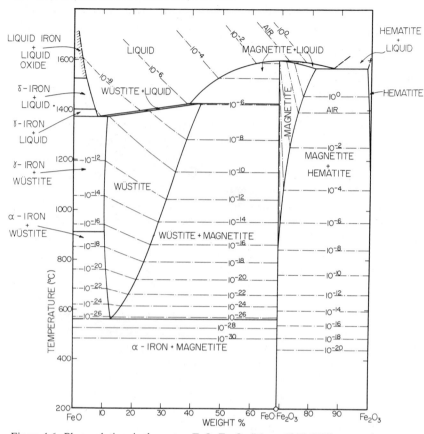

Figure 4-6. Phase relations in the system $FeO-Fe_2O_3$ (Muan, 1955, 1958) at a total pressure of 1 atm. In addition to the boundary curves between the areas of stable existence of the various condensed phases, curves of equal oxygen fugacities (in atm) have been superimposed on the diagram (light dash-dot curves).

to Fe^{3+}. The liquid phase changes its composition along the oxygen isobar in question (curve $X-c$ in Fig. 4-7). In this case, liquid and wüstite coexist in equilibrium only at one temperature, whereas in the previous case of constant total composition of the mixture these two phases co-existed over a temperature range. The selection of constant oxygen fugacity fixes the chemical potential of one of the components; hence one degree of freedom is expended, and liquid and crystals will coexist at only one temperature for each chosen oxygen fugacity. In anticipation of similar relations applicable to ternary and quaternary systems, it is in-

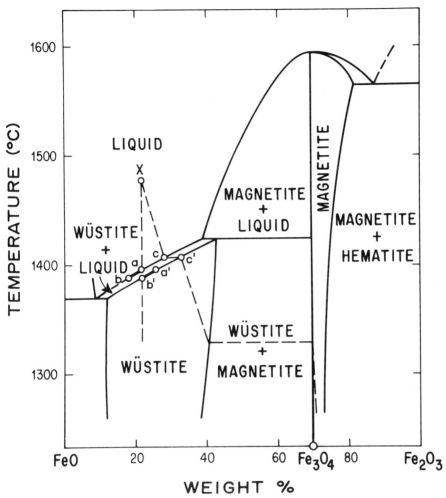

Figure 4-7. Diagram showing paths of equilibrium crystallization of a selected mixture X in the system FeO–Fe_2O_3 (Muan, 1955, 1958) under two idealized conditions, as explained in text: constant total composition of condensed phases and constant oxygen pressure.

structive to refer to this phenomenon as one of liquid "arrest" at the point of intersection between the liquidus curve and the oxygen isobar.

In addition to the two idealized conditions considered above, it is useful to consider the paths of equilibrium crystallization under yet another assumption, viz., constant ratios of species of the gas phase. In experimental studies of iron silicate systems, as well as in natural systems, the composition of the gas phase may control the oxygen fugacity of the system. This composition control may be illustrated by considering the decomposition of H_2O according to the equation

$$H_2O = H_2 + 1/2 O_2 \qquad (4\text{-}1)$$

for which the equilibrium constant may be written

$$K = \frac{f_{O_2}^{1/2} \cdot f_{H_2}}{f_{H_2O}} \qquad (4\text{-}2)$$

where f_{O_2}, f_{H_2} and f_{H_2O} are the fugacities of O_2, H_2 and H_2O, respectively. It is seen that for a given temperature (i.e., constant K), the oxygen fugacity is proportional to f_{H_2}/f_{H_2O} of the gas phase. Because the effects of

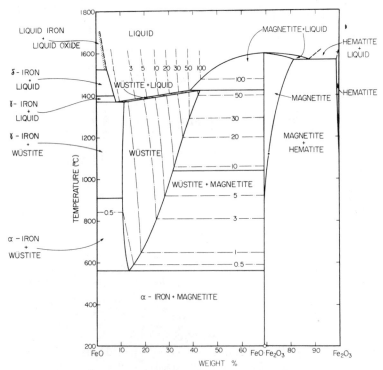

Figure 4-8. Phase relations in the system FeO–Fe$_2$O$_3$ (Muan, 1955, 1958) at a total pressure of 1 atm, showing boundary curves between areas of stable existence of various condensed phases, and superimposed curves of equal H$_2$O/H$_2$ of the coexisting gas phase (light dash-dot curves).

temperature on the equilibrium between H_2O and H_2 and between Fe_2O_3 and FeO are similar in magnitude, Fe_2O_3/FeO of the condensed phases vary relatively little with changing temperature at constant H_2O/H_2 of the gas phase. Hence, lines of constant H_2O/H_2 superimposed on the diagram of Fig. 4-8 are nearly vertical in areas where only one condensed phase is present. The directions of the crystallization paths in these areas under conditions of constant H_2O/H_2 of the gas phase, therefore, are similar to those prevailing when the process takes place at constant total composition. The curves of constant H_2O/H_2, however, cross areas of coexistence of two condensed phases as horizontal lines. Hence, during crystallization at constant H_2O/H_2 in the system Fe–O, the liquid composition, if equilibrium prevails, is "arrested" at the point of intersection between the curve of constant H_2O/H_2 and the liquidus curve, just as was the case at constant oxygen fugacity (compare Fig. 4-6).

Fractional crystallization in the system Fe–O is governed largely by the same factors considered in earlier paragraphs. In addition to fractionation caused by failure of early-formed crystals to react with liquid, however, there is the possibility of incomplete attainment of equilibrium between the gas phase and the condensed phases. The latter situation may give rise to crystallization paths intermediate between the idealized conditions discussed above. Because of the possible failure of attainment of equilibrium for both the crystal-liquid and the gas-condensed phase reactions, a particularly large variety of fractionation paths is possible in systems where the gas phase plays an important role in the equilibria.

TERNARY SYSTEMS

Examples of ternary systems presented in the following are arranged on the basis of the presence or absence of solid solubility among the phases, tendency for compound formation, and presence or absence of incongruently melting solid solutions. In addition, an example is given in which the gas phase plays an important role.

SYSTEMS WITHOUT APPRECIABLE SOLID SOLUTION AMONG THE PHASES

Systems without Compound Formation   Consider as an example the system $CaSiO_3$–$CaAl_2Si_2O_8$–SiO_2, as shown in Fig. 4-9 (Rankin and Wright, 1915; Greig, 1927). Except for the appearance of a small region of liquid immiscibility adjacent to the $CaSiO_3$–SiO_2 join, this diagram represents a ternary analog of the system $CaAl_2Si_2O_8$–$CaMgSi_2O_6$ (see Fig. 4-1). The path of change of liquid composition during crystallization, both

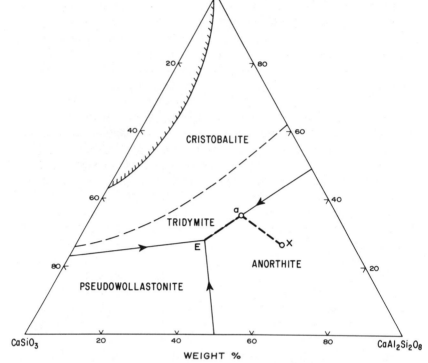

Figure 4-9. Phase diagram of the system $CaSiO_3$–$CaAl_2Si_2O_8$–SiO_2 (Rankin and Wright, 1915; Greig, 1927), showing general features of paths of crystallization, as discussed in the text. The curve with stippling on one side indicates the outline of a two-liquid region.

under equilibrium and under fractional crystallization, is represented by the curve X–a–E. If equilibrium prevails, the relative amounts of the phases during crystallization, at any given temperature, are fixed and constant in any part of the mixture. If fractionation by relative movement of liquid and crystals takes place during crystallization, however, the relative amounts of the phases, at a given temperature, may differ in different parts of the mixture, but the nature of the phases is the same as under equilibrium conditions.

Ternary Systems with Compound Formation   Crystallization paths in ternary systems with congruently melting binary or ternary compounds, or both, are derived by methods similar to those described for the simple eutectic system above. When compounds are present, the system may be divided into composition (compatibility) triangles within each of which the paths of crystallization are derived by methods resembling those described for the system $CaSiO_3$–$CaAl_2Si_2O_8$–SiO_2. For the purposes of the present discussion, the main point to note is that crystallization in this type of system always proceeds to the eutectic point at which liquid coexists with the three crystalline phases whose compositions are represented by the apices of the composition triangle in which the point representing the mixture composition is located.

As was the case in binary systems, the crystallization paths in ternary systems become more complex if one or more of the compounds melt(s) incongruently rather than congruently. An example of such relations is afforded by the system Mg_2SiO_4–$CaAl_2Si_2O_8$–SiO_2[3] (see Fig. 4-10, after Andersen, 1915; Greig, 1927).

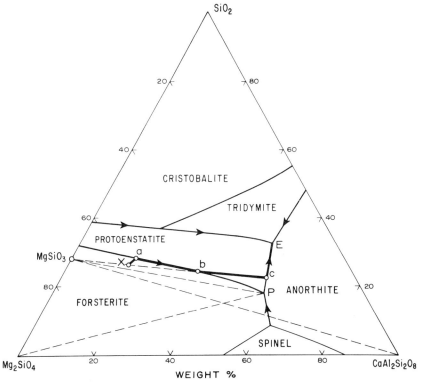

Figure 4-10. Phase diagram of the system Mg_2SiO_4–$CaAl_2Si_2O_8$–SiO_2 (Andersen, 1915; Greig, 1927), showing paths of crystallization of a selected mixture.

[3] The part of the system in which spinel is present as a phase is not ternary. The present discussion, however, will not be concerned with that part of the system.

The following discussion will be concerned primarily with liquid-solid equilibria along (or in the vicinity of) the boundary curve between the primary phase areas of forsterite (Mg_2SiO_4) and protoenstatite ($MgSiO_3$). Note first that the incongruent melting of $MgSiO_3$ in the binary system $MgO–SiO_2$ (see Fig. 4-3) is evidenced in the diagram of Fig. 4-10 by the fact that the $MgSiO_3$ composition point is outside the composition area within which protoenstatite ($MgSiO_3$) is the primary crystalline phase. Also, the tangents to the liquidus boundary curve between the forsterite and the protoenstatite primary phase fields intersects the $MgO–SiO_2$ join to the SiO_2 side of the $MgSiO_3$ composition point, rather than between the latter and the Mg_2SiO_4 composition point. Hence, one of the crystalline phases (forsterite) is dissolving while the other (protoenstatite) is crystallizing as temperature decreases and liquid compositions change from left to right along this boundary curve. With forsterite dissolving, the question of whether or not the liquid composition will follow the boundary curve all the way to the forsterite-protoenstatite-anorthite ternary peritectic (point P in Fig. 4-10) depends on how much forsterite was present as a phase when the liquid first became saturated with protoenstatite, i.e., when the liquid in its crystallization path first reached the forsterite-protoenstatite liquidus boundary curve. This behavior depends on the bulk composition of the mixture being crystallized. For instance, in mixtures within the primary phase field of forsterite located below the dashed line $MgSiO_3–P$ (but above the line $Mg_2SiO_4–P$), the liquid during equilibrium crystallization will change its composition along the forsterite-protoenstatite boundary curve all the way to the ternary peritectic point. In mixtures above the $MgSiO_3–P$ line, but below the olivine-protoenstatite boundary curve, on the other hand, forsterite will be consumed before the liquid reaches the peritectic point. In the latter case, therefore, the liquid will leave the forsterite-protoenstatite boundary curve and cross the latter field, as shown, for example, by curve $b–c$ in Fig. 4-10. Whether or not forsterite is completely consumed as crystallization proceeds along the forsterite protenstatite boundary curve, however, the fundamental reaction is the same and similar to that shown by the *reaction relation* at the peritectic temperature of the system $MgO–SiO_2$: forsterite + liquid → protoenstatite.

If equilibrium prevails during crystallization, as described above, all parts of the mass will have the same bulk composition. If, on the other hand, fractionation of phases takes place during crystallization, the bulk composition may vary from place to place within the mass. For instance, if the forsterite crystals forming in mixture X in Fig. 4-10 are partly or completely withdrawn from the liquid after their formation, the remaining mass (liquid) would have compositions represented by various stages of the equilibrium path of crystallization. It is as if the crystallization started at various points along the curve $X-a$ in Fig. 4-10, each point representing a new bulk composition of the mass being crystallized from an all-liquid state. Hence, an infinite number of paths of liquid change is possible during fractional crystallization, depending on the degree of fractionation. The multiplicity of crystallization paths during fractional crystallization is thus correlatable at least in part to the reaction relation existing between an early-formed crystalline phase (forsterite in the case considered above) and liquid to form another, "late" crystalline phase (protoenstatite in the case considered above).

The crystallization process taking place along other boundary curves in Fig. 4-10 is quite different from that described above. For instance, along the anorthite-tridymite boundary curve these two solid phases crystallize together as the liquid composition changes toward the left along the liquidus boundary curve with decreasing temperature. [Note that the tangent to the liquidus boundary curve intersects the $CaAl_2Si_2O_8-SiO_2$ join between the composition points for the two crystalline phases ($CaAl_2Si_2O_8$ and SiO_2) involved.] Bowen termed this behavior a *subtraction relation*. In this case, in contrast to that prevailing in the *reaction relation* discussed above, the compositions of successive liquids, and the nature of the crystalline phases formed, will not vary with the degree of fractionation; only the relative amounts of these phases in different parts of the crystallizing mass will vary with degree of fractionation.

The separation of crystals from liquid by relative movement of these two phases was visualized as the main mechanism of fractionation in the discussion above. Other mechanisms, however, may have similar effects. For instance, it is common that protoenstatite crystals resulting from the reaction relation form a layer on the forsterite crystals, having the effect of isolating the forsterite crystals from the liquid, and thus producing a path of liquid change characteristic of that prevailing in the absence of forsterite. Hence, the crystallization path in this case will be through the protoenstatite primary phase area and in the general direction of the protoenstatite + tridymite + anorthite ternary invariant point in Fig. 4-10.

By analogy with the binary system $MgO-SiO_2$ (Fig. 4-3) it may be concluded that in a ternary system of the type shown by $Mg_2SiO_4-CaAl_2Si_2O_8-SiO_2$ and involving an incongruently melting compound, the path of melting, while being the exact reverse of the path of equilibrium crystallization, is not the reverse of fractional crystallization. As was the case in the binary system $MgO-SiO_2$, so in the ternary system *repeated* cycles of fusion and crystallization are required in order to bring about effects similar to those produced by fractional crystallization.

Relations similar to those for an incongruently melting binary compound prevail in the case of an incongruently melting ternary compound. Again the reaction relation is operative, i.e., the early-formed crystals react with liquid to form another crystalline phase as temperature decreases. Either by partial removal of early crystals from the liquid or by formation of a surface layer on the early crystals, the reaction of the early crystals with the liquid may be curtailed, leading to fractional crystallization paths that may be derived by methods similar to those described for the system Mg_2SiO_4–$CaAl_2Si_2O_8$–SiO_2.

TERNARY SYSTEMS WITH SOLID SOLUTION AMONG THE PHASES

Consider first a system with a complete binary solid solution and no compound formation. The classic example is $NaAlSi_3O_8$–$CaAl_2Si_2O_8$–$CaMgSi_2O_6$ (see Fig. 4-11, after Bowen, 1915; Kushiro, 1972). Crystal-

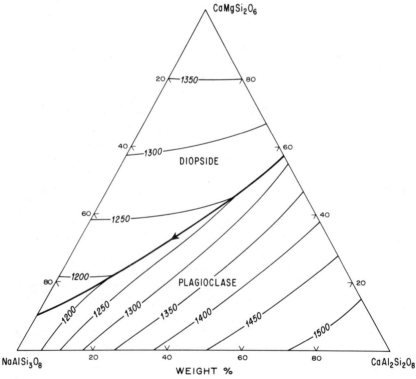

Figure 4-11. Projection of the liquidus surface of the system $NaAlSi_3O_8$–$CaAl_2Si_2O_8$–$CaMgSi_2O_6$ at a total pressure of 1 atm (Bowen, 1915; Kushiro, 1972).

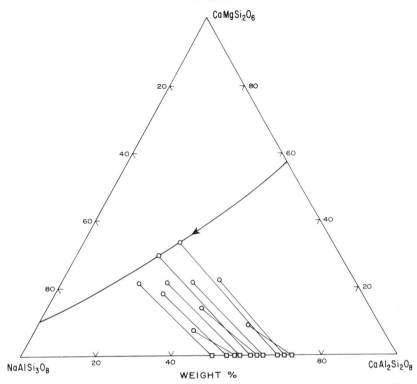

Figure 4-12. Diagram showing selected conjugation lines between coexisting crystalline and liquid phases in the system $NaAlSi_3O_8-CaAl_2Si_2O_8-CaMgSi_2O_6$ (Bowen, 1915; Kushiro, 1972).

lization within the diopside primary phase area is similar to those described above, because the composition of diopside is nearly constant.[4] Crystallization involving feldspar crystals is more complicated, however, because of the complete solid solubility between the end members $NaAlSi_3O_8$ and $CaAl_2Si_2O_8$. In the following discussion, crystallization paths will be considered under two idealized conditions: equilibrium and perfect fractional crystallization.

Assume that the interrelations between compositions of coexisting liquid and crystalline phases have been determined experimentally, and, hence, that the directions of *conjugation lines*[5] between such pairs of coexisting phases are known. A number of such lines are shown in Fig. 4-12.

---

[4] See footnote 1.

[5] Conjugation lines (= tie lines) are straight lines connecting the points representing the compositions of two phases coexisting in equilibrium.

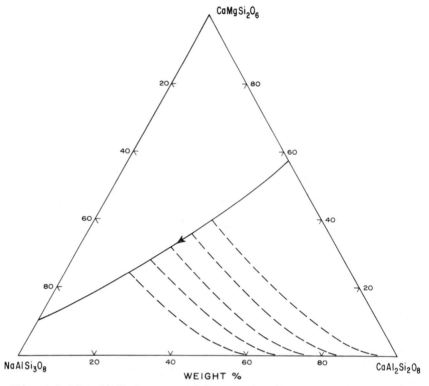

Figure 4-13. Diagram showing approximate locations of fractionation curves along the liquidus surface of the system $NaAlSi_3O_8$–$CaAl_2Si_2O_8$–$CaMgSi_2O_6$.

It is obvious from this figure that the resulting diagram tends to be rather confusing if a relatively large number of conjugation lines are drawn. A clearer picture of such liquid-solid interrelations emerges *if the intercepted arcs*[6] rather than the conjugation lines themselves are drawn, as shown in Fig. 4-13. Because there can be only one composition of the solid solution coexisting in equilibrium with a liquid of given composition at a given temperature, there can be only one direction of the conjugation line through each point representing the composition of liquid along the liquidus surface. Therefore, the intercepted arcs cannot intersect each other. Because of this relation, interpolations among these curves may be made accurately. Moreover, because an infinite number of tangents can be drawn to each arc, the family of curves shown in Fig. 4-13 sum-

[6] The intercepted arcs are the curves to which the conjugation lines are tangents.

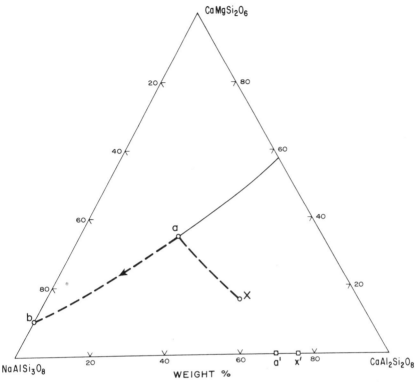

Figure 4-14. Diagram showing paths of perfect fractional crystallization in the system $NaAlSi_3O_8$–$CaAl_2Si_2O_8$–$CaMgSi_2O_6$ at a total pressure of 1 atm.

marizes in a clear and concise form a very large amount of information on liquid-solid interrelations.

Crystallization paths may be derived from these curves (see Osborn and Schairer, 1941) as follows: Consider first a mixture X shown in Fig. 4-14, as it is cooled from liquidus to solidus temperature. The first crystalline phase to appear at the liquidus temperature is represented by point x', and is determined as the intersection between the join $NaAlSi_3O_8$–$CaAl_2Si_2O_8$ and the tangent to the intercepted arc at point X. If the crystals are removed from the liquid as soon as they are formed, thus preventing, as temperature decreases, any reaction between liquid and the crystals formed at higher temperatures, the liquid composition will change along the arc X–a. Thus, these curves represent the course of liquid change in the plagioclase field if perfect fractionation prevails. These curves are, therefore, referred to as *fractionation curves*. The fractionation curves in Figs. 4-13 and 4-14 are seen to be convex to the left, and hence, their tangents rotate clockwise as crystallization proceeds with decreasing temperature. The plagioclase crystals, whose compositions are determined by the tangents to the fractionation curves, thus become richer in the $NaAlSi_3O_8$ component as the crystallization progresses. If no reaction takes place between the liquid and the crystals having separated out at the higher temperatures, the enrichment in the $NaAlSi_3O_8$ component of the phases will continue without the constraints of liquid-solid inter- action. After the liquid has reached the plagioclase-diopside boundary curve, the liquid changes its composition along this boundary curve all the way to the left-hand side of the diagram. The path of change of liquid composition under conditions of perfect fractionation is thus represented by the curve $X–a–b$, whereas the plagioclase crystals vary in composition from x' through a' to $NaAlSi_3O_8$. (For a more detailed consideration of fractional crystallization paths in this and other systems, the reader is referred to a recent paper by Morse, 1976.)

In contrast to the above, consider the path of crystallization if the cool- ing takes place so slowly that at each step of the process perfect equilibrium is maintained, i.e., liquid reacts with early crystals to adjust their composi- tion to yield solid solutions of uniform composition required for equilib- rium with the liquid at each temperature. The path of liquid descent in this case is derived on the basis of the following considerations, using as example a mixture of composition X in Fig. 4-15. As long as the liquid is in equilibrium with only one crystalline phase (in this case plagioclase of variable composition), the conjugation line connecting the points repre- senting the compositions of the two coexisting phases (liquid and crystal) must pass through the point representing the constant total composition of the mixture. The convexity to the left of the fractionation curves reflects the fact that the crystallizing plagioclase phase changes its composition to the left (becomes increasingly rich in the $NaAlSi_3O_8$ component) as temperature decreases during the crystallization. The composition of the liquid phase is thus being changed along a curve $X–c$ departing to the right of the fractionation curve in question, while the composition of the plagioclase crystals changes continuously from x' to c'. In detail, the course of liquid composition change during equilibrium crystallization is determined as follows: At any given temperature, the composition of liquids in equilibrium with plagioclase must be represented by a point on the liquidus isotherm corresponding to that temperature. At the same time, by criteria discussed above, the liquid composition must be such that the tangent to the fractionation curve through the point representing liquid composition passes through the point X representing the constant bulk composition of the crystallizing mass. Because, as previously noted,

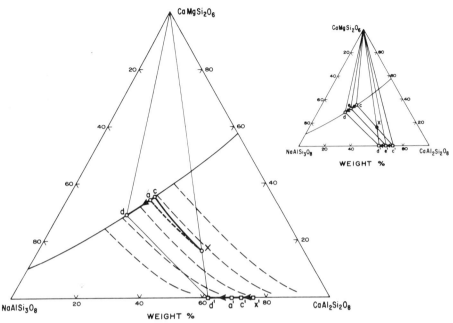

Figure 4-15. Diagram showing paths of equilibrium crystallization of a selected mixture X in the system $NaAlSi_3O_8$–$CaAl_2Si_2O_8$–$CaMgSi_2O_6$ at a total pressure of 1 atm, as discussed in the text. The smaller diagram in the upper right insert shows changes in location of the three-phase triangle plagioclase-diopside-liquid as crystallization proceeds.

fractionation curves do not cross each other, interpolations between the fractionation curves may be made and the curve of equilibrium crystallization path may be constructed accurately.

After the liquid composition has reached the plagioclase-diopside boundary curve, these two phases crystallize together, and the liquid composition changes to the left along the boundary curve. If equilibrium prevails, the plagioclase crystals continuously react with the liquid to produce crystals of uniform composition corresponding to those expressed by the directions of the conjugation lines between coexisting plagioclase and liquid along the boundary curve. During this process, the compositions of both liquid and plagioclase change to the left, i.e., toward higher contents of the $NaAlSi_3O_8$ component. When the composition of the liquid has reached point d, the composition of the plagioclase phase has reached point d', and the conjugation line between the two coexisting crystalline phases (plagioclase and diopside) passes through the point (X) representing the constant total composition of the crystallizing mass.

Hence, the mass has crystallized completely to an aggregate of these two crystalline phases.

In order to help visualize the above sequence of phase changes along the liquidus boundary curve during equilibrium crystallization, it is instructive to consider the movements with decreasing temperature of the three-phase triangle plagioclase–diopside–liquid, as shown schematically in the upper insert of Fig. 4-15. At the point where the liquid first reaches the boundary curve and the first trace of diopside appears, the three-phase triangle $(c'–Di–c)$ is in such a position that point X (representing the total composition of the mixture) is located on the plagioclase-liquid join of the triangle. The second three-phase triangle $(e'–Di–e)$ represents an intermediate case along the boundary curve where all three phases (plagioclase, diopside, and liquid) are present in finite amounts. In this case the three-phase triangle has moved to the left so that point X is within the three-phase triangle. Finally, the third three-phase triangle $(d'–Di–d)$ is located such that point X is now located on the plagioclase-diopside join of the triangle, representing the situation where the last trace of liquid is consumed, i.e., the solidus temperature. The above relations demonstrate the fact that the position of X within the three-phase triangle gives the relative amounts of the phases at any temperature where the three phases coexist in equilibrium.

Systems with Incongruently Melting Binary Solid Solution   Preceding sections dealt with liquid-solid relations attending the incongruent melting of protoenstatite ($MgSiO_3$) in binary ($MgO–SiO_2$) and ternary ($Mg_2SiO_4–CaAl_2Si_2O_8–SiO_2$)[7] systems in which there is no appreciable solid solubility. In most natural minerals and rocks, however, there is extensive solid-solution formation among the various components. An example of a ternary system of this type involving common rock-forming minerals is $Mg_2SiO_4–CaMgSi_2O_6–SiO_2$ as shown in Fig. 4-16 mainly after Bowen (1914), Kushiro (1972), and Yang and Foster (1972). Although phase relations in this system have undergone some revisions since the time of Bowen's original work, many features of the crystallization trends remain essentially as described by him. The key to an understanding of crystallization phenomena in this system is the fact that pyroxene solid solutions of compositions near $MgSiO_3$ behave similarly to the latter end member, i.e., forsterite is the primary crystalline phase on the liquidus surface. Hence, for mixtures of compositions along the pyroxene join ($MgSiO_3–CaMgSi_2O_6$) a *reaction relation* exists. The composition of the liquid

---

[7] As noted previously, the part of the system where spinel appears as a phase is not ternary.

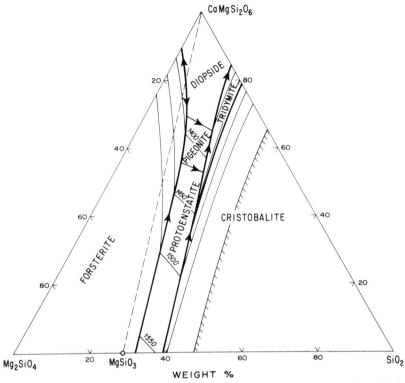

Figure 4-16. Projection of the liquidus surface of the system Mg_2SiO_4–SiO_2–$CaMgSi_2O_6$ (Bowen, 1914; Kushiro, 1972).

phase is off the binary pyroxene join, as forsterite (Mg_2SiO_4) dissolves and pyroxene crystallizes.

The derivation of a detailed quantitative account of relations in crystallization sequences involving the coexistence of two solid-solution phases and a liquid phase is very complex (Greig, unpublished work), and could not be accommodated in a relatively short chapter. An attempt is made below, however, to approximate crystallization paths involving important rock-forming solid solutions.

Approximate locations of fractionation curves along the liquidus surface are shown in Fig. 4-17. From these curves, paths of equilibrium crystallization may be derived by methods similar to those described in the preceding section. The relations are described with reference to Fig. 4-18. Take as an example the equilibrium crystallization of a mixture of

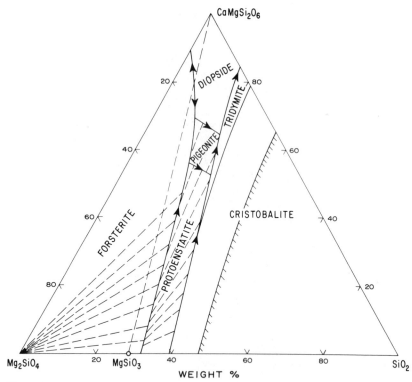

Figure 4-17. Diagram showing (as dash curves) approximate locations of liquidus fractionation curves for the system Mg_2SiO_4–SiO_2–$CaMgSi_2O_6$.

composition X in the primary field of forsterite. With forsterite of essentially constant composition (Mg_2SiO_4) crystallizing, the liquid composition changes along the straight line X–a. At the latter point the liquid has become saturated with protoenstatite of composition P_1, and the liquid composition starts changing along the forsterite-protoenstatite boundary curve. The tangent to this curve intersects the Mg_2SiO_4–SiO_2 join at a point to the SiO_2 side of the $MgSiO_3$ composition point, and thus forsterite is dissolving while protoenstatite crystallizes along this curve. As the composition of the liquid changes from a to b, the composition of the protoenstatite changes from P_1 to P_2 (as determined by the tangents to the pyroxene fractionation curves along the forsterite-pyroxene boundary curve). At the temperature of peritectic point b, pigeonite of composition pi_1 crystallizes out together with protoenstatite and the liquid is consumed, leaving as the end product of crystallization

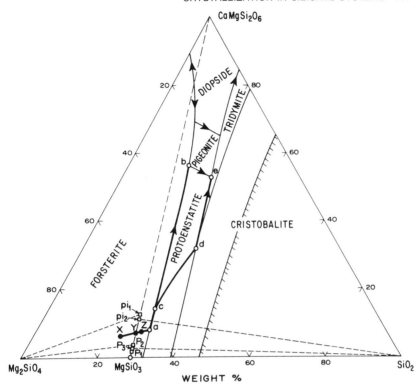

Figure 4-18. Diagram showing paths of equilibrium crystallization of selected mixtures in the system Mg_2SiO_4–SiO_2–$CaMgSi_2O_6$.

two pyroxenes (protoenstatite and pigeonite) and forsterite. The complete consumption of the liquid phase at the peritectic point is evidenced by the fact that mixture X is located within the triangle Mg_2SiO_4–P_2–pi_1.

Identical paths of change in liquid composition are followed if the composition of the mixture is that represented by point Y, located on the $MgSiO_3$–$CaMgSi_2O_6$ join. In this case, with protoenstatite crystallizing and forsterite dissolving as the liquid changes its composition along the forsterite-protoenstatite boundary curve, the liquid phase and forsterite are consumed simultaneously, at the temperature of point b, leaving as end product of equilibrium crystallization two pyroxenes (protoenstatite and pigeonite). (Note that point Y is located on the P_2–pi_1 side of the triangle Mg_2SiO_4–P_2–pi_1).

The initial steps in the crystallization of mixture Z, located within the primary phase area of forsterite but to the SiO_2 side of the pyroxene join, are identical to those described above for mixtures X and Y. In the case of mixture Z, however, forsterite rather than liquid is the first phase to be consumed at the temperature corresponding to point c. With further decrease in temperature, the liquid, therefore, leaves the boundary curve and traverses the protoenstatite field along a curved path c–d, which can be derived from the fractionation curves by methods similar to those described in detail for the system $NaAlSi_3O_8$–$CaAl_2Si_2O_8$–$CaMgSi_2O_6$ in a previous paragraph. When the composition of the liquid reaches the protoenstatite-cristobalite boundary curve at point d, the liquid has become saturated with respect to the latter phase, and protoenstatite and silica (first cristobalite, then tridymite at lower temperatures) crystallize together until the composition of the liquid has reached point e. At this point pigeonite of composition pi_2 crystallizes out together with proto-enstatite of composition P_3 and tridymite, leaving an aggregate of these three phases. (Note that mixture Z is located within the triangle P_3–pi_2–SiO_2.)

The equilibrium crystallization discussed above is an idealized condition that may be approached to various degrees but rarely is attained. Crystallization paths deviating to a larger or smaller extent from those described may, therefore, obtain in practice. The main mechanisms preventing attainment of equilibrium in the system under consideration (Mg_2SiO_4–$CaMgSi_2O_6$–SiO_2) are combinations of those examined separately for the systems Mg_2SiO_4–$CaAl_2Si_2O_8$–SiO_2 and $NaAlSi_3O_8$–$CaAl_2Si_2O_8$–$CaMgSi_2O_6$ in previous paragraphs, namely (1) failure of early crystals of forsterite to react with the liquid as the latter (either by relative movement of crystals and liquid or by formation of a reaction rim of pyroxene on the surface of the olivine crystals) becomes saturated with pyroxene and (2) failure of early crystals of pyroxene to react with the liquid so as to adjust continuously the composition of this solid solution to that required for maintaining equilibrium as temperature decreases. If perfect fractionation prevails, i.e., if there is no reaction between the liquid and the early-formed forsterite crystals as the forsterite-pyroxene boundary curve is reached, and no adjustment of pyroxene composition after the pyroxene crystals first form, the path of composition changes of the liquid follows the fractionation curves, as sketched in Fig. 4-19. For any one of the mixtures X, Y, or Z used in the previous discussion, the paths of fractional crystallization are as follows: Forsterite crystals of essentially constant composition crystallize first, and the liquid composition changes along the extension of the straight line connecting the forsterite composition point (Mg_2SiO_4) with the point representing the constant total composition of the mixture. If it is assumed that the forsterite crystals are removed from the liquid as soon as they are formed, or otherwise are prevented

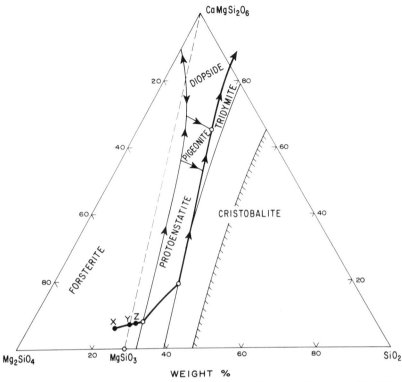

Figure 4-19. Diagram showing paths of fractional crystallization in the system Mg_2SiO_4–SiO_2–$CaMgSi_2O_6$.

from further reacting with the liquid, the liquid upon reaching the forsterite-protoenstatite boundary curve immediately leaves this boundary curve, and with protoenstatite crystallizing, the liquid crosses the protoenstatite field. The early-formed protoenstatite crystals do not adjust their composition by reaction with the liquid as temperature decreases, and the path of liquid change across the protoenstatite field is along the fractionation curve passing through the point at which the liquid reached the forsterite-protoenstatite boundary curve. The composition of the protoenstatite crystallizing out at any point along this fractionation curve is determined by the tangent to the fractionation curve at that point. After the liquid becomes saturated with silica (initially cristobalite), the liquid composition changes along the pyroxene-silica boundary curve, involving first cristobalite and protoenstatite, then

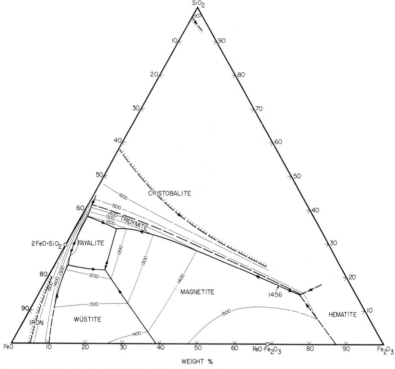

Figure 4-20. Projection of the liquidus surface of the system FeO–Fe$_2$O$_3$–SiO$_2$ at a total pressure of 1 atm (Muan, 1955, 1958).

tridymite and protoenstatite, followed by tridymite and pigeonite, and finally diopsidic pyroxene and tridymite, depending on the temperature.

TERNARY SYSTEMS IN WHICH THE GAS PHASE PLAYS
AN IMPORTANT ROLE

The relations derived for the system FeO–Fe$_2$O$_3$ (see a previous section) may be extended to ternary- and higher-component systems. A particularly suitable example of the former is FeO–Fe$_2$O$_3$–SiO$_2$.

Paths of crystallization in this system under various assumed idealized conditions may be derived by a combination of information contained in the diagrams shown in Figs. 4-20 to 4-22 (Muan, 1955, 1958). The first of these (Fig. 4-20) shows interrelations between temperature and compositions of liquids at liquidus temperatures. The second diagram (Fig. 4-21) portrays fractionation curves along the liquidus surface, and,

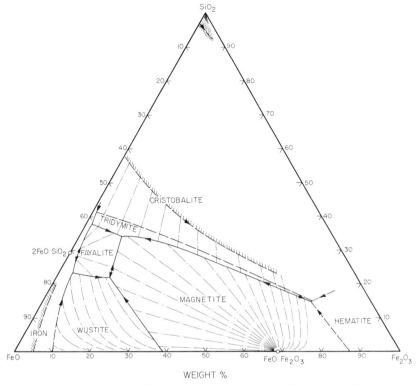

Figure 4-21. Projection of the liquidus surface of the system FeO–Fe$_2$O$_3$–SiO$_2$ showing boundary curves as heavy lines and fractionation curves as medium dash lines (Muan, 1955, 1958).

thus, illustrates interrelations of coexisting liquid and crystalline phases, as explained above. The third diagram (Fig. 4-22) shows oxygen isobars along the liquidus surface, and, hence, illustrates interrelations between compositions of condensed phases (solids and liquids) and the gas phase. The meaning of these oxygen isobars along the liquidus surface is best understood by considering the binary system FeO–Fe$_2$O$_3$ as a base and studying the effect of adding SiO$_2$ as a third component. In discussing the binary system FeO–Fe$_2$O$_3$, it was emphasized that the oxygen isobars cross areas of two coexisting condensed phases (e.g., magnetite + liquid) as horizontal lines, i.e., at constant temperature. When SiO$_2$ is added as a component, the system acquires an additional degree of freedom for the same phase assemblage (one crystalline phase plus liquid), and oxygen fugacity can vary with temperature along the liquidus surface. Each *point*,

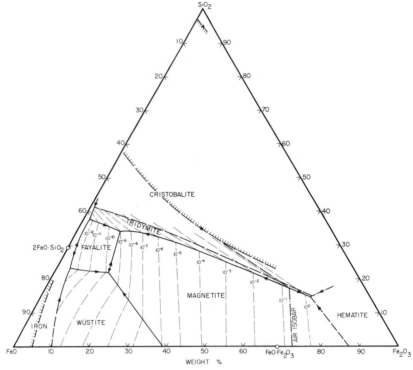

Figure 4-22.   Projection of the liquidus surface of the system $FeO–Fe_2O_3–SiO_2$ showing boundary curves as heavy lines and oxygen isobars (in atm) as light dash-dot lines (Muan, 1955, 1958).

corresponding to a given oxygen fugacity along the liquidus curve for the binary system $FeO–Fe_2O_3$, serves as the starting point for a *curve* of constant oxygen fugacity along the liquidus surface in the ternary system $FeO–Fe_2O_3–SiO_2$, as shown in Fig. 4-22.

Consider first mixture X in the primary phase area of magnetite (see Fig. 4-23) under the assumption that the total composition of condensed phases is kept constant. The liquid composition initially changes along the straight line $X–a$ as increasing amounts of magnetite separate out. (In this part of the diagram the magnetite composition may be considered constant, corresponding to the formula Fe_3O_4.) At point a the liquid has also become saturated with fayalite (Fe_2SiO_4), and with decreasing temperature, fayalite and magnetite crystallize together as the liquid changes its composition along the fayalite-magnetite boundary curve from a to b. At the latter point the liquid becomes saturated with silica (tridymite),

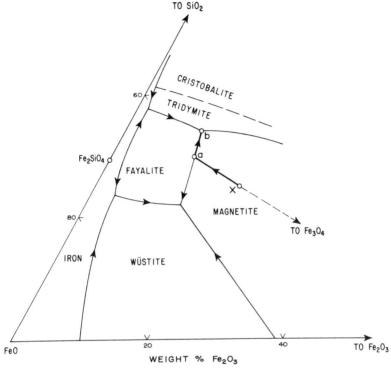

Figure 4-23. Diagram of a part of the system FeO–Fe_2O_3–SiO_2 showing paths of equilibrium crystallization of a mixture X under conditions of constant total composition of condensed phases.

which crystallizes together with fayalite and magnetite. The liquid is consumed, leaving as final product of crystallization an aggregate of the above three crystalline phases. (Note that X is within the triangle Fe_2SiO_4–Fe_3O_4–SiO_2.) During this process the oxygen fugacity changes in a manner that may be derived by tracing the path of liquid change, as described above, on the diagram showing the oxygen isobars along the liquidus surface (Fig. 4-22). It is seen that the oxygen fugacity first decreases significantly as magnetite, containing two-thirds of its iron in the trivalent state, crystallizes along the path of the liquid, X–a. When fayalite, containing essentially all its iron in the divalent state, crystallizes together with magnetite along the fayalite-magnetite boundary curve, the oxygen fugacity increases slightly.

Contrast the crystallization path described above with that resulting if the same mixture is cooled under equilibrium conditions at a constant

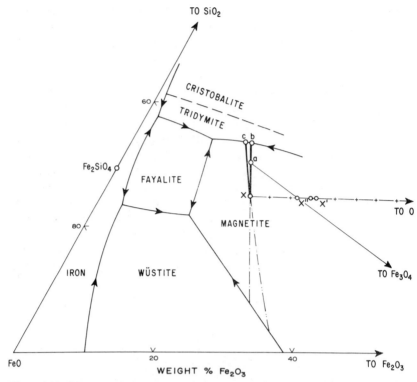

Figure 4-24. Diagram of the system $FeO-Fe_2O_3-SiO_2$ showing paths of equilibrium
crystallization under two idealized conditions, viz., constant oxygen fugacity
and constant H_2O/H_2 of the gas phase, as discussed in the text. The dash-cross
line represents changes in total composition of condensed phases during
crystallization.

oxygen fugacity corresponding to that prevailing at the liquidus temperature (10^{-7} atm). In this case (see Fig. 4-24), the liquid composition changes
along the corresponding oxygen isobar from X to b as magnetite of constant composition (Fe_3O_4) crystallizes. The total composition of the
mixture is changing during this process by virtue of the condensed phases
reacting with oxygen of the gas phase. This change, reflecting the constancy of Si/Fe and changing O/(Si + Fe) of the mixture, is described by
the "oxygen reaction line" $X–X'$ shown as a dash-cross line in Fig. 4-24
and pointing to the O apex of the triangle representing the system Fe–Si–O,
of which $FeO-Fe_2O_3-SiO_2$ is a part. In addition, the conjugation line
connecting compositions of the coexisting phases (magnetite and liquid)
in the two-phase area must pass through the point representing the total

composition of the mixture. Hence, the total composition of the mixture, at any given temperature during the crystallization, is determined as the point of intersection between these two lines. For instance, the total composition of a mixture consisting of magnetite and liquid a is represented by point X''. As temperature is lowered further and additional amounts of magnetite crystallize, the liquid composition continues to change along the oxygen isobar until point b is reached. Here the liquid has also become saturated with silica (tridymite), the liquid is consumed, and the final product of crystallization is a mixture of tridymite and magnetite. The total composition of the mixture is now the point of intersection between the oxygen reaction line and the SiO_2–Fe_3O_4 join. It is thus seen that equilibrium crystallization under constant oxygen fugacity differs strikingly, both with regard to the course of change of the liquid and with respect to the final product of crystallization, from that obtaining when the total composition of the mixture is kept constant.

As for the binary system FeO–Fe_2O_3, one may conceive of a number of alternative idealized ways of carrying out equilibrium crystallization in the system FeO–Fe_2O_3–SiO_2. One possibility is to keep the H_2O/H_2 ratio of the gas phase constant. The paths of liquid change in this case follow the curves of equal H_2O/H_2 along the liquidus surface, as shown by the curve X–c in Fig. 4-24. The interrelations among total-composition changes, and compositions of coexisting solid and liquid phases are analogous to those described above, with the curves of constant H_2O/H_2 playing the same role as did the oxygen isobars in the case described above.

Because no appreciable solid solubility and no incongruently melting phases are involved in the examples shown for the system FeO–Fe_2O_3–SiO_2, crystal–liquid fraction would not have a significant effect on the crystallization process. For instance, settling of magnetite from the liquid upon crystallization would not alter the paths of liquid composition during crystallization or the nature of the phase assemblages present; only the relative amounts of the phases would vary from place to place in the crystallizing mass as a result of such fractionation. Because reaction between the condensed phases and oxygen of the gas phase may alter the course of liquid descent and the final product of crystallization, however, the degree to which equilibrium between gas and condensed phases is attained will have a profound effect on the crystallization path. Failure of the gas phase to react with the condensed phases, even if the oxygen fugacity by some mechanism is maintained constant, may be considered a form of fractional crystallization. If equilibrium between gas and condensed phases is only partially attained, the crystallization paths would be intermediate between those shown in Figs. 4-23 and 4-24.

QUATERNARY SYSTEMS

GENERAL CONSIDERATIONS

Paths of crystallization in quaternary silicate systems can be derived by methods similar to those applied to binary and ternary systems in the preceding sections. The main difficulty in extending such relations to quaternary systems does not stem from the application of any new principles but from the complexity of illustrating the interrelations geometrically in a legible form. In going from a ternary to a quaternary system, another composition variable is added. Therefore, three-dimensional models are needed in order to express compositions within the quaternary system. The most convenient and generally accepted representation is in the form of a regular tetrahedron. Phase relations in quaternary systems are, thus, usually shown in the form of perspective drawings of such tetrahedra or in projections into selected planes through such tetrahedra.

In representing and interpreting phase relations in tetrahedral models it is important to keep in mind the restrictions imposed by Gibbs' Phase Rule, and it is particularly useful for general guidance to make comparisons with analogous ternary systems for which the relations can be represented quantitatively in the plane of the paper. For instance, a liquidus univariant situation in a ternary system is characterized by two crystalline phases coexisting with liquid along a boundary curve between two adjacent primary phase areas, whereas in a quaternary system a univariant liquidus situation is characterized by the coexistence of three crystalline phases and liquid. Similarly, the simultaneous presence of three crystalline phases and liquid corresponds to a liquidus invariant situation in a ternary system, whereas four crystalline phases coexist with liquid in a quaternary liquidus invariant situation. The general trend of liquid crystallization in quaternary systems is also similar to those of ternary systems: under equilibrium conditions the primary crystalline phase appears at the liquidus temperature, and as the temperature decreases further, increasing amounts of the primary phase separate out and the liquid composition changes along a certain path through the primary-phase volume. This path is a straight line if the crystalline phase has constant composition, and a curved path if the composition of the crystalline phase is variable (i.e., if the phase is a solid solution). When the curve representing the composition of the liquid reaches the surface between the primary-phase volume and an adjacent phase volume, the liquid becomes saturated with a second crystalline phase. With these two crystalline phases coexisting with liquid over a certain temperature range, the composition of the liquid changes along the above-mentioned boundary surface and usually stays on this surface until the path reaches a univariant curve where three primary-phase volumes join each other. This univariant curve represents the compositions of liquid in equilibrium with three crystalline phases. Depending on the characteristics of the system, and in analogy with relations discussed for ternary systems in a previous section of this chapter, all three crystalline phases may be separating out together as the liquid composition changes along the univariant curve, or one or two of the crystalline phases may be dissolving while the other(s) are crystallizing. If all three solid phases are crystallizing, the liquid composition continues to change along the univariant curve until either the liquid phase is completely consumed along this curve (possible only if extensive solid-solution formation is involved; see previous discussion of ternary systems), or until the liquid becomes saturated also with a fourth crystalline phase at a quaternary invariant point. The latter point is located where four primary-phase volumes meet. If one (or more) of the crystalline phases coexisting with liquid along a quaternary univariant curve is dissolving rather than precipitating (compare the ternary analog of this case, as discussed above), the liquid may leave the univariant curve, depending on the interrelations among the compositions of the coexisting phases and the total composition of the mixture.

The number of possible configurations of a system increases drastically as the number of components increases, and a very large number of examples would be necessary in order to derive the various possible crystallization paths in quaternary systems, even if perfect equilibrium is attained. The present discussion will be limited to one petrologically important quaternary system, $MgO-FeO-Fe_2O_3-SiO_2$. This system combines most of the features of particular interest in deriving crystallization paths: an incongruently melting compound (the metasilicate solid solution $MgSiO_3-FeSiO_3$), extensive or complete solid-solution formation between the Mg-containing and the Fe-containing end members ($Mg_2SiO_4-Fe_2SiO_4$, $MgSiO_3-FeSiO_3$, and $MgFe_2O_4-FeFe_2O_4$), and dependence of compositions of the condensed phases on the oxygen fugacity of the gas phase. If the total composition of the mixture (crystallizing mass) is kept constant, however, the paths of crystallization can be derived by the same methods as apply to systems in which the gas phase does not play an important role, as was explained in detail for the ternary analogous system $FeO-Fe_2O_3-SiO_2$. Hence, the system $MgO-FeO-Fe_2O_3-SiO_2$ also serves the purpose of illustrating paths of crystallization under conditions where the gas phase does not have to be taken into account.

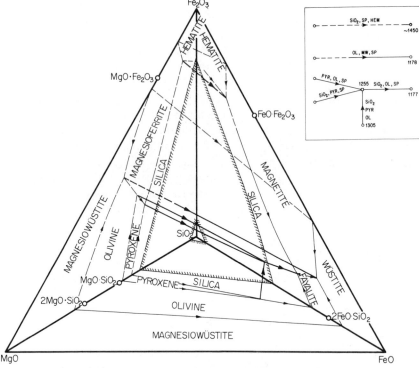

Figure 4-25. Diagram showing liquidus phase relations in the system $MgO-FeO-Fe_2O_3-SiO_2$ (Muan and Osborn, 1956, 1965). Boundary curves on the four faces of the tetrahedron are drawn as light lines, and those within the tetrahedron as heavy lines. Lines with stippling on one side show outlines of a two-liquid region. The smaller upper-right insert shows a simplified ("flow-sheet") representation of univariant liquidus curves, as explained in the text. Abbreviations: SP = spinel; HEM = hematite; OL = olivine; MW = magnesiowüstite; PYR = pyroxene.

THE SYSTEM $MgO-FeO-Fe_2O_3-SiO_2$

The phase relations of particular interest for the present discussion are portrayed in Figs. 4-25 to 4-29 (Muan and Osborn, 1956, 1965). The first of these (Fig. 4-25) is a perspective drawing showing primary-phase volumes of the various crystalline phases at liquidus temperatures, including divariant boundary surfaces between adjacent phase volumes, univariant liquidus boundary curves, and invariant points at a total pressure of 1 atm. Arrows indicate directions of decreasing temperature along

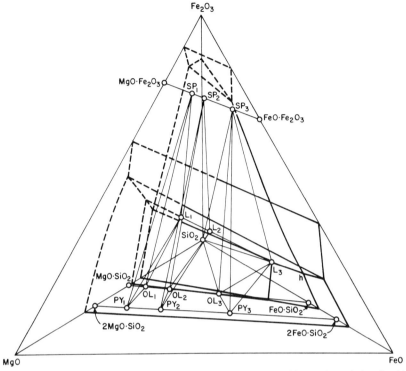

Figure 4-26. Diagram illustrating interrelations among compositions of coexisting liquid and crystalline phases in the system $MgO–FeO–Fe_2O_3–SiO_2$ (see text).

the liquidus univariant curves. It is seen that the temperature in general decreases from the MgO-rich side toward the FeO-rich side of the system.

In order to show more clearly the interrelations among the various univariant curves, and the phase assemblages and directions of decreasing temperature along these curves, it has been found useful to represent such relations in a simplified "flow sheet" (Schairer, 1942), in which the univariant curves are "spread out" in a plane without regard to their spatial orientation. A sketch showing a "flow sheet" of univariant curves in the system $MgO–FeO–Fe_2O_3–SiO_2$ is shown as an insert in Fig. 4-25 in order to introduce this concept. Such simplified, non-quantitative representations are particularly helpful for visualizing interrelations among the various curves in complex quaternary systems involving a large number of crystalline phases.

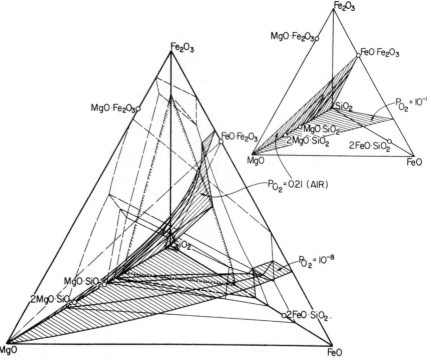

Figure 4-27. Sketch showing approximate locations and shapes of two representative
oxygen isobaric surfaces (0.21 atm and 10^{-8} atm) through the tetrahedron
representing the system $MgO–FeO–Fe_2O_3–SiO_2$, at liquidus temperatures.
The two surfaces are shown in simplified form as planes in upper right insert
(compare text).

In principle, it is possible to construct a family of fractionation *surfaces*
for the system $MgO–FeO–Fe_2O_3–SiO_2$. These surfaces would be the
quaternary analogs of the fractionation *curves* shown previously for the
ternary system $FeO–Fe_2O_3–SiO_2$. In practice, however, it is difficult to
draw legible diagrams of this type for quaternary systems. Instead, one
may portray selected conjugation lines between coexisting phases, as
shown in Fig. 4-26. It is seen that the points representing compositions
of the solid phases are located closer to the MgO side of the system than
are the composition points for the conjugate liquid phase, demonstrating
a consistent enrichment in MgO in the crystalline phases relative to the
liquid. This observation, of course, correlates with the consistent decrease
in liquidus temperatures from the MgO toward the FeO side of the
diagram.

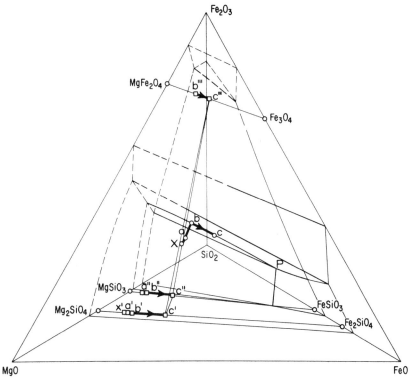

Figure 4-28. Diagram illustrating path of equilibrium crystallization of a mixture X in the system $MgO–FeO–Fe_2O_3–SiO_2$ at constant total composition of condensed phases.

Practical difficulties of a geometrical nature, similar to those pertaining to the fractionation surfaces, apply to the extension of the oxygen isobaric *curves* in the ternary system $FeO–Fe_2O_3–SiO_2$ to oxygen isobaric *surfaces* in the quaternary system $MgO–FeO–Fe_2O_3–SiO_2$. If the ternary system $FeO–Fe_2O_3–SiO_2$ is considered as a base, and MgO is added as a component, the system gains an additional degree of freedom, for a given phase assemblage. The oxygen isobaric curves on the liquidus surface of the ternary system generate oxygen isobaric surfaces in the quaternary system. These oxygen isobaric surfaces change direction abruptly as the surfaces intersect liquidus divariant surfaces or univariant curves. The oxygen isobaric surfaces, therefore, have irregular shapes rendering their quantitative reproduction in perspective drawings of three-dimensional tetrahedral models difficult. An example of an attempt to reproduce such surfaces is shown in Fig. 4-27. In order to show the trends of such relations

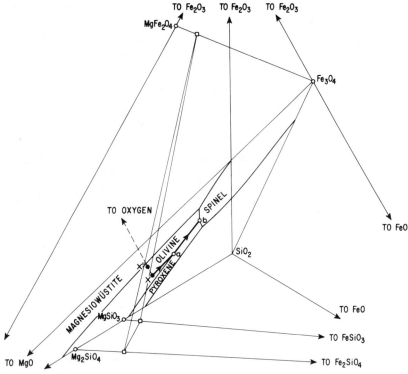

Figure 4-29. Simplified diagram presented in order to illustrate paths of equilibrium crystallization of a selected mixture X in the system $MgO–FeO–Fe_2O_3–SiO_2$ at constant oxygen pressure of the gas phase.

in more legible form, it is advantageous to use simplified representations omitting details of the curvatures of these oxygen isobaric surfaces. Examples of such simplifications are presented in the upper insert of Fig. 4-27, where the irregularly curved surfaces have been drawn as planes through the tetrahedron representing the quaternary system $MgO–FeO–Fe_2O_3–SiO_2$.

Paths of crystallization under various assumed idealized conditions for selected mixtures in the system $MgO–FeO–Fe_2O_3–SiO_2$ will be derived with reference to Figs. 4-28 and 4-29. Consider first (see Fig. 4-28) the crystallization of a mixture X at constant total composition of the mixture. Olivine of composition x' starts to crystallize at liquidus temperature, and the liquid composition changes along a slightly curved path $X–a$ through the olivine phase volume as increasing amounts of olivine of slightly changing composition separate out with decreasing temperature. The composition of the olivine at any temperature is determined by the direction of the liquid-solid conjugation line (= the tangent to the fractionation surface through the point representing the liquid composition at that temperature). When the liquid reaches the divariant surface between the olivine and pyroxene phase volumes, these two phases coexist with liquid as the latter changes its composition along a slightly curved path a–b along the divariant surface and the olivine and pyroxene crystals change their composition from a' to b' and from a'' to b'', respectively. It is to be noted that the tangents to the curve of liquid change within this surface intersect the basal plane (MgO–FeO–SiO_2) at points to the SiO_2 side of the $MgSiO_3$–$FeSiO_3$ join, i.e., outside the composition area defined by the olivine (Mg_2SiO_4–Fe_2SiO_4) and pyroxene ($MgSiO_3$–$FeSiO_3$) joins. Hence, while both crystalline phases, olivine and pyroxene, are present together with liquids along the divariant surface, one of the crystalline phases (pyroxene) is crystallizing while the other (olivine) is dissolving. A *reaction relation*, therefore, exists between the liquid and the crystalline phases, similar to that previously described for ternary systems. Whether or not the dissolving phase (olivine) is consumed (i.e., whether or not the liquid leaves the boundary surface) before a third crystalline phase appears, depends on the total composition of the mixture. If the liquid does not leave the divariant surface, a point (b) is reached at which the liquid becomes saturated also with respect to spinel ($MgFe_2O_4$–$FeFe_2O_4$ solid solution of composition b'''). As temperature decreases further, the three phases olivine, pyroxene, and spinel coexist with a liquid whose composition changes toward the right from b to c along the univariant boundary curve involving these phases. If equilibrium is maintained, the three crystalline phases will continuously change their compositions by reaction with the liquid to those required by the liquid-solid equilibrium at any particular temperature. As the crystalline phases thus increase their Fe/Mg as temperature decreases and crystallization proceeds, a situation may be reached at which the plane (c', c'', c''') formed by the composition points representing the three coexisting crystalline phases (olivine, pyroxene, spinel) passes through the point X representing the constant total composition of the mixture under consideration. When this occurs, the liquid phase has been completely consumed, i.e., the solidus temperature has been reached.

If the plane noted above (the plane formed by the composition points representing the three crystalline phases olivine, pyroxene, and spinel), as it moves in the direction of the FeO–Fe_2O_3–SiO_2 side of the diagram with decreasing temperature, does not reach the point (X) representing total composition of the mixture before the liquid composition reaches the quaternary peritectic point P, the final product of crystallization will be a four-phase assemblage of olivine, clinopyroxene, spinel, and silica (tridymite). It is seen that the "four-phase volume" made up from the above plane and the point representing liquid composition plays a role in quaternary systems similar to the "three-phase triangle" in a ternary system, and is similarly useful as a means of visualizing composition changes of the various phases coexisting in univariant equilibrium. The most striking feature of the crystallization paths discussed above is the general trend toward iron-oxide enrichment of the liquid (note that quaternary univariant liquidus curves in general have arrows pointing toward the bounding system $FeO–Fe_2O_3–SiO_2$).

If crystallization, at constant total composition of the mixture, takes place under conditions of fractionation rather than equilibrium, the trends in change of liquid composition toward the $FeO–Fe_2O_3–SiO_2$ face of the tetrahedron will be even more pronounced, as the constraining effect of liquid-crystal reaction is eliminated (compare previous discussion of the system $NaAlSi_3O_8–CaAl_2Si_2O_8–CaMgSi_2O_6$). If perfect fractionation occurs, the last trace of liquid at the end of the crystallization will have the composition of one of the eutectic points in the system $FeO–Fe_2O_3–SiO_2$.

The behavior described in the paragraphs above is in sharp contrast to that prevailing if crystallization takes place at constant oxygen fugacity. In this case, the path of equilibrium crystallization follows the oxygen isobaric surface in question. Because the oxygen isobaric surfaces in Fig. 4-27 intersect the liquidus univariant curves at a large angle, it follows that crystallization trends at constant oxygen fugacity will be radically different from those displayed when the total composition of the mixture remains constant (see above). The oxygen isobaric surfaces may be thought of as barriers preventing liquids from changing their compositions in the direction of the arrows along the univariant boundary curves. The main trends of the crystallization paths at constant oxygen fugacity are shown in simplified form in Fig. 4-29. A mixture X in the primary-phase volume of olivine is used as an example. With olivine of changing composition crystallizing, the liquid composition first changes along the oxygen isobaric surface within the olivine primary-phase volume. When the liquid reaches the divariant olivine-pyroxene liquidus boundary surface, these two crystalline phases coexist with liquids whose compositions change along a curve ($a–b$) defined by the intersection between the divariant liquidus boundary surface and the oxygen isobaric surface. Depending on the total composition of the mixture, the liquid composition may stay within the olivine-pyroxene divariant surface until the liquid is saturated with spinel ($MgFe_2O_4–FeFe_2O_4$ solid solution) at point b, or the liquid may leave the olivine-pyroxene boundary surface and move into the pyroxene phase volume (because of the *reaction relation* among olivine, pyroxene, and liquid, as explained previously). Regardless of which of these alternatives prevails with respect to details of crystallization paths, the liquid has to stay within the oxygen isobaric surface if the oxygen fugacity is kept constant and equilibrium is maintained. It is seen that the pronounced trend for an increase in FeO/MgO of the liquid during crystallization at constant total composition of the mixture is replaced by a trend toward silica enrichment in the liquid during crystallization at constant oxygen fugacity.

Various modes of fractionation may be visualized for a system crystallizing at constant oxygen fugacity. One of these would involve attainment of equilibrium between the liquid and the gas phase while perfect fractionation occurs in the solid-liquid relations. Under these circumstances the liquid composition would remain within the oxygen isobaric surface, as described above for equilibrium crystallization, but the constraining force imposed by solid-liquid equilibration within this surface would be eliminated. The paths of liquid change within this surface would be changed from those prevailing under equilibrium conditions, by virtue of the reaction relation existing among olivine, pyroxene, and liquid and by virtue of the failure of early-formed, solid-solution crystals to adjust their MgO/FeO to those required by equilibrium with the liquid phase as crystallization proceeds (compare relations discussed earlier for the ternary system $NaAlSi_3O_8$–$CaAl_2Si_2O_8$–$CaMgSi_2O_6$).

A second type of fractional crystallization would obtain if equilibrium is maintained between crystal and liquid phases but not between the condensed phases and the gas phase. In this case the paths of liquid would be similar to those observed when the crystallization takes place at constant total composition of the condensed phases.

Various combinations of partial fractionation of either of these two types may produce an infinite number of variations in paths of crystallization of iron silicate liquids even in the case of a relatively simple four-component system such as MgO–FeO–Fe_2O_3–SiO_2.

It is often desirable to represent phase relations in quaternary systems in a simplified form. This is accomplished by imposing restrictions on the system such as to keep one or more parameters constant or to minimize their variation. The system MgO–FeO–Fe_2O_3–SiO_2 lends itself particularly well to such simplifications. For instance, if the oxide and silicate phases are in equilibrium with metallic iron, the phase relations may be shown with close approximation in a projection onto the basal plane, MgO–FeO–SiO_2, of the tetrahedron representing the system (Bowen and Schairer, 1935). The triangular diagram thus obtained (see Fig. 4-30) has the appearance of a ternary system, although the relations are not strictly

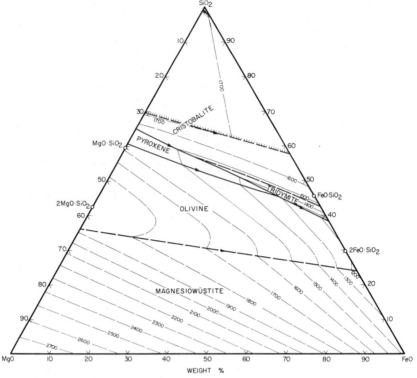

Figure 4-30. Simplified presentation of phase relations in the system MgO–iron oxide–SiO$_2$ in contact with metallic iron (Bowen and Schairer, 1935).

ternary because of the small but variable amounts of Fe$_2$O$_3$ present. Aside from this property, however, the system can be treated like a ternary system as far as derivations of crystallization paths are concerned, because the presence of metallic iron as a phase imposes a restriction on the system and thus expends one degree of freedom. The liquid compositions stay within the surface represented by the outline of the metallic iron phase volume of the MgO–FeO–Fe$_2$O$_3$–SiO$_2$ tetrahedron as long as metallic iron is present as a phase, and thus project in a uniquely defined manner onto the plane (MgO–FeO–SiO$_2$) chosen for simplified illustration. This system affords excellent examples of crystallization paths involving important rock-forming phases, and exhibits with unusual clarity the reaction relation among olivine, pyroxene, and liquids, as well as extensive solid-solution formation. Each of these factors provides the basis for significantly different crystallization paths by fractional crystallization as opposed to equilibrium crystallization. Examples of crystallization paths under various idealized conditions have been discussed extensively in previous literature (Bowen and Schairer, 1935).

SYSTEMS CONTAINING MORE THAN FOUR COMPONENTS

The phenomena discussed above for two-, three-, and four-component systems apply equally well to systems of higher order; however, the geometrical difficulties in depicting such relations in more complex systems are formidable. It is useful to make simplifications in order to represent them in reasonably legible form. One way of simplifying the representation is to hold one or more of the parameters constant.

This approach is perhaps best understood by comparison with some of the iron oxide-containing systems just discussed. Referring to Fig. 4-25, portraying liquidus phase relations for the system MgO–FeO–Fe_2O_3–SiO_2, it was shown subsequently that phase relations in contact with metallic iron may be represented in simplified form in a diagram with the *appearance* of a ternary system (Fig. 4-30). A similar approach may be used to impose constraints on multicomponent silicate systems. For instance, if the phases in a five-component iron oxide-containing system are in equilibrium with metallic iron (Fe activity = 1), the phase relations may be portrayed with close approximation in the form of a regular tetrahedron (see, for instance, the system CaO–FeO–Fe_2O_3–Al_2O_3–SiO_2 in contact with metallic iron, Schairer, 1942). If, in addition to fixing the activity of iron, the activity of one of the other components is fixed (e.g., oxide equilibria along the SiO_2 saturation surface, i.e., SiO_2 activity = 1), phase relations in the five-component system could be portrayed in a triangular diagram. Another way of imposing a constraint on a five-component system, and thus making it amenable to a representation in terms of a tetrahedral model, is to fix the oxygen fugacity of the gas phase. An example of this approach is the study by Roeder and Osborn (1966) of the system Mg_2SiO_4–"Fe_3O_4"–SiO_2–$CaSiO_3$ (a part of the five-component system CaO–MgO–FeO–Fe_2O_3–SiO_2) at an oxygen fugacity of 0.2 atm.

In order to extend the consideration of crystallization paths of silicates to systems that are fairly representative of all the main types of mineral phases predominantly present in igneous rocks, it is highly desirable to include at least the following six oxides as components: CaO, MgO, FeO, Fe_2O_3, Al_2O_3, and SiO_2. For the purpose of portraying clearly the phase relations in this highly complex six-component system, it is helpful to expend two degrees of freedom so that representation in terms of a tetrahedral model may be achieved. One way of reducing the variance of the system by one is to keep the ratio of two of the oxide components constant. If, in addition, the activity of one of the components or the product of the activities of two or more components is fixed, another degree of freedom is expended. In the case of the latter approach, it is obviously advantageous to choose a combination of components forming a common mineral species that has reasonably constant composition and with which the silicate liquid is saturated during a major part of the crystallization process. Such a phase is anorthite, $CaAl_2Si_2O_8$, which was used for this purpose by Roeder and Osborn (1966) in their study of the system Mg_2SiO_4–FeO–Fe_2O_3–SiO_2–$CaAl_2Si_2O_8$: phase relations at liquidus temperatures at a total pressure of 1 atm were determined at constant CaO/Al_2O_3 and along the anorthite saturation surface (i.e., anorthite activity = 1), as shown in Fig. 4-31. Note that the oxygen fugacity is a variable within this system. Near the base, where essentially all the iron is in the divalent state, the oxygen fugacities are very low (approximating those corresponding to equilibrium of the oxide and silicate phases with metallic iron), whereas the oxygen fugacities are high (approaching those of air, 0.2 atm) as the left-hand face (Mg_2SiO_4–SiO_2–Fe_2O_3) is approached. In applying this diagram, it is important to realize the limitations imposed by the simplified representation. For instance, the true compositions of the phases present cannot be derived quantitatively; only the ratios of Mg_2SiO_4, FeO, Fe_2O_3, and SiO_2 exclusive of the SiO_2 tied up in the $CaAl_2Si_2O_8$ component can be read off the diagram. With this limitation, however, some general trends of crystallization paths may be derived. Consider as an example a mixture in the vicinity of the univariant curve along which pyroxene, silica, magnetite (+ anorthite)[8] coexist with liquid. With these crystalline phases separating out, the liquid composition changes along the univariant boundary curve toward the invariant point Z'. If the constraints of liquid-solid equilibrium are operative, the liquid may or may not reach the invariant point, depending on the total composition of the crystallizing mixture. If the crystallization takes place under conditions of fractionation, however, the liquid not only reaches point Z', but continues its composition change along the olivine + silica + magnetite (+ anorthite)[8] liquidus boundary curve to point B', with an infinitesimal amount of liquid reaching the latter point if perfect fractionation prevails. It is thus observed that the crystallization

[8] Inasmuch as this diagram illustrates relations along the $CaAl_2Si_2O_8$ saturation surface, anorthite is always present as a phase, even though an anorthite primary-phase volume does not appear in Fig. 4-31.

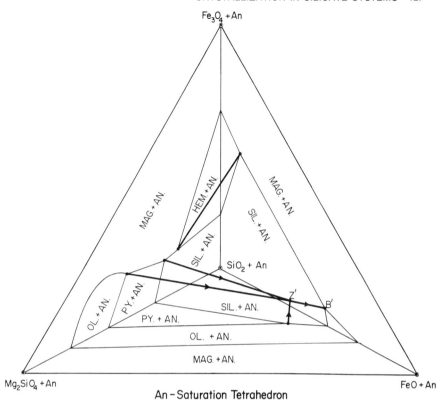

Figure 4-31. Tetrahedron representing liquidus phase relations along the anorthite-saturation volume of the system $Mg_2SiO_4-FeO-Fe_2O_3-CaAl_2Si_2O_8-SiO_2$, at a total pressure of 1 atm (Roeder and Osborn, 1966). Abbreviations: AN = anorthite; MAG = magnetite; HEM = hematite; OL = olivine; PY = pyroxene; SIL = silica (tridymite or cristobalite).

of mixtures in this system under conditions of constant total composition in general produces the effect of iron-oxide enrichment of the liquids as the crystallization proceeds, and more drastically so the larger the degree of fractionation during the crystallization.

Instead of fixing the anorthite activity at unity, as was done in the example just considered, the variance of the system may be reduced by one by fixing the oxygen fugacity of the gas phase. An example is afforded by the system $Mg_2SiO_4-\text{"}Fe_3O_4\text{"}-SiO_2-CaAl_2Si_2O_8$ (CaO/Al_2O_3 constant) at a constant oxygen fugacity of 10^{-7} atm, as shown in Fig. 4-32 after Roeder and Osborn (1966). Particular attention is drawn to the following

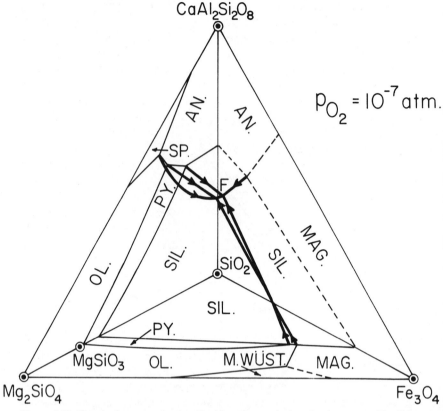

Figure 4-32. Tetrahedron representing liquidus phase relations in the system Mg_2SiO_4–FeO–Fe_2O_3–$CaAl_2Si_2O_8$–SiO_2 at constant oxygen fugacity of 10^{-7} atm and a total pressure of 1 atm (Roeder and Osborn, 1966). Abbreviations: AN = anorthite; MAG = magnetite; OL = olivine; PY = pyroxene; SP = spinel; SIL = silica (tridymite or cristobalite); MWÜST = magnesiowüstite.

relations within this tetrahedral model: As shown by the arrows (indicating directions of decreasing temperature along the univariant boundary curves), the main trend in the change of liquid compositions during crystallization is toward the invariant point F. The path along which the liquid changes its composition toward this point, including the question of whether or not the liquid actually reaches point F, depends on the total composition of the crystallizing mixture and the degree to which equilibrium crystallization as opposed to fractional crystallization obtains. As was the case in binary, ternary, and quaternary systems, establishment of equilibrium imposes a constraint on the descent of the liquid along the liquidus univariant curves, whereas perfect fractionation accelerates the change in liquid composition toward the low-temperature region of the system. Figure 4-32 also shows that the magnetite phase volume, which extends into the tetrahedron from its iron-oxide apex, forms a barrier that prevents the iron silicate liquids from becoming further enriched in iron oxide than those corresponding to the boundary surface between this magnetite primary-phase volume and the adjacent phase volumes of olivine, pyroxene, anorthite, and silica (tridymite or cristobalite). At constant oxygen fugacity, therefore, the path of crystallization is mainly toward silica enrichment.

It is seen from the above considerations that the trends observed in this complex six-component system are similar to those displayed in the relatively simple three- and four-component systems $FeO–Fe_2O_3–SiO_2$ and $MgO–FeO–Fe_2O_3–SiO_2$ previously described (see Figs. 4-23 to 4-24, and 4-28 to 4-29). This serves as a demonstration of the value and justification of using simplified model silicate systems to evaluate crystallization trends in complex systems, including natural magmas.

CONCLUDING REMARKS

A wealth of new data on oxide and silicate systems has become available since the appearance of Bowen's *The Evolution of the Igneous Rocks* (1928). Among these data is the vastly increased knowledge of the role of iron oxides and oxygen fugacities in magma crystallization. Evaluation of liquid-solid equilibria involving these parameters requires new approaches, and some of these have been treated extensively in the present chapter. Other new developments in phase-diagram interpretations, such as the treatment of crystallization paths in systems involving ternary solid solutions (see, e.g., Roeder, 1974), and a quantitative treatment of fractional crystallization involving binary solid solutions (Morse, 1976), have been left out because of space limitations. Furthermore, the relations discussed here pertain to conditions at a total pressure of 1 atm, in keeping with Bowen's original treatment. The principles discussed for a constant total pressure of 1 atm, however, apply also at higher pressures, even through the nature of the phases present at these higher pressures may be different. A discussion of crystallization paths in a pressure gradient is considered beyond the scope of the present accounts.

The advent of the electron microprobe has opened up new vistas in geosciences, and its impact on the study of liquid-solid equilibria in silicate systems is only beginning to be felt. By means of this tool, compositions of individual phases may be determined, both in natural mineral assemblages and in laboratory systems; thus much more detailed determinations of crystallization paths may be realized than was possible previously. In particular, with this new tool it is possible to determine the compositions of solid-solution phases far more conveniently and with higher accuracy than was feasible in the past. Combination of such data with improved theoretical approaches to the treatment of interrelations between liquid and solid-solution phases will undoubtedly result in major progress toward quantitative treatment of such equilibria. The geometrical basis for such interrelations has been established through a penetrating analysis by Greig (unpublished work). This analysis deals not only with the compositions of coexisting solid and liquid phases but also with the mass transport between the phases as crystallization proceeds.

The main emphasis in Bowen's original work (1928), as well as in the present chapter, has been on the *geometrical* aspects of the crystallization paths—in other words on changes in *compositional* parameters during the crystallization process. One must keep in mind that these paths are manifestations of the thermodynamic properties of the phases involved. The crystallization trends are such as to minimize the free energy of the system at each temperature. It follows that crystallization paths could be calculated if the thermodynamics of the phases involved were sufficiently well known. Modest progress in this direction has been made during the past few decades, but one is still a long way from having either the necessary data or the theoretical basis for making accurate calculations within the enormously complex multicomponent systems constituting natural magmas. It appears likely, however, that major efforts now underway in this direction will provide the basis for a more sophisticated and quantitative evaluation of magma crystallization phenomena than is possible at this time.

In addition to the evolution of a thermodynamic framework within which to evaluate crystallization processes of magmas, increasing attention is now being directed toward the phenomena involved in nucleation and crystal growth during the crystallization process. Again the possible scope of quantitative evaluations has been expanded significantly by the availability of new tools (e.g., the electron microprobe) for determining not only compositions of individual crystals, but composition gradients within these crystals. Furthermore, new instrumentation (e.g., Mössbauer spectroscopy, transmission electron microscopy) for determining fine structural and textural details in minerals greatly enhances the possibilities of progress in this area. It thus appears that new dimensions will be added to future research on phenomena involved in the crystallization of magmas.

Acknowledgments Critical reading of the various drafts of this chapter by Drs. E. F. Osborn, D. B. Stewart, and H. S. Yoder, Jr., is gratefully acknowledged.

References

Andersen, O., The system anorthite–forsterite–silica, *Am. J. Sci.* (4th series) *39*, 407–454, 1915.

Bowen, N. L., The binary system $Na_2Al_2Si_2O_8$ (nepheline, carnegieite)–$CaAl_2Si_2O_8$ (anorthite), *Am. J. Sci.* (4th series) *33*, 551–573, 1912.

———, Melting phenomenon of plagioclase feldspar, *Am. J. Sci.* (4th series) *35*, 577–599, 1913.

———, The ternary system: diopside–forsterite–silica, *Am. J. Sci.* (4th series) *38*, 207–264, 1914.

———, The crystallization of haplobasaltic, haplodioritic, and related magmas, *Am. J. Sci.* (4th series) *40*, 161–185, 1915.

———, *The Evolution of the Igneous Rocks*, Princeton University Press; Dover Publications, Inc., 332 pp., 1928 and 1956.

Bowen, N. L. and Andersen, O., The binary system $MgO–SiO_2$, *Am. J. Sci.* (4th series) *37*, 487–500, 1914.

Bowen, N. L. and Schairer, J. F., The system $MgO–FeO–SiO_2$, *Am. J. Sci.* (5th series) *29*, 151–217, 1935.

Darken, L. S. and Gurry, R. W., The system iron–oxygen. I. The wüstite field and related equilibria, *J. Am. Chem. Soc. 67*, 1398–1412, 1945.

Darken, L. S. and Gurry, R. W., The system iron–oxygen. II. Equilibrium and thermodynamics of liquid oxide and other phases, *J. Am. Chem. Soc. 68*, 798–816, 1946.

Gibbs, J. W., On the equilibrium of heterogeneous substances, *Trans. Conn. Acad. Sci. 3*, 108–248, *5*, 343–524, 1876 and 1878.

Greig, J. W., Immiscibility in silicate melts, *Am. J. Sci.* (5th series) *13*, 1–44, 133–154, 1927.

Kushiro, I., Determination of liquidus relations in synthetic silicate systems with electron probe analysis: the system forsterite–diopside–silica at 1 atmosphere, *Am. Mineral. 57*, 1260–1271, 1972.

Morse, S. A., The lever rule with fractional crystallization and fusion, *Am. J. Sci. 276*, 330–346, 1976.

Muan, A., Phase equilibria in the system $FeO–Fe_2O_3–SiO_2$, *Trans. AIME 203*, 965–976, 1955.

———, Phase equilibria at high temperatures in oxide systems involving changes in oxidation states, *Am. J. Sci. 256*, 171–207, 1958.

Muan, A. and Osborn, E. F., Phase equilibria at liquidus temperatures in the system $MgO–FeO–Fe_2O_3–SiO_2$, *J. Am. Ceram. Soc. 39*, 121–140, 1956.

Muan, A. and Osborn, E. F., *Phase Equilibria Among Oxides in Steelmaking*, Addison-Wesley Publishing Company, Reading, Mass., 236 pp., 1965.

Osborn, E. F., The system $CaSiO_3$–diopside–anorthite, *Am. J. Sci. 240*, 751–788, 1942.

Osborn, E. F. and Schairer, J. F., The ternary system pseudowollastonite–akermanite–gehlenite, *Am. J. Sci. 239*, 715–763, 1941.

Rankin, G. A. and Wright, F. E., The ternary system $CaO–Al_2O_3–SiO_2$, *Am. J. Sci.* (4th series) *39*, 1–79, 1915.

Roeder, P. L., Paths of crystallization and fusion in systems showing ternary solid solution, *Am. J. Sci. 274*, 48–60, 1974.

Roeder, P. L. and Osborn, E. F., Experimental data for the system $MgO–FeO–Fe_2O_3–CaAl_2Si_2O_8–SiO_2$ and their petrological implications, *Am. J. Sci. 264*, 428–480, 1966.

Schairer, J. F., The system $CaO–FeO–Al_2O_3–SiO_2$. I. Results of quenching experiments on five joins, *J. Am. Ceram. Soc. 25*, 241–274, 1942.

Smith, J. V., The crystal structure of protoenstatite, *Acta Cryst. 12*, 515–519, 1959.

Yang, H.-Y. and Foster, W. R., Stability of iron-free pigeonite at atmospheric pressure, *Am. Mineral. 57*, 1232–1241, 1972.

Chapter 5

THE REACTION PRINCIPLE

E. F. OSBORN

Geophysical Laboratory, Carnegie Institution of Washington, Washington, D.C.

"It will be apparent from the discussion of crystallization in typical silicate systems that a relation of liquid to crystals, characterized by reaction between them, is exceedingly common during the normal course of crystallization" (Bowen, 1928, p. 54). Bowen went on to explain that the reaction may be continuous, as in the common solid solution series, or it may be discontinuous, as when a crystalline phase that precipitated within one temperature range reacts with the liquid at lower temperatures and dissolves while one or more other phases crystallize. This feature of crystallizing silicate liquids, upon which fractional crystallization depends, was recognized by Bowen (1922) as being of such great significance in petrogenesis as to be termed the "reaction principle."

A continuous reaction series is exemplified by the plagioclase feldspars (Fig. 4-4). As crystallization of a liquid in the system proceeds, the crystals react continuously with the liquid changing their compositions in the direction of the $NaAlSi_3O_8$ component. If perfect equilibrium (complete reaction) among the phases were maintained, each crystal would be homogeneous, and all crystals at any specified temperature and pressure would have identical compositions. Complete reaction, however, cannot ordinarily be achieved during cooling in this system because of factors such as a slow rate of ionic diffusion relative to the rate of crystal growth, and separation of crystals from the main body of the liquid as by crystal settling. Reaction therefore is incomplete, as shown, for example, by a compositional zoning of the crystals. If reaction is limited, strong fractionation will produce a very small amount of residual liquid of composition close to $NaAlSi_3O_8$ and an aggregate of crystals ranging in composition from that of crystals separated at the liquidus temperature to that of nearly pure albite.

The classic example of a *discontinuous reaction* is olivine → pyroxene in the system Mg_2SiO_4–SiO_2 (Fig. 4-3). A cooling liquid of composition between Mg_2SiO_4 and $MgSiO_3$ will first precipitate olivine crystals. When the melting temperature of the incongruently melting pyroxene, $MgSiO_3$, is reached, olivine crystals react with the liquid and dissolve as pyroxene crystallizes. Crystals having this type of relation were termed by Bowen (1928, p. 55) a *reaction pair*. If equilibrium is maintained during cooling, the final liquid has the composition of point P. On the other hand, if the olivine crystals fail to dissolve, continued pyroxene precipitation will result in a change in the composition of the liquid from P to the higher SiO_2 composition of the eutectic point E. As in the case of the plagioclase feldspars, the incomplete reaction of crystals and liquid results in fractional crystallization, the liquid composition moving to higher SiO_2 contents than would be the case if equilibrium among the phases were maintained.

REACTIONS DURING FRACTIONAL CRYSTALLIZATION OF HAPLOBASALTIC LIQUIDS IN THREE-COMPONENT SYSTEMS

The reaction relations demonstrated in two-component silicate systems have been found to continue to be present as other components are added, a consideration of immense importance as phase equilibrium relations in these systems are applied to igneous rocks. This fact is illustrated by the diagrams of Fig. 5-1, in which a third compound has been added to the diagrams for the $CaAl_2Si_2O_8$–$NaAlSi_3O_8$ and Mg_2SiO_4–SiO_2 systems. The circle HB is taken as the composition of a haplobasaltic[1] liquid,

---

[1] The term, haplobasaltic, was coined by Bowen (1915) to mean simple basaltic. It referred to mixtures of diopside with various members of the plagioclase series, as illustrated by Fig. 5-1(a). Usage of haplobasaltic is herein extended to refer also to: (a) those mixtures in plagioclase-free systems that crystallize under equilibrium conditions to an aggregate of pyroxene crystals [Figs. 5-1(b), 5-1(c), and 5-3], or to pyroxene plus magnetite (Fig. 5-4), but with fractional crystallization will have olivine and silica as additional phases, and (b) simple basalt mixtures in the system illustrated by Figs. 5-6 and 5-10. In the latter, a haplobasalt will crystallize under equilibrium conditions to pyroxene, anorthite, and magnetite. Haplobasaltic mixtures are ". . . believed to be sufficiently close to basaltic and other magmas to throw considerable light on the crystallization of these natural mixtures (Bowen, 1915, p. 161)."

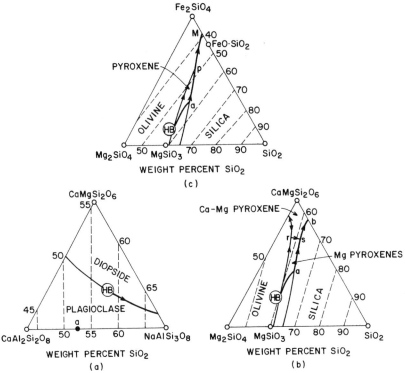

Figure 5-1. Sketch to illustrate reaction relations during crystallization in three systems having plagioclase, olivine, and pyroxene as phases (modified from Osborn, 1962a, p. 213, Fig. 2). Diagram (a) is after Bowen (1915) and Kushiro (1973); diagram (b) is after Bowen, (1914), Schairer and Yoder (1962), and Kushiro (1972); and diagram (c) is after Bowen and Schairer (1935). Circle labeled HB indicates composition of a haplobasalt. The heavy, solid line passing from the circle toward the right sideline is the course a liquid of initial haplobasaltic composition follows on extreme fractional crystallization. Lines of equal SiO_2 content are shown dashed.

and the heavy line represents the course, simplified to show only the principal relations, of fractional crystallization of HB. The following points are especially to be noted.

(1) In Fig. 5-1(a) even though diopside is present, the plagioclase solid solutions exhibit the same type of continuous reaction as in the binary system. As the composition of the liquid during fractional crystallization moves from HB toward the right along the diopside-plagioclase boundary curve, diopside precipitates along with plagioclase, the latter changing

in composition continuously from a solid solution of composition a, in equilibrium with liquid HB, to solid solutions approaching the composition of $NaAlSi_3O_8$.

(2) In Fig. 5-1(b), with $CaMgSi_2O_6$ added as a component to the Mg_2SiO_4–SiO_2 binary system, the Mg olivine → Mg pyroxene[2] reaction relation existing at the binary reaction point continues along the olivine-pyroxene boundary curve in the ternary system.[3] An Mg pyroxene → Ca-Mg pyroxene reaction is also present, occurring at point s in Fig. 5-1(b) (Schairer and Yoder, 1962; Kushiro, 1972). The composition of the fractionating liquid [heavy line, Fig. 5-1(b)] moves from the olivine-pyroxene boundary curve across the pyroxene field to a, continues along the pyroxene-silica boundary curve to s, and then moves from s through b. At s, Mg pyroxene ceases to precipitate as Ca-Mg pyroxene and tridymite crystallize. A *discontinuous reaction series* for these phase relations can therefore be written, Mg-olivine → Mg pyroxenes → Ca-Mg pyroxenes.

(3) In Fig. 5-1(c), with Fe_2SiO_4 added as a component to the Mg_2SiO_4–SiO_2 binary system, the Mg olivine → Mg pyroxene reaction relation occurs along the olivine-pyroxene boundary curve at the high-MgO end of this boundary curve. During fractional crystallization of liquid HB, the composition of the liquid leaves the olivine-pyroxene boundary curve,

---

[2] The general term, Mg pyroxene, is used herein following Bowen's usage (Bowen, 1928, p. 60, Table II). Mg pyroxene includes the iron-free pigeonite of Yang and Foster (1972), and the Ca-poor pyroxene of Kushiro (1972).

[3] For purposes of this discussion, the three systems shown in Fig. 5-1 can be considered ternary and the bounding systems binary. Actually, ternary phase relations, in the range of temperature between the liquidus and the solidus, do not exist for all compositions in any of the three triangular joins of Fig. 5-1, if the compounds shown at the apices of each triangle are considered to be the three components in each case. For example, in Fig. 5-1(a), the diopside is aluminous (Osborn, 1942). Consequently, both the diopside crystals and a liquid phase in equilibrium with diopside cannot be represented in terms of the three components. Furthermore, the albite crystals in equilibrium with liquids in mixtures along the join $CaMgSi_2O_6$–$NaAlSi_3O_8$ are calcic (Schairer and Yoder, 1960), and, therefore, the composition of these liquids cannot be represented in terms of the two compounds, $CaMgSi_2O_6$ and $NaAlSi_3O_8$. In Fig. 5-1(b), relations are ternary if CaO, MgO, and SiO_2 are selected as components, but are not entirely ternary for the component selection used. The non-ternary nature can be seen from the fact that the liquid phase coexisting with diopside in mixtures whose compositions lie along the Mg_2SiO_4–$CaMgSi_2O_6$ and SiO_2–$CaMgSi_2O_6$ joins has a composition lying within the triangle CaO–MgO–SiO_2, but outside of the triangular join of Fig. 5-1(b) (Kushiro, 1972). For the system of Fig. 5-1(c), all liquids contain small amounts of Fe^{3+} (Bowen and Schairer, 1935), and hence liquid compositions all lie outside of this triangular join and within the tetrahedron representing the system MgO–FeO–Fe_2O_3–SiO_2. Departures from true ternary relations for the triangular joins of Fig. 5-1 can be neglected, however, for purposes of this discussion.

crosses the pyroxene field to a, and moves along the pyroxene-silica boundary curve to p. At p, pyroxene ceases to crystallize as an Fe-rich olivine and tridymite crystallize, causing the liquid composition to change from p toward the minimum liquidus point M. A *discontinuous reaction series* is illustrated by the fractionation curve in Fig. 5-1(c) as follows: Mg olivines → Mg-Fe pyroxenes → Fe olivines.

(4) Fractional crystallization in haplobasaltic systems does not always produce final liquids with compositions higher in SiO_2. In Fig. 5-1(a), liquids are enriched in SiO_2, in Fig. 5-1(b) the final liquid has nearly the same SiO_2 content as the original haplobasalt, and in Fig. 5-1(c) the SiO_2 content of the liquid decreases from about 55% to 40%.

(5) As more components are added, crystal-liquid reactions become of increasing importance during crystallization. Note, for example, that eutectic, or eutectic-like minimum liquidus points with only a small amount of crystal-liquid reaction, occur in two of the three bounding binary systems in Fig. 5-1(a) and (c), and in all three of the bounding systems in Fig. 5-1(b), whereas there are no eutectic or eutectic-like points in any of the ternary systems. Furthermore, an Mg pyroxene crystallizes along with silica from the eutectic liquid in the binary system, Mg_2SiO_4–SiO_2 (Fig. 4-3), whereas an Mg pyroxene reacts with the liquid at point s [Fig. 5-1(b)] and dissolves as a Ca-Mg pyroxene and silica crystallize; and the Mg-Fe pyroxene of Fig. 5-1(c) reacts with liquid p and dissolves, as an Fe-rich olivine and silica crystallize. Eutectic relations have thus become an unimportant aspect of crystallization in progressing from two- to three-component systems. It is concluded that continuous and discontinuous reaction relations are the major determinants of courses of crystallization in silicate systems that can serve as reliable models of the common magmas cooling in the earth's crust.

REACTIONS DURING FRACTIONAL CRYSTALLIZATION OF HAPLOBASALTIC LIQUIDS IN THE SYSTEM CaO–MgO–Al_2O_3–SiO_2

Reaction in more complex silicate systems can be visualized by the use of the tetrahedral diagrams of Figs. 5-2, 5-3, and 5-4. The tetrahedron of Fig. 5-2 is used to illustrate reactions among the common rock-forming minerals, spinel, olivine, plagioclase, and pyroxene, as they occur at atmospheric pressure in the iron-free system, CaO–MgO–Al_2O_3–SiO_2. An internal tetrahedron is shown in Fig. 5-2, having as apices An, Di, Fo, and Sp. Phase relations are indicated for the two compatibility triangles, An–Fo–Sp and An–Fo–Di, that form the back faces of this internal tetrahedron. Heavy lines are boundary curves on these two triangular joins. The boundary curve along which anorthite, forsterite, spinel, and liquid

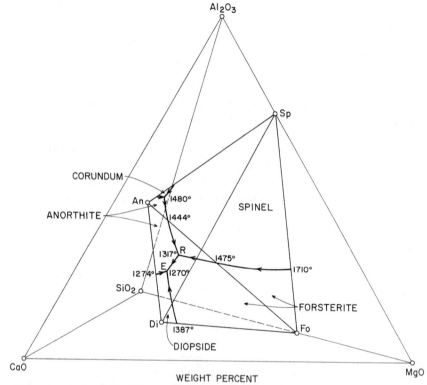

Figure 5-2. Sketch to illustrate liquidus phase relations for the two compatibility triangles,
An–Fo–Sp, after DeVries and Osborn (1957) and An–Fo–Di, after Osborn and
Tait (1952). These triangular joins are shown as forming two faces of the
tetrahedron, An–Di–Fo–Sp, within the tetrahedron representing the quaternary
system $CaO–MgO–Al_2O_3–SiO_2$. Heavy lines are boundary curves on the two
triangular joins, An–Fo–Sp and An–Fo–Di. Arrows on the curves indicate
direction of decreasing temperature. Numbers are liquidus temperatures in °C.
Abbreviations: An = $CaO \cdot Al_2O_3 \cdot 2SiO_2$,   Fo = $2MgO \cdot SiO_2$,   Sp =
$MgO \cdot Al_2O_3$, Di = $CaO \cdot MgO \cdot 2SiO_2$.

coexist lies outside of (does not intersect) the An–Fo–Sp triangle, and is
shown in Fig. 5-2 only as the piercing point, R, in the lower triangle
An–Fo–Di. A maximum temperature on this boundary curve of 1320°
(Osborn and Tait, 1952) occurs at a point on the SiO_2 side of R where the
downward extension of the An–Fo–Sp plane intersects the boundary
curve. Inasmuch as the composition of the liquid in equilibrium with
anorthite, forsterite, and spinel lies outside the An–Fo–Sp triangle, one

of the phases, spinel, must dissolve with decreasing temperature, as anorthite and forsterite crystallize, if equilibrium is maintained among the phases.

To illustrate reaction relations during crystallization of a liquid, take a mixture within the triangular area, An–Fo–R, of the lower compatibility triangle, An–Fo–Di. Spinel will crystallize as the primary or secondary phase from any cooling liquid having a composition within the area An–Fo–R. While spinel is crystallizing, the liquid composition moves out of the An–Fo–Di plane along a path leading to a point on the anorthite-forsterite-spinel boundary curve having a higher SiO_2 content than R. When the liquid composition reaches this boundary curve, spinel reacts with the liquid and dissolves, as forsterite and anorthite crystallize. If equilibrium is maintained, the liquid composition changes along the anorthite-forsterite-spinel boundary curve toward R, attaining the composition of R just as solution of spinel is completed. The liquid composition then moves to E as anorthite and forsterite continue to crystallize with decreasing temperature. At E, diopside, anorthite, and forsterite crystallize until the liquid is consumed.[4] If, however, equilibrium is not maintained and spinel fails to react with the liquid on the anorthite-forsterite-spinel boundary curve, the liquid composition, with continuing anorthite and forsterite precipitation, leaves the boundary curve and moves along the anorthite-forsterite boundary surface to either the diopside-anorthite-forsterite boundary curve[5] at a point on the SiO_2-rich side of E, or the Mg pyroxene-anorthite-forsterite boundary curve, depending on the initial composition of the mixture. If the former, then with continued cooling, diopside, anorthite, and forsterite crystallize, and the liquid composition moves to the quaternary reaction point [analogous to the ternary reaction point, r, of Fig. 5-1(b)] at which diopside, anorthite, forsterite, and Mg pyroxene coexist in equilibrium with liquid. At

---

[4] Because the diopside is aluminous rather than pure $CaMgSi_2O_6$ (Osborn, 1942), the liquid composition does not remain constant during crystallization of anorthite, forsterite, and diopside, but with decreasing temperature moves a short distance along the An–Fo–Di boundary curve to a higher SiO_2 content than point E, before the liquid is consumed.

[5] Quaternary boundary curves, i.e., liquidus univariant curves along which three crystalline phases coexist with liquid, are not shown in Fig. 5-2. Quaternary boundary curves referred to in this discussion are, however, analogous to the ternary boundary curves shown in Fig. 5-1(b), with anorthite added as a coexisting phase. For example, the dopside-anorthite-Mg pyroxene boundary curve of the quaternary system $CaO–MgO–Al_2O_3–SiO_2$ is analogous to the diopside-Mg pyroxene boundary curve, $r–s$ of Fig. 5-1(b). The general nature of courses of crystallization for mixtures whose compositions lie within the tetrahedron, An–Fo–Di–SiO_2 of Fig. 2 can, therefore, be understood from relations shown in Fig. 5-1(b).

this point, forsterite reacts with the liquid and dissolves, as the other crystalline phases precipitate, if equilibrium is maintained. But if forsterite fails to react with the liquid, the liquid composition moves along the diopside-anorthite-Mg pyroxene boundary curve to the quaternary invariant point [analogous to the ternary reaction point, s, of Fig. 5-1(b)] at which diopside, Mg pyroxene, tridymite, and anorthite coexist with the liquid, If, on the other hand, the initial composition of the liquid is such that the fractionation curve intersects the Mg pyroxene-forsterite-anorthite boundary curve instead of the diopside-forsterite-anorthite boundary curve[6] then with perfect fractionation the liquid composition leaves the boundary curve and follows a fractionation curve across the Mg pyroxene-anorthite boundary surface to either the Mg pyroxene-diopside-anorthite, or Mg pyroxene-tridymite-anorthite boundary curve, and then moves along one of these boundary curves to the diopside-Mg pyroxene-tridymite-anorthite quaternary invariant point. This point is probably a reaction point, since the diopside-Mg pyroxene-tridymite-liquid point in the $CaO-MgO-SiO_2$ system [point s in Fig. 5-1(b)] has been shown to be of this nature (Schairer and Yoder, 1962), with Mg pyroxene reacting with the liquid to dissolve as diopside and tridymite crystallize. If Mg pyroxene fails to react with the liquid at the quaternary reaction point, the liquid composition leaves this point and proceeds along the tridymite-diopside-anorthite boundary curve toward the quaternary eutectic point at which tridymite, diopside, anorthite, and wollastonite coexist with liquid. Thus, during fractional crystallization of a mixture having a composition within the triangular area An–Fo–R of Fig. 5-2, a succession of crystalline phase assemblages may precipitate as follows: spinel + forsterite or spinel + anorthite → forsterite + anorthite → forsterite + diopside + anorthite or Mg pyroxene + anorthite → Mg pyroxene + diopside + anorthite   or   Mg   pyroxene + tridymite + anorthite → diopside + tridymite + anorthite → diopside + tridymite + wollastonite + anorthite. Spinel, forsterite, and Mg pyroxene each in turn crystallize over a temperature range and then cease to crystallize. A *discontinuous reaction series* may be written as follows: spinel → olivine → Mg pyroxene → Ca-Mg pyroxene. Inasmuch as anorthite is one of the early phases, along with spinel and olivine, and continues to be a phase throughout the entire fractional crystallization sequence, if Na_2O were added to the initial liquid (a composition within the triangle An–Fo–R of Fig. 5-2) to bring the composition closer to that of a basaltic liquid,

---

[6] In general, fractionation curves for liquids within An-Fo-R (Fig. 5-2) will intersect the Mg pyroxene-forsterite-anorthite boundary curve for initial liquid compositions close to the An–Fo join and will intersect the diopside-forsterite-anorthite boundary curve for initial liquid compositions close to R (cf. Osborn and Tait, 1952, p. 427).

the continuous series of plagioclase feldspars would precipitate along with the members of the discontinuous series, in analogy with the course of fractional crystallization illustrated in Fig. 5-1(a).

REACTIONS DURING FRACTIONAL CRYSTALLIZATION OF HAPLOBASALTIC LIQUIDS IN THE SYSTEM $CaO-MgO-FeO-SiO_2$

In the tetrahedron of Fig. 5-3, the component $CaSiO_3$ has been added to the $Mg_2SiO_4-FeO-SiO_2$ diagram to extend phase relations shown in the Ca-free system of Fig. 5-1(c). The related tetrahedron for $Mg_2SiO_4-Fe_3O_4-CaSiO_3-SiO_2$, to be discussed below, is presented here in Fig. 5-4 for comparative purposes. The point X in Fig. 5-3 represents a haplo-basaltic composition projected upward from the base and located in the olivine primary phase volume close to the olivine-Mg-Fe pyroxenes boundary surface. Liquidus phase relations for the base of the tetrahedron, after Bowen and Schairer (1935), and for the right face, after Bowen, Schairer, and Posnjak (1933) are those for liquids in equilibrium with metallic iron projected onto the FeO plane by computing all iron oxide as FeO. Relations shown on the left face are principally after Bowen (1914), Schairer and Bowen (1942), Schairer and Yoder (1962), and Kushiro (1972). The phase relations shown for the interior of the tetra-hedron represent an arrangement of boundary curves that seems reason-able on the basis of phase relations known for the three faces of the tetrahedron. At R, as heat is withdrawn, an Mg-Fe pyroxene dissolves, as an Fe olivine, a Ca-Mg-Fe pyroxene, and tridymite crystallize; and at P, a Ca-Mg-Fe pyroxene dissolves as Fe olivine, tridymite, and Fe wollastonite crystallize. A liquid of initial composition, X, fractionally crystallizing, may have a very complex course of crystallization, but in general will change in composition upward and to the right in the tetra-hedron, through the Mg-Fe pyroxene primary phase volume to the two-pyroxene boundary surface, $r-s-R$. The two pyroxenes coprecipitate as the liquid composition moves along this surface to the two-pyroxene-tridymite boundary curve, $s-R$. At this boundary curve, Mg-Fe pyroxene ceases to crystallize as tridymite joins the Ca-Mg-Fe pyroxene in precipi-tating, and the liquid composition then moves along the Ca-Mg-Fe pyroxene-tridymite boundary surface to the boundary curve $R-P$. Along $R-P$, Ca-Mg-Fe pyroxene, Fe olivine, and tridymite crystallize as the liquid composition moves to P. At P, Ca-Mg-Fe pyroxene ceases to crys-tallize, and Fe wollastonite, Fe olivine, and tridymite coprecipitate as the liquid composition moves down along $P-S$ toward S. The *discontinuous*

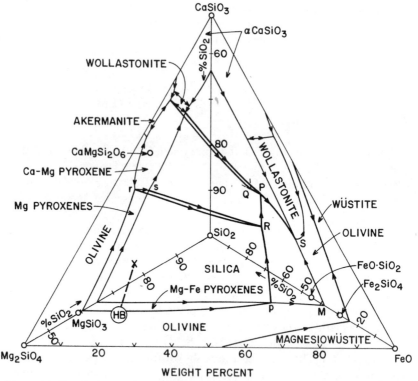

Figure 5-3. Tetrahedron to illustrate liquidus phase relations in the system, Mg₂SiO₄–FeO–CaSiO₃–SiO₂ (modified from Osborn, 1962a, p. 216, Fig. 3). The left face is principally after Bowen (1914), Schairer and Bowen (1942), Schairer and Yoder (1962), and Kushiro (1972). The base is after Bowen and Schairer (1935). The right face is after Bowen, Schairer, and Posnjak (1933). Arrows on boundary curves indicate direction of falling temperature. Positions of boundary curves within the tetrahedron have not been determined experimentally, but are drawn to indicate approximate phase relations judged to exist on the basis of those that have been determined for the three faces of the tetrahedron. Four invariant points at which four oxide crystalline phases are in equilibrium with metallic iron, liquid, and gas exist in the tetrahedron. In order to avoid an unnecessary confusion of lines, the invariant point at which αCaSiO₃ is one of the phases has been omitted, as well as the boundary curve passing through the tetrahedron along which αCaSiO₃, wollastonite, and tridymite coexist with liquid. The position of a liquidus invariant point Q, at which akermanite, wollastonite, Ca–Mg pyroxene, and olivine coexist is shown, but two of the boundary curves intersecting at this point are omitted except for the two short, dashed lines indicating their presence. The invariant points R and P are reaction points of relevance to crystallization of haplobasaltic liquids, as discussed in the text. Point X indicates the composition of a haplobasaltic liquid in the tetrahedron, analogous to HB on the base and in Fig. 5-1(c).

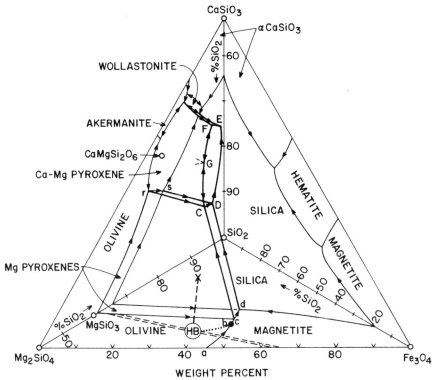

Figure 5-4. Tetrahedron to illustrate liquidus phase relations in the system Mg_2SiO_4–Fe_3O_4–$CaSiO_3$–SiO_2 at a constant oxygen partial pressure of $10^{-0.7}$ atm (modified from Osborn, 1962a, b, p. 217, Fig. 4). The left face is principally after Bowen (1914), Schairer and Bowen (1942), Schairer and Yoder (1962), and Kushiro (1972). The base is after Muan and Osborn (1956) and may be seen as a part of the plane, MgO–Fe_3O_4–SiO_2, in the tetrahedron of Fig. 5-8. The right face is after Phillips and Muan (1959). The arrangement of boundary curves within the tetrahedron is sketched on the basis of the phase relations experimentally determined for the left face and base. To minimize confusion of lines, boundary curves extending from the right face and front face into the tetrahedron are not shown. At invariant point G, at which akermanite, Ca-Mg pyroxene, olivine, and either magnetite or hematite coexist with liquid and gas, boundary curves extending into the tetrahedron from invariant points in the left and front faces are shown only as two short, dashed lines. Invariant points, and associated boundary curves, having $\alpha CaSiO_3$ as a phase have been omitted. Points HB and X represent haplobasaltic liquid compositions on the base and in the tetrahedron, respectively. Dashed lines on the base are tie lines joining coexisting pyroxene and magnetite compositions in equilibrium with liquids c and d (after Osborn 1959, p. 626, Fig. 8). Arrows on boundary curves represent direc-directions of decreasing temperature.

reaction series illustrated by the Ca-free system of Fig. 5-1(c) is thus expanded in Fig. 5-3 by adding two additional members, Ca-Mg-Fe pyroxenes and Fe wollastonites.

REACTION SERIES

In the preceding discussion, the basis has been presented for believing that reaction series exist and play an important role during the fractional crystallization of subalkaline basaltic magma at low pressures. There are several continuous reaction series, notably the plagioclase feldspars, the olivines, the pyroxenes, and the spinels. Discontinuous reactions among phases of importance in basaltic magmas start with spinel and Mg olivine as discussed in connection with Fig. 5-2; and if the fractionation occurs at constant total composition, approximated for the phase relations shown in Fig. 5-3, a *discontinuous reaction series* occurs of the nature of Mg olivine → Mg-Fe pyroxene → Ca-Mg-Fe pyroxene → Fe olivine + Fe wollastonite. Bowen (1922) introduced a diagram presenting the complicated reactions and trends occurring during the fractional crystallization of a basaltic magma, as a simple picture of two converging reaction series, one discontinuous and the other the plagioclase continuous series, with spinel as an early, coprecipitating phase. He called this arrangement of minerals "Reaction Series in Subalkaline Rocks," and stated that "Neither rigid accuracy nor finality is claimed. It is regarded merely as a framework upon which others may build, making such modifications and additions as may be found necessary" (Bowen, 1922, p. 190). Later he stated (Bowen, 1928, p. 60), "The matter is really too complex to be presented in such simple form. Nevertheless the simplicity, while somewhat misleading, may prove of service in presenting the subject in concrete form." In a continuation of Bowen's method of presentation there are shown in Fig. 5-5 two such reaction series schemes. These show, in each case, the discontinuous reaction series of ferromagnesian minerals and the continuous reaction series of the plagioclase feldspars as two converging lines. This schematic convergence of these two series was Bowen's method of indicating an actual tendency for compositional convergence. He stated, referring to his Table 2 (Bowen, 1928, p. 60) that is similar to Fig. 5-5(b): "Beginning at the upper end of the series in the more basic mixtures we have at first two distinct reaction series, the continuous series of the plagioclases and the discontinuous series, olivines-pyroxenes-amphiboles, etc. As we descend in these series, however, they become less distinct, in the aluminous pyroxenes and amphiboles a certain amount of interlocking begins and they finally merge into a single series."

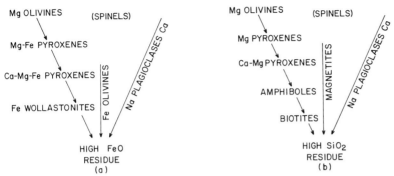

Figure 5-5. Two groups of reaction series for subalkaline igneous rocks at low pressure (after Osborn, 1962a, p. 221, Fig. 5). Group (a) is related to Fig. 5-3 and to fractional crystalization of basaltic magma as a closed system. Group (b), modified after Bowen (1922), is related to Fig. 5-4 and to fractional crystallization of basaltic magma under oxygen-buffered conditions. For each group, the line on the right represents the continuous series of plagioclase solid solutions from calcic at the top to sodic at the bottom.

The reaction series of Fig. 5-5(a) apply to the case discussed above in connection with Fig. 5-3. The reaction series of Fig. 5-5(b), to be discussed later with reference to Fig. 5-4, apply where the fractionating basaltic magma is oxygen-buffered. Bowen did not describe a reaction series such as that shown in Fig. 5-5(a). His presentation of this matter was prior to the classic experimental work (Bowen and Schairer, 1932) that determined phase relations in the system FeO–SiO_2, and that led to studies of other iron oxide systems, particularly MgO–FeO–SiO_2 (Bowen and Schairer, 1935), CaO–FeO–SiO_2 (Bowen, Schairer, and Posnjak, 1933), and MgO–FeO–Fe_2O_3–SiO_2 (Muan and Osborn, 1956). Bowen's reaction series scheme was similar to that shown in Fig. 5-5(b). From experimental, geologic, and petrologic work since publication of his book, the relations shown in Fig. 5-5(a) have been established and represent an additional scheme.

    The reaction series shown in Fig. 5-5(a) apply to subalkaline basaltic magma intruded into the upper crust and fractionally crystallizing as a closed system. Of the bodies formed in this manner, and that have been studied, the one exhibiting the most extreme course of fractional crystallization is the Skaergaard intrusion of East Greenland (Wager and Deer, 1939). The major features of the reaction series of Fig. 5-5(a) are illustrated by the sequence of crystalline aggregates formed during solidification of the Skaergaard magma. The Fe wollastonites, shown as a late, iron-rich phase in the reaction series (Fig. 5-5a), were not found by Wager and Deer

(1939), but these authors noted that this phase undoubtedly had been present in the fayalite ferrogabbros, and is now represented by pyroxene of the hedenbergite-clinoferrosilite solid solution series formed by inversion of Fe wollastonite solid solutions on cooling.

DISCONTINUOUS PRECIPITATION OF OLIVINE AND SPINEL PHASES

As indicated in Fig. 5-5, three solid solution series are simultaneously present in the early stages of fractional crystallization of subalkaline basaltic magma at low pressure: Mg olivines, spinels, and Ca plagioclases. Crystals of any one of these three can be the first or second phase to precipitate, depending on the initial composition of the liquid, but based on relations shown in Fig. 5-2, spinel cannot be the third phase. That is, if plagioclase and olivine are crystallizing without spinel, spinel will not appear, and if spinel is crystallizing along with either olivine or plagioclase, spinel will cease to crystallize when the third crystalline phase appears. Both the olivines and the spinels that form early in the fractional crystallization sequence will later cease to be precipitating phases. They may reappear as later phases in the fractional crystallization of basaltic magma, however, because the phase relations are such that a gap may exist between the precipitation of early Mg-rich olivines and spinels and the later Fe-rich members of the two series.

The gap in olivine precipitation was noted earlier in the discussion of Fig. 5-1(c). The gap in spinel precipitation is illustrated, for the case of an oxygen-buffered system, by the phase relations shown in Figs. 5-6 and 5-7. Although spinel solid solution compositions exist as a continuous series in equilibrium with liquids along the boundary curve $b-P$ (Fig. 5-6), spinel dissolves rather than precipitates as the liquid composition with decreasing temperature moves along $b-P$ toward P. Therefore, when the composition of a liquid precipitating olivine and spinel or anorthite and spinel reaches the curve $b-P$, spinel precipitation ceases, and with *fractional crystallization*, the liquid composition immediately leaves the curve $b-P$, and moves across the olivine-anorthite boundary surface $b-p-P$, to the olivine-anorthite-pyroxene boundary curve $p-P$. Olivine has a reaction relation with liquid along this boundary curve, and hence, with continued fractional crystallization the liquid composition leaves the boundary curve $p-P$, and crosses the pyroxene-anorthite-boundary surface $p-e-E-P$, following a fractionation curve either to $P-E$ or to $e-E$. On reaching $P-E$ or $e-E$, the liquid composition moves to E. A spinel, designated as magnetite in Fig. 5-6, and commonly containing more than 10 mol % Al_2O_3

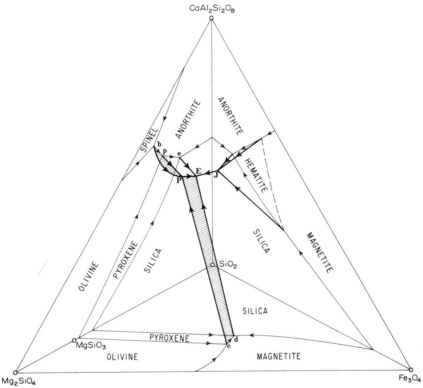

Figure 5-6. Tetrahedron representing liquidus phase relations in the Mg_2SiO_4–FeO–Fe_2O_3–$CaAl_2Si_2O_8$–SiO_2 system at a constant oxygen partial pressure (P_{O_2}) of $10^{-0.7}$ atm, after Roeder and Osborn (1966). Left face of tetrahedron is after Andersen (1915), base is after Muan and Osborn (1956), and right face is after Roeder and Osborn (1966). Heavy solid lines within the tetrahedron represent liquid compositions in equilibrium with three crystalline phases and a gas phase of $P_{O_2} = 10^{-0.7}$ atm. Arrows indicate the direction of decreasing temperature. Points P, E, and J are liquid compositions coexisting with four crystalline phases and a gas phase of $P_{O_2} = 10^{-0.7}$ atm. To improve readability of the diagram, the surface along which olivine and anorthite coexist with liquid and gas, b–p–P, and the surface along which magnetite and pyroxene coexist with liquid and gas, P–E–d–c, have been ruled. (With permission of the *American Journal of Science*.)

(Roeder and Osborn, 1966, p. 476, Table 7), reappears as a phase when the liquid composition reaches P–E or E.

The course of *equilibrium crystallization* is different from that just described, because a liquid composition on reaching the curve b–P does not immediately leave this curve to cross the surface b–p–P, but moves along *b–P* toward *P*. As the liquid composition, with equilibrium crystallization, moves along *b–P*, spinel reacts with the liquid, changes in composition and dissolves, and olivine and anorthite crystallize. Whether or not all of the spinel phase dissolves before the liquid composition reaches the reaction point *P* depends on the bulk composition of the mixture. If all of the spinel does not dissolve along *b–P*, the liquid composition will reach *P*, at which point spinel again crystallizes. Thus, under equilibrium conditions, even though there is a gap in spinel crystallization, spinel continues to be present as a phase in the mixture, increasing in amount as the liquid composition moves to *b–P*, decreasing in amount as the liquid composition moves along *b–P* to *P*, and then increasing again in amount when the liquid composition reaches *P*. At *P*, spinel, pyroxene, and anorthite crystallize as olivine dissolves. If the composition of the mixture is such that all of the spinel dissolves along *b–P*, the liquid composition leaves *b–P* at the point of disappearance of spinel, and follows an equilibrium curve across the olivine-anorthite surface *b–p–P*, to the *p–P* boundary curve. The liquid composition then moves along *p–P* toward *P* as olivine dissolves and anorthite and pyroxene crystallize. The liquid composition will move along *p–P* to *P* if the olivine is not entirely dissolved. If olivine is entirely dissolved along *p–P*, the liquid composition will leave *p–P* at the point of disappearance of olivine and follow an equilibrium curve across the pyroxene-anorthite surface *p–e–E–P* to *P–E*, or to *e–E*, the precise course depending on the bulk composition of the mixture. Spinel again crystallizes as the liquid composition moves along *P–E*, or on reaching *E* after moving along *e–E*. For haplobasaltic liquids, therefore, that have initial compositions such that their crystallization curves intersect the boundary curve *b–P* (Fig. 5-6), a gap exists in spinel crystallization. Haplobasaltic liquids within the tetrahedron of Fig. 5-6 may, however, have compositions such that a gap in spinel crystallization does not occur during crystallization. These liquids will be higher in iron oxide and lower in MgO than those having a gap in their spinel crystallization, and their courses of crystallization will follow along the spinel-olivine surface to the *c–P* boundary curve rather than to *b–P* (Fig. 5-6). After intersecting *c–P*, spinel crystallization will continue, as the temperature decreases.

These two examples of haplobasaltic liquids in the system illustrated by the tetrahedra in Figs. 5-6 and 5-7, are analogous to the two basaltic liquids whose crystallization paths were studied by Hill and Roeder (1974). They found that a gap in spinel precipitation during crystallization does not occur for one basaltic liquid studied, the one having the higher iron oxide and lower MgO content, but does occur for the other, an olivine tholeiite. A gap in spinel precipitation during crystallization of the olivine

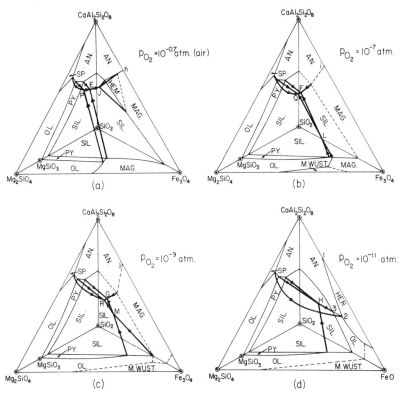

Figure 5-7. Tetrahedra representing liquidus phase relations in the Mg_2SiO_4–FeO–Fe_2O_3–$CaAl_2Si_2O_8$–SiO_2 system at four different oxygen partial pressures, after Roeder and Osborn (1966, p. 445, Fig. 10). For plotting purposes, iron oxide is calculated as Fe_3O_4 in (a), (b), and (c) and as FeO in (d). Figure (a) is redrawn from Fig. 5-6 for ready comparison with (b), (c), and (d).
(a) $P_{O_2} = 10^{-0.7}$ atm (air), (b) $P_{O_2} = 10^{-7}$ atm, (c) $P_{O_2} = 10^{-9}$ atm, (d) $P_{O_2} = 10^{-11}$ atm. Abbreviations: AN = anorthite, SP = spinel, HEM = hematite, OL = olivine, PY = pyroxene, SIL = silica, M. WÜST = magnesiowüstite, MAG = magnetite, HER = hercynite. (With permission of the *American Journal of Science*.)

tholeiite occurs at an oxygen fugacity of 10^{-9} to 10^{-11} atm in the temperature range of about 1160° to 1120° (Hill and Roeder, 1974, p. 718, Fig. 2). Such a gap does not occur at higher oxygen fugacities. An explanation for occurrence of the gap only at the lower oxygen fugacities can be derived from phase relations for the model system MgO–FeO–Fe_2O_3–$CaAl_2Si_2O_8$–SiO_2 as shown in Fig. 5-7, where phase relations are illustrated for four different conditions of oxygen partial pressure (P_{O_2}). Figure 5-7(a) is that for the high P_{O_2} of air, as in Fig. 5-6. Figures 5-7(b), 5-7(c), and 5-7(d) are for the much lower oxygen partial pressures of 10^{-7} atm, 10^{-9} atm, and 10^{-11} atm, respectively. The composition of a haplobasalt may be such that, as olivine and spinel crystallize at a high P_{O_2}, the liquid composition moves to the spinel-olivine-pyroxene boundary curve (c–P of Fig. 5-6). There is, in that case, no gap in spinel precipitation, as explained in the above discussion, because the liquid composition during crystallization intersects the spinel-olivine-pyroxene rather than the spinel-olivine-anorthite boundary curve. As P_{O_2} is reduced, however, the spinel-olivine-anorthite boundary curve extends farther out into the tetrahedron, i.e., to higher iron oxide and lower MgO contents [from P in Fig. 5-7(a), to Q in Fig. 5-7(b), to R in Fig. 5-7(c), and to 2 in Fig. 5-7(d)]. As a consequence, the crystallization curve for the same haplobasaltic mixture may, at a low P_{O_2}, intersect the spinel-olivine-anorthite boundary curve rather than the spinel-olivine-pyroxene boundary curve. A gap in spinel crystallization will then occur.

MAGNETITE PRECIPITATION DURING FRACTIONAL CRYSTALLIZATION OF HAPLOBASALTIC LIQUIDS UNDER CONDITIONS OF CONSTANT TOTAL COMPOSITION

The phase relations shown in Fig. 5-3 are those for silicate liquids and crystals in equilibrium with metallic iron at 1 atm total pressure. Magnetite is not a phase. In a basaltic magma crystallizing within the earth's crust, on the other hand, metallic iron is not a phase[7] but magnetite ordinarily does appear as a phase. The oxygen partial pressure is thus lower for the system shown in Fig. 5-3 than for basaltic magma, and this must be taken into consideration in applying the phase relations shown in Fig. 5-3 to magmas intruding the earth's crust. The phase relations of Fig. 5-3 serve, however, as a useful guide in interpreting reactions during fractional crystallization of basaltic magma treated as a system closed to oxygen (constant total composition), just as the phase relations do in the system MgO–FeO–SiO$_2$ (Bowen and Schairer, 1935). With closed system crystallization, at initial oxygen partial pressures similar to those for basaltic magma, and, therefore, higher than those obtaining in Fig. 5-3, magnetite precipitation occurs.

[7] Metallic iron occurs in a basalt in Greenland, but its origin is attributed to reduction of iron oxide by included organic matter (see, e.g., Vaasjoki, 1965; Melson and Switzer, 1966) or to its being xenolithic material from the mantle (see, e.g., Bird and Weathers, 1977).

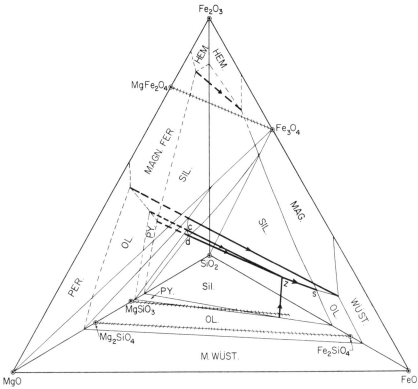

Figure 5-8. Tetrahedron to illustrate liquidus phase relations in the quaternary system MgO–FeO–Fe$_2$O$_3$–SiO$_2$ (from Roeder and Osborn, 1966, p. 431, Fig. 1, after Muan and Osborn, 1956). Phase relations on the base are after Bowen and Schairer (1935), on the right face are after Muan (1955), and on the left face are inferred. Univariant boundary curves on three of the faces of the tetrahedron are shown by light lines. The heavy lines represent univariant curves within the tetrahedron along which three crystalline phases, one liquid, and vapor are in equilibrium. Arrows indicate the direction of decreasing temperature. Lines are solid if experimentally determined, dashed if inferred. Composition joins for three solid solution series are shown by cross-hatched lines. Phase boundary curves on the triangular plane MgO-Fe$_3$O$_4$-SiO$_2$ passing through the tetrahedron are shown by light lines. Abbreviations: HEM = hematite, MAGN. FER = magnesioferrite, SIL = silica, MAG = magnetite, PER = periclase, OL = olivine, PY = pyroxene, WÜST = wustite, M. WÜST = magnesiowüstite. (With permission of the *American Journal of Science* and the *Journal of the American Ceramic Society*.)

The relation of oxygen partial pressure to magnetite precipitation is illustrated by phase relations in the system $MgO–FeO–Fe_2O_3–SiO_2$. This system is represented by a tetrahedron in Fig. 5-8, after Muan and Osborn (1956), where liquidus phase relations on the base are those for liquids in equilibrium with metallic iron, as in Fig. 5-3. Oxygen partial pressures at liquidus temperatures increase from the base toward the Fe_2O_3 apex (Fig. 5-8). A haplobasaltic composition (not shown) lies above the base in the olivine primary phase volume, close to the olivine-pyroxene boundary surface. During fractional crystallization of this haplobasaltic liquid at constant total composition, the liquid composition moves upward in the tetrahedron to the pyroxene-magnetite boundary surface, $c–d–z$, moves along this surface to z, and then moves along the boundary curve from z to s. Magnetite is a phase precipitating from the liquid continually after the liquid composition reaches the pyroxene-magnetite surface, and its precipitation does not result in the liquid composition's moving toward higher SiO_2 contents as happens in an oxygen-buffered system, e.g., Figs. 5-4 and 5-7(a, b, and c). Magnetite precipitation, however, may be considered as a controlling regulator of the maximum Fe_2O_3/FeO of the liquid (Fig. 5-8), because the liquid composition during crystallization cannot move above the lower surface of the magnetite volume to enter that volume. The magnetite-olivine and magnetite-pyroxene boundary surfaces act as a "roof" that liquid compositions reach but cannot penetrate. During fractional crystallization at constant total composition, the liquid composition "slides" down this roof toward s.

REACTIONS DURING FRACTIONAL CRYSTALLIZATION OF OXYGEN-BUFFERED, HAPLOBASALTIC LIQUIDS IN THE SYSTEM CaO–MgO–Iron Oxide–SiO_2

Reaction series of a different nature from those illustrated in Fig. 5-5(a) were visualized by Bowen (1922, 1928) as operative for the common subalkaline igneous rocks. They are illustrated in Fig. 5-5(b) modified from Bowen's original sketch principally by the addition of magnetites. In the scheme shown in Fig. 5-5(b), water plays a role, and the residual liquid is high in SiO_2. This type of reaction series results from fractional crystallization of a magma when the oxygen partial pressures are buffered so that they do not drop as rapidly during fractional crystallization as they do in the system closed to oxygen (Osborn, 1959, 1962, 1969a, 1976; Roeder and Osborn, 1966; Presnall, 1966). Oxygen is added to the system during crystallization (Osborn, 1959, pp. 621, 626, 638, 642). The phase relations of Fig. 5-4, where oxygen partial pressure is constant, are applicable as a guide to the reaction relations among the anhydrous ferromagnesian phases of Fig. 5-5(b).[8] The left face of the tetrahedron (Fig. 5-4) is the same as that for the tetrahedron in Fig. 5-3. Liquidus phase relations for the right face are after Phillips and Muan (1959), and for the base, after Muan and Osborn (1956). Dashed lines on the base are tie-lines joining compositions of the two pyroxenes and two magnetites in equilibrium with liquids c and d. The compositions of saturated haplobasaltic liquids on the base of the tetrahedron lie within the zone bounded by the dashed lines. Liquids in this zone will, under equilibrium conditions, crystallize completely to an assemblage of pyroxene and magnetite (see Osborn, 1959, p. 626). On addition of $CaSiO_3$ to the base, to form the tetrahedron of Fig. 5-4, a haplobasaltic composition HB, on the base, will move upward to X. Point X lies in the olivine primary phase volume, and within a triangular, planar zone (not shown) of liquid compositions, analogous to that within the dashed lines on the base. This zone has two plane, triangular faces. The edges of one face are the tie-lines joining compositions of the Mg pyroxene, Ca-Mg pyroxene, and magnetite that coexist in equilibrium with liquid C, and for the other face the edges are similar tie-lines joining compositions of these crystalline phases coexisting in equilibrium with liquid D. This zone of liquid compositions contains the saturated haplobasaltic liquids for the tetrahedron, inasmuch as all liquids in the zone crystallize under equilibrium conditions to form an aggregate of only the two pyroxenes and magnetite, i.e., without either olivine or a silica phase being present.

With fractional crystallization of liquid HB (Fig. 5-4), olivine precipitation will result in the liquid composition's following a fractionation curve (dotted line) to the olivine-magnetite boundary curve at b, along this curve to c, as olivine and magnetite coprecipitate, and then from c to d as pyroxene and magnetite coprecipitate. A fractionating haplobasaltic liquid X will have a similar course of crystallization with, however, a

---

[8] The relations as shown in Fig. 5-4 are for a constant oxygen partial pressure only, but the same type of phase relations, with magnetite precipitation occurring to prevent continuing iron enrichment of a fractionally crystallizing haplobasaltic liquid, have been shown to exist with oxygen buffering over a range of CO_2–CO gas mixtures (Muan and Osborn, 1956) and CO_2–H_2 gas mixtures (Roeder and Osborn, 1966; Presnall, 1966; Nafziger, 1970), and by buffers in the range of fayalite-magnetite-quartz to magnetite-hematite as can be inferred from Fig. 4 of Osborn (1969a, p. 315). In nature, the crustal country rocks can act as the oxide buffering mixture if the fractionally crystallizing magma contains H_2O (Osborn, 1959, p. 643; 1969a, p. 316).

Ca-Mg pyroxene joining an Mg pyroxene and magnetite in precipitating from liquids along the boundary curve $C–D$. Magnetite precipitation thus plays a very significant role in determining the course of fractional crystallization of a haplobasaltic liquid in the system of Fig. 5-4.

In moving from HB to b (Fig. 5-4), the liquid composition increases strongly in iron oxide (cf. Osborn, 1959, p. 627; Presnall, 1966, p. 773) and in $(FeO + Fe_2O_3)/(FeO + Fe_2O_3 + MgO)$, with little or no increase in silica. Along $b–c–d$, the trend of liquid composition is one of strong silica enrichment with declining iron oxide content and only slowly rising $(FeO + Fe_2O_3)/(FeO + Fe_2O_3 + MgO)$ (cf. Osborn, 1959, p. 629, Fig. 9). Curves of $FeO + Fe_2O_3$ vs. SiO_2 or of $(FeO + Fe_2O_3)/(FeO + Fe_2O_3 + MgO)$ vs. SiO_2, for the liquid phase during fractional crystallization, thus have a discontinuity in slope at the point of initial precipitation of magnetite, and for the latter type of curves, this discontinuity in slope has been referred to as the Sp corner (Osborn, 1976).

The reaction relations among the silicates as derived from Fig. 5-4 may be written: Mg olivines → Mg pyroxenes → Ca-Mg pyroxenes. At the lower temperatures of the latter part of the crystallization of a magma, amphiboles[9] and biotites may succeed pyroxenes, as shown in the original reaction series of Bowen (1922, 1928), and in Fig. 5-5(b). These hydrous minerals are more characteristic of coarse-grained rocks, where crystallization and reactions among the phases continue to lower temperatures, than of volcanic rocks. In the latter, amphiboles and biotites commonly do not appear as stable phases, and this is true for those rock suites used as examples in Fig. 5-9.

---

[9] Amphibole has been found not to be a stable phase in andesites and subalkaline basalts at liquidus or near-liquidus temperatures, even though the magmas have a water content of several percent. For example, the andesite flows at Paricutin volcano, Mexico, were at a temperature of approximately 1200° on extrusion (Zies, 1946). Eggler (1972) estimated that at a water content of 2.2%, the temperature at which silicate phases now present in the lava as megaphenocrysts would crystallize from a Paricutin andesite is 1110° ± 40°. The liquidus temperature at a water content of 4.5% and a total pressure of 5kbar is approximately 1050° (Eggler and Burnham, 1973) for both Paricutin and Mount Hood, Oregon, andesites. The temperature of stability of amphibote in andesite and basalt increases initially with increasing water pressure, but reaches a maximum that for both the Paricutin and the Mount Hood andesites was found to be about 950° (Allen, et al., 1972; Eggler, 1972; Eggler and Burnham, 1973). For a high-alumina basalt from Medicine Lake, California, Yoder and Tilley (1962, p. 451, Fig. 28) found that the liquidus temperature decreased from about 1250° for the anhydrous material to about 1090° at a water pressure of 5 kbar, and that at this water pressure, the maximum temperature of stability of amphibole was about 1025°. Amphibole does not, therefore, appear as a stable phase in basalts and andesites until an advanced stage in their crystallization. It follows that amphibole is an unlikely precipitating phase during fractional crystallization of basaltic magma to produce an andesite, and that amphibolites are an unlikely source of andesitic or basaltic magma by partial melting.

SIGNIFICANCE OF CURVES ILLUSTRATING
$(FeO + Fe_2O_3)/(FeO + Fe_2O_3 + MgO)$ vs.
SiO_2 FOR VOLCANIC ROCK SUITES

For certain subalkaline volcanic rock suites, an Sp corner has been found to be characteristic of curves illustrating $(FeO + Fe_2O_3)/(FeO + Fe_2O_3 + MgO)$ vs. SiO_2. Examples are shown in Fig. 5-9. The curves, Th, S, T, and M, representing rock suites from Thingmuli volcano, Iceland (Carmichael, 1964), Santorini volcano, Greece, (Nicholls, 1971), Talasea Peninsula, New Britain, Papua New Guinea (Lowder and Carmichael, 1970), and Mashu volcano, Japan (Katsui *et al.*, 1975), respectively, all have a corner, and as noted by the authors, the corner is followed by a strong silica-enrichment trend along which there is conspicuous magnetite crystallization from the liquid. This feature is analogous to the fractional crystallization curves of haplobasalts HB and X (Fig. 5-4). Similarly shaped curves develop on plotting analyses of many other volcanic rock suites, five additional examples of which are shown in Fig. 5-9. Presumably the corner on each curve coincides with the stage of beginning of major magnetite crystallization in a series of liquids produced by fractional crystallization of an oxygen-buffered basaltic magma. Magnetite precipitation, therefore, would seem to play a key role in developing the series of extrusive-rock compositions characteristic of many volcanic areas, and, hence, is appropriately shown in the reaction series diagram of Fig. 5-5(b).[10]

For the central group of curves of Fig. 5-9 (*Fl*, *F*, *N*, *U*, *T*, *M*, and *S*, representing rock suites from Flores Group, Fiji, Nicaragua, Umnak, Talasea, Mashu, and Santorini, respectively), the general arrangement suggests that the higher the silica content of the liquid during the early, steeply rising part of the curve, the greater the $(FeO + Fe_2O_3)/(FeO + Fe_2O_3 + MgO)$ at the corner. This progression in location of the corner can be explained by reference to phase relations for the system Mg_2SiO_4-Fe_3O_4-SiO_2 as shown on the base of the tetrahedron in Fig. 5-4. If the composition of the initial crystallizing liquid is moved from HB along

---

[10] It was argued by Taylor *et al.* (1969a, 1969b) that an andesite cannot form from a high-Al basalt by fractional crystallization involving magnetite precipitation, as proposed by Osborn (1959), because the vanadium abundance in andesite as compared with high-Al basalt is too great for magnetite removal from the high-Al basalt magma in the amount required to produce the derivative andesite magma. In arriving at this conclusion the authors apparently assumed that the amount of andesite derived from basalt would be 100% of the original basalt instead of an actual amount of about 30%. When recalculated (Osborn, 1969b) on the realistic basis of about 30% of andesite being derived from high-Al basalt by fractional crystallization, the data of Taylor *et al.* are in accord with a fractional crystallization origin of the andesite as proposed (Osborn, 1959).

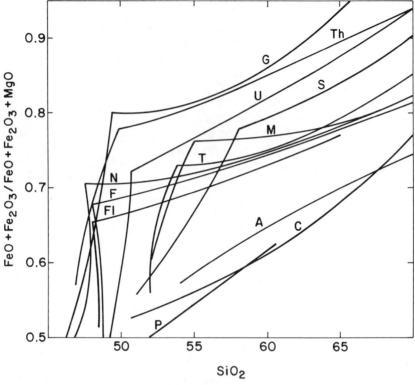

WEIGHT PERCENT

Figure 5-9. Curves representing $(FeO + Fe_2O_3)/(FeO + Fe_2O_3 + MgO)$ vs. SiO_2 for volcanic rock suites. Abbreviations: G = Galápagos Islands [plot of analyses reported by McBirney and Aoki (1966)]; Th = Thingmuli volcano, Iceland, after Carmichael (1964); U = Northeast Umnak Island, Aleutian Islands, Alaska [plot of analyses reported by Byers (1961)]; S = Santorini volcano, Cyclades, Greece [curve for the Main Series, after Nicholls (1971)]; M = Mashu volcano, East Hokkaido, Japan, after Katsui, Ando, and Inaba (1975); T = Talasea Peninsula, New Britain, Papua New Guinea [plot of analyses reported by Lowder and Carmichael (1970)]; N = Nicaragua [plot of analyses reported by Williams (1952), McBirney and Williams (1965), Williams and McBirney (1969) and Ui (1972)]; F = Vitu Levu, Fiji [plot of analyses reported by Gill (1970)]; Fl = Flores Group, Indonesia [plot of analyses reported by Newman van Padang (1951)]; A = Aegean [curve for lavas of the western South Aegean Arc, after Nicholls (1971)]; C = Cascades, western United States, after Osborn (1959); P = Parícutin volcano, Mexico [plot of analyses reported by Wilcox (1954), Foshag and Gonzáles (1956), and Tilley, Yoder, and Schairer (1966)]. Analyses for G, U, N, F, Fl, and P were taken from the data base of F. Chayes at the Geophysical Laboratory, by routine application of his information system, RKNFSYS.

the base of the tetrahedron toward the SiO_2 apex (Fig. 5-4), the point at which its fractionation curve reaches the magnetite field boundary will shift from b through c and on toward d. Liquid compositions continually increase both in SiO_2 and in $(FeO + Fe_2O_3)/(FeO + Fe_2O_3 + MgO)$ along the boundary curve from b to d. Hence, the composition of the corner on the fractionation curve for haplobasaltic liquids in Fig. 5-4 increases in SiO_2 and in $(FeO + Fe_2O_3)/(FeO + Fe_2O_3 + MgO)$ as the composition of the initial liquid increases in SiO_2.

The curves for Thingmuli (Th) and Galápagos (G), on the other hand, have a very high $(FeO + Fe_2O_3)/(FeO + Fe_2O_3 + MgO)$ at their corners, even though the silica content of the first segment of the curve is low. A possible explanation for this apparent anomaly is that the Thingmuli and Galápagos magma series formed at lower oxygen fugacities than the others. With reference to phase relations on the base of the tetrahedron of Fig. 5-4, the boundary curve, $a-c-d$, will shift to the right, i.e., toward the Fe_3O_4 apex, if the oxygen fugacity at which the system is buffered is decreased (cf. Muan and Osborn, 1956, p. 133, Fig. 11, and the shift in the magnetite-pyroxene-anorthite boundary curve to higher iron oxide contents with decreasing P_{O_2} as shown in Fig. 5-7). The composition of the fractionating liquid at the stage of initial precipitation of magnetite will, therefore, be higher in iron oxide and in $(FeO + Fe_2O_3)/(FeO + Fe_2O_3 + MgO)$ the lower the prevailing oxygen fugacity. Phase relations in the model system of Fig. 5-4 thus suggest that the two principal factors determining the position of corners on the curves in Fig. 5-9 are silica content of the parent basaltic magma and oxygen fugacity during fractional crystallization. The differing silica contents of the parent basaltic magmas may be related to differing degrees of fractionation during an early high-pressure phase of differentiation, as discussed below.

The curves (Fig. 5-9) for the Cascades (C), Aegean (A), and Parícutin (P) volcanic rock suites are examples of curves for $(FeO + Fe_2O_3)/(FeO + Fe_2O_3 + MgO)$ vs. SiO_2 that do not possess a corner. Absence of a corner, and the overall low $(FeO + Fe_2O_3)/(FeO + Fe_2O_3 + MgO)$ for these three curves, may signify that the fractionating magma systems represented by them were oxygen buffered at higher oxygen fugacities than were magmas represented by the other curves of Fig. 5-9. To illustrate these relations with a model, and with reference to phase relations shown for the base of the tetrahedron of Fig. 5-4, the magnetite-silicate boundary curve, $a-b-c-d$ (Fig. 5-4), moves to the left, i.e., to lower iron oxide contents with increase in oxygen fugacity (cf. Muan and Osborn, 1956, p. 133, Fig. 11). If the shift of the boundary curve to the left is sufficient to place the composition of a haplobasaltic liquid on the magnetite-silicate boundary curve (cf. Osborn, 1959, p. 626, Fig. 8, point n), magnetite will be a primary phase and no Sp corner will appear on the curves of $(FeO + Fe_2O_3)/(FeO + Fe_2O_3 + MgO)$ vs. SiO_2 for the fractionating liquid. It is possible, however, that the absence of a corner on curves A, C, and P (Fig. 5-9), indicative of continuous precipitation of spinel during fractional crystallization, should be attributed largely to a high-pressure effect that changes positions of phase boundaries (see below). In this view, curves A, C, and P of Fig. 5-9 represent magma suites formed largely at high pressure (e.g., 10 kbar); curves T, M, and S (Fig. 5-9) represent suites formed by fractionation initially at high pressure, followed by low pressure; and curves on the far left (G, Th, and N, Fig. 5-9) represent suites formed by fractionation entirely at low pressure.

That magnetite was continuously precipitating from the Cascades, Parícutin, and Aegean liquids, during at least the basalt to basaltic andesite stage of development of the magma series, is suggested by results of phase equilibrium studies of specimens of high-alumina basalt and basaltic andesite from the Crater Lake and Mount Jefferson districts of the Cascades region, Oregon (Osborn and Watson, 1977). In that experimental work, the powdered and dried samples were run in sealed capsules of both 95Pt-5Au and 50Ag-50Pd composition, using a gas-media, high-pressure apparatus of Yoder (1950). Runs were of eight hours' duration. Although a moderate loss of iron to the capsule occurred, the loss was believed not sufficient to affect the general phase relations. Results of the runs were similar for the two types of capsules. At a pressure of 1 kbar, magnetite was a phase at, or within a small temperature interval below, the liquidus temperatures. Presumably, therefore, magnetite was a precipitating phase, along with silicate crystals, in these basalt and basaltic-andesite magmas when they had moved to within 3 to 4 km of the surface. Preliminary data, however, show that when these Cascades rock compositions are subjected to a pressure of 10 kbar, magnetite is not a liquidus or near-liquidus phase, but appears at a temperature some $30°$ to $40°$ below the liquidus. This difference in phase relations at the two pressures can be understood by reference to Fig. 5-10. If a liquid in equilibrium with magnetite at low pressure is subjected to high pressure, the liquid may no longer be in equilibrium with magnetite, for the phase boundaries have shifted in response to the pressure increase. Taking the composition of liquid P (Fig. 5-10) as an example, this liquid composition coexists with magnetite, olivine, pyroxene, and anorthite at a pressure of 1 atm, but at a pressure of 10 kbar, a liquid of composition P is in equilibrium only with pyroxene, for the liquid composition lies within the pyroxene primary phase volume at 10 kbar. With crystallization of liquid P at 10 kbar, the composition of the liquid phase will change in response to precipitation of pyroxene crystals so that magnetite will appear only at a sub-liquidus temperature.

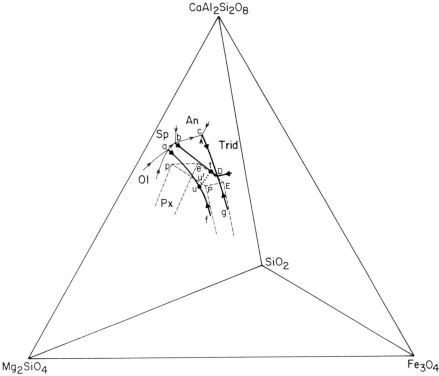

Figure 5-10. Sketch (after Osborn, 1976) to illustrate phase relations in a part of the system Mg_2SiO_4–Fe_3O_4–$CaAl_2Si_2O_8$–SiO_2 at a pressure of 10 kbar (solid lines) compared to those at a pressure of 1 atm (light dashed lines). Data at 10 kbar for compositions within the tetrahedron were obtained from runs in a gas-media, high-pressure apparatus of Yoder (1950), with the charges sealed in 95Pt-5Au capsules. Prior to encapsulation, the samples were equilibrated in air at the temperature to be used in the high-pressure run. Data at 1 atm were obtained at a constant oxygen partial pressure of $10^{-0.7}$ atm (Roeder and Osborn, 1966). Phase relations shown for the left face of the tetrahedron (light, solid lines) are after Kushiro (1972). The dotted curve u–u'–t is a fractionation curve for mixture u at 10 kbar. Point u' lies midway along u–t. The phase relations are shown on a larger scale than is the tetrahedron in order to minimize confusion of lines (cf. Fig. 5-6). Abbreviations: Ol = olivine, Px = pyroxene, Trid = tridymite or other silica phase, An = anorthite, Sp = spinel group phase. (With permission of the Department of Mineralogy, Petrology, Geology; National Technical University of Athens.)

In a parent Cascades magma fractionally crystallizing at a pressure of 10 kbar, magnetite could very well have been a crystallizing phase, but precipitating from a liquid having a composition and temperature different from that of the liquid precipitating magnetite at a lower pressure.

On the basis of results of high-pressure experiments on a Cascades andesite specimen from Mount Hood, Oregon, Eggler and Burnham (1973) concluded that fractional crystallization with magnetite precipitation cannot be the explanation for the compositional trends of the Cascades volcanic rock series. They found that, for this specimen, plagioclase is the liquidus phase if the mixture is dry or contains 2% H_2O, at all pressures up to the 10 kbar maximum pressure employed. At higher H_2O contents, plagioclase may be displaced as the liquidus phase by pyroxene, but even at 4.7% H_2O, plagioclase remains the liquidus phase up to a total pressure of about 9 kbar. Magnetite is not a liquidus or near-liquidus phase, and indeed appears only at a temperature about 100° below the liquidus. Inasmuch as magnetite is present only at temperatures much below the liquidus for this rock specimen, under any conditions of pressure, oxygen fugacity, or water content that might reasonably exist, Eggler and Burnham (1973, p. 2517) conclude: "Because differentiation of basalt melts to andesite must involve iron-rich oxide phase subtraction, such fractionation models appear unreasonable." The question to be answered, however, is whether a spinel group phase was in equilibrium with, and precipitating from the *liquid phase of the magma*, not whether a spinel group phase is in equilibrium with a liquid made by laboratory melting of the whole rock. In this connection, a statement by Bowen is appropriate. In his discussion of the glassy rocks he stated (Bowen, 1928, p. 125), "They are the only rocks of which we can say with complete confidence that they correspond in composition with a liquid. . . . Porphyritic rocks and granular phanerites often correspond very closely in composition with the liquid from which they formed, but in many instances there is great departure of such rocks from the composition of any liquid."

Abundant, zoned phenocrysts of plagioclase are present in the Mount Hood andesite. In describing these plagioclase phenocrysts, Wise states (1969, p. 977), "The cores of most of these phenocrysts have compositions between An_{45} and An_{48}. Phenocrysts in many flows show a short reversal in the zoning, crystallizing An_{55} to An_{65} following resorption. Groundmass plagioclase begins crystallizing An_{38-40} but is zoned to An_{25} where crystallization was nearly complete." The bulk composition of the plagioclase phenocrysts was, therefore, not that of crystals in equilibrium with the liquid. Furthermore, because plagioclase has a density much closer to that of the liquid than pyroxene or magnetite, differential settling of phenocrysts may have occurred, with the proportion of plagioclase crystals remaining suspended in the liquid, relative to that of pyroxene or magnetite, being larger than would be the case if the three types of crystals had the same tendency to sink. Because of these factors, the composition and liquidus temperature of this mixture are different from the composition of the liquid phase of the magma and from the temperature of the magma, respectively. The plagioclase liquidus temperature of the andesite, determined in the laboratory, is undoubtedly higher than was the temperature of the magma, and could be much higher in view of the large amount of plagioclase phenocrysts present. Because the magmatic liquid phase had a different composition and was at a lower temperature than the liquid that is formed in the laboratory by the equilibrium melting of the whole rock, the results of the studies of Eggler and Burnham (1973) may not be applicable to the question of whether or not magnetite was precipitating from the liquid phase of the andesite magma.

REACTIONS DURING FRACTIONAL CRYSTALLIZATION OF HAPLOBASALTIC LIQUIDS AT HIGHER PRESSURES

The reaction series illustrated in Fig. 5-5 apply to fractionation of basaltic magma at low pressures, below ~ 8 kbar. At higher pressures, reaction relations are significantly different. Olivine and plagioclase may not be compatible phases (Kushiro and Yoder, 1966; Cohen et al., 1967), whereas pyroxene may coexist with the early-precipitating spinels, as shown by Kushiro (1972) for the system Mg_2SiO_4–$CaAl_2Si_2O_8$–SiO_2 at a pressure of 10 kbar. Reactions at pressures greater than ~ 8 kbar but less than that at which garnet becomes a phase, ~ 20 kbar, (MacGregor, 1965; Cohen et al., 1967), can be inferred from phase equilibrium studies of the system Mg_2SiO_4–FeO–Fe_2O_3–$CaAl_2Si_2O_8$–SiO_2.[11] The tetrahedron of Fig. 5-6, after Roeder and Osborn (1966) is a representation of the phase relations at 1 atm total pressure and at a constant oxygen partial pressure of $10^{-0.7}$ atm. The curve b–P (Fig. 5-6) is the spinel-forsterite-anorthite boundary curve, discussed earlier in connection with Fig. 5-2, extending in Fig. 5-6 from the left face into the tetrahedron. As the liquid composition changes along b–P with decreasing temperature, spinel reacts with the liquid to dissolve as anorthite and olivine crystallize. Pyroxene does not coexist in equilibrium with these spinels, but only with the high-Fe spinels (magnetites) in equilibrium with liquids along the surface P–E–d–c. Along the curve p–P with decreasing temperature, olivine reacts with the liquid and dissolves as pyroxene and anorthite crystallize. At P, olivine dissolves as pyroxene, magnetite, and anorthite crystallize. Along the

[11] At still higher pressures, and for the much more complex system represented by basaltic magma, other reactions between crystals and liquid occur. These will not be discussed, for the author believes that the examples given in this chapter adequately illustrate the reaction principle.

curve e–E, pyroxene, anorthite, and tridymite coprecipitate as the liquid composition moves toward E. At E, pyroxene, magnetite, anorthite, and tridymite coprecipitate as heat is withdrawn.[12]

Figure 5-10 is a preliminary equilibrium diagram, after Osborn (1976), illustrating phase relations among spinel, olivine, pyroxene, anorthite, tridymite, and liquid at approximately the same oxygen partial pressure as in Fig. 5-6 ($10^{-0.7}$ atm) and at a pressure of 10 kbar, for comparison of relations with those at 1 atm (Fig. 5-6). In Fig. 5-10, heavy lines represent boundary curves at 10 kbar within the tetrahedron. Light solid lines, after Kushiro (1972), represent boundary curves at 10 kbar on the left face. Dashed lines are boundary curves at 1 atm outlining the upper part of the pyroxene primary phase volume, from Fig. 5-6. The phase relations are shown on a larger scale than is the tetrahedron in order to minimize confusion of lines. The incompatibility of anorthite and olivine in this system, at 10 kbar, is shown by the fact that their primary phase volumes do not have a common bounding surface (Fig. 5-10). The boundary curve, p–P, along which anorthite and forsterite coexist with pyroxene and liquid at 1 atm (Fig. 5-6) has been replaced at 10 kbar by the two boundary curves a–f and b–D (Fig. 5-10). Liquid coexists with olivine, pyroxene, and spinel along a–f, and with anorthite, pyroxene, and spinel along b–D. Both of the boundary curves, a–f and b–D, are believed to be reaction curves on the basis of preliminary data (Osborn, 1976). With decreasing temperature, olivine reacts with the liquid along a–f, and spinel reacts with the liquid along b–D. It will be noted that pyroxene is compatible with all spinel compositions at 10 kbar (Fig. 5-10), the common pyroxene-spinel surface being a–b–D–g–f. Thus, for haplobasaltic compositions in the system of Figs. 5-6 and 5-10, the order in which olivine and spinel react out is reversed at 10 kbar from that at low pressures. The order of disappearance from the precipitating aggregate, with fractional crystallization, is spinel followed by olivine at 1 atm (Figs. 5-2 and 5-6) and olivine followed by spinel at 10 kbar (Fig. 5-10). *A high pressure, discontinuous reaction series* that can be derived from relations shown in Fig. 5-10 is: olivine → spinel → pyroxene.

The reaction point P of Fig. 5-6 is not present at high pressure (Fig. 5-10), but a temperature minimum is present on the boundary curve a–f at

---

[12] At both P and E (Fig. 5-6), four crystalline phases coexist with liquid and gas over a range of temperature of 10° or more. They are not, therefore, invariant points. For purposes of this discussion, however, P and E can be considered as peritectic and eutectic points, respectively, at a P_{O_2} of $10^{-0.7}$ atm and a total pressure of 1 atm. The compositions and temperatures of these liquidus points are listed in Table 9 of Roeder and Osborn (1966, p. 478). For further discussion related to phase relations in this system, the reader is referred to the preceding chapter, particularly Figs. 4-31 and 4-32.

u (Osborn, 1976). The composition of a primary haplobasaltic liquid that forms on partial melting of a haploperidotite, at a pressure of 10 kbar, is at a point on a–f in the vicinity of the minimum, u, its precise composition depending on the composition of the haploperidotite and on the degree of melting. A primary haplobasaltic liquid of a composition u (Fig. 5-10), in equilibrium with olivine, spinel, and pyroxene, will, if oxygen buffered, have a fractional crystallization curve of the nature of u–u'–t, as pyroxene and spinel coprecipitate.[13] Spinel will cease to crystallize and anorthite will begin to crystallize when the liquid composition reaches the boundary curve b–D at t.

Along the fractionation curve u–u'–t, the liquid becomes enriched in both the SiO_2 and the $CaAl_2Si_2O_8$ components. A liquid, u', located on the dotted curve midway between u and t, has the composition of a high-alumina, haplobasaltic andesite in equilibrium with spinel and pyroxene at a pressure of 10 kbar. At a pressure of 1 atm, however, u' lies within the anorthite primary phase volume, and, hence, at low pressure anorthite is the liquidus phase for the composition u' instead of spinel and pyroxene. Therefore, if liquid u' is moved from a 10 kbar region to one of low pressure, anorthite will precipitate on cooling, followed by olivine or pyroxene, but without spinel. Cascades basalts and basaltic andesites might, therefore, in analogy with liquid u' have spinel as a liquidus phase during fractionation at high pressure, but not have a spinel group phase as phenocrysts in the magma during a continuing cooling and fractionation period at lower pressure, prior to extrusion at the surface.[14]

Based on a comparison of relations shown in Figs. 5-6 and 5-10, the sequence of mineral assemblages for fractional crystallization of basaltic magma, first at high pressure (~ 8 to ~ 20 kbar), followed by continuing fractionation at low pressure, is as follows: (high pressure) spinel + olivine $\rightarrow$ spinel + pyroxene $\rightarrow$ (low pressure) plagioclase + olivine $\rightarrow$ plagioclase + pyroxene $\rightarrow$ plagioclase + pyroxene + magnetite. The positions of the curves of Fig. 5-9 that have their origins at relatively high SiO_2 contents, and that also have a corner, can possibly be explained by

[13] Because the pyroxene and the spinel that precipitate during the crystallization of mixture u are both moderately high in alumina, their compositions, and consequently also that of the fractionating liquid, depart significantly from compositions that can be represented in terms of the apices of the tetrahedron. For illustrating fractional crystallization, therefore, the diagram of Fig. 5-10 can serve only as a guide to indicate the general nature of the course of crystallization.

[14] Some basalts and basaltic andesites of the Cascades province, northern California, did not have magnetite as a crystallizing phase at the time of extrusion (Carmichael and Nicholls, 1967, p. 4682). These lavas may be cases where spinel was a crystallizing phase at high pressures, but did not continue as a crystallizing phase at low pressure (cf. Osborn, 1969a, p. 318).

such a two-stage process. The Talasea and Mashu curves, for example, have starting liquid compositions of about 52% SiO_2, as compared with about 47% SiO_2 for Galapagos, Thingmuli, and Nicaragua. A shift in liquid composition to the higher SiO_2 content of about 52% may have occurred during a high-pressure crystallization stage at which a spinel group phase was precipitating along with pyroxene, whereas the curves shown on Fig. 5-9 for Talasea and Mashu may represent the second stage, that of the low-pressure fractionation: plagioclase + olivine → plagioclase + pyroxene → plagioclase + pyroxene + magnetite. Because spinel is a continuing liquidus phase as the liquid composition changes from u to t along the high-pressure crystallization curve, u–u'–t (Fig. 5-10), a plot of $(FeO + Fe_2O_3)/(FeO + Fe_2O_3 + MgO)$ vs. SiO_2 for liquid compositions along this curve will show no Sp corner. These relations suggest, by analogy, that fractional crystallization at high pressure with spinel and pyroxene as precipitating phases continued until an advanced stage for the Cascades, Parícutin and western South Aegean Arc magma suites. Continuing precipitation of spinel would explain the absence of a corner and the low $(FeO + Fe_2O_3)/(FeO + Fe_2O_3 + MgO)$ for their curves (Fig. 5-9). For the Talasea and Mashu suites, on the other hand, high-pressure crystallization with spinel as a phase ceased at a liquid composition of about 52% SiO_2. A gap in spinel precipitation then occurred, with spinel reappearing as a phase during low-pressure fractional crystallization, at the liquid compositions represented by the corners on the curves (Fig. 5-9).

When the fractionation curve of u reaches t (Fig. 5-10), spinel ceases to precipitate, and with continuing withdrawal of heat, anorthite and pyroxene precipitate as the liquid follows a fractionation curve across the anorthite-pyroxene boundary surface to a point on the boundary curve c–D and then changes composition down this curve to D as anorthite, pyroxene, and a silica phase precipitate. At D, with continuing withdrawal of heat, an iron-rich spinel precipitates along with pyroxene, anorthite, and a silica phase. With this course of fractional crystallization, therefore, a gap in spinel precipitation occurs, but the gap is over ranges of SiO_2 content of the fractionating liquid, and of spinel compositions, different from those for fractional crystallization at 1 atm. Those liquids that do not coexist with spinel at 10 kbar are of medium to high SiO_2 content, as compared with a low SiO_2 content for liquids that do not coexist with spinel at a pressure of 1 atm. If, however, the fractionation curve for haplobasaltic liquid u is such that the liquid composition moves to a point on the boundary curve between g and D, there will be no gap in spinel precipitation. The liquid composition, on reaching g–D, will move along this curve toward D as pyroxene, spinel, and silica crystallize. At D,

Table 5-1. Successive assemblages precipitating at high, medium, and low pressures from an oxygen-buffered, fractionally crystallizing, haplobasaltic liquid in the system Mg_2SiO_4-FeO-Fe_2O_3-$CaAl_2Si_2O_8$-SiO_2.

High pressure	Medium pressure	Low pressure
spinel + olivine	spinel + olivine	spinel + olivine or anorthite
↓	↓	↓
spinel + pyroxene	spinel + pyroxene	olivine + anorthite
	↓	↓
	pyroxene + anorthite	pyroxene + anorthite
	↓	
spinel + pyroxene + silica	pyroxene + silica +.anorthite	pyroxene + magnetite + anorthite
	pyroxene + silica + magnetite + anorthite	

anorthite appears. With this course of crystallization, a spinel group phase precipitates continuously, and silica appears as a phase before anorthite. This sequence of phase assemblages will occur particularly as pressure is increased beyond 10 kbar, causing the primary phase volume of anorthite to decrease in size, and shifting D toward the left face of the tetrahedron. Successive crystalline phase assemblages precipitating during fractional crystallization of a haplobasaltic liquid of composition u in the system Mg_2SiO_4–FeO–Fe_2O_3–$CaAl_2Si_2O_8$–SiO_2 (Figs. 5-6 and 5-10) are illustrated in Table 5-1 for high pressure ($\sim$10 to 20 kbar), medium pressure ($\sim$8 to 10 kbar), and low pressure ($\gtrsim$8 kbar).

CONCLUSION

In the foregoing, the reaction principle of Bowen has been illustrated and discussed by examining phase relations in silicate systems more simple in composition than their natural analogs. These systems have been studied in the laboratory and have phase relations that will, in general, be of a type similar to those in the much more complex natural magmas. Reaction relations between crystals and liquid are commonplace and extensive in these silicate systems. Because of these relations, fractional crystallization will occur and may be extreme during the slow cooling of haplobasaltic liquids. During fractionation, certain crystalline phases precipitate only over a range of temperature, and therefore, unlike eutectic crystallization, will not be continuously precipitating phases. Residual liquids may be much higher or much lower in FeO or in SiO_2 than the starting liquid. By analogy with silicate systems studied in the laboratory, successive liquids produced by fractional crystallization of natural magmas, and separated from time to time from the crystallizing parent liquid, can form a series of related rocks, such as the calc-alkaline suite. Successive precipitating crystal fractions from a cooling, basaltic magma can form layered gabbroic and ultramafic bodies. Bowen's sense of the importance of the reaction between crystals and liquid has been borne out by the continuing experimental work of the past fifty years.

References

Allen, J. C., P. J. Modreski, C. Haygood, and A. L. Boettcher, The role of water in the mantle of the earth: The stability of amphiboles and micas, *Proc. 24th Int. Geol. Congr., Sec. 2*, 231–240, 1972.

Andersen, O., The system anorthite–forsterite–silica, *Am. J. Sci., 39*, 407–454, 1915.

Bird, J. M., and M. S. Weathers, Native iron occurrences of Disko Island, Greenland, *J. Geol., 85*, 359–371, 1977.

Bowen, N. L., The ternary system: diopside–forsterite–silica, *Am. J. Sci., 38*, 207–264, 1914.

———, The crystallization of haplobasaltic, haplodioritic and related magmas, *Am. J. Sci., 40*, 161–185, 1915.

———, The reaction principle in petrogenesis, *J. Geol., 30*, 177–198, 1922.

———, *The Evolution of the Igneous Rocks*, Princeton University Press, Princeton, New Jersey, 334 pp., 1928.

Bowen, N. L., J. F. Schairer, and E. Posnjak, The system $CaO-FeO-SiO_2$, *Am. J. Sci., 26*, 193–284, 1933.

Bowen, N. L., and J. F. Schairer, The system $MgO-FeO-SiO_2$, *Am. J. Sci., 26*, 151–217, 1935.

Byers, F. M., Volcanic suites, Umnak and Bogoslof Islands, Aleutian Islands, Alaska, *Geol. Soc. Am. Bull., 72*, 93–128, 1961.

Carmichael, I. S. E., The petrology of Thingmuli, a Tertiary volcano in eastern Iceland, *J. Petrol., 5*, 435–460, 1964.

Carmichael, I. S. E., and J. Nicholls, Iron-titanium oxides and oxygen fugacities in volcanic rocks, *J. Geophys. Res., 72*, 4665–4687, 1967.

Cohen, L. H., K. Ito, and G. C. Kennedy, Melting and phase relations in an anhydrous basalt to 40 kilobars, *Am. J. Sci., 265*, 475–518, 1967.

DeVrics, R. C., and E. F. Osborn, Phase equilibria in high-alumina part of the system $CaO-MgO-Al_2O_3-SiO_2$, *J. Am. Ceram. Soc., 40*, 6–15, 1957.

Eggler, D. H., Water-saturated and undersaturated melting relations in a Paricutin andesite and an estimate of water content in the natural magma, *Contr. Mineral. Petrol., 34*, 261–271 (1972).

Eggler, D. H., and C. W. Burnham, Crystallization and fractionation trends in the system andesite–$H_2O-CO_2-O_2$ at pressures to 10 kb, *Geol. Soc. Am. Bull., 84*, 2517–2532, 1973.

Foshag, W. F., and J. Gonzáles, Birth and development of Parícutin volcano, Mexico, *U.S. Geol. Surv. Bull., 965-D*, 355–489, 1956.

Gill, J. B., Geochemistry of Vitu Levu, Fiji, and its evolution as an island arc, *Contrib. Mineral. Petrol.*, *27*, 179–203, 1970.

Hill, R. E., and P. L. Roeder, The crystallization of spinel from basaltic liquid as a function of oxygen fugacity, *J. Geol.*, *82*, 709–729, 1974.

Katsui, Y., S. Ando, and K. Inaba, Formation and magmatic evolution of Mashu volcano, East Hokkaido, Japan, *J. Fac. Sci. Hokkaido Univ.*, *Ser. 4*, *16*, 533–552, 1975.

Kushiro, I., Determination of liquidus relations in synthetic silicate systems with electron probe analysis: the system forsterite–diopside–silica at 1 atmosphere, *Am. Mineral.*, *57*, 1260–1271, 1972.

———, The system Mg_2SiO_4–$CaAl_2Si_2O_8$–SiO_2 at 10 kbar, *Carnegie Institution of Washington Yearbook*, *71*, 357–359, 1972.

———, The system diopside–anorthite–albite: determination of composition of coexisting phases, *Carnegie Institution of Washington Yearbook*, *72*, 502–507, 1973.

Kushiro, I., and H. S. Yoder, Jr., Anorthite-forsterite and anorthite-enstatite reactions and their bearing on the basalt-eclogite transformation, *J. Petrol.*, *7*, 337–362, 1966.

Lowder, G. G., and I. S. E. Carmichael, The volcanoes and caldera of Talasea, New Britain: geology and petrology, *Geol. Soc. Am. Bull.*, *81*, 17–38, 1970.

MacGregor, I. D., Stability fields of spinel and garnet peridotites in the synthetic system MgO–CaO–Al_2O_3–SiO_2, *Carnegie Institution of Washington Yearbook*, *64*, 126–134, 1965.

McBirney, A. R., and K. Aoki, Petrology of the Galápagos Islands, in *The Galápagos; Proceedings of the Symposia of the Galápagos International Scientific Project*, R. I. Bowman, ed., University of California Press, Berkeley and Los Angeles, pp. 71–77, 1966.

McBirney, A. R., and H. Williams, Volcanic history of Nicaragua, *Univ. Calif. Publ. Geol. Sci.*, *55*, 65 pp., 1965.

Melson, W. G., and G. Switzer, Plagioclase-spinel-graphite xenoliths in metallic iron-bearing basalts, Disko Island, Greenland, *Am. Mineral*, *51*, 664–676, 1966.

Muan, A., Phase equilibria in the system FeO–Fe_2O_3–SiO_2, *Trans. AIME*, *203*, 965–976, 1955.

Muan, A., and E. F. Osborn, Phase equilibria at liquidus temperatures in the system MgO–FeO–Fe_2O_3–SiO_2, *J. Am. Ceram. Soc.*, *39*, 121–140, 1956.

Nafziger, R. H., The join diopside–iron oxide–silica and its relation to the join diopside–forsterite–iron oxide–silica, *Am. Mineral.*, *55.* 2042–2052, 1970.

Neumann van Padang, M., *Catalogue of the Active Volcanoes of the World, Part I, Indonesia*, International Volcanological Association, Naples, pp. 161–213, 1951.

Nicholls, I. A., Petrology of Santorini volcano, Cyclades, Greece, *J. Petrol.*, *12*, 67–119, 1971.

Osborn, E. F., The system $CaSiO_3$–diopside–anorthite, *Am. J. Sci.*, *240*, 751–788, 1942.

———, Role of oxygen pressure in the crystallization and differentiation of basaltic magma, *Am. J. Sci.*, *257*, 609–647, 1959.

————, Reaction series for subalkaline igneous rocks based on different oxygen pressure conditions, *Am. Mineral.*, *47*, 211–226, 1962a.

————, Addendum note to "Reaction series for subalkaline igneous rocks based on different oxygen pressure conditions," *Am. Mineral.*, *47*, 1480–1481, 1962b.

————, The complementariness of orogenic andesite and alpine peridotite, *Geochim. Cosmochim. Acta*, *33*, 307–324, 1969a.

————, Genetic significance of V and Ni content of andesites: Comments on a paper by Taylor, Kaye, White, Duncan and Ewart, *Geochim. Cosmochim. Acta*, *33*, 1553–1554, 1969b.

————, Origin of calc-alkali magma series of Santorini volcano type in the light of recent experimental phase equilibrium studies, *Proceedings of the International Congress on Thermal Waters, Geothermal Energy and Volcanism of the Mediterranean Area, Athens, Greece, October 1976*, *3*, 154–167, 1976.

Osborn, E. F., and D. B. Tait, The system diopside–forsterite–anorthite, *Am. J. Sci., Bowen Vol.*, 413–433, 1952.

Osborn, E. F., and E. B. Watson, Studies of phase relations in subalkaline volcanic rock series, *Carnegie Institution of Washington Yearbook*, *76*, 472–478, 1977.

Phillips, B., and A. Muan, Phase equilibria in the system CaO–iron oxide–SiO_2 in air, *J. Am. Ceram. Soc.*, *42*, 413–423, 1959.

Presnall, D. C., The join forsterite–diopside–iron oxide and its bearing on the crystallization of basaltic and ultramafic magmas, *Am. J. Sci.*, *264*, 753–809, 1966.

Roeder, P. L., and E. F. Osborn, Experimental data for the system MgO–FeO–Fe_2O_3–$CaAl_2Si_2O_8$–SiO_2 and their petrologic implications, *Am. J. Sci.*, *264*, 428–480, 1966.

Schairer, J. F., and N. L. Bowen, The binary system $CaSiO_3$–diopside and the relations between $CaSiO_3$ and akermanite, *Am. J. Sci.*, *240*, 725–742, 1942.

Schairer, J. F., and H. S. Yoder, Jr., The nature of residual liquids from crystallization, with data on the system nepheline–diopside–silica, *Am. J. Sci., Bradley Vol.*, *258-A*, 273–283, 1960.

Schairer, J. F., and H. S. Yoder, Jr., The system diopside–enstatite–silica, *Carnegie Institution of Washington Yearbook*, *61*, 75–82, 1962.

Taylor, S. R., M. Kaye, A. J. R. White, A. R. Duncan, and E. Ewart, Genetic significance of Co, Cr, Ni, Sc and V content of andesites, *Geochim. Cosmochim. Acta*, *33*, 275–286, 1969a.

Taylor, S. R., M. Kaye, A. J. R. White, A. R. Duncan, and E. Ewart, Genetic significance of V and Ni content of andesites: Reply to Prof. E. F. Osborn, *Geochim. Cosmochim. Acta*, *33*, 1555–1557, 1969b.

Tilley, C. E., H. S. Yoder, Jr., and J. F. Schairer, Melting relations of volcanic rock series, *Carnegie Institution of Washington Yearbook*, *65*, 260–269, 1966.

Ui, T., Recent volcanism in Masaya-Granada area, Nicaragua, *Bull. Volcanol.*, *36*, 174–190, 1972.

Vaasjoki, O., On basalt rocks with native iron in Disko, West Greenland, *C. R. Soc. Geol. Finlande*, *37*, 85–97, 1965.

Wager, L. R., and W. A. Deer, Geological investigations in East Greenland, Part III.

The petrology of the Skaergaard intrusion, Kangerdlugssuaq, East Greenland, *Medd. Groenl.*, *105*, 335 pp., 1939.

Wilcox, R. E., Petrology of Paricutin volcano, Mexico, *U.S. Geol. Surv. Bull.*, *965-C*, 281–353, 1954.

Williams, H., The great eruption of Coseguina, Nicaragua, in 1935 with notes on the Nicaraguan volcanic chain, *Univ. Calif. Publ. Geol. Sci.*, *29*, 21–46, 1952.

Williams, H., and A. R. McBirney, Volcanic history of Honduras, *Univ. Calif. Publ. Geol. Sci.*, *85*, 101 pp., 1969.

Wise, W. S., Geology and petrology of the Mt. Hood area: A study of high Cascade volcanism, *Geol. Soc. Am. Bull.*, *80*, 969–1006, 1969.

Yang, H. Y., and W. R. Foster, Stability of iron-free pigeonite at atmospheric pressure, *Am. Mineral.*, *57*, 1232–1241, 1972.

Yoder, H. S., Jr., High-low quartz inversion up to 10,000 bars, *Trans. Am. Geophys. Union*, *31*, 827–835, 1950.

Yoder, H. S., Jr., and C. E. Tilley, Origin of basalt magmas: an experimental study of natural and synthetic rock systems, *J. Petrol.*, *3*, 342–532, 1962.

Zies, E. G., Temperature measurements at Paricutin volcano, *Trans. Am. Geophys. Union*, *27*, 178–180, 1946.

Chapter 6

FRACTIONAL CRYSTALLIZATION
OF BASALTIC MAGMA

IKUO KUSHIRO

Geological Institute, University of Tokyo, Tokyo, Japan and Geophysical Laboratory,
Carnegie Institution of Washington, Washington, D.C.

INTRODUCTION

Bowen emphasized that the fractional crystallization of basaltic magma
is a fundamental process in the generation of igneous rocks varying widely
in chemical and mineralogical composition, including most of the major
igneous rock types. He proposed a scheme of fractional crystallization
of basaltic magma that is still essential to the understanding of the mechanisms by which the diversity of igneous rocks is produced; however,
application has proved to be more limited than Bowen postulated. First,
the concept of basaltic magma has changed. Although different magma
types had been recognized in the 1920's in Mull (Bailey *et al.*, 1924), it was
not certain whether such different magma types were of worldwide significance. Now, at least three different magma types are recognized, each
producing a different series of rocks by fractional crystallization. Bowen's
scheme covers part but not all of these fractional crystallization trends.
Second, some major magma types such as granitic or rhyolitic magma
and calc-alkaline andesitic magma may not be products of fractional
crystallization of basaltic magma. The change in concept of basaltic
magma, the recognition of different rock series, and doubt as to the role
of fractional crystallization in the genesis of some magmas are based on
extensive and detailed observations of many well-differentiated layered

© 1979 by Princeton University Press
The Evolution of the Igneous Rocks: Fiftieth Anniversary Perspectives
0-691-08223-5/79/0171-33$01.65/0 (cloth)
0-691-08224-3/79/0171-33$01.65/0 (paperback)
For copying information, see copyright page

intrusions, particularly the Skaergaard intrusion, and on studies of extrusive rocks in many different areas, particularly in Hawaii and young orogenic belts. Third, the fractional crystallization of basaltic magma is thought to take place in most cases at considerable depths, at least not near the surface. The liquid-crystal equilibria among common rock-forming silicates established at 1 atm, therefore, are not always applicable to the problem of fractional crystallization of basaltic magmas at depths. The depth of generation of magma is deeper than the level where fractional crystallization takes place. Since the 1950's great progress has been made in the techniques of experimental petrology, especially for experiments at high pressures and in controlled atmospheres. The phase equilibrium relations, involving major rock-forming silicates, applied by Bowen to the fractional crystallization of basaltic magma undergo significant changes at high pressures and under different fugacities of volatile components such as oxygen, water, and carbon dioxide. Some of the conclusions reached by Bowen must be significantly modified to accommodate these new experimental results obtained under conditions believed to be closer to those existing during the generation of natural magmas. It must be emphasized, however, that the principle of fractional crystallization as the main process in the development of igneous rock series established by Bowen is still valid, and his approach in applying experimental results to igneous petrogenesis has been a guide for later studies in petrology.

This paper comprises the following sections: (1) changes in concept of basaltic magma types, (2) reaction relation of olivine with liquid under varying physicochemical conditions and its significance in magma genesis, (3) modification of the courses of fractional crystallization of basaltic magma, (4) crystallization of the major rock-forming silicate minerals pyroxene, plagioclase, hornblende, and biotite from magmas and their influence on the trend of fractional crystallization, and (5) fractional crystallization of basaltic magma at high pressures.

BASALT MAGMA TYPES

Basaltic magmas on the surface of the earth have a wide range of chemical composition. On the basis of many studies of eruptive rocks, notably those by Bailey et al. (1924), Kennedy (1933), and Tilley (1950), at least three different basaltic magma types[1] have been distinguished. Each type

---

[1] Bailey et al. (1924) stated that the concept of magma type is based upon composition alone and differs from the concept of rock type which takes into account texture as well. Each basalt magma type has been defined on the basis of the compositions of a group of basalts and related rocks that approximate those of magmas and have certain compositional characteristics.

exhibits a unique line of descent during fractional crystallization according to the present interpretation of field associations.

Bowen was aware of different basaltic magma types occurring in Mull. He demonstrated by means of an addition-subtraction diagram that the non-porphyritic central magma type could be derived from the plateau magma type by separation of olivine, diopsidic clinopyroxene, calcic plagioclase, and magnetite (Bowen, 1928, pp. 74–78). In this demonstration he presumed the plateau magma type had a reaction relation of olivine with the liquid. This assumption is, however, not acceptable (Tilley, 1950, p. 45). Subsequent work by Kennedy (1933) and Tilley (1950) has shown that two of the magma types recognized in Mull are of worldwide significance. The two types corresponding to the non-porphyritic central and the plateau magma types were later called tholeiitic and alkali olivine basalt magma types, respectively (Kennedy, 1933; Tilley, 1950). The generalization of Bowen's argument that tholeiitic magma is derived from alkali olivine basalt magma by the subtraction of early crystallizing minerals has been shown to be of doubtful validity at low pressures, at least in Hawaii (Kuno *et al.*, 1957). According to Kuno's calculation, about 13 wt.% nepheline must crystallize with diopsidic augite, olivine, iron oxide, and calcic plagioclase to produce a magma of average tholeiite composition (average of 32 Hawaiian tholeiites) from a magma of average alkali olivine basalt composition (average of 7 Hawaiian alkali olivine basalts). Kuno argued that crystallization of nepheline from basaltic magma in the early stage is quite unlikely, a conclusion opposite to that reached by Bowen (1928, pp. 77–78). Alkali olivine basalt is higher in both K_2O and Na_2O contents than tholeiite in Hawaii as well as in many other localities, whereas the Hebridean plateau magma type used in the Bowen's calculation is lower in alkalies than the non-porphyritic central magma type. These differences are the reasons for the two entirely different conclusions. In other words, the composition of plateau magma type used by Bowen is not representative of the composition of alkali olivine basalt magma type. Later, Yoder and Tilley (1962) suggested on the basis of iron-free model silicate systems that the chemical compositions of alkali olivine basalt and tholeiitic magma types are separated by a thermal divide[2] at or near 1 atm, and therefore, one of the two magma types cannot be derived from the other by crystal extraction under near-surface

[2] A thermal divide (Yoder and Tilley, 1962, p. 398) is a liquidus temperature maximum, or a locus of such maxima, analogous to a topographic divide. It is a temperature barrier that the liquid composition cannot cross through crystallization or melting of the solid phase(s) with which it is in equilibrium, and liquids on either side of the divide will move away from it compositionally as they crystallize on cooling. The thermal divide, for example, is a point in a binary system, a line in a ternary system, and a surface in a quaternary system.

conditions. The essential difference in chemical composition between these two magma types must, therefore, develop at high pressures, probably in the upper mantle. The derivation of these magma types at high pressures will be discussed in a later section of this chapter.

A further complication arises from evidence that tholeiitic and alkali olivine basalt magmas both vary considerably in composition. Tholeiite ranges from silica-undersaturated olivine tholeiite to silica-oversaturated quartz tholeiite, and alkali basalt ranges from weakly silica-undersaturated olivine basalt to strongly silica-undersaturated basanites. Highly silica-undersaturated varieties such as olivine nephelinite, olivine melilitite, and kimberlite have affinities with alkali basalts. Recent experimental studies at high pressures suggest that some of these highly silica-undersaturated alkali basic rocks are formed independently in the upper mantle under special chemical conditions.

REACTION RELATION OF OLIVINE WITH LIQUID

Bowen (1928, pp. 70–72) pointed out that because of the presence of a reaction relation of olivine with liquid, olivine separates from the liquid in amounts greater than its actual "stoichiometric proportion" thereby producing free silica in the late stages of fractional crystallization. In his theory of the genesis of igneous rocks, Bowen placed great importance on the reaction relation of olivine, and stated," . . . some petrologists liked to think that high pressure or the presence of volatile components destroys the reaction relation, but this brief review of the actual relations displayed in rocks did not justify their hopes." Bowen's argument was indeed correct for the subalkaline basalts crystallized at or near the surface, which often show evidence of reaction relation of olivine in the form of a reaction rim of pyroxene around olivine crystals and the disappearance of olivine at a certain stage of crystallization.

Since the 1960's many experimental studies related to the reaction relation of olivine have been carried out at high pressures and in the presence of volatile components. The effect of pressure on the reaction relation of forsterite to the liquid was first studied by Boyd and England (1961). They showed that the incongruent melting of enstatite (protoenstatite) to forsterite + liquid persists only to about 3 kbar under dry conditions (Fig. 6-1). It is implied that the reaction relation of forsterite to liquid does not hold at pressures higher than about 3 kbar in the system MgO–SiO_2. Even in more complex systems, the reaction relation of olivine to quartz-normative liquid does not persist to high pressures. O'Hara (1968)

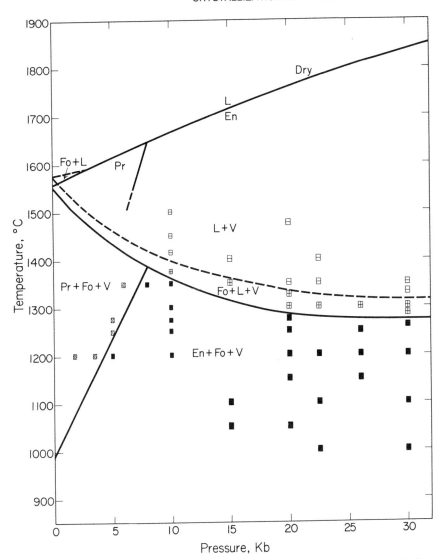

Figure 6-1. Melting relations for $MgSiO_3$ composition under both anhydrous and H_2O-saturated conditions (after Boyd, England, and Davis, 1964, and Kushiro, Yoder, and Nishikawa, 1968, respectively). Under H_2O-saturated conditions, Fo reacts with liquid to produce En (or Pr) at the boundary between the regions Fo + L + V and En + Fo + V (or Pr + Fo + V).

reviewed the reaction relation of olivine with liquid in various iron-free synthetic systems both at 1 atm and at high pressures, and realized that the reaction relationship between olivine and silica-oversaturated basic liquids is suppressed under anhydrous conditions by pressures of about 5 kbar.

A reaction relation of olivine with a hypersthene- and olivine-normative liquid, however, persists to much higher pressures. For example, in the system $CaO-MgO-Al_2O_3-SiO_2$, the reaction relation of olivine with liquid persists to about 28 kbar (Kushiro and Yoder, 1974). The reactions in this case are forsterite + liquid $\rightleftharpoons$ clinopyroxene + orthopyroxene + spinel (10–26 kbar) and forsterite + liquid $\rightleftharpoons$ clinopyroxene + orthopyroxene + garnet (26–28 kbar). The liquids involved in these reactions are not directly relevant to the generation of quartz-normative liquid or rocks with free silica, but the reactions eliminate olivine from the liquids at high pressures. They are considered important to the genesis of olivine-free eclogite, the high-pressure equivalent of basalt, in the upper mantle through the partial melting of peridotites, a possible source rock for magmas of basaltic composition.

In the presence of volatile components the phase relations involving the reaction relation of olivine undergo significant changes. Under hydrous conditions, the reaction relation of olivine to quartz-normative liquid persists to much higher pressures than under dry conditions. For example, in the system $MgO-SiO_2-H_2O$, the reaction relation of olivine persists to at least 30 kbar (Kushiro et al., 1968) (Fig. 6-1); in the system $CaO-MgO-Al_2O_3-SiO_2-H_2O$, it persists to at least 10 kbar (Yoder and Chinner, 1960); and in the system $Na_2O-CaO-MgO-Al_2O_3-SiO_2-H_2O$, it persists to about 25 kbar (Kushiro, 1972). These experimental results indicate that a quartz-normative liquid can be produced at considerable depths (at least to 80 km) in the presence of H_2O from olivine-normative liquids by fractional crystallization or from olivine-bearing rocks by fractional fusion. Thus, although the mechanism of generation of quartz-normative liquid by separation of olivine would not be important under anhydrous conditions at the site of magma generation, it may still be important under hydrous conditions.

Carbon dioxide is another important volatile component in magma. Its effect on the phase relations involving the reaction relation of olivine was determined by Eggler (1975, 1976), who found that in the system $MgO-SiO_2-H_2O-CO_2$, the reaction relation of forsterite to the quartz-normative liquid disappears when the mole fraction of CO_2 in the CO_2-H_2O vapor exceeds 0.45 at 20 kbar, as shown in Fig. 6-2. In more complex systems, the liquidus field of olivine is reduced relative to that of Ca-poor pyroxene in the presence of CO_2, indicating that the reaction relation of olivine would tend to disappear in natural magmas if CO_2 is present.

Thus, the possibility of generation of quartz-normative magma by separation of olivine probably decreases when CO_2 is present in the magma.

COURSE OF FRACTIONAL CRYSTALLIZATION OF BASALTIC MAGMA

In his discussion on the course of fractional crystallization of basaltic magma, Bowen (1928, pp. 79–80) emphasized the enrichment of the residual liquid in silica and alkali feldspar components and did not consider significant the enrichment in iron or ferriferous pyroxene component. On

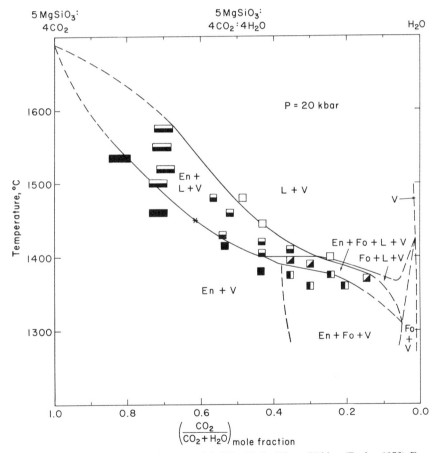

Figure 6-2. Phase relations in the system $MgSiO_3$–H_2O–CO_2 at 20 kbar (Eggler, 1975). Fo reacts with liquid to produce En in the region En + Fo + L + V.

the other hand, Fenner (1929) emphasized the enrichment of the residual liquid in iron rather than silica during such crystallization. Bowen's argument was valid for the fractional crystallization of calc-alkalic magmas, and in all the variation diagrams shown by Bowen, the total iron content of the suites of rocks plotted decreases with decreasing MgO. As discussed in the previous chapter by Osborn, the sequence of minerals in the discontinuous reaction series by Bowen also represents that of the calc-alkalic rock series. It should be realized that the tholeiitic magma had not as yet been recognized.

In 1939, Wager and Deer published their detailed study on the Skaergaard intrusion, where fractional crystallization of basaltic magma was strikingly demonstrated. The residual liquid here was remarkably enriched in iron, presumably by the process of fractional crystallization. Total iron oxide increased from about 10 wt.% in the most primitive magma to about 22 wt.% in the residual liquid when 98 wt.% of the magma was believed to have solidified, whereas MgO decreased from 8.6 to 1.3 wt.%, and silica decreased until about 90 wt.% of the magma was solidified (Wager and Brown, 1967) (Fig. 6-3). These features did not support Bowen's argument, although continuous enrichment of the alkali feldspar component and sudden increase in silica in the final stage of fractional crystallization occurred in this intrusion.

After the detailed studies of the Skaergaard intrusion had been carried out, similar fractional crystallization of tholeiitic magma was recognized in many differentiated intrusive bodies, such as the Karroo dolerite sills (Walker and Poldervaart, 1949; Poldervaart, 1944), the Palisade diabase sill (F. Walker, 1940; K. R. Walker, 1969), the Dillsburg sills (Hotz, 1953), the Beaver Bay gabbro (Muir, 1954), and the Stillwater complex (Hess, 1960). These intrusions show moderate to strong iron-enrichment during fractional crystallization, although the degree of iron-enrichment is not as strong as in the Skaergaard intrusion. Iron enrichment during the crystallization of basaltic magma was, therefore, found to be common for a certain type of basaltic magma, now identified as a tholeiitic magma type.

The trends of fractional crystallization of different basaltic magmas have also been investigated in detail in many volcanic rock suites. The studies of Kuno (1950, 1959, 1960) in the Japanese Islands are perhaps the most comprehensive. Kuno (1950) distinguished two different rock series, for the first time, on the basis of groundmass pyroxene: the "pigeonitic rock series" that contains pigeonite in the groundmass, and the "hypersthenic rock series" that contains orthopyroxene in the groundmass. Kuno (1968) correlated the pigeonitic rock series with the tholeiitic rock series and the hypersthenic rock series with the calc-alkalic rock

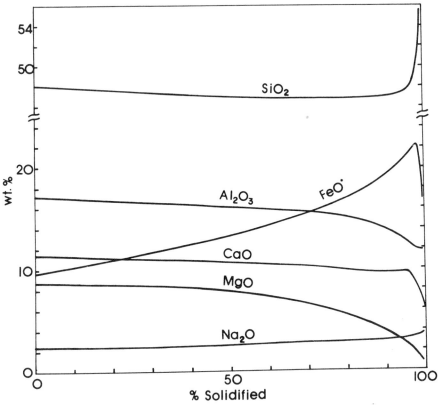

Figure 6-3. Oxide variation against percentage solidified in the Skaergaard intrusion (Wager and Brown, 1967). The SiO$_2$ content (in wt. %) of the liquid decreases slightly whereas total iron (FeO*) increases until about 90% of the magma is solidified. The composition of the first liquid is represented by that of the chilled margin.

series[3], although there are some objections to his generalization (for example, Yoder and Tilley, 1962, pp. 353–354). It was demonstrated by Kuno that the volcanic rocks of the calc-alkalic rock series are relatively low in iron and do not exhibit iron enrichment, whereas those of the tholeiitic rock series are relatively high in iron and exhibit iron enrichment, as shown in the AMF diagram (Fig. 6-4). These fractionation trends are

---

[3] Kuno (1960, 1968) distinguished high-alumina basalt series from the other two rock series. Rocks of this series, according to him, belong mostly to pigeonitic and partly to hypersthenic rock series.

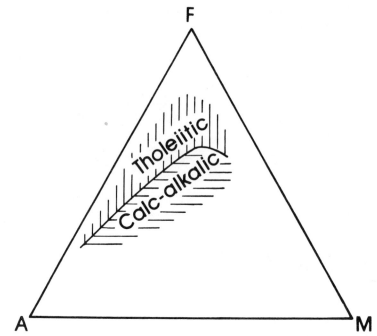

Figure 6-4. AFM ($Na_2O + K_2O$–$FeO + Fe_2O_3$–MgO) diagram showing the trends of tholeiitic and calc-alkali rock series in the Izu-Hakone region (Kuno, 1968).

also observed within relatively thick lava flows, which often contain segregation veins formed from residual liquids. Kuno (1965) showed that the lavas of the tholeiitic rock series segregated iron-rich residual liquids, whereas those of calc-alkalic rock series segregated silica-rich residual liquids with no iron enrichment. Examples are shown in Fig. 6-5.

The difference between tholeiitic and calc-alkalic trends of fractional crystallization was explained elegantly by Osborn (1959) based on the difference in oxygen fugacity during the crystallization of magma. The principle of his argument is illustrated in the system MgO–FeO–Fe_2O_3–SiO_2 under controlled P_{O_2} (Fig. 6-6). If the system is open to oxygen so that P_{O_2} is kept constant and high during the crystallization, magnetite (magnesioferrite) crystallizes from liquid (m) with olivine and subsequently with pyroxene. The residual liquid is depleted in iron oxides and enriched in silica, as shown by a trend $m \rightarrow a \rightarrow b$ in Fig. 6-6. On the other hand, if the system is low in P_{O_2}, olivine alone crystallizes from the same liquid and the residual liquid is enriched in iron oxides and depleted

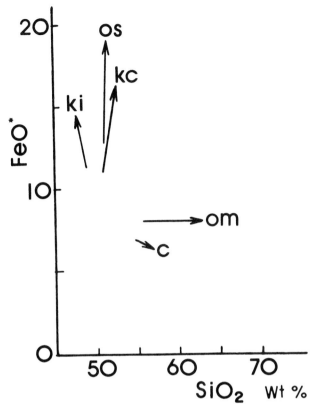

Figure 6-5. SiO$_2$-total iron (as FeO*) variation diagram for some differentiated lava flows of tholeiitic and calc-alkalic magmas (Kuno, 1965).

slightly in silica, as shown by a trend $m \to n \to o$. The variations in SiO$_2$ and iron oxides for these two different courses of crystallization are shown in Fig. 6-7 with those for the Skaergaard intrusion and the calc-alkalic rock suite from Cascades. The trend of fractional crystallization of the Skaergaard intrusion is similar to that in the synthetic system under low P_{O_2} conditions, and the trend of the Cascades calc-alkalic rock suite is similar to that under constant and high P_{O_2} conditions. Some alkalic rock suites (for example, Oslo alkalic rocks) exhibit trends similar to those of calc-alkalic rock suites (Yoder and Tilley, 1962, Fig. 20). The studies of Osborn (1959) appear to have resolved the well-known Bowen-Fenner debate on whether or not the residual liquid is enriched in iron during the fractional crystallization of basaltic magma.

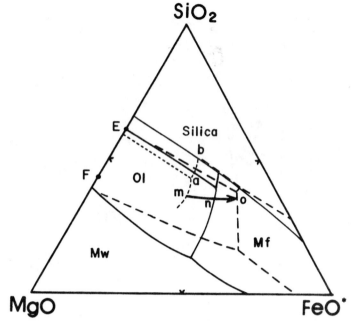

Figure 6-6. Phase relations in the system MgO–FeO–Fe$_2$O$_3$–SiO$_2$ at different f_{O_2} (after Muan and Osborn, 1956). Dotted lines, heavy lines, and dashed lines are liquidus boundaries at 1 atm P_{O_2}, at P_{O_2} equal to that of air (i.e., 0.21 atm), and at P_{O_2} for a constant CO$_2$:CO ratio of 132, respectively. Crystallization course $m \rightarrow a \rightarrow b$ is for constant P_{O_2} (i.e., 1 atm) and another course $m \rightarrow n \rightarrow o$ is for maximum fractionation at P_{O_2} for CO$_2$:CO of 132 (after Osborn, 1976).

There are, however, some objections to Osborn's discussion of the origin of calc-alkalic andesite (T. H. Green and Ringwood, 1968; Ringwood, 1975; Cawthorn and O'Hara, 1976). The major argument arises from the absence of extensive near-surface crystallization of magnetite in basalt, basaltic andesite, and andesite of calc-alkalic rock series, and no magnetite-rich cumulates have been found in orogenic belts. In addition, experiments on natural andesite under hydrous conditions indicate that magnetite would not be stable at high temperatures above 2 kbar P_{H_2O} (Eggler and Burnham, 1973), and at oxygen fugacities lower than 10^{-6} bar at 1 atm (Bigger, 1974), which seems to be an unrealistically high oxygen fugacity. Taylor et al. (1969) also argued from the vanadium contents that the maximum amount of magnetite that could separate from the high-alumina basalt magma to produce calc-alkalic andesite magma is less than 2 percent.

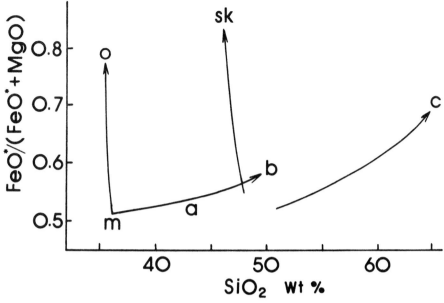

Figure 6-7. Variations in SiO_2 and $(FeO + Fe_2O_3)/(FeO + Fe_2O_3 + MgO)$ of liquid during the crystallization of the Skaergaard intrusion (Sk), the Cascades calc-alkalic rock suites (C), and mixture m of Fig. 6-6, under two different conditions.

Osborn (1969) pointed out, however, that the calculation by Taylor *et al.* (1969) is not correct and that the amount of magnetite in a gabbro cumulate that would form in deriving an andesite from a high-alumina basalt is similar to the actual amount of magnetite (3.3 wt.%) in Nockolds' (1954) average gabbro. Osborn (1976) also pointed out that magnetite does appear and causes the change in fractional crystallization trend to calc-alkalic directions in some orogenic volcanic rock series, such as in Talasea, New Britain; Mashu, Hokkaido, Japan; and Santorini, Cyclades, Greece.

Recently, Irvine (1976) proposed a new interpretation of the origin of the above two different crystallization trends (i.e., iron-enrichment trend and silica-enrichment trend) on the basis of the effect of metastable liquid immiscibility in silicate melt. He demonstrated that in the system Mg_2SiO_4–Fe_2SiO_4–$CaAl_2Si_2O_8$–$KAlSi_3O_8$–SiO_2 at 1 atm, the field of crystallization of olivine and Ca-poor pyroxene is suppressed relative to that of silica mineral in the absence of $KAlSi_3O_8$ component, whereas the field of silica mineral is suppressed greatly relative to that of olivine and pyroxene in the presence of appreciable amounts of $KAlSi_3O_8$

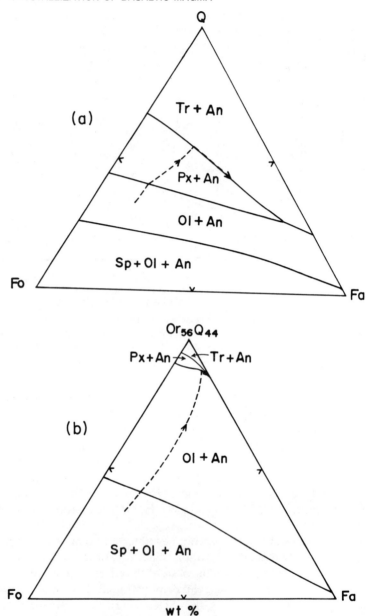

Figure 6-8. Liquidus relations of the anorthite saturation surface of the join Mg_2SiO_4–Fe_2SiO_4–$CaAl_2Si_2O_8$–SiO_2 (a) and the join Mg_2SiO_4–Fe_2SiO_4–$CaAl_2Si_2O_8$–($KAlSi_3O_8$ 56 · SiO_2 44 in mole %) (b) at 1 atm. Dashed curve is one of the liquidus fractionation curves (Irvine, 1976).

component (Fig. 6-8). Consequently, the liquids precipitating olivine, Ca-poor pyroxene, and anorthite cannot fractionate toward SiO_2 beyond certain limits in the absence of or in the presence of only small amounts of $KAlSi_3O_8$, whereas the same liquids fractionate to very silica-rich compositions in the presence of appreciable amounts of $KAlSi_3O_8$ (Fig. 6-8). Irvine (1976) attributed these changes to the suppression of the metastable liquid immiscibility region by the addition of the $KAlSi_3O_8$ component. Whatever the reason is, it is significant that the field of crystallization of olivine greatly expands and that of silica mineral is suppressed by the addition of the $KAlSi_3O_8$ component, thereby shifting the trend of fractional crystallization toward silica-enrichment. According to the regularities on the shift of the liquidus boundary found by Kushiro (1975), H_2O and $NaAlSi_3O_8$ would have similar effects. Calc-alkaline magmas are believed to be enriched in these components as well, whereas tholeiitic magmas, which show strong iron-enrichment, appear to be depleted in these components.

Bowen (1928, p. 90) raised the question why the magma of noritic gabbro could crystallize, under certain conditions, the minerals of noritic gabbro, and, under other conditions, give the series of rocks, diorite, quartz diorite, and granite. In both cases the late-crystallizing residuum has similar granitic compositions. In the noritic gabbro, fractionation was brought about by zoning mainly in plagioclase, and the product was the characteristic gabbroid rock with granophyric mesostasis. The intermediate stages were, however, different; i.e., augite gabbro in the one case, and diorite and quartz diorite in the other. Bowen attributed these differences to the difference in the degree of fractionation; the crystallization of the noritic gabbro was characterized by high fractionation (low reaction), whereas the trend leading to separate masses of diorite and quartz diorite and granite was characterized by low fractionation (high reaction).

This important suggestion does not seem to have been fully taken into account in the subsequent studies of fractional crystallization trends in basaltic magma. In distinguishing these trends, degree of iron enrichment has usually been taken as the major factor. For example, the diversity in trend of fractional crystallization with different degrees of iron enrichment is shown in the AFM diagram (Fig. 6-9). The difference in these trends was interpreted by some as due to the difference in stage for the initiation of crystallization of iron oxide minerals (e.g., Kuno, 1968). In the high-iron concentration types, crystallization of iron oxides is delayed relative to mafic silicate minerals, so that iron is enriched in the residual liquids, whereas in the moderate to low-iron concentration types, iron

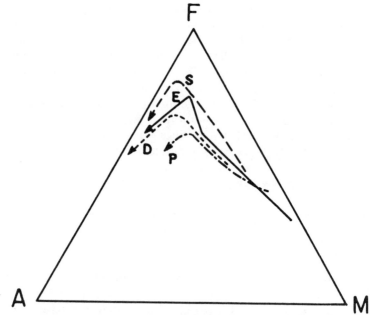

Figure 6-9. AFM diagram showing the trends of fractional crystallization with different degrees of iron enrichment. *S*, Skaergaard intrusion (Wager and Deer, 1939; Wager and Brown, 1967); *E*, Elephant's Head dike and New Amalfi sheet (Walker and Poldervaart, 1949); *D*, Dillsburg sill (Hotz, 1953); and *P*, Palisade diabase sill (Walker, 1940).

oxides crystallize early relative to mafic silicates, and the residual liquids are depleted in iron. The trends of fractional crystallization of tholeiitic series differ even in the very early stages when the crystallization of iron oxides is not extensive, however, as indicated by the fact that no, or very rare, phenocrysts of iron oxides are found in tholeiitic basalts.

The simplest explanation based on Bowen's suggestion is that the difference in degree of iron enrichment is due to the difference in degree of fractionation of basaltic magma. The high iron-enrichment trends such as those of the Skaergaard intrusion and the New Amalfi, South Africa, sheet and the associated Elephant's Head dike are caused by high fractionation (low reaction of mafic silicates with liquid), whereas the moderate to low-iron enrichment trends such as those of the Palisade diabase sill and pigeonitic rock series in the Izu-Hakone region, Huzi and in west Scotland are caused by low fractionation (high reaction) (Fig. 6-9). There is a complete gradation from the high iron-enrichment trend to the low iron-enrichment trend within the tholeiitic series that may depend wholly on the degree of fractionation.

The differences between the tholeiitic series and the calc-alkalic series, however, cannot be explained by the difference in the degree of fractionation. Even by low fractionation of mafic silicates, the residual liquids must be enriched in iron, and, therefore, the calc-alkalic trend cannot be produced by the fractionation of only mafic silicates. All the major mafic silicates including amphibole have lower Fe/Mg than the coexisting liquids. Thus, Bowen's original suggestion cannot be generalized to explain the difference in the fractionation trends between the tholeiitic and the calc-alkali series. Other factors such as separation of iron oxides must be taken into consideration to account for the calc-alkali trends.

CRYSTALLIZATION OF PYROXENE AND PLAGIOCLASE

Bowen (1928, pp. 64–69) argued, on the basis of observations of natural rocks as well as of the relations in simple silicate systems, that plagioclase and pyroxene separate at nearly the same time in the early stages of crystallization of basaltic magma. This generalization has been illustrated by the results of melting experiments on ten different basaltic rocks at 1 atm by Yoder and Tilley (1962, p. 382), which are given in Fig. 6-10. In eight basaltic rocks plagioclase and clinopyroxene (with orthopyroxene in two cases) begin to crystallize within only 20°. Two exceptions to the generalization include a high-alumina basalt from Medicine Lake, California, and an olivine nephelinite from Oahu, Hawaii (the latter specimen does not include plagioclase and should be excluded from consideration). It is evident that Bowen's generalization is applicable to the crystallization of most basaltic magmas at or near 1 atm. The results given in Fig. 6-10 are for equilibrium crystallization, and if incomplete reaction between olivine and liquid occurs, the temperature of crystallization of plagioclase may be slightly lowered relative to that of pyroxene. It is quite unlikely, however, that this minor variation significantly affects Bowen's conclusion.

A new problem is raised, however, by the "exceptional" Medicine Lake high-alumina basalt. This basalt (18.28 wt.% Al_2O_3) is non-porphyritic, and therefore, the high Al_2O_3 content (and hence the normative plagioclase component) would not be due to plagioclase accumulation. A high-alumina basalt from Parícutin also shows a similar crystallization sequence: plagioclase begins to crystallize at 1208°C, and clinopyroxene and orthopyroxene both begin to crystallize at 1145°C

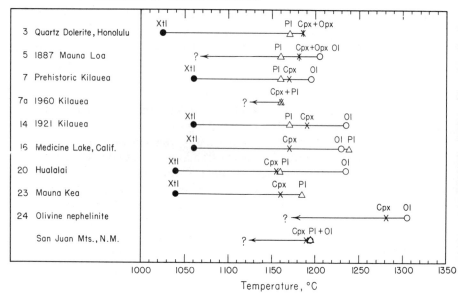

Figure 6-10. Experimentally determined crystallization sequence of minerals at 1 atm for ten different basaltic rocks (Yoder and Tilley, 1962).

(Tilley *et al.*, 1966). A plagioclase tholeiite (T-87) from the Mid-Atlantic Ridge that contains only 0.7 volume % plagioclase phenocryst (Shido *et al.*, 1971) exhibits a crystallization sequence and temperature of plagioclase and clinopyroxene crystallization similar to those of the Parícutin high-alumina basalt (Tilley *et al.*, 1972). Calc-alkalic andesite often exhibits an unusually high temperature of crystallization of plagioclase at 1 atm (Yoder, 1969). It would appear from these experimental results that crystallization of plagioclase at temperatures considerably higher than that of pyroxene at 1 atm is not exceptional, but may be common for some varieties of basalts.

As mentioned above, the high crystallization temperature of plagioclase for the Medicine Lake basalt is probably not due to plagioclase accumulation. Such a high crystallization temperature of plagioclase at 1 atm for high-alumina basalts and calc-alkalic andesites can be best explained by the presence of H_2O at the site of origin or separation of magma (Yoder, 1969). In the presence of H_2O, the cotectic boundary between plagioclase and pyroxene shifts toward plagioclase with increasing pressure, as shown in the system diopside–anorthite–H_2O (Fig. 6-11). The magma saturated with both plagioclase and clinopyroxene at 10 kbar P_{H_2O} is, therefore, much more enriched in the plagioclase component than that at 1 atm. If the composition of such a hydrous magma is treated at 1 atm in the absence of H_2O, plagioclase will crystallize alone at temperatures much higher than those of the plagioclase-pyroxene boundary.

FORMATION OF BIOTITE

Bowen (1928, pp. 83–85) described the formation of biotite and quartz in the late stage of fractional crystallization of basaltic magma as due to the enrichment of the residual liquid in alkali feldspar component as well as in volatile constituents such as H_2O, CO_2, S, and Cl. He considered the extraction of biotite to be one of the processes whereby free quartz may be developed during the crystallization of basaltic magma. Enrichment of the alkali feldspar component in the residual liquids is observed in all the differentiated intrusive bodies, and is a general trend of fractional crystallization of basaltic magma. Biotite does not always crystallize in

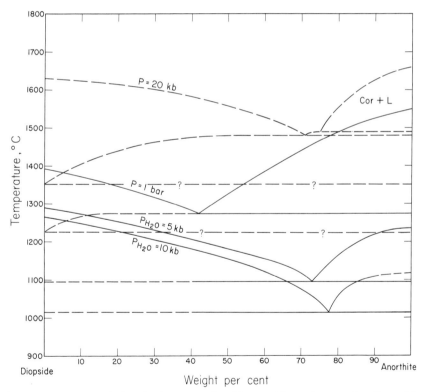

Figure 6-11. Phase relations in the system diopside–anorthite at 1 atm and at P_{H_2O} of 5 and 10 kbar, and an estimate at 20 kbar (Yoder, 1969).

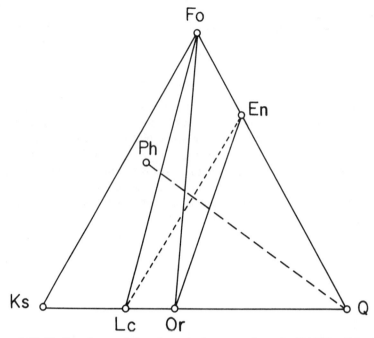

Figure 6-12. Tie lines for coexisting phases in the system forsterite–KAlSiO$_4$–SiO$_2$. The En–Lc join is stable at temperatures higher than about 1000°C at 1 atm, and the Ph–SiO$_2$ join is stable at temperatures lower than about 820°C (Luth, 1967).

the late differentiates of basaltic magmas, however, and it is observed mostly in the late differentiates of the calc-alkalic rock series and in some differentiates of the alkalic rock series. It would appear that volatile constituents, especially H$_2$O, do not concentrate significantly in the residual liquids of tholeiitic magmas that are originally low in volatiles, whereas H$_2$O concentrates in the residual liquids of calc-alkalic and alkalic magmas that are relatively high in volatiles. Furthermore, the formation of both biotite and quartz does not depend on the concentration of the orthoclase component and volatile constituents in the residual magmas, but depends on the stabilization of the join biotite-quartz, which breaks the joins olivine–leucite–H$_2$O, olivine–sanidine(Or)–H$_2$O[4] and hypersthene–sanidine–H$_2$O, as illustrated in Fig. 6-12. If the join biotite-quartz is stable and no K-bearing amphibole is stable, both biotite and quartz

---

[4] The olivine–sanidine assemblage is found in some alkalic rocks; however, in the iron-free system, the join forsterite–sanidine is not stable relative to the join enstatite–leucite at temperatures above 1000°C (Luth, 1967).

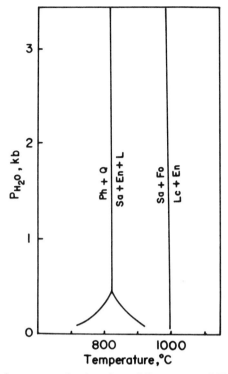

Figure 6-13. Univariant curves showing the stability ranges of Ph + Q and Sa + Fo assemblages (Luth, 1967).

form upon complete crystallization of basaltic magma even if the amount of the orthoclase component is very small.

The reaction

$$3MgSiO_3 + KAlSi_3O_8 + H_2O \rightleftharpoons KMg_3AlSi_3O_{10}(OH)_2 + 3SiO_2$$

has been suggested (Luth, 1967) and the inferred univariant curve of this reaction under $P_{H_2O} = P_{total}$ is shown in Fig. 6-13 (low-temperature limb of curve for PH + Q assemblage). On the lower-temperature side of the univariant curve, the phlogopite-quartz assemblage is stable.

The curve is shifted by the addition of other components, especially iron, and by changes in the partial pressure of H_2O; hence it cannot be applied directly to determine the conditions of formation of the biotite–quartz assemblage. If the partial pressure of H_2O in the system is low, the univariant curve is shifted to the low-temperature side and the formation of biotite and quartz is limited to the lower-temperature range. The presence of iron would also lower the upper stability limit of the biotite-quartz assemblage.

In the late differentiates of tholeiitic magma, iron-rich, Ca-poor pyroxene coexists with K-feldspar or anorthoclase (e.g., in dacite of the pigeonitic rock series in Hakone described by Kuno, 1950), indicating that the temperature and the partial pressure of H_2O during the crystallization of tholeiitic magma are such that biotite and quartz are unstable. On the other hand, during the crystallization of calc-alkalic magmas the partial pressure of H_2O is high and the crystallization temperature is low, resulting in the stabilization of the biotite-quartz assemblage.

In summary, the residual liquid is enriched in alkali feldspar components and silica in the late stage of fractional crystallization of basaltic magmas, even in the tholeiitic series; however, formation of the biotite and quartz assemblage is not a necessary consequence. That assemblage depends on the partial pressure of H_2O and the temperature during crystallization of the magma.

SEPARATION OF HORNBLENDE

In criticizing the deductions of Asklund (1923), Bowen (1928, pp. 86-91) demonstrated that in the Stavsjö gabbro complex, quartz diorite can be derived from noritic gabbro by the subtraction of plagioclase, pyroxene, and hornblende. Furthermore, the formation of hornblende at some stage in the crystallization of the noritic gabbro would yield an increased amount of free quartz and, therefore, of possible quartzose differentiate. The argument is supported by evidence from a recent study of the Guadalupe igneous complex, California, in which Best and Mercy (1967) showed that a series of gabbro, diorite, granodiorite, and granite was produced by separation of amphibole.

Yoder and Tilley (1962) made a series of experiments on four representative basalts at pressures up to 10 kbar under H_2O-saturated conditions. They found that the highest crystallization temperature of amphibole rises with increasing pressure in the presence of an excess of H_2O and may become the primary phase at pressures above 10.5 to 12.6 kbar depending on the bulk composition. These results indicate that the fractionation of amphibole may become significant at deeper levels under hydrous conditions. Later work shows that crystallization of amphibole is limited to pressures less than 30 kbar, as summarized by Yoder (1976. p. 23).

Application of amphibole fractionation to the genesis of calc-alkalic andesite magma has been made by several investigators (Tilley, 1950; T. H. Green and Ringwood, 1968; Holloway and Burnham 1972; Cawthorn and O'Hara, 1976; Allen, Boettcher and Marland, 1975).[5] Detailed discussion of this problem has recently been given by Cawthorn and O'Hara (1976), who reviewed the phase equilibria in the system $Na_2O–CaO–MgO–Al_2O_3–SiO_2–H_2O$ and the natural materials at pressures between 2 and 10 kbar. They showed that amphibole as well as olivine and clinopyroxene is the near-liquidus phase in basic and intermediate magma compositions and that the liquid approaches an andesitic composition through fractional crystallization. They concluded from these results that calc-alkalic andesite is produced from hydrous basaltic magmas by fractionation of amphibole. One problem that should be resolved before accepting this hypothesis concerns iron-enrichment during the fractional crystallization of amphibole. Holloway and Burnham (1972) have made controlled experiments on tholeiitic basalt and showed that amphibole is stable up to about 970°C at 2 kbar and 1060°C at 8 kbar under the P_{O_2} of the Ni–NiO buffer. They also showed that all the amphiboles analyzed have much lower Fe/Mg than the coexisting glass,[6] indicating that the liquid becomes iron-rich by separation of amphibole.[7] Separation of amphibole alone is, therefore, not effective in generating a calc-alkalic trend with no iron enrichment.

---

[5] T. H. Green and Ringwood (1968) have suggested, on the basis of hydrous experiments on natural basaltic and andesitic compositions, that calc-alkalic andesite may be derived by fractional crystallization of hydrous basalt magma involving separation of amphibole. They showed that amphibole crystallizing near the liquidus temperature of 9–10 kbar is subsilicic and has higher Fe/Mg than the coexisting pyroxenes. The experiments on which the discussion is based, however, are not well controlled (i.e., with respect to loss of iron to platinum capsules and loss of water from unsealed capsules).

[6] The composition of glass may not necessarily represent that of an equilibrium liquid because of the effect of quench crystals, particularly when the amount of liquid is small (e.g., Cawthorn and O'Hara, 1976). In Holloway and Burnham's experiments the amount of glass coexisting with amphibole was up to 35% and the quench problem may not be serious. The discrepancy in Fe-Mg partitioning for olivine and liquid between the data of Holloway and Burnham and those of Roeder and Emslie (1970) obtained at 1 atm and at temperatures between 1150° and 1300°C for basaltic compositions would probably be due to the difference in temperature, pressure, and chemical composition of liquid (particularly H_2O).

[7] Ringwood (1975) picked one of the analyses of amphiboles obtained by Holloway and Burnham and compared its Fe/Mg (0.69) with the Fe/Mg of the original basalt (0.64), believing that separation of such amphibole would produce the calc-alkalic trend (no iron enrichment). Comparison of the Fe/Mg of amphibole with that of the original basalt, however, is not meaningful. The Fe/Mg of the coexisting amphibole and liquid must be compared.

FRACTIONAL CRYSTALLIZATION OF BASALTIC MAGMA AT HIGH PRESSURES

Since 1960, large amounts of data on equilibrium phase relations at high pressures have been obtained that are relevant to defining courses of fractional crystallization of basaltic magma at high pressures. Yoder and Tilley (1962) were the first to demonstrate differences between the crystallization trends of basaltic magma at low pressures and high pressures (Fig. 6-14a and b). At low pressures, thermal divides involving olivine and plagioclase constitute a barrier to hypersthene-normative (tholeiitic) liquids fractionating toward the nepheline-normative (alkali basaltic) compositions, or, nepheline-normative liquids fractionating toward the hypersthene-normative compositions. At high pressures, however, the low-pressure thermal divides become unstable and are replaced by thermal divides involving pyroxenes (and garnet), which control the fractionation trends. For example, in the system $Mg_2SiO_4-NaAlSiO_4-SiO_2$, the join enstatite–jadeite becomes stable in place of the join forsterite–albite at pressures above about 30 kbar, and hypersthene-normative liquids can fractionate into the nepheline-normative region (Fig. 6-14b). In fact, this change can even occur at lower pressures because there is a reaction relation of forsterite + albite $\rightleftharpoons$ enstatite + liquid below 30 kbar (Kushiro, 1968). Nevertheless, the studies by Yoder and Tilley (1962) indicated that an olivine- and hypersthene-normative (olivine tholeiitic) magma that yields quartz-normative residual magmas by fractional crystallization at low pressures can produce nepheline-normative magmas at high pressures in the absence of H_2O and CO_2.

O'Hara (1965) argued that most of the extrusive basalts do not represent the primary basalt magmas but are the residual liquids of advanced crystal fractionation. This view is supported by Yoder (1976, pp. 145–146). O'Hara emphasized the continuous fractional crystallization of magmas during their ascent from the depth of magma generation to the surface; that is, the fractional crystallization of magma takes place under polybaric conditions. He also considered the possibility that the ascent of magma is "interrupted" at different depths, and described the possible trends of fractional crystallization in such cases. If magmas are interrupted in their ascent at high pressures, the compositional variation of residual liquid is controlled by the fractional crystallization of eclogite (O'Hara and Yoder, 1967), with the residual liquids trending toward the compositions of potassic mafic lavas, kimberlite, and carbonatite. If magmas are interrupted in their ascent at intermediate pressures, the residual liquids formed by fractional crystallization of spinel–lherzolite and spinel–wehrlite assemblages are strongly silica-undersaturated nephelinitic

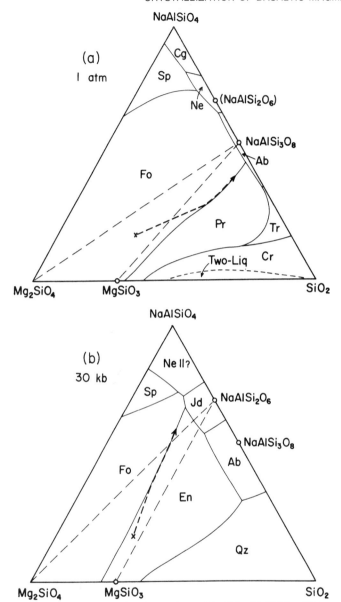

Figure 6-14. Liquidus relations in the system Mg_2SiO_4–$NaAlSiO_4$–SiO_2 at 1 atm
(a) (Schairer and Yoder, 1961) and those at 30 kbar (b) constructed on the basis
of data by Kushiro (1968) and Bell and Roseboom (1969). Dashed lines indicate
crystallization trends of an olivine tholeiitic composition at 1 atm and 30 kbar.
Albite is unstable below 1200°C at 30 kbar.

compositions. If magmas are interrupted in their ascent at low pressures, the residual liquids through fractional crystallization of the plagioclase–peridotite assemblage range from alkali olivine basaltic through high-alumina basaltic to quartz tholeiitic with decreasing depth. This study first suggested the role of fractional crystallization at various depths on the generation of magmas of wide compositional ranges.

The fractional crystallization of basaltic magmas at high pressures has also been studied in natural rock systems. (e.g., Yoder and Tilley, 1962; D. H. Green and Ringwood, 1964, 1967; Ito and Kennedy, 1968; Thompson, 1972). D. H. Green and Ringwood (1964, 1967) made a series of high-pressure experiments on four different synthetic compositions simulating analyzed natural basalts and determined the liquidus phases at various pressures for each one. The compositions selected were based on those of basaltic lavas that were erupted rapidly and, therefore, considered least fractionated during their ascent (i.e., olivine tholeiite, olivine basalt, picrite, and alkali basalt). On the basis of these data Green and Ringwood proposed a scheme for genesis of basaltic magmas in which fractional crystallization of olivine tholeiite magma (about 22% normative olivine) yields quartz tholeiite magma at 1 atm, high-alumina olivine tholeiite magma at about 9 kbar, and nepheline-normative alkali olivine basaltic magma at 13–18 kbar.[8] Their experiments on natural basaltic compositions also demonstrated that the trends of fractional crystallization of basaltic magma at high pressures are very different from those at 1 atm.

The hypotheses and schemes proposed in the above studies will, no doubt, undergo further changes; however, these studies all show that the effects of pressure on magmas formed solely by fractional crystallization may give rise to much more compositional diversity than was thought possible by Bowen fifty years ago. In addition to the effects of pressure, volatile components such as H_2O and CO_2 further widen the possible compositional range of the derivative magmas formed by fractional crystallization. Recent experimental studies have demonstrated that the composition of liquid coexisting with both olivine and pyroxene changes drastically with changes in volatile components and pressure (e.g., Eggler, 1975, 1976; Kushiro, 1972, 1975; Mysen and Boettcher, 1975; Wyllie and Huang, 1976). For example, liquid coexisting with forsterite, enstatite, and diopsidic pyroxene varies from a silica-rich andesitic composition

---

[8] This fractional crystallization scheme, particularly orthopyroxene fractionation, was later criticized (O'Hara, 1968; Kushiro, 1969) on the grounds that the primary phase volume of olivine expands relative to that of orthopyroxene with decreasing pressure. Liquids formed by partial melting of peridotitic upper mantle must be within the olivine volume at lower pressures and cannot fractionate by separation of orthopyroxene alone during their ascent, except at pressures less than about 5 kbar.

in the presence of a pure H_2O or H_2O-rich gas phase to a silica-poor kimberlitic, or even carbonatitic composition, in the presence of a CO_2 or CO_2-rich gas phase. Figure 6-15 illustrates the shift of the forsterite-enstatite liquidus boundary in the presence of excess CO_2 in the system $CaO-MgO-SiO_2$. Liquids with silica contents between the above two extreme compositions can be produced by changing the ratio of CO_2 to H_2O in the gas phase. Magmas ranging widely in composition can, therefore, be produced by fractional crystallization involving separation of olivine and pyroxene in the presence of gas phases of different CO_2/H_2O.

Fractional crystallization of magma may be more efficient at high pressures than at low pressures. The viscosities of basaltic and andesitic melts have recently been measured at high pressures (Kushiro, Yoder, and Mysen, 1976), and it was found that the viscosity of melt of Kilauea 1921 olivine tholeiite, for example, decreases from about 107 poise at 1 atm to about 8 poise at 30 kbar along its liquidus (Fig. 6-16). This large decrease reflects the tendency of viscosity to decrease both with increasing temperature at constant pressure and with increasing pressure at constant temperature, although the effect of pressure is not large. Kushiro *et al.* (1976) also determined the viscosity of Crater Lake calc-alkalic andesitic melt at high pressures under anhydrous conditions, and demonstrated that the viscosity decreases from about 5000 poise at 1 atm to about 900

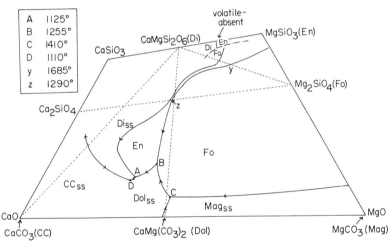

Figure 6-15. Liquidus boundaries on the CO_2 saturation surface of the system $CaO-MgO-SiO_2-CO_2$ at 30 kbar (solid curves) determined by Eggler (1976). The Fo–En liquidus boundary on the H_2O saturation surface of the system $CaO-MgO-SiO_2-H_2O$ at 20 kbar lies above the join Di–En, as determined by Kushiro (1969).

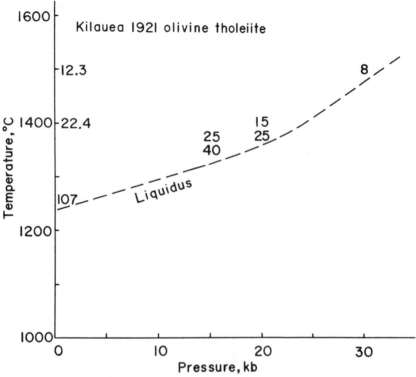

Figure 6-16. Viscosity of melt of Kilauea 1921 olivine tholeiite (Kushiro *et al.*, 1976). Numbers are viscosities in poise. Viscosities at 1 atm have been determined by T. Murase and A. R. McBirney (unpublished data).

poise at 20 kbar along the liquidus. Addition of 4 wt.% H_2O reduces the viscosity of the same melt by a factor of about twenty at 15 kbar.

Because of lower viscosity of basaltic melt at high pressures, separation of crystals from the melt should be more efficient. On the other hand, density of melt increases with increasing pressure and this effect reduces the efficiency of crystal separation (although it is much smaller than the effect of the viscosity decrease). For example, a pyroxene crystal, 2 mm in diameter, sinks in a basaltic melt with a velocity of 0.24 m/hr at 1 atm, whereas the same pyroxene crystal sinks with a velocity of 2.5 m/hr at 30 kbar. A garnet crystal, 2 mm in diameter, sinks with a velocity 4.0 m/hr in a basaltic melt at 30 kbar.

The velocity of ascent of magma may become greater with decreasing the viscosity. If so, the chance of fractional crystallization becomes smaller with increasing depth. Alkali basalts containing peridotite nodules probably had the least chance for fractional crystallization. The presence of peridotite nodules indicates that fractional crystallization due to separation of crystals was not effective in these alkali basalt magmas during their ascent, at least after inclusion of the nodules. The velocity of ascent of magma, however, also depends on other factors such as the physical properties of surrounding rocks, and the velocity of ascent of magma may not be linearly related to its viscosity.

CONCLUDING REMARKS

Significant revisions or modifications have been made to Bowen's scheme of fractional crystallization of basaltic magma, due to (1) the recognition of different magma types of worldwide significance, (2) the consequences of detailed studies of many differentiated layered intrusions and volcanic rock series in various localities, and (3) the results of extensive experimental studies at high pressures and under controlled fugacities of oxygen, H_2O, and CO_2. The concept of fractional crystallization, however, has not been changed and is, no doubt, of fundamental importance in the study of the petrogenesis of igneous rocks. The process of fractional crystallization can actually be seen in many differentiated magma bodies. Even in lava flows, small-scale fractional crystallization can be observed as zonal textures of minerals and as segregation veins and fractionated liquid in the interstices of the groundmass. The problem is whether or not fractional crystallization of basaltic magma is essential to the genesis of major magmas. The diversity of igneous rocks in both chemical and mineralogical composition can certainly be explained to a considerable extent by the process of fractional crystallization of basaltic magma. The diversity in basaltic magma, however, cannot be explained by near-surface fractional crystallization (Yoder and Tilley, 1962). The diversity of basaltic magma must, therefore, be generated for the most part at high pressures. The processes suggested for the generation of different basaltic magma types are partial melting or fractional fusion (Presnall, 1969) of the upper mantle materials under different physicochemical conditions, and fractional crystallization at different depths of a "more primitive magma" that is rarely seen on the surface. Recent experimental studies in various silicate systems with H_2O and CO_2 at high pressures indicate that liquids coexisting with olivine and pyroxene vary from silica-saturated compositions to very silica-poor carbonatitic compositions with changing H_2O/CO_2 of the gas phase. Magmas of a wide range in composition can, therefore, be produced by partial melting of the upper mantle materials (e.g., spinel- or garnet–lherzolite) in the presence of gas phases having different CO_2/H_2O. In addition, the compositional range of liquid is widened as pressure changes. The compositional range of magma that would be produced by partial melting is probably as wide as that produced by fractional crystallization. Unfortunately, the process of partial melting is rarely observed directly, except as shown in small scale in some peridotites such as those in Ronda, South Spain (Dickey, 1970), in some high grade metamorphic rocks, and in some xenoliths.

Although the writer agrees that fractional crystallization is an important igneous process, he also assigns a fundamental role to partial melting or fractional fusion under different physicochemical conditions in the generation of diversity in chemical composition of magma. The composition of magmas formed by partial melting may or may not change by subsequent fractional crystallization during their ascent toward the surface.

Acknowledgments The author wishes to thank Drs. F. Chayes, T. N. Irvine, E. F. Osborn, D. Velde, and H. S. Yoder, Jr. for their reviews of the manuscript.

References

Allen, J. C., A. L. Boettcher, and G. Marland, Amphiboles in andesite and basalt: I. Stability as a function of $P-T-f_{O_2}$, *Am. Mineral. 60*, 1069–1085, 1975.

Asklund, B., Granites and associated basic rocks of the Stavsjö area, *Sveriges Geol. Undersök. Arsbok, 17, No. 6*, 1923.

Bailey, E. B., C. T. Clough, W. B. Wright, J. E. Richey, and G. V. Wilson, Tertiary and post-tertiary geology of Mull, Loch Aline, and Oban. *Mem. Geol. Survey Scot.*, 445 pp., 1924.

Bell, P. M., and E. H. Roseboom, Melting relationships of jadeite and albite to 45 kilobars with comments on melting diagrams of binary systems at high pressures, *Min. Soc. Amer.*, Special Paper No. 2, 151–161, 1969.

Best, M. G., and E. L. P. Mercy, Composition and crystallization of mafic minerals in the Guadalupe igneous complex, California, *Am. Mineral., 52*, 436–474, 1967.

Biggar, G. M., Phase equilibria studies of the chilled margins of some layered intrusions, *Contr. Mineral. Petrol., 46*, 159–167, 1974.

Bowen, N. L., *The Evolution of the Igneous Rocks*, Princeton University Press, Princeton, New Jersey, 334 pp., 1928.

Boyd, F. R., and J. L. England, Melting of silicates at high pressures, *Carnegie Institution of Washington Yearbook, 60*, 113–125, 1961.

Boyd, F. R., J. L. England, and B. T. C. Davis, Effects of pressure on the melting and polymorphism of enstatite, $MgSiO_3$. *J. Geophys. Res.*, 69, 2101–2109, 1964.

Cawthorn, R. G., and M. J. O'Hara, Amphibole fractionation in calc-alkaline magma genesis, *Am. J. Sci.*, *276*, 309–329, 1976.

Dickey, J. S., Partial fusion in Alpine-type peridotites: Serrania de la Ronda and other examples, *Min. Soc. Amer.*, Special Paper No. 3, 33–49, 1970.

Eggler, D. H., CO_2 as a volatile component of the mantle: The system Mg_2SiO_4–SiO_2–H_2O–CO_2, *Phys. Chem Earth*, *9*, 869–881, 1975.

———, Does CO_2 cause partial melting in the low-velocity layer of the mantle?, *Geology*, *4*, 69–72, 1976.

Eggler, D. H., and C. W. Burnham, Crystallization and fractionation trends in the system andesite–H_2O–CO_2–O_2 at pressures to 10 kb, *Geol. Soc. Am. Bull.*, *84*, 2517–2532, 1973.

Fenner, C. N., The crystallization of basalts, *Am. J. Sci.*, 5th ser., *18*, 225–253, 1929.

Green, D. H., and A. E. Ringwood, Fractionation of basalt magmas at high pressures, *Nature*, *201*, 1276–1279, 1964.

Green, D. H., and A. E. Ringwood, The genesis of basaltic magmas, *Contr. Mineral. Petrol.*, *15*, 103–190, 1967.

Green, T. H., and A. E. Ringwood, Genesis of calc-alkaline igneous rock suite, *Contr. Mineral. Petrol.*, *18*, 105–162, 1968.

Hess, H. H., Stillwater igneous complex, Montana: A quantitative mineralogical study, *Geol. Soc. Am. Mem. 80*, 230 pp., 1960.

Holloway, J. R., and C. W. Burnham, Melting relations of basalt with equilibrium water pressure less than total pressure, *J. Petrol.*, *13*, 1–29, 1972.

Hotz, P. E., Petrology of granophyre in diabase near Dillsburg, Pennsylvania, *Geol. Soc. Am. Bull.*, *64*, 675–704, 1953.

Irvine, T. N., Metastable liquid immiscibility and MgO–FeO–SiO_2 fractionation patterns in the system Mg_2SiO_4–Fe_2SiO_4–$CaAl_2Si_2O_8$–$KAlSi_3O_8$–SiO_2, *Carnegie Institution of Washington Yearbook*, 75, 597–611, 1976.

Ito, K. and G. C. Kennedy, Melting and phase relations in the plane tholeiite–lherzolite–nepheline basanite to 40 kilobars with geological implications, *Contrib. Min. Petrol.*, *19*, 177–211, 1968.

Kennedy, W. Q., Trends of differentiation in basaltic magmas, *Am. J. Sci.*, 5th ser., *25*, 239–256, 1933.

Kuno, H., Petrology of Hakone Volcano and the adjacent areas, Japan, *Geol. Soc. Am. Bull. 61*, 957–1020, 1950.

———, Origin of Cenozoic petrographic provinces of Japan and surrounding areas, *Bull. Volcanol.*, ser. II, *20*, 37–76, 1959.

———, High-alumina basalt, *J. Petrol.*, *1*, 121–145, 1960.

———, Fractionation trends of basalt magmas in lava flows, *J. Petrol.*, *6*, 302–321, 1965.

———, Differentiation of basalt magmas, in *Basalts: The Poldervaart Treatise on Rocks of Basaltic Composition*, H. H. Hess and A. Poldervaart, editors, Interscience, John Wiley & Sons, New York, 623–688, 1968.

Kuno, H., K. Yamasaki, C. Iida, and K. Nagashima, Differentiation of Hawaiian magmas, *Japn. J. Geol. Geogr.*, *28*, 179–218, 1957.

Kushiro, I., Compositions of magmas formed by partial zone melting in the earth's upper mantle, *J. Geophys. Res.*, *73*, 619–634, 1968.

————, Discussion of paper "The origin of basaltic and nephelinitic magmas in the earth's mantle," *Tectonophys.*, *7*, 427–436, 1969.

————, Effect of water on the composition of magmas formed at high pressures, *J. Petrol.*, *13*, 311–334, 1972.

————, On the nature of silicate melt and its significance in magma genesis: regularities in the shift of the liquidus boundaries involving olivine, pyroxene, and silica minerals, *Am. J. Sci.*, *275*, 411–431, 1975.

Kushiro, I., and H. S. Yoder, Formation of eclogite from garnet lherzolite: Liquidus relations in a portion of the system $MgSiO_3$–$CaSiO_3$–Al_2O_3 at high pressures, *Carnegie Institution of Washington Yearbook*, *73*, 266–269, 1974.

Kushiro, I., H. S. Yoder, and B. O. Mysen, Viscosities of basalt and andesite melts at high pressures, *J. Geophys. Res.*, *81*, 6351–6356, 1976.

Kushiro, I., H. S. Yoder, and M. Nishikawa, Effect of water on the melting of enstatite, *Geol. Soc. Am. Bull.*, *79*, 1685–1692, 1968.

Luth, W. C., Studies in the system $KAlSiO_4$–Mg_2SiO_4–SiO_2–H_2O. I, Inferred phase relation and petrologic applications, *J. Petrol.*, *8*, 372–416, 1967.

Muan, A. and E. F. Osborn, Phase equilibria at liquidus temperatures in the system MgO–FeO–Fe_2O_3–SiO_2, *J. Am. Ceram. Soc.*, *39*, 121–140, 1956.

Muir, I. D., Crystallization of pyroxenes in an iron-rich diabase from Minnesota, *Mineral. Mag.*, *30*, 376–388, 1954.

Mysen, B. O., and A. L. Boettcher, Melting of a hydrous mantle: II, Geochemistry of crystals and liquids formed by anatexis of mantle periodotite at high pressures and high temperatures as a function of controlled activities of water, hydrogen and carbon dioxide, *J. Petrol.*, *16*, 549–500, 1975.

Nockolds, S. R., Average chemical compositions of some igneous rocks, *Geol. Soc. Am. Bull*, *65*, 1007–1032, 1954.

O'Hara, M. J., Primary magmas and origin of basalts, *Scot. J. Geol.*, *1*, 19–40, 1965.

————, The bearing of phase equilibria studies in synthetic and natural systems on the origin and evolution of basic and ultrabasic rocks, *Earth Sci. Rev.*, *4*, 69–133, 1968.

O'Hara, M. J., and H. S. Yoder, Formation and fractionation of basic magmas at high pressures, *Scot. J. Geol.*, *3*, 67–117, 1967.

Osborn, E. F., Role of oxygen pressure in the crystallization and differentiation of basaltic magma, *Am. J. Sci.*, *257*, 609–647, 1959.

————, Genetic significance of V and Ni content of andesites: Comments on a paper by Taylor, Kaye, White, Duncan and Ewart, *Geochim. Cosmochim. Acta*, *33*, 1553–1554, 1969.

————, Origin of calc-alkali magma series of Santorini volcano type in the light of recent experimental phase equilibrium studies, *Proceedings of the International Congress on Thermal Waters, Geothermal Energy and Volcanism of the Mediterranean Area, Athens, Greece, October 1976*, *3*, 154–167, 1976.

Poldervaart, A., The petrology of the Elephant's Head dike and New Amalfi sheet (Mataliele), *Roy. Soc. S. Africa, Trans.*, *30*, pt. 2, 85–120, 1944.

Presnall, D. C., Geometrical analysis of partial fusion, *Am. J. Sci.*, *267*, 1178–1194, 1969.

Ringwood, A. E., *Composition and Petrology of the Earth's Mantle*, McGraw-Hill, 618 pp., 1975.

Shido, F., A. Miyashiro, and M. Ewing, Crystallization of abyssal tholeiites, *Contr. Mineral. Petrol.*, *31*, 251–266, 1971.

Schairer, J. F., and H. S. Yoder, Crystallization in the system nepheline–forsterite–silica at one atmosphere pressure, *Carnegie Institution of Washington Yearbook*, *60*, 141–144, 1961.

Taylor, S. R., M. Kaye, A. J. R. White, A. R. Duncan, and E. Ewart, Genetic significance of Co, Cr, Ni, Sc and V content of andesites. *Geochim. Cosmochim. Acta*, 33, 275–286, 1969.

Thompson, R. N., Melting behavior of two Snake River lavas at pressures up to 35 kb, *Carnegie Institution of Washington Yearbook*, *71*, 406–410, 1972.

Tilley, C. E., Some aspects of magmatic evolution, *Quart. J. Geol. Soc. London*, *106*, 37–61, 1950.

Tilley, C. E., R. N. Thompson, and P. A. Lovenbury, Melting relations of some oceanic basalts, *Geol. J.*, *8*, 59–64, 1972.

Tilley, C. E., H. S. Yoder, and J. F. Schairer, Melting relations of volcanic rock series, *Carnegie Institution of Washington Yearbook*, *65*, 260–268, 1966.

Wager, L. R. and G. M. Brown, *Layered Igneous Rocks*, Oliver & Boyd, Edinburgh, London, 588 pp., 1967.

Wager, L. R. and W. A. Deer, The petrology of the Skaergaard intrusion, Kangerdlugssuaq, East Greenland, *Meddl. om Grønland*, *105*, No. 4, pt. 3, 1–352, 1939.

Walker, F., Differentiation of the Palisade diabase, New Jersey, *Geol. Soc. Am. Bull.*, *51*, 1059–1105, 1940.

Walker, F. and A. Poldervaart, Karroo dolerites of Union of South Africa, *Geol. Soc. Am. Bull.*, *60*, 591–706, 1949.

Walker, K. R., The Palisades Sill, New Jersey—A reinvestigation, *Geol. Soc. Am.*, Special Paper 111, 178 pp., 1969.

Wyllie, P. J., and W-L. Huang, Carbonation and melting relations in the system $CaO-MgO-SiO_2-CO_2$ at mantle pressures with geophysical and petrological applications, *Contrib. Mineral. Petrol.*, *54*, 79–107, 1976.

Yoder, H. S., Jr., Calcalkalic andesites: Experimental data bearing on the origin of their assumed characteristics, in *Proc. Andesite Conf.*, A. R. McBirney, editor, *State of Oregon, Dep. Geol. Mineral Ind. Publ.*, 77–89, 1969.

——, *Generation of Basaltic Magma*, National Academy of Sciences, 265 pp., 1976.

Yoder, H. S., Jr., and G. A. Chinner, Grossular–pyrope–water system at 10,000 bars, *Carnegie Institution of Washington Yearbook*, *59*, 78–81, 1960.

Yoder, H. S., Jr., and C. E. Tilley, Origin of basalt magmas: An experimental study of natural and synthetic rock systems, *J. Petrol.*, *3*, 342–532, 1962.

Chapter 7

THE LIQUID LINE OF DESCENT AND VARIATION DIAGRAMS

RAY E. WILCOX

U.S. Geological Survey, Denver, Colorado,

In pointing out the effectiveness of the variation diagram for illustrating chemical relationships among members of an igneous rock series, Bowen (1928) particularly emphasized it as a tool in exploring the kinds and quantities of mineral phases that might have been subtracted from an evolving magma in bringing about the observed relationships in an evolutionary series by fractional crystallization. The "liquid line of descent"[1] refers to the evolutionary trend in composition of the liquid phase of the magma that results from differential withdrawal of chemical constituents from the liquid during growth of crystals. This trend is apart from the changing *bulk* composition of the magma that results from fractional crystallization; that is, from differential movement of the growing crystals in respect to the liquid, or from assimilation of xenolithic material or admixture of yet another magma. In this chapter the variation diagram will be applied to the elucidation not only of the fundamentally important liquid lines of descent, but also to the changing bulk compositions of magmas, more directly accessible to the investigator through bulk analyses of members of the corresponding series of erupted rocks.

A recent major advance has been the development of the electron microprobe, by which compositions of groundmass and individual phenocryst phases of each member of the series may be determined. A further major step forward in technique has been the extension of the

© 1979 by Princeton University Press
The Evolution of the Igneous Rocks: Fiftieth Anniversary Perspectives
0-691-08223-5/79/0205-28$01.40/0 (cloth)
0-691-08224-3/79/0205-28$01.40/0 (paperback)
For copying information, see copyright page

graphical methods to those of their computer analogs, by which analysis of the relationships within a rock series can be carried out in exhaustive detail. In this modern approach, however, most of the fundamental precepts and strictures outlined by Bowen (1928) for the graphical method still hold, and the computer analysis must be built around a framework of the observed petrographic and spatial relationships of the rock series.

INFERRING MAGMA COMPOSITION FROM ROCK ANALYSIS

A glassy or fine-grained volcanic rock may be regarded as a quenched sample of magma, and its chemical analysis may be taken as a reasonable representation of the composition of the magma, neglecting possible loss of volatiles (see Chapter 8). On a larger scale, an assemblage of volcanic rocks of similar geologic age in a limited geographic area is commonly taken as representing a series of samplings from a common magma or magmatic system at successive stages of its evolution. Elucidation of the evolutionary development requires (1) field mapping, (2) petrographic analysis, and (3) chemical analysis of rock specimens selected on the basis of those results.

To reconstruct the liquid line of descent, certain restrictions are imposed on the choice of material for chemical analysis. The most important of these is that the material analyzed should be glassy, or sufficiently fine grained to offer assurance that its contained crystals have not moved differentially with respect to the liquid before quenching (Bowen, 1928, p. 94) If large crystals (phenocrysts) are present in the sample there is the possibility that they may have immigrated from another portion of the magma or that the proportion of their several mineral species may have changed owing to differential gravitative action, so that a whole-rock analysis would not represent the composition of any actual *liquid* in the evolutionary series. This simple precept, second nature to the experimental petrologist working with quenched samples in the laboratory, unfortunately has not been taken seriously in many studies of natural igneous rock series, to the detriment of the conclusions reached concerning the liquid line of descent.

These constraints were special handicaps when only wet chemical methods were available, for an appreciable amount of material was

---

[1] Also called the "path of crystallization" or simply the "liquid path."

necessary for rock analysis. The porphyritic rock sample had to be left out of consideration entirely unless the effect of the phenocrysts could be eliminated, either by physical removal before analysis or by recalculating the whole-rock analysis as phenocryst-free, using the data from modal analysis under the microscope. Neither approach was felicitous, for both were tedious and subject to error. This difficulty may now be avoided for some porphyritic volcanic rocks, at least for major chemical constituents, by use of the electron microprobe with which analysis can be made of specific small zones of the groundmass (glassy or finely crystalline) without interference from the phenocrysts. Furthermore, the individual phenocrysts can be analyzed, to provide a complete picture of the phase assemblage. Inclusions of glass commonly present in phenocrysts have been investigated by Anderson (1974) and Watson (1976) on the assumption that they represent trapped samples of previous stages in the liquid line of descent. Although the potential of the electron microprobe in this application appears great, inhomogeneities in the groundmass of some rocks and in the glass inclusions, as well as compositional zoning in the phenocrysts, introduce complications that cause difficulties but at the same time are informative.

Numerous inquiries into magmatic evolution have been based on analyses of the coarse-grained rocks of plutonic igneous series, on the assumption that the variations in bulk chemical composition represent liquid lines of descent. There is much doubt, however, that the bulk rock composition represents any magmatic liquid on the liquid line of descent or even a magma with suspended crystals on the bulk line of descent. With slow cooling there is every opportunity for differential gravitation of the one or several species of growing crystals out of their parental liquid, and for their proportions in the final rock to change radically from those present as they crystallized in the parent liquid. Increased opportunity also exists for late-stage alteration of already formed crystals and for differential transfer of more volatile constituents. As extreme cases, one may cite the cumulate rocks (Chapter 9) in which crystal sorting and accumulation has been the dominant process and the actual parental liquid of these crystals is not represented, even in the interstitial material.

VARIATION DIAGRAMS

A variation diagram is a useful device for investigating the origin of differences in chemical composition between members of a rock series presumed to represent stages in the evolution of a magma. The diagram may be a plot of constituents, groups of constituents, or ratios of constituents against each other, based on the chemical analyses of the individual rock samples of the series. In addition to providing a graphic portrayal of the chemistry of the rock series as a whole, such a diagram serves as a basis for testing processes of evolution (Bowen, 1928, p. 92; Wright, 1974). With the aid of the diagram one can simulate graphically the subtraction of mineral phases (crystal fractionation, liquid unmixing) or addition of rock materials or magmas (assimilation) to ascertain how closely these operations might be able to account for the observed chemical differences between members of the rock series.

Whereas a prominent advantage of the variation diagram is the graphic picture it provides of the chemical changes taking place in the magma, it has the disadvantage that only a small number of separate materials can be subtracted or added at one time in following the changing composition. Fortunately, the modern computer can handle a relatively large number of end members in a similar manner to the variation diagram and, when applied with due regard for petrographic and thermodynamic constraints, considerably extends the range of tests that can be made on igneous rock series. In this chapter, examples of the application of variation diagrams will be taken up first, followed by several examples of analogous computer applications.

REPRESENTATION OF ROCK SERIES

Simplest of the many types of chemical variation diagrams is a plot of the percentage of one oxide constituent as ordinate against another as abscissa for each analysis of the rock series. A best-fit line drawn through the resulting series of points, either by eye or by least-squares fitting, focuses attention on the general relationship between those two oxides.[2] The relations of several oxides, or for that matter, of all oxides of the analyses, can be shown similarly by selecting one oxide as abscissa and plotting each of the others against it. To avoid clutter, a different zero point may be used for the different ordinate oxides; alternatively, each oxide may be shown in a separate tier.

In the silica-variation diagram (Iddings, 1892; Harker, 1909), often called the Harker diagram, silica is taken as the abscissa. As an example, the well-known Katmai, Alaska, rock series (Fenner, 1926) is shown on Fig. 7-1. Plotting silica values on the abscissa has had a wide appeal, both because the large range of silica percentages in many rock series spreads

---

[2] Pearce (1968) points out, however, that one is actually dealing here with a ternary system, namely, the two selected oxides and the sum of the remaining oxides.

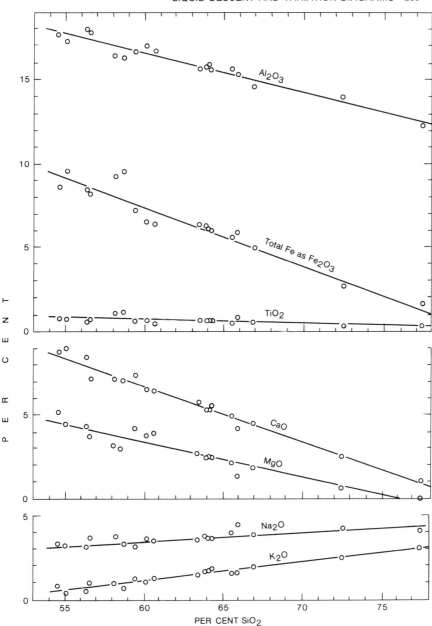

Figure 7-1. Silica-variation diagram of Katmai, Alaska, rock series (after Fenner, 1926).

the diagram laterally for ease of reading without necessity for scale change, and because it would seem to fit so naturally into the popular concept of progressive evolution of a relatively silica-poor parent magma towards a more silica-rich magma. It should not be necessary, however, to emphasize that mere plotting of a series of rock analyses on this type of diagram does not magically make the rocks a "differentiation series" nor the diagram a "differentiation diagram." Rather, the diagram is only a plot of analyses in order of increasing silica and, in itself, carries no compelling implications of evolutionary development.

It may be noted here that a negative slope of the best-fit line through the points of a particular oxide plotted against silica should not be taken directly as indicating a negative correlation of the absolute amounts of that oxide in the evolving magma with respect to silica. Chayes (1960, 1964) points out that, because silica is the dominant oxide and commonly has a wide range of percentages in igneous rock suites, an apparent negative tendency is introduced in the other oxide plots in the diagram, constructed as it is from a data array having constant sums (here 100 wt. percent). This negative tendency may, in some cases, produce a negative slope for a constituent on the silica variation diagram, whereas statistically the correlation between absolute amounts is actually slightly positive.[3] Whereas this property may handicap the use of silica-variation diagrams in statistical studies of interelement trends, it will not affect their main use in the present inquiry, namely, graphical testing of compositional changes in the magma resulting from subtraction or addition of specific mineral phases or phase assemblages.

In a rock series exhibiting only a narrow range of silica percentages, an oxide other than silica may serve better as abscissa. For basaltic suites, Wright (1974) recommends MgO with its relatively wide range, presumably due mainly to progressive fractional crystallization and addition of magnesium-rich olivine. (See also Murata and Richter, 1966; Wright and Peck, 1978.)

Other types of chemical variation diagrams designed to illustrate or to emphasize certain genetic aspects of igneous rock suites employ composite abscissas; that is, the abscissa values are derived by an arithmetic operation involving several constituents of the analyses. The abscissa of the

---

[3] In an attempt to circumvent this difficulty, Pearce (1968, 1970) plots "Al_2O_3" molar ratios of oxide constituents in a variation diagram with moles SiO_2/moles Al_2O_3 as abscissa, on the assumption that Al_2O_3 remains constant in absolute mass in the unit magma during during fractional crystallization. Subtraction of olivine (of whatever Fo-content) leads to a trend line of a characteristic slope; subtraction of pyroxene to another characteristic slope. Pearce does not discuss the common situation in which Al_2O_3 does not remain constant, for instance when an Al-bearing mineral such as plagioclase (or for that matter, aluminous pyroxene) is also involved in the fractionation.

Larsen diagram (Larsen, 1938; Nockolds, 1941, pp. 493 and 510), for instance, is the quantity $(1/3 \; SiO_2 + K_2O - FeO - MgO - CaO)$. The abscissa of the Thornton and Tuttle (1960) diagram is the "Differentiation Index," the sum of the felsic normative minerals, $Q + Or + Ab + Ne + Kp + Lc$. In some diagrams both abscissa and ordinates are composite, as for instance in the Niggli diagram (Burri, 1964), formerly called the "Differentiation Diagram." Here the four ordinates, *al*, *fm*, *c*, and *alk*, as well as the abscissa *si*, are two or more oxide constituents combined and recast to 100 percent. Although graphical addition and subtraction tests of the type to be used later in this chapter could be performed on these diagrams, they would require that the added or subtracted materials also be recast in terms of the composite abscissa (and ordinates in the case of the Niggli diagram). In addition to being costly in time, these operations may obscure the details of interelement relations and unnecessarily complicate their interpretation.

Mutual relations among three selected oxides or groups of oxides from the analyses of a series of closely related rocks may be shown in a triangular variation diagram. The values of the chosen trio are plotted after recasting to total 100 percent. Figure 7-2(a) is such a plot for MgO,

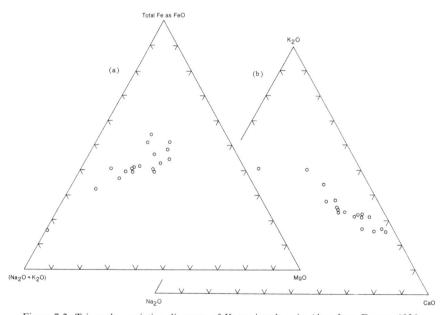

Figure 7-2. Triangular variation diagrams of Katmai rock series (data from Fenner, 1926, Table 1). (a) MgO–Total Fe as FeO–$(Na_2O + K_2O)$. (b) CaO–Na_2O–K_2O.

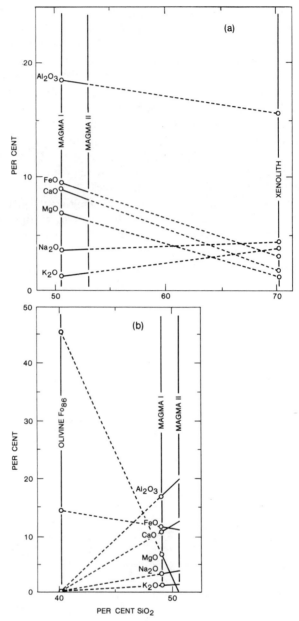

Figure 7-3. Simulated trends in silica-variation diagram. (a) Trends produced by addition of xenolith (Table 7-1). (b) Trends produced by subtraction of olivine, Fo_{86}.

total iron as FeO, and the sum of the alkalies, $Na_2O + K_2O$, for the Katmai series; Fig. 7-2(b) is of CaO, Na_2O, and K_2O for the same series. The triangular diagram serves to bring into focus groups of constituents considered to dominate the course of evolution. But, by excluding other major constituents of the analyses, it provides only a partial picture, and where analytical error is relatively large in respect to the total percentage of the selected constituents, the recast percentages may plot in a wide scatter or give misleading trends.

REPRESENTATION OF CHANGING MAGMA COMPOSITION

On a variation diagram, one can follow the changes of a magma's chemical composition brought about by addition or subtraction of material of known composition. Figure 7-3(a) shows an example of addition (assimilation) of a material by a magma. With the compositions of the two materials (Table 7-1) plotted on the silica variation diagrams, the percentage of a particular oxide is seen to change along a straight line from the oxide plot for the initial magma directly *towards* the oxide plot for the added material. At a given intermediate silica percentage of the resultant magma, the percentages of the other oxides may be read off as the ordinates directly above the abscissa value of SiO_2.

For subtraction of a substance, for instance by crystallization and removal of a mineral phase of fixed composition, the example of Fig. 7-3(b) shows the percentage of a particular oxide changing along a straight line from the oxide plot of the original magma directly *away* from the oxide plot of the subtracted material. When a given silica percentage has been reached in the residual magma, the percentages of the other oxides may likewise be read off as ordinates above the SiO_2 abscissa value. On the assumption that the composition of the subtracted phase does not change,

Table 7-1. Chemical compositions of materials involved in operations of Figure 7-3

	(a) Addition			(b) Subtraction		
	Magma I	+ Xenolith	= Magma II	Magma I	− Fo$_{86}$	= Magma II
SiO_2	50.6	70.2	53.1	49.2	40.1	50.6
Al_2O_3	18.6	15.7	18.2	17.0	—	19.6
Total Fe as FeO	9.5	3.0	8.7	11.7	14.1	11.3
MgO	7.4	1.3	6.6	6.6	45.8	0.6
CaO	9.0	1.8	8.1	10.8	—	12.5
Na_2O	3.6	4.3	3.7	3.4	—	3.9
K_2O	1.3	3.7	1.6	1.3	—	1.5
Total	100.0	100.0	100.0	100.0	100.0	100.0

it is obvious that the process could not continue beyond the point at which the most rapidly depleted oxide constituent [MgO in Fig. 7-3(b)] projects to zero.

The quantities of materials involved in these transactions are determined very simply by application of the "method of moments" (lever principle) to the compositional changes read off the diagram. Any particular oxide could be used, but it is usually convenient to use the one plotted as abscissa. For addition of material (assimilation):

$$\text{Wt. xenolith added} = \frac{(\%SiO_2)_{\text{magma II}} - (\%SiO_2)_{\text{magma I}}}{(\%SiO_2)_{\text{xen.}} - (\%SiO_2)_{\text{magma I}}}$$

$$\times \text{ Wt. of magma I.}$$

In the example of Fig. 7-2(a), assuming a starting weight of 100 gms for magma I,

$$\text{Wt. xenolith added} = \frac{53.1 - 50.6}{70.2 - 50.6} \times 100 \text{ gms} = 12.8 \text{ gms.}$$

For subtraction of material (crystal fractionation):

$$\text{Wt. phenocryst subtracted} = \frac{(\%SiO_2)_{\text{magma II}} - (\%SiO_2)_{\text{magma I}}}{(\%SiO_2)_{\text{magma II}} - (\%SiO_2)_{\text{phen.}}}$$

$$\times \text{ Wt. of magma I.}$$

In the example of Fig. 7-2(b),

$$\text{Wt. olivine subtracted} = \frac{50.6 - 49.2}{50.6 - 40.1} \times 100 \text{ gms} = 13.3 \text{ gms.}$$

In the case of crystal fractionation in a natural magma, the subtraction of just one crystalline phase of constant composition would, of course, not be expected to continue indefinitely. Eventually another phase joins the first, whereupon the directions of the trends of oxide percentages in successive residual liquids may be radically altered. Furthermore, with gradually lowering temperature and change of composition of the magma, most of the crystallizing phases themselves change progressively in composition in accordance with the reaction principle (Chapter 5). The result is that, rather than produce straight trend lines on the silica-variation diagram, the trends of the other oxide percentages show flexures or curvature.

A further factor in controlling the trends in the composition of the successive residual liquids is the degree to which equilibrium between liquid and growing crystals has been maintained before the crystals are removed from contact with the liquid. The material that is finally removed from the crystallizing magma if equilibrium has been maintained is different from the material that results if the growing crystals have been mantled by subsequent members of their isomorphous series. Such modifications of the liquid line of descent are illustrated by Bowen (1928, pp. 96–110) using examples from the simpler laboratory-investigated silicate systems of three and four components. In the ternary feldspar system, for instance, non-equilibrium crystallization leads to more sodic mantles over early calcic cores. Eventual physical removal of these composite crystals from the magma leaves a liquid higher in SiO_2, thus more strongly fractionated than would have been the case had the early-formed plagioclase equilibrated completely with the liquid before its removal. In an analogous manner, differing oxide trends in the residual liquids may also be expected in the fractional crystallization of ferromagnesian solid solutions. These behaviors, easily demonstrated in systems of few components, must also be present to some degree in the complex systems of natural magmas and must act similarly to modify the trends of the liquid line of descent.

With simple assimilation—that is, with progressive dissolution of a homogeneous foreign material by a liquid, unaccompanied by fractional crystallization—the resulting trends in the variation diagram should be straight lines. Where the compositions of the successive portions of added materials change, however, departures from the straight-line trends may be expected. The kinds of materials that could be assimilated (dissolved) by magmatic liquids depend on the available heat energy along with the relative positions of the magma in the reaction series (see Chapters 5 and 10).

During the 1920's and 1930's, when there was strong debate over fractional crystallization as a dominant process in producing the diversity of igneous rocks, curvature of oxide trends in the variation diagram was taken by some as evidence that the particular rock series had indeed originated through fractional crystallization. In the classical controversy on the igneous rocks of the Katmai magmatic province, for example, Fenner (1926, p. 704) used eighteen analyzed specimens for a silica-variation diagram, on which he drew best-fit straight lines through data points of the other oxides (Fig. 7-1). The departures of data points from these straight lines were not enough, in his opinion, to be significant, and he concluded therefore that, with straight trends, fractional crystallization could not have been a prominent process in the origin of the Katmai rocks.[4]

Bowen (1928, pp. 116–118), in reexamining the data for the Katmai rocks, saw as particularly significant the departures of the plotted points from the straight lines (in some cases more than 10 percent of the oxide percentages) and detected a certain amount of curvature in the series as a whole. Going further, he applied the strictures referred to above, eliminating all specimens except those that were holohyaline, aphanitic or only sparsely porphyritic. Smooth trend lines through the oxide points of the five remaining specimens were definitely curved (Fig. 7-4), and were taken as an argument for an origin by fractional crystallization.[5]

The appropriateness of the Katmai series for use as a test of magmatic evolution may be questioned on several additional bases: the samples were taken from a fairly large geographic area (some 700 km²); only sketchy reconnaissance mapping was available on which to infer the geologic relations between samples; and their ages were only imperfectly known and ranged widely from Jurassic to Holocene. Despite these weaknesses and the doubt cast by Bowen on the pertinence of most samples, later authors (e.g., Thornton and Tuttle, 1960), have used the analyses of all eighteen samples of the Katmai suite as collectively illustrating a liquid line of descent. Objections similar to those listed here for the Katmai series apply as well to other rock suites that have been introduced in support of one or another magmatic process (Harker, 1909; Bailey et al., 1924, p. 26; Larsen, 1948; Nockolds and Allen, 1953; Nockolds and Mitchell, 1948). From rock series of such complex origins and hazy magmatic relations, one can infer little else than their general chemical type and the obvious but unhelpful general rule that as their members become more siliceous they also become less cafemic. And this holds true for almost any series of igneous rocks, even a "series" whose individual members come from widely separated geologic areas! The prime requisites for initial testing of mechanisms of origin are volcanic rock series of simple geologic setting and narrowly prescribed spatial and age limits. Several series are discussed below, in some cases less with the

---

[4] Subsequently, Fenner (1948) suggested that these apparent straight-line trends, as well as those of the Gardiner River series (see below) might be the result of liquid unmixing at high temperature followed by incomplete remixing of the conjugate pairs before final consolidation.

[5] It is not clear why so much importance was attached here to curvature of the liquid path as a unique indicator of origin by fractional crystallization, when at other points Bowen (1928, pp. 185, 191) remarked that, except for distance traversed, the course of liquid compositions must be the same during assimilation as during fractional crystallization.

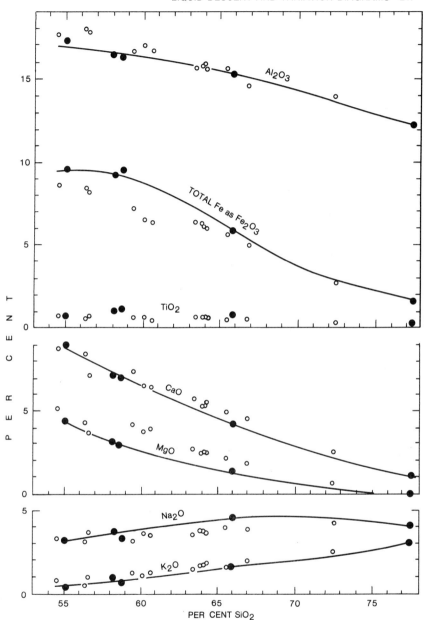

Figure 7-4. Silica-variation diagram of the Katmai rock series (after Bowen, 1928). Analyses of aphanitic or near-aphanitic rocks are represented by filled circles.

aim of delineating their actual liquid lines of descent than of testing the abilities of various magmatic processes to have produced the observed bulk relationships.

QUANTITATIVE GRAPHICAL TESTING OF RIVAL MAGMATIC PROCESSES

One rock series that has the simple setting desired for testing alternative hypotheses is that of the Gardiner River rhyolite-basalt complex in Yellowstone Park, Wyoming (Fenner, 1938, 1944; Wilcox, 1944). It crops out in contact with other formations over an area less than half a kilometer in diameter; the rhyolite intimately intrudes basalt and includes masses of basalt that show all stages of disintegration in the rhyolite; and inhomogeneity is present at all scales, even in hand specimen and thin section (Fenner, 1938, pls. 7 and 8). The complete range of bulk compositions from basalt to rhyolite is documented by bulk chemical analyses of eighteen specimens (Fenner, 1938, Table 6). In the silica-variation diagram, reproduced here as Fig. 7-5, the close adherence of the data points to straight-line variation between the extremes of composition is readily apparent. In this straight-line relationship Fenner saw a denial of the efficacy of fractional crystallization, and without considering simple assimilation, he postulated an origin through pneumatolytic metasomatism of basalt country rock by a surface rhyolitic lava flow. Based on detailed mapping and petrographic study, Wilcox (1944) concluded that the straight-line character of the Gardiner River complex was consistent with an origin by simple cross-assimilation—that is, by mixing of fluid rhyolitic lava with simultaneously erupted fluid basaltic lava, each at its respective liquidus temperature.

An instructive illustration of the liquid line of descent resulting from straightforward fractional crystallization is furnished by the detailed studies of samples taken from drill holes during the solidification of the 1963 lava lake formed in Alae pit crater, Hawaii (Wright and Fiske, 1971; Wright and Peck, 1978). As the crust thickened, drill holes were deepened to sample both the newly crystallized rock and the glass of the interstitial liquid that had oozed by filter pressing from the walls of the hole. Samples were studied in thin section, and selected specimens were chemically analyzed. The MgO-variation diagram of the compositions of undifferentiated lava and successive oozes (Fig. 7-6) gives a picture of the liquid line of descent and a striking confirmation in a natural setting of trend changes due to onset of separation of several crystalline phases as predicted by laboratory experiment. As MgO decreases with separation of Mg-rich

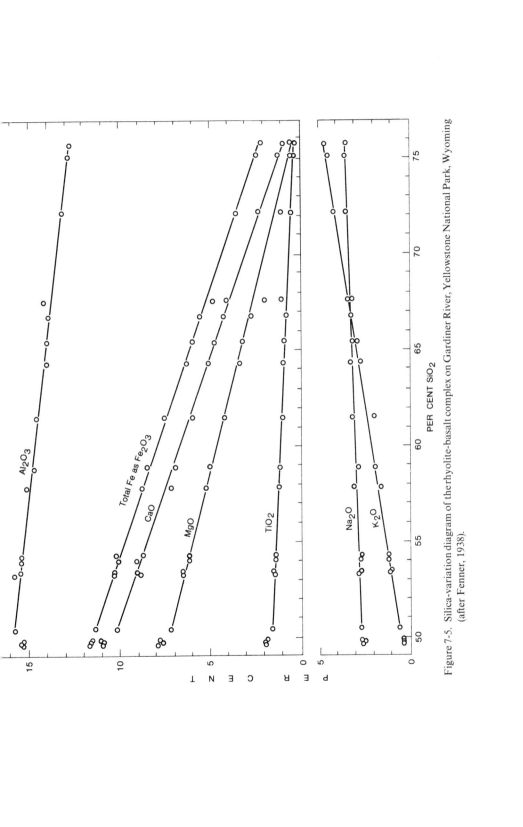

Figure 7-5. Silica-variation diagram of the rhyolite-basalt complex on Gardiner River, Yellowstone National Park, Wyoming (after Fenner, 1938).

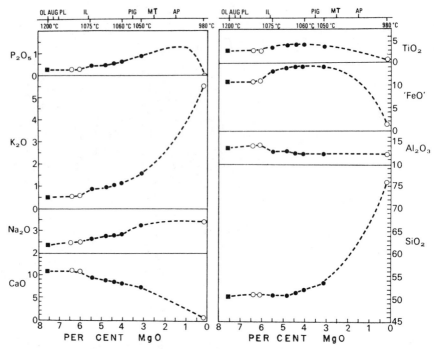

Figure 7-6. Magnesia-variation diagram showing the liquid line of descent for Alae lava lake, Hawaii. Squares represent bulk composition; open circles represent analyzed glasses; solid circles represent filterpressed oozes. Temperatures of magma and first appearances of minerals are shown at the top. (Wright and Fiske, 1971; Wright and Peck, 1978).

olivine, there is initially little change in relative percentages of other oxide constituents. After the appearance of augite and plagioclase, however, and with continued decrease in MgO, the relative percentage of CaO decreases while the alkalies increase and iron increases to a low maximum. After the appearance of magnetite, iron decreases strongly, while silica and K_2O increase strongly and Na_2O only slightly. Throughout there is little change in the relative percentage of Al_2O_3.

A more complex example, but nevertheless well suited from the standpoint of simplicity of geologic setting desired for meaningful tests with the variation diagram, is the rock series from lavas successively erupted from Parícutin Volcano, Mexico during 1943–1953 (Wilcox, 1954). There can be little doubt that all samples of these lavas are quenched products from the same magma chamber, inasmuch as the eruption was continuous and came from a single master feeder conduit. The early erupted lavas were olivine-bearing basaltic andesites of around 55 percent SiO_2[6], and subsequently erupted lavas were in general progressively more siliceous. The final lavas to be erupted were hypersthene andesites of about 60 percent SiO_2. Total amount of lava and pyroclastic ejecta erupted during the nine years of activity was estimated to have been about 3,560 million metric tons (Fries, 1953), equivalent to about 1.4 km^3 of molten magma (Wilcox, 1954, p. 289). In the early Parícutin lavas the sparse mega-phenocrysts were olivine Fo_{75-80} containing inclusions of scattered tiny opaque oxide crystals, presumed to be magnetite. These were associated in some specimens with plagioclase An_{70-75} megaphenocrysts. In the lavas of 1945–1947, the olivine megaphenocrysts carried rims of fine-grained orthopyroxene, indicating reaction with the liquid. Such relics of olivine were occasionally present in lavas erupted after 1947, along with sparse megaphenocrysts of orthopyroxene En_{60-75} containing small amounts of tiny opaque oxide crystals.

The chemical setting of this close-knit lava series in relation to its phenocrysts and xenoliths is shown in the silica-variation diagram of Fig. 7-7. The solid bands, drawn to include the plotted points of the twenty-two analyzed lavas, may be regarded as approximating a series of magmatic liquids, inasmuch as so few phenocrysts are present in the lavas. From the plotted compositions of the representative phenocrysts and xenolithic materials on the diagram, it is apparent that the observed compositional differences through the series cannot be accounted for by addition or subtraction of any one of these alone. The question remains whether combinations of these materials as end members in various addition or subtraction operations might account for the trends.

Graphic procedures for testing the adequacy of selected combinations are illustrated in Fig. 7-8, extending the approach used in Fig. 7-3. For a combination of two end members (in addition to the starting magma) two oxides are used as "controls." In effect, the graphical operation adds or subtracts just enough of each constituent to or from the beginning lava (magma) to match the observed percentages of the two control oxides in the supposed derivative lava (magma). The merit of the operation is judged by the closeness of match of the other oxides, that is, by the small-ness of the residuals. In Fig. 7-8(a), using SiO_2 and Al_2O_3 as controls, a test is made of the transition from Lava No. 1 to Lava No. 11 of the Parícutin series by subtracting olivine Fo_{80} and plagioclase An_{70}, whereby the residuals in 'FeO', CaO, and MgO are found to be about 1.2, 0.8, and 0.3 percent, respectively.

---

[6] Another sample of early lava analyzed by J. H. Scoon (Tilley *et al.*, 1967, p. 264, Table 5, Anal. 3; Nockolds and Allen, 1953, Table 8, Anal. 7) was slightly more basic, showing 54.08 percent silica.

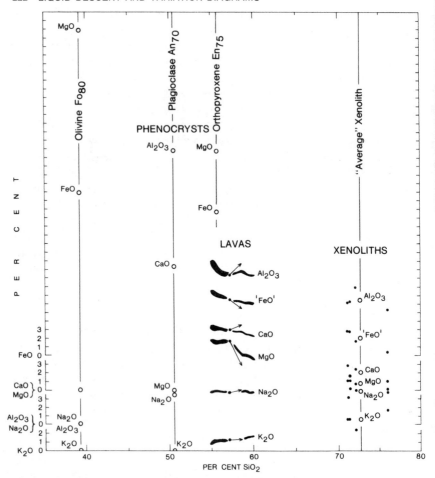

Figure 7-7. Silica-variation diagram of erupted materials of Parícutin volcano, Mexico. Blacked-in areas include plotted lava compositions, with 'FeO' representing total iron recast as FeO. Open circles show compositions of xenoliths and megaphenocrysts (except for opaque oxide, which plots off the diagram). Arrows indicate trends that would be produced by subtracting olivine Fo_{80} from an intermediate lava. (After Wilcox, 1954).

For testing a combination of three end members (in addition to the starting magma), three oxides are used as controls, and the graphical procedure is complicated by the necessity for a trial-and-error method to arrive at the proportions leading to a match of all three control oxides. In Fig. 7-8(b), using SiO_2, Al_2O_3, and MgO as controls, a test is made of

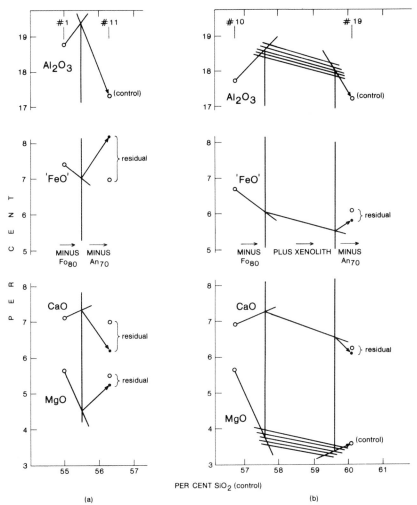

Figure 7-8. Silica-variation diagrams showing manner of testing adequacies of additions and subtractions of materials to account for compositional differences between members of the Parícutin lava series. Open circles represent compositions of lavas. 'FeO' represents total iron recast as FeO. (a) Subtraction of phenocrystic olivine Fo_{80} and plagioclase An_{70} from Lava No. 1 to produce a lava of same SiO_2 and Al_2O_3 contents as Lava No. 11. (b) Subtraction of Fo_{80} and An_{70} from Lava No. 10 with simultaneous addition of "average" xenolith to produce a lava of same SiO_2, Al_2O_3, and MgO contents as Lava No. 19. (After Wilcox, 1954).

the transition of Lava No. 10 to Lava No. 19 by subtracting Fo_{80} and An_{70} and simultaneously adding xenolithic materials. In this case, the residuals in 'FeO' and CaO are found to be about 0.3 and 0.2 percent, respectively. For testing more than three end members in addition to the starting magma, the graphical method becomes unreasonably cumbersome. (This number can be handled readily by modern computer methods, however, and examples will be described briefly in the next section.)

After the results of graphic tests of the Parícutin lava series were summarized (Wilcox, 1954, p. 336), fractional crystallization involving just olivine and plagioclase did not appear to be acceptable because of large overall residuals. The combination of olivine, plagioclase, and magnetite gave much lower residuals, but was regarded as petrologically infeasible because of the large proportion of magnetite (one-quarter to one-half as much magnetite as olivine) that would have to have been removed.[7] Lowest residuals were obtained in graphical tests of the combination "average xenolith," olivine, and plagioclase. When the thermal requirements for incorporation of the implied amount of xenolithic material were taken into account, it was concluded that assimilation combined with fractional crystallization was a possible mode of origin for the series.

It does not necessarily follow, of course, from the differing compositions of the successively erupted lavas that the total amount of change in composition had been accomplished in the magma as a whole merely during the nine years of the eruption. Rather, it is suggested (Wilcox, 1954, p. 346) that compositional gradients may have been set up in this part of the magma by these processes, in large part prior to onset of the eruption.

COMPUTER ANALOGS OF GRAPHICAL TESTS

In modes analogous to those described above for variation diagrams, the computer can be used to extend considerably the range for exploring possible combinations in origins of igneous rock series by mixing or unmixing, for it is less limited in the number of phases and in the variants within phases that can be conveniently handled. The proportions of

---

[7] Osborn (1962, p. 223), however, saw no difficulty in the amount of magnetite involved here, and thus no objection to an origin of the Parícutin lava series by fractional crystallization of these three mineral phases. Electron microprobe analyses by Ulmer and Drory (1974 and *in press*) reveal that the supposed magnetite microphenecrysts are chromian spinels with up to 27 percent Cr_2O_3.

added or subtracted end members are calculated so as to provide the closest correspondence to the observed compositions of the respective samples of the rock series. As with the graphical methods, the merit of a selected set of end members is judged by overall closeness of approach together with the geologic plausibility of the calculated proportions. (In respect to closeness of approach, however, a perfect correspondence becomes inevitable when the number of end members equals the number of oxide variables; therefore, this cannot be taken as an indication of a valid combination of end members.)

Bryan (1969) employed matrix algebra and least-squares approximation as a computerized analog of the graphical method used by Wilcox (1954) for the Parícutin suite, described above. Testing the favored xenolith-plagioclase-olivine combination, he found close correspondences not only for the major oxides, but also for the alkalies and minor oxides, which had not been tested graphically by Wilcox. In another example, Bryan *et al.* (1969) tested subtraction of olivine, plagioclase, and augite to account for observed compositional differences between sequential lavas of the 1960 eruption of Kilauea volcano, Hawaii. Again, agreement was found with results of previous tests on a magnesia-variation diagram (Murata and Richter, 1966), even though the phenocrysts were assumed, perhaps unrealistically, to have the same composition in successive stages.

Wright and Fiske (1971, p. 46 and Table 15) showed by trial-and-error calculations that the relatively silica-rich (55%) pumice of Yellow Cone on the southwest rift of Kilauea, Hawaii could have originated by mixing of a hot undifferentiated magnesium-rich magma (represented by nearby lava flows) with an already strongly differentiated magma. To avoid the tedium of the trial-and-error approach, Wright and Doherty (1970) used a combination of linear programming and conventional least-squares techniques to show that the composition of the late 1955 lava of Kilauea volcano could be closely matched by mixtures of appropriate amounts of olivine and two magmas, represented by the lava of 1952 and the still-to-be-erupted lava of 1961. As an example of a liquid line of descent calculation, they showed that the lava of the 1955 flank eruption of Kilauea could have been derived from the magma represented by a prehistoric lava from Cone Crater on the southwest rift by removal of olivine, pyroxene, plagioclase, and ilmenite. Wright (1973) also proposed magma mixing to account for the compositions of the lavas erupted from Kilauea during November and December 1959.

In another approach, Miesch (1976b) applied an extended form of Q-mode factor (vector) analysis to test hypotheses of assimilation, magma mixing, and fractional crystallization. The details of the method are described by Miesch (1976a,b), and computer programs are given by

Klovan and Miesch (1976) and by Miesch (1976c). At the outset, a factor-variance diagram is derived that indicates the number, but not the identities, of end numbers that would be required to account for any portion of the variability in each oxide of the suite of analyzed igneous rocks. This diagram can also indicate whether a "unique" end member is involved; for instance, one consisting of only a single oxide constituent. Petrologically plausible end members, for example, those of composition observed in phenocrysts, are then tested for mathematical adequacy, and combinations of these are used to determine the required mixing proportions.

Data from four igneous rock suites of progressively greater complexity were taken from the literature by Miesch (1976b) to illustrate applications of the method. The first test suite was that of the rhyolite-basalt complex on Gardiner River, Wyoming (Fenner, 1938, 1944; Wilcox, 1944), graphical treatment of which has already been discussed above. The factor-variance diagram (Fig. 7-9) showed that only two end members (still undefined) could account for 97 percent or more of all the chemical variations in six of the major oxides and for 85 percent of the variation of Na_2O. The two end members would account, however, for only 28 percent of the variability in Fe_2O_3 and very little of the variability in H_2O^+. The Q-mode vector diagram (Fig. 7-10) showed that the roles of the two dominant end members could be adequately filled by the extreme rock types of the analyzed suite, basalt and rhyolite; in other words, the main process of origin could have been simple mixing of these two materials in varying proportions. To account for the remaining Na_2O variations, as well as for the variations of Fe_2O_3 and H_2O^+, oxidation, weathering, and post-emplacement volatile transport are suggested.[8]

As a second example, Miesch (1976b) chose a granitoid pluton on Snake Creek in the southern Snake range, described by Lee and Van Loenen (1971). The rocks of this pluton are coarse-grained and range from quartz monzonite of about 76 percent SiO_2 to granodiorite of about 63 percent SiO_2. The variations in compositions were attributed by Lee and Van Loenen to assimilative contamination of a granitic magma by sedimentary country rock, specifically the associated Pioche Shale formation of Cambrian age; scattered dark enclaves (as low as 57 percent SiO_2), of finer grain than the host granodiorite, are regarded as partly digested relics of the shale.

The factor-variance diagram for the rocks of the Snake Greek pluton indicated that mixtures of two end members (still unidentified) could

---

[8] A likely alternative would be that the ferric iron and water as well as the soda could have come in by volatile transport, the iron perhaps deposited as hydrated $FeCl_3$, then converted to a hydrated ferric oxide complex.

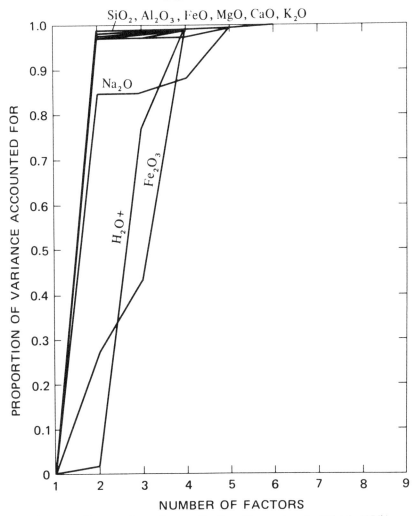

Figure 7-9. Factor-variance diagram, Gardiner River rock series (Miesch, 1976b).

account for 70 percent or more of the variabilities in most constituents, but for only about 1 percent of the variability in Na_2O and 25 percent in H_2O^+. The Q-mode vector diagram in turn indicated that, whereas a material near the composition of the most siliceous granites could have been one major end member, the other end member could not have been the Pioche Shale, as Lee and Van Loenen (1971) had suggested, nor, for that matter, could it have been any of the analyzed associated sedimentary

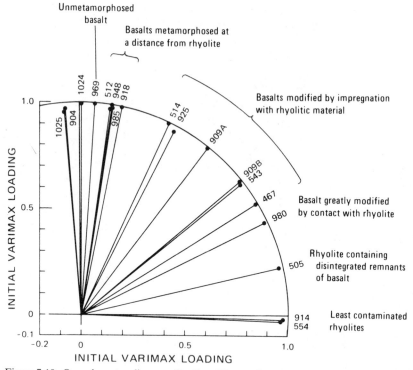

Figure 7-10. Q-mode vector diagram, Gardiner River rock series (Miesch, 1976b). Numbers at ends of vectors are the sample numbers of Fenner (1938). Notes on vectors give interpretations of Fenner.

units or combinations thereof. What did appear as plausible for the second major end member was material of the composition of the darkest of the enclaves. The major compositional variations through the pluton thus would seem best accounted for by simple mixing of granitic magma with material of the composition of, or perhaps slightly more basic than, that of the dark enclaves, whatever may have been the latter's origin.

Miesch (1976b) also reexamined the already mentioned compositional variations of lavas of the 1959 summit eruption of Kilauea volcano, Hawaii, which Wright (1973) on the basis of linear programming and least-squares analysis, had accounted for by mixing of five end members: two magmas and addition (not subtraction) of two olivines (Fa_{10} and Fa_{30}) and chromite. The initial factor-variance diagram showed that the number of end members needed to account for all the variations was indeed five, but that three would be sufficient to account well for nearly all of the variability in seven major oxides, except for 23 percent and 29 percent of the variability in FeO and Fe_2O_3, respectively, and 15 percent in K_2O. Post-eruptive oxidation could account for the FeO/Fe_2O_3 relations. From the Q-mode vector diagram (a stereogram to represent the three dimensions imposed by the three end members involved), Miesch concluded that the variations (excluding those of Fe_2O_3/FeO, and of K_2O) could have been produced from a parent magma of composition near that of the oceanite parent of Hawaiian lavas suggested by Macdonald (1968, pp. 502, 511), from which varying amounts of the following two groups of minerals were subtracted: (1) plagioclase, clinopyroxene, and olivine Fa_{15}; and (2) plagioclase, clinopyroxene, and olivine Fa_{24}.[9]

As a final example, Miesch (1976b) examined the classical plutonic layered series of the Skaergaard (Wager and Brown, 1967) but was unable to reach clearcut conclusions. Whereas the factor-variance diagram indicated that five end members would account for more than 85 percent of the variations of each oxide, no combination of five geologically plausible end members could be found. Only by going to nine end members could a geologically acceptable solution be derived. Because nine was the number of analyzed oxides employed for the sample array, however, the perfect agreement between observed and reproduced compositions of the samples was mathematically inevitable and could not be crosschecked.

In the Skaergaard example, one is perhaps confronted with variables too complex for even the Q-mode factor methods to handle. With the exception of the one sample of the chilled border zone that was taken to represent the parent magma, each analyzed Skaergaard sample consists predominantly of crystals that had separated from another part of the magma, perhaps of different composition even than that of the minor amounts of liquid trapped between crystals. It may be that the examination of the physical-chemical magmatic process is handicapped, if not totally frustrated, due to different mechanisms and rates of segregation in which the relative proportions of the different mineral species accumulating in any one sample changed greatly from the proportions in which they separated from magma. There is, in addition, the possibility that new batches of magma, different in composition from that of the lone sample from the border zone chosen as the parent, were introduced into the system during the buildup of the layered deposit. With so many complications, comparisons between bulk chemical compositions of samples of the cumulate layered series may lead to less helpful conclusions concerning

---

[9] The petrologic acceptability of the two groups is not discussed.

the magmatic line of descent involved in their production than would comparisons from sample to sample of the composition of each individual crystal phase.

In conclusion it is emphasized that the purpose of these tests, whether carried out by computer methods or on the variation diagram, is to reveal combinations of end members that, by mixing with or unmixing from a chosen member of the rock series, satisfy the chemical relationships to other members of the series. All conceivable combinations of end members must be tested.[10] It should be kept in mind that reduction of total residuals brought about by introduction of an additional end member into the mixing tests may be merely a mathematical artifact unrelated to progress toward a solution of the origin of the rock series. This aspect is apparent in the graphical operations, described above, where for each additional end member one must preempt as control yet another oxide, residual of which then becomes zero. The ultimate expression of this tendency is seen in computer tests when the number of end members becomes equal to the number of oxides in the array of chemical analyses, and a perfect, but perhaps meaningless, correspondence is attained. Finally, each combination of end members that is found by these tests to be mathematically satisfying must be carefully examined further to ascertain its petrologic and thermodynamic acceptability before one can regard it as a viable candidate to account for the evolution of the rock series.

Acknowledgments   Grateful acknowledgment is made to Drs. A. R. McBirney, A. T. Miesch, T. L. Wright, and H. S. Yoder, Jr., for review of the manuscript and helpful suggestions.

References

Anderson, A. T., Evidence for a picritic, volatile-rich magma beneath Mt. Shasta, California, *J. Petrol.*, *15*, 243–267, 1974.
Bailey, E. B., Clough, C. D., Wright, W. B., Richey, J. E., Wilson, G. V., and Thomas, H. H., Tertiary and post-Tertiary geology of Mull, Loch Aline, and Oban, *Scotland Geological Survey Mem.*, 445 pp., 1924.
Bowen, N. L., *The Evolution of the Igneous Rocks*, Princeton University Press, 334 pp., 1928.

---

[10] To facilitate participation of the petrologist in exploring the many possibilities, Miesch (1976c) provides a series of interactive computer programs for the extended form of Q-mode factor analysis.

Bryan, W. B., Materials balance in igneous rock suites, *Carnegie Institution of Washington Yearbook*, *67*, 241–243, 1969.

Bryan, W. B., Finger, L. W., and Chayes, F., Estimating proportions in petrographic mixing equations by least-squares approximation, *Science*, *163*, 926–927, 1969.

Burri, Conrad, *Petrochemical calculations based on equivalents* (*Methods of Paul Niggli*). Israel Program for Sci. Translations, Jerusalem, 295 pp., 1964.

Chayes, Felix, On correlation between variables of constant sum, *J. Geophys. Res.*, *65*, 4185–4193, 1960.

———, Variance-covariance relations in some published Harker diagrams of volcanic suites, *J. Petrol.*, *5*, 219–237, 1964.

Fenner, C. N., The Katmai magmatic province, *J. Geology*, *34*, 675–772, 1926.

———, Contact relations between rhyolite and basalt on Gardiner River, Yellowstone Park, Wyoming, *Geol. Soc. America Bull.*, *49*, 1441–1484, 1938.

———, Rhyolite-basalt complex on Gardiner River, Yellowstone Park, Wyoming: a discussion, *Geol. Soc America Bull.*, *55*, 1081–1096, 1944.

———, Immiscibilities of igneous magmas, *Am. J. Sci.*, *246*, 465–502, 1948.

Fries, Carl, Jr., Volumes and weights of pyroclastic material, lava, and water erupted by Parícutin Volcano, Michoacán, Mexico, *Am. Geophys. Union Trans.*, *34*, 603–616, 1953.

Harker, A., *The Natural History of the Igneous Rocks*, New York, 384 pp., 1909.

Iddings, J. P., The origin of igneous rocks, *Phil. Soc. Washington Bull.*, *12*, 89–213, 1892.

Klovan, J. E., and Miesch, A. T., Extended CABFAC and QMODEL computer programs for Q-mode factor analysis of compositional data, *Computers and Geosciences*, *1*, 161–178, 1976.

Larsen, E. S., Some new variation diagrams for groups of igneous rocks, *J. Geology*, *46*, 505–520, 1938.

Larsen, E. S., Jr., Batholith and associated rocks of Corona, Elsinore, and San Luis Rey quadrangles, southern California, *Geol. Soc. America, Mem. 29*, 182 pp., 1948.

Lee, D. E., and Van Loenen, R. E., Hybrid granitoid rocks of the southern Snake Range, Nevada, *U.S. Geol. Survey Prof. Paper 668*, 48 pp., 1971.

Macdonald, G. A., Composition and origin of Hawaiian lavas, in *Studies in Volcanology*, R. R. Coats, R. L. Hay, C. A. Anderson, eds., *Geol. Soc. Am. Mem. 116*, 477–522, 1968.

Miesch, A. T., Q-mode factor analysis of compositional data, *Computers and Geosciences*, *1*, 147–159, 1976a.

———, Q-mode factor analysis of geochemical and petrologic data matrices with constant-row sums, *U.S. Geol. Survey Prof. Paper 574-G*, 47 pp., 1976b.

———, Interactive computer programs for petrologic modeling with extended Q-mode factor analysis, *Computers and Geosciences*, *2*, 439–492, 1976c.

Murata, K. J., and Richter, D. H., Chemistry of the lavas of the 1959–60 eruptions of Kilauea Volcano, Hawaii, *U.S. Geol. Survey Prof. Paper 537–A*, 26 pp., 1966.

Nockolds, S. R., The Garabal Hill-Glen Fyne igneous complex, *Geol. Soc. London, Quart. Jour.*, *96*, 451–511, 1941.

Nockolds, S. R., and Allen, R. J., The geochemistry of some igneous rock series, *Geochim. Cosmochim. Acta*, *4*, 105–142, 1953.

Nockolds, S. R., and Mitchell, R. L., The geochemistry of some Caledonian plutonic rocks, *Royal Soc. Edinburgh, Trans.*, *61*, 533–575, 1948.

Osborn, E. F., Reaction series for subalkaline igneous rocks based on different oxygen pressure conditions, *Am. Mineral.*, *47*, 211–226, 1962.

Pearce, T. H., A contribution to the theory of the variation diagram, *Contrib. Min. and Petrol.*, *19*, 142–157, 1968.

———, Chemical variations in the Palisades Sill, *J. Petrol.*, *11*, 15–32, 1970.

Thornton, C. P., and Tuttle, O. F., Chemistry of igneous rocks; pt. I, Differentiation Index. *Am. J. Sci.*, *258*, 664–684, 1960.

Tilley, C. E., Yoder, H. S., Jr., and Schairer, J. F., Melting relations of volcanic rock series, *Carnegie Institution of Washington Yearbook*, *65*, 260–269, 1967.

Ulmer, Gene C., and Drory, A., Parícutin andesite: two new lines of evidence, *Geol. Soc. Am., Abstracts*, *6*, 1069, 1974 (also *Geochim. Cosmochim. Acta*, in press).

Wager, L. R., and Brown, G. M., *Layered Igneous Rocks*, Freeman, San Francisco, 588 pp., 1967.

Watson, E. B., Glass inclusions as samples of early magmatic liquid: determinative method and application to a South Atlantic basalt, *J. Volcanic and Geothermal Res.*, *1*, 73–84, 1976.

Wilcox, R. E., Rhyolite-basalt complex on Gardiner River, Yellowstone Park, Wyoming, *Geol. Soc. America Bull.*, *55*, 1047–1080, 1944.

———, Petrology of Parícutin Volcano, Mexico, *U.S. Geol. Survey Bull. 965-C*, 281–353, 1954.

Wright, T. L., Magma mixing as illustrated by the 1959 eruption, Kilauea Volcano, Hawaii, *Geol. Soc. America Bull.*, *84*, 849–858, 1973.

———, Presentation and interpretation of chemical data for igneous rocks, *Contrib. Min. Petrol.*, *48*, 233–248, 1974.

———, and Doherty, P. C., A linear programming and least squares computer method for solving petrologic mixing problems, *Geol. Soc. America Bull.*, *81*, 1995–2008, 1970.

———, and Fiske, R. S., Origin of the differentiated and hybrid lavas of Kilauea Volcano, Hawaii, *J. Petrol.*, *12*, 1–65, 1971.

———, and Peck, D. L., Solidification of the Alae lava lake, Hawaii, Chapter C. Crystallization and differentiation of the Alae magma, *U.S. Geol. Survey Prof. Paper 935-C*, 20 pp., 1978.

Chapter 8

GLASS AND THE GLASSY ROCKS

I.S.E. CARMICHAEL

Department of Geology and Geophysics, University of California, Berkeley, California

INTRODUCTION

In his compelling advocacy of the role of crystallization-differentiation as the principal process controlling the diversity of igneous rocks, Bowen sought to use the compositions of glassy rocks as the best evidence for a liquid line of descent. Obviously, a glassy rock resulted from congealed magma, and he was concerned to show that the range of composition represented by naturally occurring glasses was more restricted than that of the crystalline rocks. Of those which fell outside the glassy range, factors other than crystallization of almost crystal-free liquids were presumed to be at work in modifying their composition—perhaps accumulation of crystals or subsequent alteration by reaction with percolating water at near-solidus temperatures.

In the last half-century the emphasis on the study of glassy rocks has changed. Petrologists are no longer certain that glassy rocks represent the pristine composition of magmas, because the capacity for oxygen exchange at low temperatures from percolating water is now well documented (Taylor, 1968). Because the exchange of cooling intrusive silicic magma with its rock-water envelope is virtually ubiquitous, glassy rocks, judiciously chosen, can be used to elucidate the crystallization paths of granitic magma. Indeed, the best information available on the range of liquidus temperatures for silicic magmas comes from their glass equivalents, for glass completely devoid of crystals is rare. But Bowen's

The Evolution of the Igneous Rocks: Fiftieth Anniversary Perspectives
0-691-08223-5/79/0233-12$00.60/0 (cloth)
0-691-08224-3/79/0233-12$00.60/0 (paperback)
For copying information, see copyright page

emphasis remains entirely valid—unaltered or fresh glass still provides the best evidence for the composition and type of the early crystallizing phases, for the temperatures at which they equilibrate, and for the intensive variables, such as pressure or the fugacity of volatile components, which are characteristic of silicic magmas. All too often this evidence of the initial conditions of the magma is obliterated during the slow cooling and reaction of large intrusive bodies.

Before examining some of the more intriguing aspects of petrology illustrated by the glassy rocks, it is appropriate to discuss the nature of the glass itself and the kinetics of the transformation that has preserved this metastable, and hence reactive, substance for perhaps fifty million years on earth and for almost four billion years on the moon.

PROPERTIES OF GLASS

Some liquids when cooled rapidly below their equilibrium crystallization temperature T_e may fail to crystallize, and instead form amorphous solids that lack the long range crystallographic arrangement of atoms characteristic of the crystalline state. These solids are commonly called glass, and result from the cooling of supercooled liquids, but the two states, supercooled liquid and glass, are not identical.

The reversible transformation, liquid $\rightleftarrows$ crystals, at T_e entails a discontinuous change in the first-order thermodynamic functions, enthalpy, entropy, and volume; silicate liquids crystallize with a decrease of about 5–15 percent of their volume, and with an evolution of heat (heat of fusion) and, therefore, necessarily with a decrease in entropy. The fusion or crystallization transition is, therefore, a first-order transformation. When a glass-forming liquid, typically polymeric,[1] is cooled metastably below T_e, no discontinuity is observed at T_e in the first-order thermodynamic properties, nor in the heat capacity at constant pressure C_p or in the thermal expansion α (see Fig. 8-1), both of which are second-order thermodynamic functions. Continued cooling of the supercooled liquid eventually results in a discontinuity in C_p and in the thermal expansion, and the temperature of this discontinuity is called the glass transformation temperature T_g (Fig. 8-1).

The transformation, supercooled liquid $\rightleftarrows$ glass, is second order but

---

[1] The characteristically small entropy of fusion (per gram atom) of glass-forming compounds is indicative of a high degree of order in the liquid state.

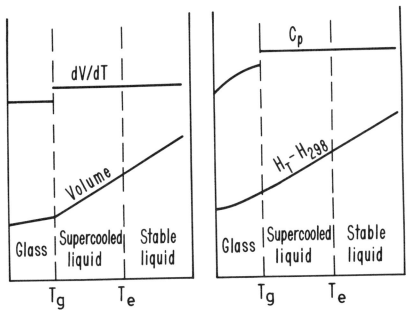

Figure 8-1. Change in properties as a function of temperature. T_e represents the equilibrium temperature, T_g the glass transformation temperature.

T_g is itself a function of the cooling rate of supercooled liquid. There is a limiting value of T_g, namely T_0, that is the temperature of a thermodynamic second-order equilibrium transformation, which will be achieved by increasingly slow cooling of the supercooled liquid. The value of T_0 is the thermodynamic low-temperature limit to the supercooled liquid state; however, both glass and supercooled liquid are metastable with respect to the crystalline state. The temperature at which the viscosity of the supercooled liquid exceeds 10^{13} poise has been used widely in the glass industry as a definition of T_g, and is close to the temperature of the operationally defined T_g given above.

Measurements of C_p for many silicate glasses have led investigators to conclude that at T_g the heat capacity, on a gram atom basis, is approximately equal to $3R$. This observation may be interpreted in classical terms of an atom being represented by a harmonic oscillator, which will have zero vibrational energy at $0°K$, but which will rise to a maximum value of kT (Boltzmann's constant times the absolute temperature) for each degree of vibrational freedom at high temperature. Thus, for an atom to vibrate in three mutually perpendicular directions, the vibrational energy per mole at high temperatures is given by

$$E = 3RT$$

so that the corresponding heat capacity at constant volume is

$$C_v = \left(\frac{\partial E}{\partial T}\right)_v = 3R.$$

Although measurements of heat capacity are usually made at constant pressure (C_p), the difference between C_p and C_v for glasses is usually quite small and equal to $TV\alpha^2/\beta$; this difference in most glasses is not significantly greater than the errors of the measurements. The glass transformation temperature may be interpreted as that temperature at which appreciable translational immobility of the atoms or particles sets in.

Bacon (1977) has measured the heat content of a variety of silicate glasses from which C_p is derived, and found that one naturally occurring anhydrous silicic lava has a T_g that may be as high as 808°C, where the cooling rate is determined by the design of the calorimeter. The calculated viscosity for the same glass shows that it exceeds 10^{13} poise at 735°C, and because water decreases the viscosity of silicate liquids, it may be presumed that T_g decreases with increasing amounts of water in solution. Thus, water-rich silicic magmas at 735°–750°C will precipitate phases in equilibrium with liquid, rather than with glass.

As an illustration of the change in properties at the glass transformation temperature, the requisite data are set out in Table 8-1 for $NaAlSi_3O_8$ and $KAlSi_3O_8$ compositions. The properties of glass are, therefore, quite distinct from those of the corresponding *stable* liquids, particularly with regard to their heat capacity and thermal expansion. It is these properties

Table 8-1. Properties of Feldspathic Glasses and Liquids

	$NaAlSi_3O_8$	$KAlSi_3O_8$
T_g °C	763	905
T_e °C	1118	1520
α_{liquid}, deg^{-1}	45×10^{-6}	42×10^{-6}
α_{glass}, deg^{-1}	22.2×10^{-6}	18.3×10^{-6}
$C_{p\,glass}$ at T_g, cal. mole^{-1} deg^{-1}	72.27	74.74
$C_{p\,liquid}$, cal. mole^{-1} deg^{-1}	85.69	83.56

Data taken from Arndt and Haberle (1973) and Kelley (1960); the values for $C_{p\,liquid}$ are calculated from the partial molar heat capacities given by Carmichael *et al.* (1977) and are independent of temperature.

of glass that have often in the past been used, wrongly, for calculations of fusion curves, solubility curves, and adiabatic gradients for silicate liquids.

KINETICS OF GLASS FORMATION

It is a matter of common observation that naturally occurring glass is found forming the carapace of large silicic lava flows, as dikes or sheets, and as the principal component of the pumice and vast ash flow eruptions. It is less frequent, and less well developed, in basic rocks, and the most common occurrence seems to be as the skin of deep sea lava flows. The formation of glass is in some way related to viscosity and cooling rate; the inability of experimental geochemists to quench certain liquid compositions to glass, for example carbonate and water-rich basic liquids, is in accord with these conclusions. It is appropriate, therefore, to inquire into the conditions that favor the nucleation and growth of crystals.

Consider a cooling liquid in which the crystals are few in number and grow with isotropic growth rates, or in other words, as spheres. Based on the assumption of no induction time or time dependence of nucleation, the volume of a single crystallite originating at time t is given by

$$v_{crystallite} = 4/3\pi y^3 t^3, \qquad (8\text{-}1)$$

where y is the isotopic growth rate (cm sec^{-1}) and t is the time in seconds.

In the body of the liquid the number of nuclei generated in time dt is given by $BV_{liquid}dt$ where B is the nucleation rate (cm^{-3} sec^{-1}) and V_{liquid} is the volume of the liquid. Initially the volume of nuclei will be very small and so the total number of crystallites will be very much less than the remaining liquid. Thus, in time dt the volume of crystallites will be given by

$$dV_{crystallites} = v_{crystallite}BV_{liquid}dt. \qquad (8\text{-}2)$$

Because there are so few crystallites, the volume of the liquid (V_{liquid}) is essentially equal to the volume V of the whole body. The total volume of crystallites in time t is, therefore,

$$V_{crystallites} = 4/3\pi V \int_{t=0}^{t=t} By^3 t^3 dt,$$

and in the simplest case, both B and y can be considered to be independent of time so that

$$V_{crystallites}/V = X_{crystallites} = \pi/3\, y^3 Bt^4, \qquad (8\text{-}3)$$

or the fraction of crystallites increases with the fourth power of time.

A more involved treatment (e.g., Christian, 1965) for the development of Equation 8-3 results in an equation of the form

$$X_{crystallites} = 1 - \exp\left(-\pi/3y^3 Bt^4\right), \qquad (8\text{-}4)$$

which reduces to Equation 8-3 as $t \to 0$.

Before the volume fraction of crystallites generated in a given time can be calculated (and these must be kept to a small number if glass is to be preserved), the temperature dependence of y, the rate at which the crystal-liquid interface advances, and B, the rate at which nuclei are generated, must be considered. The reader is referred to Christian (1965) for a detailed treatment of what follows; for the present a simple treatment will suffice.

For the rate of advance of the crystal-liquid interface in thermally activated growth (controlled by rate of heat loss rather than by diffusion), y can be related to the entropy of fusion, and also to the viscosity through the Stokes-Einstein relationship. Thus, the growth rate is given approximately by

$$y \simeq (a/\eta)(\Delta S_m \Delta T) \qquad (8\text{-}5)$$

where η is the viscosity, a is a complex factor involving particle size, site availability on the growing crystal, and thickness of the interface or jump distance, ΔS_m is the entropy of fusion per mole, and ΔT is equal to $T_e - T$, or the undercooling.

On the simplest assumption of homogeneous nucleation, the classical nucleation rate can be represented as

$$B = N\nu \exp\left(-EB/RT\right) \qquad (8\text{-}6)$$

where N is the number of particles in unit volume, ν is a frequency, and E_B is an energy that will include a term resulting from the heat of fusion and another contribution that is often taken to be given by the activation energy for diffusion (e.g., Christian, 1965).

If Equation 8-5 and Equation 8-6 are substituted in Equation 8-4, and the conditions for a crystallite volume fraction of 10^{-6}, for example, are calculated as a function of temperature and time, a T–T–T (time–temperature–transformation) plot with the familiar 'C'-shape is obtained. Uhlmann (1972) has calculated a T–T–T diagram for SiO_2, on the assumption that a crystallite volume fraction in excess of 10^{-6} would be readily observable, and therefore the bulk material would no longer be glassy; his diagram is reproduced in Fig. 8-2. The nose of the curve results from the competition of the rate of nuclei formation, which increases with falling temperature, with the rate of crystal growth, which decreases with falling temperature as the mobility of particles decreases. If, for example, liquid SiO_2 is quenched to 1550°C, then it would take approxi-

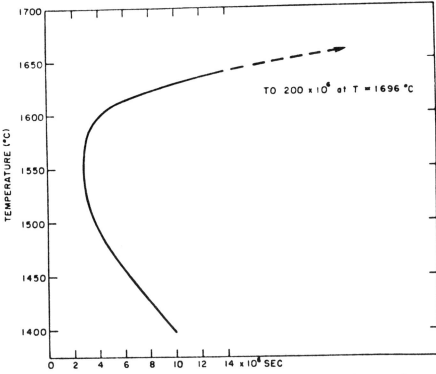

Figure 8-2. Time–temperature–transformation curve for SiO_2 corresponding to a volume fraction crystallized of 10^{-6} (after Uhlmann, 1972).

mately 3×10^6 sec for a crystallite fraction of 10^{-6} to develop. It is readily appreciated that if the same SiO_2 liquid were quenched to 25°C, the time to develop such a small volume of crystallites would be enormously long. Scherer, Hopper, and Uhlmann (1972) have provided data on two lunar basalt compositions, and their calculated $T–T–T$ curves lie well to the left of the SiO_2 curve on the time axis, in accord with the lower viscosities.

It is clear from the curve in Fig. 8-2 that in order to avoid the nose of the curve, a cooling rate for SiO_2 liquid from T_e is of the order of 7×10^{-5} deg sec^{-1}. Uhlmann (1972) has suggested that the thickness of a sample that can be found without detectable crystallites ($X_{crystallites} = 10^{-6}$) should be of the order of

$$y = (\kappa t_n)^{1/2} \qquad (8\text{-}7)$$

if heat transfer processes at the surface of the sample are ignored. In Equation 8-7, y is the thickness, κ is the thermal diffusivity ($10^{-1} - 10^{-2}$

cm^2 sec^{-1}), and t_n is the time at the nose of the $T-T-T$ curve; the thickness calculated is about 550 cm.

From this general treatment it is clear that liquids with very low viscosities may be unquenchable to a glass within experimentally achievable cooling rates; moreover, as basaltic liquids have lower viscosities than silicic liquids, glass associated with basalts will be more infrequent and form thinner flows than the glass of extremely viscous silicic lavas. As Bowen suggested, the absence of glass of ultrabasic composition may be a kinetic phenomenon but the absence of anorthosite glass, with its high viscosity, may indicate an absence of anorthosite magma;[2] however, as most of these bodies are pre-Cambrian it is not likely that glass would be preserved.

TEMPERATURES OF SILICIC GLASSES

The determination experimentally of the liquidus temperatures of nearly anhydrous silicic liquids is very difficult in view of their very slow crystallite nucleation and growth rates, so that the evidence obtained from natural glasses, although subject to the same slow processes, represents the best evidence available. It takes a large amount of experimental effort to determine the effect of all possible components on liquidus temperatures, but only nature provides the resultant effect of these, and the evidence is usually preserved only in the glassy rocks.

Many silicic obsidians contain only very small amounts of crystals, sometimes only 0.1 percent by volume, although it is rare to find glassy rocks completely devoid of crystals. Often this crystalline assemblage includes iron-titanium oxides, a magnetite solid solution co-existing with an ilmenite solid solution, and as their composition is temperature dependent, the temperature of their equilibration may be obtained from the calibration curves of Buddington and Lindsley (1964). It is reasonable to assume that these equilibration temperatures will be close petrologically to the liquidus temperature of each rock, because the total volume of crystals may be minute.

As the iron-titanium oxides are extremely susceptible to oxidation and are almost never preserved in their pristine state in slowly cooled intrusive silicic rocks, temperatures obtained from the glassy rocks are the best estimates of the near-liquidus temperatures of this type of magma. De-

---

[2] Editor's note: The kenningite of Von Eckermann (*Geol. Fören. i Stockh. Förh.*, 60, 243–284, 1938) is believed to be representative of water-rich anorthosite magma.

pending on the composition of the rhyolitic obsidian, temperatures have been recorded in the range of 735°-920°C (Carmichael, Turner, and Verhoogen, 1974), the lower values being for magma with a notable concentration of water and the higher temperatures for magma with insignificant amounts of water. A temperature range of 200° for the same general magma type places severe constraints on the conditions of its generation. If silicic magma is the product of crystallization differentiation of a basaltic parent, then its liquidus temperature cannot be far removed from the solidus temperature of the parent. This necessary but not sufficient evidence can be used to conclude that the low-temperature silicic magmas typically found in continental areas, often as voluminous ash flow deposits, are not directly related to cooling basic magma, but are perhaps the product of crustal fusion. It takes detailed mineralogic and geochemical studies on glasses (e.g., Hildreth, 1977) to substantiate such a proposition, but it is now clear that the key to the origin of granitic magma is to be found in their glassy surface equivalents, where all the intricate patterns of crystal composition are preserved by quenching during the eruptive process.

WATER IN GLASSES

At room temperature, glass is metastable compared with the crystalline state, and as the metastability may amout to almost 10 kcal per mole, it is to be expected that the alteration of glass will be a relatively fast and extensive process. Hydrous glasses called pitchstones, often with characteristic perlite textures, are one result of this reactivity. Many of these pitchstones, which must have erupted with only a small amount of water (0.09–0.29%), now contain large amounts of water, perhaps 2–7 percent weight. Their oxygen isotopic ratios indicate that this water is essentially meteoric (Taylor, 1968), and that it has been incorporated at low temperatures; indeed, the rate of water incorporation has been used as a dating method for obsidians.

With such a large-scale exchange of oxygen, which forms perhaps 95 percent by volume of the obsidians, it is inconceivable that all the other constituents will be preserved in their initial, prehydration concentrations. Several studies (Noble et al., 1967) have shown that this is the case, and there may be substantial changes in the concentration of the constituents, compared to their aphanitic crystalline equivalents. Thus, the view that all glassy rocks represent the composition of congealed magma should perhaps be restricted to those rocks which contain only small amounts of water.

But this process of incorporation of meteoric water may occur before the silicic magma was erupted, presumably by infiltration and diffusion through the liquid at shallow depths in the magma chamber (Lipman and Friedman, 1975). Pumice and glass resulting from the eruption of such magma will have low water concentrations reflecting exsolution during eruption, but as time passes, it may incorporate percolating meteoric water on the surface. Only isotopic geochemists may be able to disentangle the magnitude of the two effects, but it seems clear that water diffusing through a silicic magma prior to eruption will itself redistribute elements throughout the liquid column, and to this must be added the components that the water contained at the time it entered the magma body. It may become a very difficult task to determine the initial composition at the site of generation of a huge body of silicic magma. The complexity of the problem emphasizes the dynamic nature of magma generation, ascent, and eruption, and the glass represents an end product of many competing effects, almost all of which are time dependent.

A CONTRAST BETWEEN GLASS AND THE CRYSTALLINE EQUIVALENT

One of the well-documented contrasts between glassy rocks and their finely crystalline equivalents is to be found in the peralkaline lavas. One of the characteristics of these lavas is the high concentration of chloride, and representative analyses of Cl (in ppm) are as follows:

Glassy pantellerites	5600
Crystalline pantellerites	400
Glassy trachytes	1700
Crystalline trachytes	2100

Pantellerites are glassy rocks that contain quartz, whereas the trachytes may contain nepheline; the contrasting chloride concentration in pantelleritic rocks and the silica-poor trachytes is of interest here. It is apparent that Cl is retained as trachytic liquids crystallize, but the opposite is true for the pantellerites, and the crystalline equivalents contain less than a a tenth of the amount in the glass.

All these rocks are very poor in water, and Cl can be shown (Zies, 1960) not to be a superficial contaminant, so that the concentration of Cl in the glassy rocks probably represents the original amount in the magma. Whether or not Cl is retained in the crystalline equivalents substantially depends on the stability of the mineral sodalite, which Stormer and Carmichael (1971) showed not to be stable in liquids whose activities of silica were high enough to precipitate quartz. The Cl, and accompanying Na, were therefore lost during the crystallization process of pantellerites, and the lack of conformity between glass and the crystalline equivalent is readily explained. An analogous approach was taken by Anderson (1974) who analyzed the glass inclusions trapped in the early crystallizing phases of basaltic magma. These inclusions were also greatly enriched in Cl and S compared to the average sub-aerial crystalline basalt, and this emphasis on the composition of glass as representing more closely the original magma parallels the interest of Bowen.

While these rather brief comments on glassy rocks serve to show a little of why and how they are studied, the data obtained must be considered in relation to the crystalline rocks with which glassy rocks occur. In the case of the deep-sea basalts, the glassy skin may preserve the initial concentrations of the volatile components that were lost by degassing in sub-aerial basaltic eruptions. But in older rocks, it is likely that the crystalline varieties have undergone less alteration than their glasses, for the potential for change, represented by the free-energy change of any reaction, is enormously enhanced in the equivalent glass.

References

Anderson, A. T., Chlorine, sulfur and water in magmas and oceans, *Geol. Soc. Amer. Bull.*, *85*, 1485–1492, 1974.

Arndt, J., and F. Haberle, Thermal expansion and glass transition temperatures of synthetic glasses of plagioclase-like compositions, *Contrib. Min. Petrol.*, *39*, 175–183, 1973.

Bacon, C. R., High temperature heat content and heat capacity of silicate glasses: experimental determination and a model for calculation, *Am. J. Sci.*, *277*, 109–135, 1977.

Buddington, A. F., and D. H. Lindsley, Iron-titanium oxide minerals and synthetic equivalents, *J. Petrol.*, *5*, 310–357, 1964.

Carmichael, I. S. E., F. J. Turner, and J. Verhoogen, *Igneous Petrology*, McGraw-Hill, New York, 739 pp., 1974.

Carmichael, I. S. E., J. Nicholls, F. J. Spera, B. J. Wood, and S. A. Nelson, High-temperature properties of silicate liquids: applications to the equilibration and ascent of basic magma, *Phil. Trans. Roy. Soc. Lond.*, *A286*, 373–431, 1977.

Christian, J. W., *The Theory of Transformations in Metals and Alloys*, Pergamon, New York, 973 pp., 1965.

Hildreth, E. W., The magma chamber of the Bishop Tuff: gradients in temperature and composition. Unpublished Ph.D. thesis, Univ. Calif., Berkeley, 1977.

Kelley, K. K., Contributions to the data on theoretical metallurgy. XIII. High-temperature heat content, heat capacity and entropy data for the elements and inorganic compounds, *U.S. Bur. Mines. Bull. 584*, 232 pp. 1960.

Lipman, P. W., and I. Friedman, Interaction of meteoric water with magma: an oxygen-isotope study of ash flow sheets in southern Nevada, *Geol. Soc. Am. Bull., 86*, 695–702, 1975.

Noble, D. C., V. C. Smith, and L. C. Peck, Loss of halogens from crystallized and glassy silicic volcanic rocks. *Geochim. Cosmochim. Acta, 31*, 215–224, 1967.

Scherer, G., R. W. Hopper, and D. R. Uhlmann, Crystallization behavior and glass formation of selected lunar compositions, 3rd Lunar Sci. Conf., *Geochim. Cosmochim. Acta*, Suppl. 3, *3*, 2627–2637, 1972.

Stormer, J. C., and I. S. E. Carmichael, The free-energy of sodalite and the behavior of chloride, fluoride and sulphate in silicate magmas, *Am. Mineral., 56*, 292–306, 1971.

Taylor, H. P., The oxygen isotope geochemistry of igneous rocks, *Contrib. Mineral. Petrol., 19*, 1–71, 1968.

Uhlmann, D. R., A kinetic treatment of glass formation, *J. Noncrystalline Solids, 7*, 337–348, 1972.

Zies, E. G. Chemical analyses of two pantellerites. *J. Petrol., 1*, 304–308, 1960.

Chapter 9

ROCKS WHOSE COMPOSITION IS DETERMINED BY CRYSTAL ACCUMULATION AND SORTING

T. N. IRVINE

Geophysical Laboratory, Carnegie Institution of Washington, Washington, D.C.

INTRODUCTION

To this point in his book, Bowen had been primarily concerned with establishing that fractional crystallization was a principal process—indeed in his view, *the* principal process—in the compositional diversification of magmatic liquids. Implied in this concept is that there are large volumes of igneous rocks representing accumulations of the fractionated crystals, and as a next step Bowen undertook to search for evidence of such rocks. He concentrated particularly on the Scottish Hebrides, this being an area that was petrologically well documented for the times and with which he was personally familiar. He first examined the Central magma type of Mull, a basaltic suite that had been described by Bailey *et al.* (1924) as having two principal varieties, one non-porphyritic, the other characterized throughout by plagioclase phenocrysts. Bowen showed that in the chemical analyses of the porphyritic rocks, there was a direct correlation between the amount of the normative plagioclase and its An proportion, and he found indications of the same relationship in basalts from elsewhere in the world, using analyses compiled by Washington (1917). He cited this relationship as evidence that the plagioclase phenocrysts had been fractionated and accumulated.

Next Bowen turned to the olivine-rich rocks and gave a relatively detailed description of the peridotitic dikes of Skye, followed by a discussion of olivine basalts in which he presented geochemical evidence that olivine is crystallized in substantial amounts from quartz-normative liquids. In all these sections, he acknowledged that the plagioclase and olivine phenocrysts in the lavas, and the abundant olivine in the dikes, had probably once been in solution in magmatic liquid, but he could find no evidence of ultra-aluminous or ultrabasic liquids exceptionally rich in the normative components of these minerals. His conclusion, therefore, was that the plagioclase and olivine had been concentrated from much larger volumes of liquids that were essentially basaltic.

Bowen then gave passing attention to rocks formed by accumulation of pyroxene and hornblende, and completed the chapter with brief comments on banded gabbro and anorthosite and on magmatic ore deposits of chromite, ilmenite, and sulfides. In each case he expressed the view that the rocks in question had formed by accumulation of particular phases—pyroxene, hornblende, or plagioclase, chromite, ilmenite, or immiscible sulfide liquid—from silicate liquids that were not of especially unusual composition.

The title of the chapter stemmed from prior discussions in which Bowen had concluded that plagioclase and the mafic minerals in basaltic magmas generally crystallize simultaneously along cotectic liquidus boundaries. He believed, therefore, that the process of their individual enrichment involved not only *crystal accumulation*, such that the minerals were separated from their parental liquid, but also marked *crystal sorting*, whereby different minerals were mechanically separated from one another. Thus, in the example that he particularly stressed, plagioclase was concentrated in some parts of the magma body, and olivine in others. Bowen pointedly avoided a discussion of the physical aspects of the inferred sorting process but indicated an opinion that the effects of gravity must be dominant. It is of interest also that he was uncertain (p. 136) as to whether plagioclase would float or sink. This question, which was first raised by Darwin (1844), is still not completely answered today, even though it is critical to some of the principal concepts that have evolved in the past fifty years as to mechanisms of crystal fractionation (see later section on crystal sorting).

Bowen also proposed a mechanism of emplacement for the Skye peridotite dikes, involving a suspension of olivine crystals in basaltic liquid, that verges closely on the process now known as flow differentiation whereby the solids in a suspension are segregated away from the walls of fractures or pipes through which the suspension is injected (Baragar,

1960; Smith and Kapp, 1963; Bhattacharji and Smith 1964; Simkin, 1967; Gibb, 1968; Komar, 1972a,b). In describing his mechanism, however, Bowen somewhat unfortunately used the term "composite intrusion," but perhaps only for want of better terminology, because it is clear that he did not mean two or more discrete injections of magma as usually distinguishes composite intrusions (cf. Gary, McAfee, and Wolf, 1972). What Bowen visualized was a process whereby some relatively crystal-free magmatic liquid that had just been strained from the olivine suspension was injected into fractures immediately in advance of the crystal-laden magma where it served as a lubricant for the main intrusion. This process would seem just as plausible as that of pure flow differentiation, although the latter no doubt also was significant in the emplacement of the Skye dikes (Gibb, 1968).

Since 1928, studies of porphyritic lavas and dike rocks mentioning crystal accumulation and sorting are literally countless, and these processes have also come to be widely regarded as fundamental in the formation of layered intrusions. Layered intrusions were brought to prominence by Wager and Deer (1939) through their classic memoir on the Skaergaard intrusion in East Greenland, and since the mid-1950's there have appeared (1) descriptions of numerous major occurrences (e.g., Brown, 1956; Hess, 1960; Worst, 1960; Jackson, 1961; Smith, 1962; Rossman, 1963; Morse, 1969; Emslie, 1968, 1970; Naldrett and Mason, 1968; Naldrett et al., 1970; MacRae, 1969; Irvine and Baragar, 1972; Irvine, 1974, 1975c); (2) four symposium volumes (Fisher et al., 1963; Isachsen, 1969; Wilson, 1969; Visser and von Gruenewaldt, 1970) and one book (Wyllie, 1967) partly devoted to layered intrusions; and (3) a major book specifically devoted to the subject by Wager and Brown (1968). The characteristics of ultramafic rocks formed by crystal accumulation have been reviewed by Jackson (1971).

Bowen's view that basalt is the most primitive magmatic liquid appeared for many years to be valid, but truly ultrabasic liquids called komatiites, rich in normative olivine and pyroxene, have recently been identified in lavas, principally in Archean terranes (e.g., Viljoen and Viljoen, 1969; Pyke, Naldrett, and Eckstrand, 1973; Arndt, Naldrett, and Pyke, 1977). Also, high-pressure melting experiments indicate that primary melts in the upper mantle should commonly be picritic (Green and Ringwood, 1967; Ito and Kennedy, 1967, 1968; O'Hara and Yoder, 1967), and it has been suggested that many basaltic magmas are derivative from this type of melt (O'Hara, 1965, 1968). Several types of high alumina basalt have been recognized (Kuno, 1960; Yoder and Tilley, 1962; Engel and Engel, 1964), but as yet no truly ultra-aluminous liquids have been reported, although such liquids might be associated with the large anorthositic complexes (cf. Emslie, 1971, 1973).

In the present chapter, attention is essentially restricted to rocks formed through accumulation of olivine, the pyroxenes, and plagioclase in tholeiitic magmas. Also, the emphasis is more on compositional variations associated with crystal fractionation and accumulation than on those due to crystal sorting. The reason in part is that the former variations are more amenable to systematic analysis, but also it appears that to demonstrate that crystal sorting has occurred on more than just a local scale, it is necessary first to eliminate the possibility that the features in question might be due only to fractional crystallization. In the organization of the chapter, some general principles are first outlined on experimental and theoretical grounds; then their application is illustrated in a few selected rock suites; and finally the possibilities of crystal sorting are discussed. Particular attention is given to the problem of defining parental liquid compositions.

COMPOSITIONAL EFFECTS OF CRYSTAL FRACTIONATION AND ACCUMULATION: PRINCIPLES AND THEORY

Features Relating to Crystallization Order

Figure 9-1 illustrates the common liquidus relations of olivine, the pyroxenes, plagioclase, and the silica minerals in anhydrous tholeiitic magmas at low pressures. The main diagram is modeled after a host of simple silicate systems, several of them investigated by Bowen, and the phase boundaries have been drawn to conform as closely as possible with experimental data on the melting relations of natural basalts (see figure caption for details). The quantitative relations can change considerably, however, depending on certain compositional characteristics of the liquids not specificially distinguished in the coordinates, and they also change with pressure. Thus, it can generally be expected that with increased alkalies or H_2O, the liquidus volumes of olivine, clinopyroxene (augite), and orthopyroxene (or pigeonite) will all expand at the expense of the plagioclase volume, and the olivine liquidus will also expand into the region of the orthopyroxene volume while the latter shifts to more siliceous compositions. By contrast, with increased Fe^{2+}/Mg in the liquid, the plagioclase volume increases, and the olivine volume contracts relative to the pyroxene liquidi. With increased pressure, the pyroxene volumes

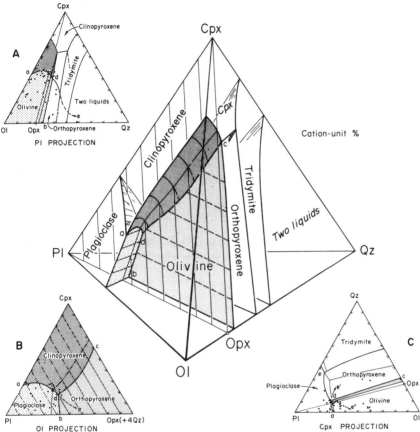

Figure 9-1. Phase diagram model of the "system" Ol–Cpx–Pl–Qz (olivine–clinopyroxene–plagioclase–quartz) illustrating the general liquidus relations of subalkaline basaltic and related magmas at low pressures. Revised from Irvine (1970a). The phase boundaries are modeled after relations in numerous simple silicate systems and have been drawn in accord with experimental data on the 1-atm melting relations of natural basalts represented by the data points in the projections. In the Ol projection (B), bounding surfaces of the olivine liquidus volume are displayed on the join Cpx–Pl–Opx, and compositions with Qz in the norm are projected onto this plane by adding Ol in accordance with the equation

$$SiO_2 + (Mg,Fe)_2SiO_4 = 2(Mg,Fe)SiO_3,$$

which is equivalent in cation units to

$$1 \text{ Qz} + 3 \text{ Ol} = 4 \text{ Opx}.$$

Thus the coordinates of the projection are Cpx:Pl:Opx(+4 Qz). See Irvine (1970a) for sources of data and further details on construction methods and norm conventions.

increase relative to those of both olivine and plagioclase, and a spinel volume develops between the liquidi of plagioclase and olivine, completely intervening at about 8–10 kbar.[1]

Relations in the silica (Qz)-rich region are still not well known, but it is apparent that they depend strongly on the extent of the two-liquid region (Irvine, 1976). With high Mg/Fe^{2+} and moderate concentrations of alkalies, there is probably a continuous one-liquid region from the 5-phase "point" d (where basaltic to andesitic liquids coexist with olivine, two pyroxenes, and plagioclase) through to a comparable point located approximately at e in the Cpx projection (Fig. 9-1C), representing rhyolitic liquid in equilibrium with a silica mineral. But Hawaiian tholeiites, which are relatively poor in alkalies, differentiate on trends such that they occasionally intersect a two-liquid solvus (Roedder and Weiblen, 1971), apparently somewhere near e' in the Cpx projection (Fig. 9-1C); and the K-poor tholeiitic liquid of the Skaergaard intrusion probably never entered the Qz-normative region, its well-known trend of marked iron-enrichment leading instead to an intersection with a much expanded two-liquid solvus (McBirney and Nakamura, 1974) near e'' in the Cpx projection. These trends are documented in later sections.

A principal observation to be made from the relations in Fig. 9-1 is that they include eighteen different orders of crystallization for just olivine, the two pyroxenes, and plagioclase. These orders are listed in Table 9-1, where it is seen that olivine is the first mineral in six orders, plagioclase in five, orthopyroxene in two, and clinopyroxene in five. (The six orders with olivine first are readily identified in the Ol projection in that each of the three liquidus boundaries describing the olivine saturation surface can be approached from either side.) If the minerals are fractionally crystallized, then each order will yield a specific sequence of crystalline products, also listed in Table 9-1. Many of these sequences have been identified in the stratigraphic successions of accumulated (cumulus) minerals characteristic of layered intrusions (Irvine, 1970a; Jackson,

---

[1] The various effects of the parameters mentioned in this paragraph are exemplified in the following systems:

Alkalies: Di–An–Ab (Bowen, 1915); Fo–An–Ab (Schairer and Yoder, 1967); Fo–Lc–Qz (Schairer, 1954); Fo–Ne–Qz (Schairer and Yoder, 1961); Di–Ne–Qz (Schairer and Yoder, 1960); Fo–Fa–An–Or–Qz (Irvine, 1976).

H_2O: Di–An–H_2O (Yoder, 1965).

Fe^{2+}/Mg: MgO–FeO–SiO_2 (Bowen and Schairer, 1935); Fo–Di–FeO–Fe_2O_3–Qz (Presnall, 1966); Fo–Fa–An–Or–Qz (Irvine, 1976).

Pressure: Fo–An, En–An (Kushiro and Yoder, 1966); Fo–Di–An–Qz (Presnall *et al.*, 1978).

Table 9-1. Possible orders of crystallization for the minerals olivine, clinopyroxene, orthopyroxene, and plagioclase, based on the liquidus relations in Fig. 9-1, and the cumulate sequences that would derive from these orders under ideal conditions of crystal fractionation. After Irvine (1970a, Table 2).

Crystallization Order[1]	Cumulate Sequence[2]
1. Ol; Cpx; Pl; Opx	Ol; Ol + Cpx; Ol + Cpx + Pl; Cpx + Pl + Opx
2. Ol; Cpx; Opx; Pl	Ol; Ol + Cpx; Cpx + Opx; Cpx + Opx + Pl
3. Ol; Opx; Cpx; Pl	Ol; Opx; Opx + Cpx; Opx + Cpx + Pl
4. Ol; Opx; Pl; Cpx	Ol; Opx; Opx + Pl; Opx + Pl + Cpx
5. Ol; Pl; Cpx; Opx	Ol; Ol + Pl; Ol + Pl + Cpx; Pl + Cpx + Opx
6. Ol; Pl; Opx; Cpx	Ol; Ol + Pl; Pl + Opx; Pl + Opx + Cpx
7. Pl; Ol; Cpx; Opx	Pl; Pl + Ol; Pl + Ol + Cpx; Pl + Cpx + Opx
8. Pl; Cpx; Ol; Opx	Pl; Pl + Cpx; Pl + Cpx + Ol; Pl + Cpx + Opx
9. Pl; Ol; Opx; Cpx	Pl; Pl + Ol; Pl + Opx; Pl + Opx + Cpx
10. Pl; Cpx; Opx	Pl; Pl + Cpx; Pl + Cpx + Opx
11. Pl; Opx; Cpx	Pl; Pl + Opx; Pl + Opx + Cpx
12. Opx; Cpx; Pl	Opx; Opx + Cpx; Opx + Cpx + Pl
13. Opx; Pl; Cpx	Opx; Opx + Pl; Opx + Pl + Cpx
14. Cpx; Ol; Opx; Pl	Cpx; Cpx + Ol; Cpx + Opx; Cpx + Opx + Pl
15. Cpx; Ol; Pl; Opx	Cpx; Cpx + Ol; Cpx + Ol + Pl; Cpx + Pl + Opx
16. Cpx; Pl; Ol; Opx	Cpx; Cpx + Pl; Cpx + Pl + Ol; Cpx + Pl + Opx
17. Cpx; Pl; Opx	Cpx; Cpx + Pl; Cpx + Pl + Opx
18. Cpx; Opx; Pl	Cpx; Cpx + Opx; Cpx + Opx + Pl

[1] Symbols as follows: Ol, olivine; Cpx, clinopyroxene (augite); Opx, orthopyroxene (or pigeonite); Pl, plagioclase. It is assumed that olivine and clinopyroxene and olivine and plagioclase have cotectic relations, and that olivine reacts to orthopyroxene. The possibility of simultaneous appearance of phases is neglected.

[2] Within each cumulate sequence the minerals are listed in the same order as their overall order of crystallization, rather than in their probable order of abundance, to make the derivation of the sequence more apparent. The expected orders of abundance for the minerals when coprecipitated under ideal conditions of crystal fractionation are: plagioclase, clinopyroxene, olivine; and plagioclase, clinopyroxene, orthopyroxene. The advent of simultaneous appearance of minerals would result in gaps in the sequences listed here.

1970), and so they are termed *cumulate sequences*. It is notable also that all of the crystallization orders in Table 9-1 can be derived from liquids plotting immediately around the 5-phase "point" *d*. Thus, the crystallization characteristics of these liquids can variously be changed by only small changes in their compositions or in the locations of the phase boundaries.

It is also possible from the relations in Fig. 9-1 (at least in principle) to predict the modal proportions in which two or more of the minerals will coprecipitate. This information is of importance both in defining fractionation trends and in deciding whether crystal sorting has occurred. The modal proportions are indicated by lines called *control lines* drawn as tangents to the liquid path. From path segments where only one mineral is precipitating, the control lines extend to the mineral composition(s) in equilibrium with the liquid. Where more than one mineral is precipitating, the intersection of the control lines with the mineral composition join defines the relative proportions in which the minerals are currently forming.

Minerals other than the four considered in Table 9-1 are also fractionated from tholeiitic magmas. They are perhaps best identified on the basis of their occurrence as cumulus phases in layered intrusions. In these bodies, cumulus chromite typically occurs in small amounts with olivine precipitated at early stages, and cumulus magnetite, ilmenite, and apatite commonly accompany the pyroxenes and plagioclase formed at later stages. Hornblende does not seem to have been reported as a cumulus phase in tholeiitic intrusions, but it commonly occurs in reaction relation with the pyroxenes and plagioclase and so may be fractionated in some circumstances. Immiscible liquids, especially sulfide liquids but also silica-rich types, probably also occur as fractionated phases.

In view of the variety of possibilities indicated above, it will be apparent that to prove that a rock or group of rocks has formed by crystal sorting rather than by fractional crystallization may frequently be impossible.

Magnesium-Iron Fractionation

General Theory   There is probably no chemical parameter that has been more widely used to monitor the effects of crystal fractionation in igneous rocks than the ratio of magnesium to ferrous iron. Attention here will be concentrated on the effects of olivine fractionation on this ratio, in which case its variations are conveniently examined on the basis of the distribution coefficient:

$$K_D = (X_{Mg}{}^L/X_{Fe^{2+}}{}^L)/(X_{Mg}{}^{Ol}/X_{Fe^{2+}}{}^{Ol}) \tag{9-1}$$

where $X_{Mg} = Mg/(Mg + Fe^{2+})$ and $X_{Fe^{2+}} = Fe^{2+}/(Mg + Fe^{2+}) = 1 - X_{Mg}$ in coexisting liquid and olivine as indicated by the superscripts L and Ol respectively. Experimentally determined equilibrium values for this coefficient appear to be practically independent of temperature and pressure and generally fall in the range 0.30–0.36 for melt compositions encompassing terrestrial basaltic and ultrabasic liquids and lunar basalts (Roedder and Emslie, 1970; Longhi and Walker, 1975; O'Hara, Saunders, and Mercy, 1975; N. T. Arndt, pers. comm., 1976). Values for haplo-basaltic liquids in the system Fo–Fa–An–Or–Qz are typically 0.30–0.32, although there is an increase to 0.36 (and probably higher) for melts of low Mg/Fe^{2+} on the brink of liquid immiscibility (Irvine, 1976, Fig. 61). For present purposes, K_D is assumed to be constant. A value of 0.31 was used for numerical calculations summarized in Table 9-2 involving liquids of relatively high Mg/Fe^{2+}; a value of 0.33 was used in preparing Fig. 9-4 where the diagram covers the full range of Mg/Fe^{2+}.

Given a constant K_D, the cation fraction of $Mg + Fe^{2+}$ that is transferred from liquid to olivine during fractional crystallization can be related to $Mg/(Mg + Fe^{2+})$ in the residual liquid by the following Rayleigh-type fractionation equation (Irvine, 1977a):

$$F_{MgFe}{}^{Ol} = 1 - \left[\left(\frac{X^0{}_{Fe^{2+}}{}^L}{X_{Fe^{2+}}{}^L} \right) \left(\frac{X_{Mg}{}^L}{X^0{}_{Mg}{}^L} \right)^{K_D} \right]^{1/(1-K_D)} \tag{9-2}$$

where the zero superscript distinguishes an initial or primitive composition, in this case for the liquid. A similar relationship between $F_{MgFe}{}^{Ol}$ and olivine composition exists through Equation 9-1. Curves that describe these relationships are illustrated and explained in Fig. 9-2.

If the cation fraction of $Mg + Fe^{2+}$ in the initial liquid is known, then the proportion of liquid solidified, expressed as a fraction of its total initial content of cations, is given by

$$F_S = F_{MgFe}{}^{Ol}(N^0{}_{MgFe}{}^L/N_{MgFe}{}^{Ol}) \tag{9-3}$$

where N_{MgFe} denotes the cation fraction of $Mg + Fe^{2+}$ (=0.67 for the olivine).

More commonly, however, the initial liquid is the unknown. The above equations can be applied to this problem if the composition of a derivative liquid from which only olivine has been fractionated is available, and if the composition of the "first" or most magnesian olivine precipitated in the system under study is known or can be inferred. A case in point might involve the comparison of the chilled margin of a layered intrusion (an assumed derivative liquid) with the most magnesian cumulus olivine in the body. From the olivine composition, it is inferred on the basis of Equation 9-1 that

$$X^0{}_{Mg}{}^L = R^0{}_{Ol}K_D/(1 + R^0{}_{Ol}K_D) \tag{9-4}$$

where $R^0{}_{Ol} = X^0{}_{Mg}{}^{Ol}/X^0{}_{Fe^{2+}}{}^{Ol}$. Then by substituting $X^0{}_{Mg}{}^L$ into Equation 9-2 along with $X_{Mg}{}^L$ for the derivative liquid, a value of $F_{MgFe}{}^{Ol}$ is obtained corresponding to the fractionation stage of the derivative liquid. It is then

Table 9-2. Assumed derivative and calculated primitive liquid compositions and comparative data for Kilauea, Mauna Loa, the Muskox intrusion, abyssal tholeiites, and the Skaergaard intrusion. Ferrous-ferric ratios have been adjusted so that Fe^{3+}/Ti (atomic) = 0.5, and most analyses are normalized to 100% without H_2O and CO_2. D.L. = derivative liquid; P.L. = primitive liquid.

	Kilauea 1959					Kilauea Prehistoric		Mauna Loa Historic		Muskox Intrusion		Abyssal Tholeiites			Skaergaard Intrusion
	1 D.L.	2 P.L.	3 Glass	4	5	6 D.L.	7 P.L.	8 D.L.	9 P.L.	10 D.L.	11 P.L.	12 D.L.	13 P.L.	14 Glass	15 Chill
SiO_2	50.18	48.37	49.20	48.21	47.92	51.23	49.18	52.18	50.04	51.08	49.20	48.93	47.70	47.92	48.55
Al_2O_3	13.79	11.36	12.77	11.37	10.75	14.14	11.66	14.17	11.68	13.65	11.98	16.13	13.72	17.89	17.39
Fe_2O_3	1.38	1.14	1.50	1.50	1.08	1.22	1.01	1.03	0.85	0.53	0.48	0.37	0.31	0.52	0.59
FeO	10.01	10.25	10.05	10.18	10.65	9.76	10.19	10.01	10.25	9.72	10.02	9.18	9.17	9.28	9.19
MgO	7.00	14.07	10.00	13.94	15.43	7.00	13.97	7.00	14.07	9.78	14.88	9.93	15.80	8.67	8.70
CaO	11.63	9.58	10.75	9.74	9.33	11.22	9.25	10.58	8.72	11.30	9.78	11.92	10.14	11.77	11.49
Na_2O	2.31	1.90	2.12	1.89	1.79	2.30	1.90	2.29	1.89	1.81	1.56	2.43	2.07	2.59	2.39
K_2O	0.60	0.49	0.51	0.44	0.44	0.41	0.34	0.37	0.30	0.64	0.56	0.15	0.12	0.05	0.25
TiO_2	2.77	2.28	2.57	2.24	2.16	2.44	2.01	2.07	1.71	1.07	0.93	0.74	0.63	1.03	1.18
MnO	0.17	0.18[1]	0.17	0.18	0.18	0.17	0.18[2]	0.17	0.18[1]	0.17	0.18	0.13	0.16[1]	0.17	0.16
P_2O_5	0.28	0.23	0.25	0.22	0.23	0.22	0.17	0.21	0.17	0.09	0.08	0.05	0.04	0.07	0.10
Cr_2O_3	0.06[2]	0.14[3]	—	—	0.14[3]	0.06[2]	0.14[3]	0.06[2]	0.14[3]	0.08	0.24	0.06	0.13[3]	0.05	—
total	100.18	99.99	99.89	99.91	100.10	100.17	100.00	100.14	100.00	99.92	99.89	100.02	99.99	100.01	99.99
$100\,Mg/(Mg+Fe^{2+})$	55.5	69.5	63.9	70.9	72.1	55.6	69.5	55.5	69.5	64.2	72.6	65.8	74.7	62.5	62.8
$100\,Al/(Mg+Fe^{2+})$	86.4	45.3	64.6	45.8	39.7	88.8	46.7	88.8	46.5	70.9	46.2	84.6	51.7	102.0	99.2
Equilibrium Olivine, $K_D = 0.31$		88.00					88.00		88.00		89.50		90.50		

[1] Equilibrated olivine assumed to contain 0.18% MnO.

[2] Cr_2O_3 assumed to be 0.06 wt % in accord with data from Wright (1973).

[3] Calculated on the assumption that equilibrated olivine is accompanied by 0.5 cation unit % $(Mg, Fe^{2+})Cr_2O_4$ as indicated by data on associated picritic rocks.

Column Headings

1. Average summit lava for the 1959 Kilauea eruption, as normalized to 7 wt % MgO by Wright (1972, Table 14).
2. Primitive 1959 Kilauea liquid, calculated from 1 by method described in text.
3. Most basic glass analyzed from 1959 Kilauea eruption, from Murata and Richter (1966, Table 8, sample s-5g).
4. Weighted average composition of lava from the first phase of the 1959 eruption of Kilauea Iki, from Murata and Richter (1966, Table 8).
5. Weighted average composition for all lava from the 1959 eruption of Kilauea, from Wright (1973, Table 6).
6. Average prehistoric Kilauea olivine-controlled lava, as normalized to 7 wt % MgO by Wright (1972, Table 14).
7. Primitive prehistoric Kilauea liquid, calculated from 6 by method described in text.
8. Average historic Mauna Loa olivine-controlled liquid, as normalized to 7 wt % MgO by Wright (1972, Table 14).
9. Primitive historic Mauna Loa liquid, calculated from 8 by method described in text.
10. Average of two analyses of Muskox chilled margin gabbro, from Smith and Kapp (1963).
11. Primitive Muskox liquid, calculated by method described in text.
12. Average olivine-phyric abyssal tholeiite from DSDP Leg 37, from data by Blanchard *et al.* (1976, Table 2).
13. Primitive abyssal tholeiite liquid, calculated from 12 by method described in text.
14. Abyssal tholeiite glass En−70−25−6−70 from Mathez (1976, Table 1).
15. Skaergaard chilled margin gabbro, from Wager (1960).

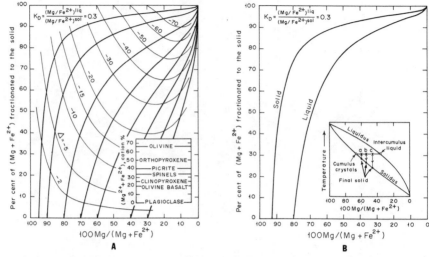

Figure 9-2. Curves illustrating the variation of $Mg/(Mg + Fe^{2+})$ of a solid such as olivine and its parental liquid during fractional crystallization in a system closed to oxygen, based on the indicated distribution coefficient and Equations 9-1 and 9-2 in the text. Diagram from Irvine (1974) with permission of the Geological Society of America. The heavy curves in A describe the composition of the fractionated solid when the first crystals have the compositions given by the intercepts with the horizontal axis. The light curves with Δ-labels indicate the compositional difference between the crystals precipitated as fractionation progresses and the first crystals. In the case of solids crystallizing from their own melt, the vertical scale is equivalent to percent solidification. In magmas, however, the liquid generally has a different content of $Mg + Fe^{2+}$ than the minerals that crystallize from it (see inset in A), and more than one mineral may precipitate at a time; thus to obtain percent solidification, the vertical scale must by multiplied by the ratio:

$$\frac{\text{percent } Mg + Fe^{2+} \text{ in the initial liquid}}{\text{percent } Mg + Fe^{2+} \text{ in the fractionated solids,}}$$

where the denominator is a constant. If the denominator changes, for example, because of the appearance of a new mineral on the liquidus, then it is necessary to start over in terms of percent solidification of the *remaining* liquid, using the new value of the ratio above and the fractionation curve appropriate to the solids forming by that stage.

Diagram B shows a paired set of solid and liquid curves, and the inset illustrates schematically how variations in the amount of interstitial (intercumulus) liquid can affect the composition of the solid that is finally crystallized. In igneous rocks, the latter effect is dependent not just on the ratio of solid to liquid, but on

$$\frac{\text{percent liquid} \times \text{percent } Mg + Fe^{2+} \text{ in the liquid}}{\text{percent solid} \times \text{percent } Mg + Fe^{2+} \text{ in the solid,}}$$

and the relationship may be further complicated by the formation of interstitial Fe-rich minerals such as magnetite.

recalled that

$$N^0_{MgFe}{}^L = F_S N_{MgFe}{}^{Ol} + (1 - F_S) N_{MgFe}{}^L \qquad (9\text{-}5)$$

Combination of this equation with Equation 9-3 gives

$$F_S = \frac{N_{MgFe}{}^L F_{MgFe}{}^{Ol}}{N_{MgFe}{}^{Ol}(1 - F_{MgFe}{}^{Ol}) + N_{MgFe}{}^L F_{MgFe}{}^{Ol}}. \qquad (9\text{-}6)$$

The complete composition of the liquid is then obtained by adding forsterite and fayalite to the derivative liquid in proportions appropriate to satisfy $X^0_{MgFe}{}^L$ and F_S from Equations 9-4 and 9-6.

Imperfect Separation of Crystals and Liquid A complication commonly encountered in studying fractionation effects in igneous rocks is that the crystals and liquid were not perfectly separated during the fractionation event. This possibility reflects not only on the bulk composition of the rocks but also on the compositions of the constituent minerals (and of any liquid that might have been quenched as glass) if the crystals and liquid still associated after fractionation continued to equilibrate during the final stages of solidification. A simple illustration of the importance of this effect in determining the compositions of the cumulus mafic minerals in layered intrusions when they react with differing proportions of intercumulus liquid is shown in the inset diagram in Fig. 9-2B.

A method of evaluating this effect is to compare the Mg/Fe^{2+} variations in the rock series with variations of some constituent that reflects as closely as possible the proportion of liquid. There are many elements in magmatic liquids that are effectively excluded from olivine, and aluminum has been found to be particularly useful for present considerations. Its variations will be compared with variations of Mg/Fe^{2+} in graphs of $100\ Al/(Mg + Fe^{2+})$ vs. $100\ Mg/(Mg + Fe^{2+})$ like those illustrating analyses of the lavas of Kilauea and Mauna Loa volcanoes, Hawaii, in Fig. 9-3. A reciprocal relationship with respect to $Mg + Fe^{2+}$ is used on both coordinates in order to obtain a graph in which mixtures of the compositions represented by any two points will plot on a straight line between those points. This property is valuable when considering processes involving the fractionation of crystals from liquid (or vice versa) and the mixing of various combinations of crystals and liquids.[2]

[2] It is a property that is common to any rectilinear plot of chemical ratios in which both coordinates have the same denominator. An extreme example is the plot advocated by Pearce (1968, 1970) in which only one constituent is represented in the denominator. At the other extreme are the simple plots of oxides, such as the Harker diagram, where the denominator traditionally is not specified because it is the total of the concentrations of all constituents in the analyzed samples.

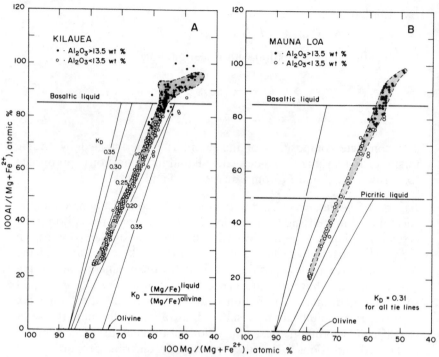

Figure 9-3.   Plots of $100Al/(Mg + Fe^{2+})$ vs. $100Mg/(Mg + Fe^{2+})$ in the lavas of Kilauea and Mauna Loa volcanoes, Hawaii. Data mainly from Murata and Richter (1966), Wright and Fiske (1971), and Wright (1972). Other sources are given by Irvine (1970a, Fig. 14). Comparative tie lines are shown for olivine and model basaltic and picritic liquids, based on the indicated K_D values. Equilibrium values for K_D generally fall in the range 0.30–0.35.

With regard to the data in Fig. 9-3, the observation frequently has been made that the compositional variations of the lavas of Kilauea and Mauna Loa can be largely explained simply on the basis of mixing the basaltic liquids typical of the volcanoes with olivine Fo_{85-87} (e.g., Powers, 1935; Macdonald, 1949; Murata and Richter, 1966; Wright, 1972). The relationships displayed in Fig. 9-3 are in accord with this observation. In addition, however, it is seen that the K_D values that would characterize such mixtures would fall in the range 0.20–0.25 and so would be significantly different from the equilibrium values enumerated above. On the other hand, as illustrated in Fig. 9-3B, the main spreads of data points are roughly parallel to equilibrium tie lines for olivine and picritic liquid. It is evident, therefore, that the trends may reflect fractionation of this type of magma and, hence, that the graphs are also useful for assessing possible parental liquid compositions.

The variations of an "excluded" element such as Al in liquids from which only olivine is being fractionated can be modeled on the basis of partitioning coefficients of the form $D = C_{Ex}^{O1}/C_{Ex}^{L}$, where C_{Ex}^{O1} and C_{Ex}^{L} denote the concentrations of the excluded element Ex in the olivine and liquid at any stage of fractionation. If D is constant, then the variation of Ex is described by the well-known Rayleigh equation:

$$C_{Ex}^{L} = C_{Ex}^{0 L}(1 - F_S)^{D-1} \qquad (9\text{-}7)$$

Thus, by substituting first Equation 9-2 into 9-3 and then 9-3 into 9-7, an equation can be obtained that relates $Mg/(Mg + Fe^{2+})$ and the concentration of Ex in the fractionated liquid to $Mg/(Mg + Fe^{2+})$ and the concentrations of $Mg + Fe^{2+}$ and Ex in the original liquid, independently of percent solidification. This equation describes liquidus fractionation curves of the type introduced into the petrological literature by Bowen (1941) and Osborn and Schairer (1941). The actual substitution of equations just outlined can be conveniently done in a computer program designed to calculate the curves, so the rather cumbersome result will not be given here.

A set of liquidus fractionation curves for $K_D = 0.33$ and $D = 0$ is shown in Fig. 9-4. Specific processes are indicated for three hypothetical parental liquids to show the difference between paths of equilibrium and fractional crystallization and to illustrate certain effects of blending liquids at different stages of differentiation. Solid-liquid tie lines are tangent to the fractionation curves; equilibrium mixtures of crystals and liquid should, of course, plot along the tie lines. If the fractionation curve appropriate to the magmatic series or system under study can be established on the basis of the composition of a fractionated liquid (such as FL_2), then the tangency point of the tie line from the most magnesian olivine precipitated in the system (e.g., FS_0) defines $Ex/(Mg + Fe^{2+})$ in the initial liquid (as at PL_1). In the foregoing theory, this relationship is generalized to cover all constituents of the derivative liquid.

If the fractionated olivine is not completely separated from its parental liquid and continues to equilibrate with the part that has not been removed, then its composition will evolve to approximately the bulk $Mg/(Mg + Fe^{2+})$ of the mixture as indicated by the location of the bulk mixture composition along the olivine-liquid tie line. (The exact $Mg/(Mg + Fe^{2+})$ developed in the olivine will depend on the kind and amount of the other mafic minerals such as the pyroxenes and magnetite that may crystallize during the final stages of solidification.) Accordingly, if the mixture composition is variable, there may develop a correlation between the composition of the olivine and its amount—the same relationship that was noted by Bowen for plagioclase. On the other hand, as may be seen from Fig. 9-4, the composition and trend of the main body of liquid will not be affected by imperfect removal of liquid from the fractionated crystals; the only effect is that the amount of the main body of liquid is proportionately reduced.

Illustrative Models More specific illustrations of the way in which the fractionation curves relate to the compositions of rocks and liquids developed during fractional crystallization are given in model diagrams in Fig. 9-5. In diagram A, the liquid is fractionated according to the crystallization order: olivine; clinopyroxene; plagioclase. The initial

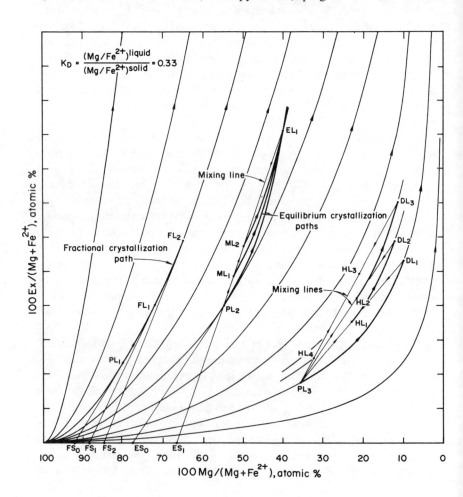

liquid plots at a, and the first olivine is Fo_{90}. As solidification proceeds, the liquid composition moves along the fractionation curve and progressively more iron-rich olivine is precipitated. Clinopyroxene appears at b, at which point the olivine being formed is about Fo_{81}. With crystallization of the pyroxene, a slightly different fractionation curve is followed, its trend depending additionally on the K_D for pyroxene-liquid equilibrium, the Al content of the pyroxene, and the modal ratio of pyroxene to olivine in the fractionated crystals. The values assumed for these quantities in the model are typical for basaltic magmas. The new curve is followed to c, where plagioclase appears. Precipitation of the feldspar prevents further marked increase of $Al/(Mg + Fe^{2+})$ in the liquid; consequently, the next segment of liquid path cuts sharply across the general trend of the olivine

---

Figure 9-4. Liquidus fractionation curves on a plot of $100Ex/(Mg + Fe^{2+})$ vs. $100Mg/(Mg + Fe^{2+})$ illustrating the relationship between the concentration of an "excluded" element Ex and the variation of Mg/Fe^{2+} in a liquid from which olivine is fractionated, based on the indicated K_D value. Specific events are illustrated for three "parental liquids." Liquid PL_1 undergoes simple fractional crystallization through FL_1 to FL_2, yielding a succession of olivine compositions from FS_0 to FS_2. Note that the olivine-liquid tie lines are always tangent to the fractionation curve. By contrast, when liquid PL_2 undergoes equilibrium crystallization to EL_1, the tie lines always pass through the initial composition point, and the liquid moves across the fractionation curves becoming disproportionately enriched in E as compared with the change in Mg/Fe^{2+}. However, if liquid EL_1 undergoes batch fractionation, it can then be blended with more of PL_2 to give liquids such as ML_1 and ML_2, with equilibrium crystallization paths that also pass through EL_1. Repetition of this process in such a way that the crystallization always progresses to about the same point could, therefore, yield large amounts of fractionated crystals and interstitial liquid of approximately constant composition. Such a process, carried out on an incremental scale has been suggested as an explanation of the relatively constant composition of cumulus olivine and clinopyroxene through large stratigraphic thicknesses in the Duke Island ultramafic complex in southeastern Alaska (Irvine, 1974); and it may be the reason for the lack of marked or systematic trends of Mg/Fe^{2+} in the mafic minerals and of Ab/An in plagioclase through thick successions of gabbroic cumulates in the Crillon-La Perouse intrusion in southeastern Alaska (Rossman, 1963) and in the Axelgold intrusion in British Columbia (Irvine, 1975c). Parental liquid PL_3 undergoes fractional crystallization, yielding the derivative liquid DL_1, which is then mixed with more PL_3 to give the hybrid liquid HL_1. This liquid is partially fractionated and then combined with more PL_3 to give HL_2, and so on. This process would yield fractionation products exhibiting cyclic variations in Mg/Fe^{2+} like those observed in the Muskox intrusion (see Fig. 9-10). The hybrid liquids may also become disproportionately enriched in Ex (as in going from HL_1 to HL_2 to HL_3; cf. O'Hara, 1977) but this possibility is critically dependent on the mixing proportions—e.g., HL_4 is a hybrid liquid on the same mixing line as HL_3, but it is impoverished in Ex as compared with HL_2, rather than enriched.

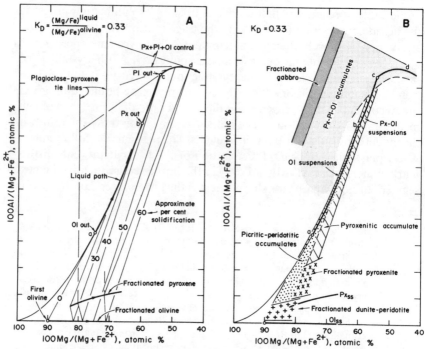

Figure 9-5. A. Crystallization model on a plot of $100Al/(Mg + Fe^{2+})$ vs. 100 Mg/$(Mg + Fe^{2+})$, based on the liquidus fractionation curves in Fig. 9-4, for a liquid that fractionates according to the crystallization order: olivine; clinopyroxene; plagioclase. Plagioclase solid solutions effectively plot at infinity on the vertical scale, hence pyroxene-plagioclase tie lines (and olivine-plagioclase tie lines not shown) are essentially vertical. The olivine-liquid compositions satisfy the K_D value of 0.33; the pyroxene-olivine relations are approximately those observed in nature.

B. Inferred composition fields for the different kinds of rocks that might be produced by accumulating the fractionated minerals in the model in A with different proportions of contemporary liquid. For further explanation, see text.

and pyroxene fractionation curves. The curvature shown for this path segment is only schematic, but an indication of similar curvature can be seen in the Kilauea data in Fig. 9-3A. The intersections of control lines from the curved path with the tie lines from the mafic minerals to plagioclase define the bulk compositions of the crystals formed during this stage of fractionation.

In Fig. 9-5B, general compositional fields have been outlined for rocks theoretically formed by combining the fractionated crystals in the model in 9-5A with different proportions of contemporary liquid. A simple terminology has been introduced to describe the rocks, based on the ratio of crystals to liquid. Thus, if the crystals were accumulated with only small amounts of liquid trapped in the pore spaces between them, *fractionated* dunite or peridotite, pyroxenite, and gabbro, like the cumulates of layered intrusions, are produced. But if the separation of crystals and liquid was less complete—as might generally be expected in flow differentiation, for example—then rocks that are here termed *accumulates* are formed. If the crystals are only slightly concentrated, or if the liquid simply undergoes partial crystallization with no fractionation, then the magma is effectively a *suspension* such as might be quenched as a porphyritic volcanic rock. It is emphasized, however, that this terminology is not intended to be formal. The divisions are obviously arbitrary, and it may be desirable to modify or change the definitions, depending on the circumstances—e.g., a suspension in one environment might be similar to an accumulate in another. Detailed examples cannot be illustrated here, but it is the author's experience that distribution patterns like those in Fig. 9-5B are common to differentiated ultramafic-gabbroic complexes and that the models are a useful guide in interpreting the patterns.

EFFECTS OF LIQUID BLENDING

In principle, the effects of crystallization order and Mg/Fe^{2+} fractionation (and also other fractionation effects) can be considerably modified and diversified simply through the blending of batches of liquid of the same parentage at different stages of differentiation. Such a process can readily be visualized as occurring in subvolcanic intrusions and passageways through eruptive processes whereby fresh magma is repeatedly added to residues of earlier liquids that have been differentiated by partial crystal-lization. This possibility is indicated, for example, in the Muskox intrusion in the Canadian Northwest Territories (Irvine and Smith, 1967) and in Kilauea volcano (Wright and Fiske, 1971; Wright, 1973).

Some possible effects of the above type of liquid blending on Mg/Fe^{2+} variations are illustrated in Figs. 9-4 and 9-6C. In Fig. 9-4, the repeated addition of parental liquid PL_3 to its derivatives results in a cyclic repeti-tion of Mg/Fe^{2+} variation and at the same time the concentration of the excluded element Ex in the hybrid liquids may be relatively increased, depending on the mixing proportions. In Fig. 9-6C, the combination of fresh liquid P_1 with daughter liquid D_2 from which some plagioclase has been fractionated results in a hybrid H_1 featuring olivine on the liquidus at relatively high Fe^{2+}/Mg.

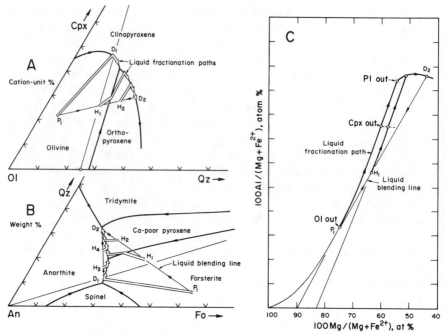

Figure 9-6. Model diagrams illustrating some possible effects of blending liquids of the same parentage at different stages of differentiation. (A) Pl projection from Fig. 9-1; (B) System Fo–An–Qz modified from Irvine (1975b, Fig. 66); (C) Al–Mg–Fe^{2+} diagram, after Fig. 9-5. See text for explanation.

Possible effects on crystallization order are modeled in Fig. 9-6, A and B. In diagram A, the mixing of primitive liquid P_1 in which clinopyroxene is next to crystallize after olivine with daughter liquid D_2 yields hybrids in which orthopyroxene precedes clinopyroxene. Similarly in diagram B, the combination of an initial liquid P_1 in which plagioclase follows olivine with a derivative D_2 yields blends in which orthopyroxene precedes plagioclase. A particularly intriguing possibility concerns derivative liquids D_1 and D_2 in Fig. 9-6B. Through this combination, liquids in which plagioclase is the first mineral to crystallize can be generated from a liquid in which olivine was first. This possibility, based on the system Fo–An–Qz, may be significant to the origin of anorthositic layers in certain ultramafic-gabbroic intrusions (Irvine, 1975b).

Orthopyroxene crystallization can also be advanced relative to the crystallization of clinopyroxene and plagioclase in basic magmas with olivine on the liquidus by contamination with granitic liquid such as might be melted from siliceous country rocks. (In each diagram in Fig. 9-6, liquid D_2 is on a boundary that ultimately leads to a granitic composition.)

SOME NATURAL EXAMPLES OF CRYSTAL FRACTIONATION AND ACCUMULATION: OBSERVATIONS AND INTERPRETATIONS

In choosing the examples considered below, an attempt has been made to match volcanic series with layered intrusions with respect to their crystallization characterisitics in order to compare rocks that once were dominantly liquid with approximate genetic counterparts formed mainly from accumulated crystals. The choice also illustrates differences between the matched pairs that apparently reflect important differences in parental liquid composition. The lavas of Kilauea and Mauna Loa are matched with the Muskox intrusion in contrast with abyssal tholeiites and the Skaergaard intrusion.

KILAUEA AND MAUNA LOA VOLCANOES

Differentiation Characteristics The lavas of Kilauea and Mauna Loa are of particular interest because they exhibit large chemical variations that can be closely correlated with the occurrence and abundance of different types of phenocrysts. The variations in Mg/Fe^{2+} were illustrated in Fig. 9-3; the main normative variations are illustrated in Figs. 9-7 and 9-8 on the projections introduced in Fig. 9-1.

As noted above, the main compositional variations in the lavas are related to the abundance of olivine phenocrysts. This feature is evident in Figs. 9-7 and 9-8 in the large spread of points along the indicated olivine control lines. Wright (1972) has distinguished several olivine control lines for Kilauean lavas, depending on their age, based on detailed study of their chemistry; but only two such lines are shown on the present generalized plots, corresponding essentially to historic (H) and prehistoric (P) lavas. Small amounts of chromite generally acompany the olivine (Wright, 1973).

The lavas of the two volcanoes differ significantly in their crystallization characteristics. As seen in Fig. 9-7, the olivine control lines for Kilauea lead to the clinopyroxene liquidus, and it is evident that some of the lavas have differentiated according to crystallization order 1 in Table 9-1 (Ol; Cpx; Pl; Opx). The Mauna Loan control line, on the other hand, leads practically to the 5-phase point d, indicating the crystallization

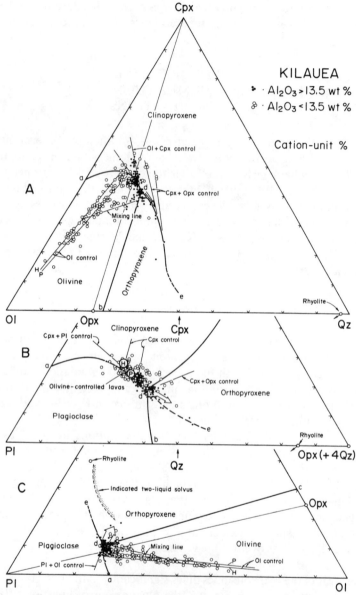

Figure 9-7. Plots of analyses of the lavas of Kilauea volcano on the projections derived in
Fig. 9-1. Sources of data as in Fig. 9-3. Different symbols are used for rocks
containing more or less than 13.5 wt % Al$_2$O$_3$ because this division generally
distinguishes lavas in which there has been appreciable plagioclase fractionation
from those in which olivine and pyroxene fractionation are dominant. Two
olivine control lines are shown, corresponding in general to historic and prehistoric lavas (cf. Wright, 1973). In plot B, most of these lavas plot in the two
circled areas, and it is apparent that they have the crystallization order: Ol;
Cpx; Pl; Opx. For further explanation, see text.

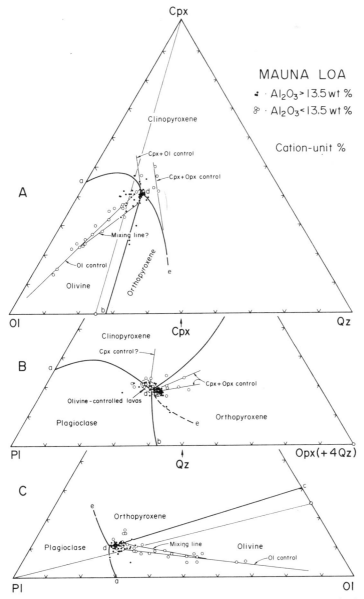

Figure 9-8. Plots of the analyses of the lavas of Mauna Loa volcano on the projections in Fig. 9-1. Sources of data as in Fig. 9-3. Note that the olivine-controlled lavas essentially project to the 5-phase "point" *d* in plot B.

order: Ol; Cpx + Opx + Pl (see also Jameison, 1970; Irvine, 1970a, Fig. 14). These relations accord with the parageneses of the lavas as described by Macdonald (1949), Tilley (1960, 1961), Wright and Weiblen (1968), Wright and Fiske (1971), Wright (1972), and many others.[3]

Also notable in Fig. 9-7C is the splay of points in the area of the Cpx + Ol control lines. This distribution appears to reflect the accumulation of olivine and augite phenocrysts in gradually changing proportions as would be precipitated along the curved olivine-clinopyroxene cotectic. Similar indications of joint accumulation of hypersthene and augite can be seen in the data from both Kilauea and Mauna Loa as denoted by the Opx + Cpx control lines in Figs. 9-7C and 9-8C.

As mentioned earlier, Wright and Fiske (1971) have cited evidence of magma mixing and hybridization in Kilauean lavas such as would indicate the combination of new, olivine-rich liquid with a residue of earlier liquid enriched in silica by differentiation in shallow reservoirs. Schematic mixing lines have been drawn in Figs. 9-7 and 9-8 to indicate this type of possibility and its potential in explaining some of the otherwise anomalous features of the data-point distributions. It is emphasized, however, that these lines are based on the present plots and are not the same as those suggested by Wright and Fiske.

The fractionation trend of the lavas enriched in silica beyond the olivine liquidus is not well defined, but the Kilauea data warrant brief attention. In the Pl projection in Fig. 9-7A their trend is directed strongly away from Cpx, and if continued, would pass through the base of the triangle into the corundum-normative region where most rhyolites plot. Wright and Fiske (1971, Fig. 3) have graphed analyses of liquids collected at temperature from Kilauea Alae lava lake that show a sudden change in the interval from 1050°C to 980°C from a basic composition with about 53 wt% SiO_2, 14% total iron as "FeO," and 1.5% K_2O, to a rhyolitic composition with about 76% SiO_2, 1.2% "FeO," and 5.6% K_2O. The general relations of these liquids are shown in Fig. 9-7. The trend of the basic liquids is not obviously directed toward the rhyolite, however, and it appears either that liquid immiscibility has intervened or that the liquid trend was sharply deflected on approaching a two-liquid solvus (cf. Irvine, 1975a). Wright and Fiske do not mention immiscibility, but Roedder and Weiblen (1971)

---

[3] The close correlation in Figs. 9-7 and 9-8 between the variations in the lavas and the experimentally defined reference liquidus boundaries is not entirely coincidental. The reference boundaries are largely based on the melting relations of lavas from Kilauea and Mauna Loa and so, in effect, define only the liquid trends of these lavas on general cotectic surfaces that would appear as bands in the projections. These bands might variously be contoured for Mg/Fe^{2+}, An/Ab, etc., and accordingly other liquids could have different trends along them.

have recorded immiscibility between "high-Fe" and "high-Si" liquids in basalt collected at 1020°C from Kilauea Makaopuhi lava lake.

Primitive Liquid Compositions Two important observations to be made from Figs. 9-7 and 9-8 are (1) despite the fractionation and concentration of the phenocryst minerals, the whole-rock variations generally follow the liquid path, and (2) the modal proportions of the phenocryst minerals appear generally to reflect only the contemporary fractionation stage of the liquid. In particular, it seems clear that olivine was the only mineral (other than minor chromite) that was fractionated and accumulated in most of the lavas plotting along the olivine control lines. The compositional differences indicated by these lines would appear, therefore, to have developed before the olivine was fractionated—for example, as Wright (1972) has suggested, during melting of the magmas in the upper mantle. Thus, as emphasized by Murata and Richter (1966), the problem of defining primitive liquid compositions for the volcanoes is primarily a matter of determining how much olivine was fractionated from the olivine-controlled liquids. The Al–Mg–Fe^{2+} relations of the lavas (Fig. 9-3) will now be further examined in this regard.

In Fig. 9-9, the Al–Mg–Fe^{2+} relations are interpreted in terms of the models in Figs. 9-5 and 9-7C. The main trends have been fitted with olivine fractionation curves, using average data for the olivine-controlled lavas from Wright (1972) for derivative liquid compositions. Points marking the appearance of the pyroxenes and plagioclase were selected by matching the data-point distributions in Fig. 9-3 with those in Figs. 9-7 and 9-8. The fit for the Kilauean lavas leaves little room for alternatives, and although that for Mauna Loa could be improved by using a slightly different derivative liquid composition, it was considered better to use Wright's average than arbitrarily to select some specific analysis as being more representative of the liquid.

As explained in the theory section, given the olivine fractionation curve for a magmatic system, the composition of the most primitive liquid recorded in the system can be estimated on the basis of the composition of the most magnesian olivine precipitated in the magma. The most magnesian olivine reported for Kilauean lavas is Fo$_{87.8}$ (Murata and Richter, 1966), and this composition is about the most magnesian that can be inferred from the analyses of the lavas on the basis of the tie-line relations in Fig. 9-9A. For $K_D = 0.31$, such olivine would be in equilibrium with liquid on the Kilauea fractionation curve having $100Mg/(Mg + Fe^{2+}) = 69.5$ and $100Al/(Mg + Fe^{2+}) = 45.3$. Essentially the same values can be inferred for these ratios in Mauna Loa "primitive liquid" from the relations in Fig. 9-9B. Complete major element compositions

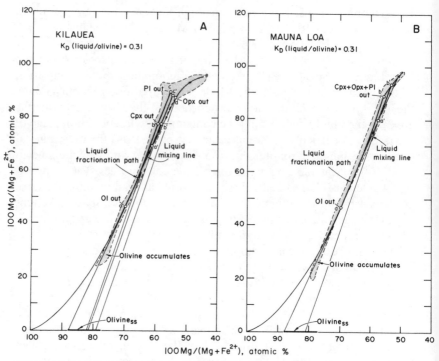

Figure 9-9. Interpretations of the compositional variations of the lavas of Kilauea and Mauna Loa shown in Fig. 9-3, based on the models in Fig. 9-5.

for liquids are listed in Table 9-2.[4] The liquids could be termed picritic tholeiites. As may be seen from comparative data in the table, they are only slightly more basic than the most basic glass found in Kilauea lavas, and they are similar to average lava compositions for the 1959 eruption of Kilauea calculated by Murata and Richter (1966) and Wright (1973) by weighting analytical data for different batches of magma in accordance with their size.

---

[4] The uncertainties in the present estimates are obviously large, and they are not readily estimated because the method by which they are derived involves subjective judgements. The choice of fractionation curve, for example, is not so much a matter of statistically fitting the data as it is a case of selecting the curve that is most satisfactory in terms of the physical chemistry of the problem. A large uncertainty also stems from decisions concerning the oxidation state of iron. Several investigators (e.g., Murata and Richter, 1966) have noted that Kilauea lavas are oxidized by exposure to the atmosphere during eruption, and some correction or adjustment for this alteration is necessary. In preparing Fig. 9-3, two preliminary plots were made for each volcano. For the one plot, only Fe_2O_3 in excess of 20% of the

MUSKOX INTRUSION

General Features   The Muskox intrusion presents an interesting comparison with Kilauea and Mauna Loa in its chemical differentiation and also, to some extent, in its physical history. The intrusion was originally described by Smith (1962) and a more recent general review was given by Irvine and Baragar (1972). A summary description is useful for present purposes.

As exposed, the intrusion appears as a markedly elongate, north-northwesterly trending body, about 120 km long. Over the southern half of this length it consists simply of a segmented vertical dike, 150–500 m wide, known as the "Feeder dike." This unit is composed mainly of bronzite gabbro but generally also contains one or two internal zones of picrite in parallel with its length. Some of the gabbro is very fine grained or "chilled" along the dike walls, and the picrite variously contains from 15% to 50% granular olivine in a gabbroic matrix.

The main body of the intrusion is developed to the north of the feeder dike but apparently extended southward over the dike before erosion. The main body is funnelform in cross section with lower walls dipping inward, generally at 25°–35°. It plunges gently to the north (at about 5°) and consequently its outcrop width gradually increases northward, reaching a maximum of about 11 km where the body finally disappears beneath its roof rocks and younger cover. The intrusion does not end at this point, however, but can be traced on to the north in aeromagnetic maps for another 30 km, over which distance there develops a major gravity anomaly that peaks some 60 km to the north and ultimately extends for more than 230 km. The exposed rocks, therefore, appear to represent only the southern extremity of a much larger igneous complex.

Within the main body of the intrusion, the footwall contacts are each lined by a marginal zone 130–230 m thick. In general, these zones grade

---

total iron was converted to FeO. This adjustment is essentially a minimal correction on only the most oxidized samples. For the other plot, the adjustment was extreme—all Fe_2O_3 was converted to FeO. With the extreme adjustment, the scatter of points along the main trend lines was reduced to about a third, and this seemed clearly to be an improvement. To be more realistic, however, the final plots were made with Fe^{3+}/Ti set to 0.5 on an atomic basis, with the rest of the iron as FeO. The premise was that Fe^{3+} and Ti are fractionated in much the same way, and the ratio value 0.5 was about the lowest recorded in any of the original analyses. With this adjustment, as it turned out, the scatter of data points was reduced slightly more. The same adjustment has been made in preparing all other plots, except the graphs for the Muskox analyses. The Muskox analyses have been adjusted on an individual basis, depending on the petrography of the rocks, but in general Fe_2O_3 was reduced to a relatively low level.

inward or upward from bronzite gabbro at the contact (also locally chilled) through picrite and feldspathic peridotite to peridotite. In combination, they form a trough in which has accumulated a layered series about 1800 m thick comprised of 42 mappable layers of 18 different rock types. The layers range in thickness from 3 m to 350 m and, in general, are sharply defined. They are inclined to the north at about 5° and typically are continuous between the marginal zones. Some can be traced in outcrop for more than 25 km, and from drill hole intersections it is apparent that many have areal extents in excess of 250 km². The succession of rock types in the layered series ranges generally from dunite at the base through peridotite and various pyroxenites and gabbros to granophyric gabbro at the top. Certain rock types tend to be repeated in specific sequences, however, and the series as a whole has been divided into 25 repetitive divisions called *cyclic units*, a term introduced by Jackson (1961) to describe similarly repeated stratigraphic divisions in the Stillwater complex in Montana. The top of the layered series is gradational with an irregular roof zone of granophyric rocks that locally is heavily charged with xenoliths of various siliceous roof rocks.

The differentiation of the intrusion is most systematic within the cyclic units. Three general classes of these units are recognized, distinguished in principle by different successions of rock types. These successions reflect different crystallization orders, as summarized in Table 9-3, and it has been shown that the repetition of units is the result of repeated influx of fresh magma into the exposed part of the intrusion while the layered series was accumulating (Irvine and Smith, 1967; Irvine, 1970a). The feeder dike does not cut the layer series, however, and does not appear to have been an access route for magma after the initial formation of the intrusion. The later additions are believed, therefore, to have come from the north from a major feeder system or magma reservoir at depth in the area of the large

Table 9-3. Rock layer sequences and crystallization orders characteristic of the principal cyclic units in the Muskox intrusion.

Class	Rock Sequence	Crystallization Order
I.	Dunite; Olivine clinopyroxenite; Olivine gabbro	Ol; Cpx; Pl; Opx
II.	Dunite; Olivine clinopyroxenite; Websterite	Ol; Cpx; Opx; Pl
III.	Peridotite; Orthopyroxenite; Websterite; Two-pyroxene gabbro	Ol; Opx; Cpx; Pl

Abbreviations: Ol, olivine; Cpx, clinopyroxene; Opx, orthopyroxene; Pl, plagioclase.

gravity anomaly, the new magma having flowed southward through the intrusion in the space between the accumulating layers and the roof rocks. There is evidence that residual magma was displaced in this process, and most probably it was pushed on to the south and then to the surface as volcanic eruptions, the products of which have since been removed by erosion (Irvine and Smith, 1967). This history may be compared with that of Kilauea volcano where it has been found that magma rises from the mantle to a reservoir 3–4 km beneath the summit and then in part feeds laterally outward along rift zones to be erupted from fissures as distant as 35 km on the flanks of the volcano (e.g., Wright and Fiske, 1971).

Compositional Variations Relating to Crystallization Order The crystallization orders listed for the three classes of cyclic units in Table 9-3 correspond respectively to orders 1, 2, and 3 in Table 9-1. The sequence in which the classes are numbered, from I to III, is essentially the sequence in which they occur in the layered series; thus it is evident that the differences between them developed because orthopyroxene progressively advanced in the crystallization order of the magma. The crystallization order of class I is the same as that observed in Kilauean lavas; the earlier appearance of orthopyroxene in classes II and III may be compared with the relatively early appearance of hypersthene in Mauna Loan magma.

Figure 9-10 illustrates modal data and Mg/Fe^{2+} variations in drill-core sections through cyclic units assigned to classes I and III. Chemical data from the ultramafic rocks in these two kinds of units are shown in normative projections in Fig. 9-11. Class II has only limited development and so is not considered.

From the data in Fig. 9-10A, it is seen that in cyclic unit 7, which is the type class I unit, olivine was the first mineral to precipitate and subsequently was joined, first by clinopyroxene and then by plagioclase, thus giving the rock succession dunite–olivine clinopyroxenite–olivine gabbro. Orthopyroxene is present as a minor post-cumulus phase formed from trapped intercumulus (interstitial) liquid, but did not appear as a cumulus mineral, apparently because the crystallization course of the main body of magma was terminated by an influx of fresh magma before orthopyroxene saturation was reached. Similarly in cyclic unit 6, in which dunite is overlain only by olivine clinopyroxenite, it is apparent that accumulation was terminated before plagioclase could precipitate; and in units 4 and 5, which consist only of dunite and are defined on the basis of repeated chemical trends such as those illustrated for Mg/Fe^{2+}, it may be inferred that fresh magma was introduced before even clinopyroxene could crystallize. The layered series includes six other cyclic units like unit 6 and at least five others like units 4 and 5. Chromite occurs as a minor cumulus

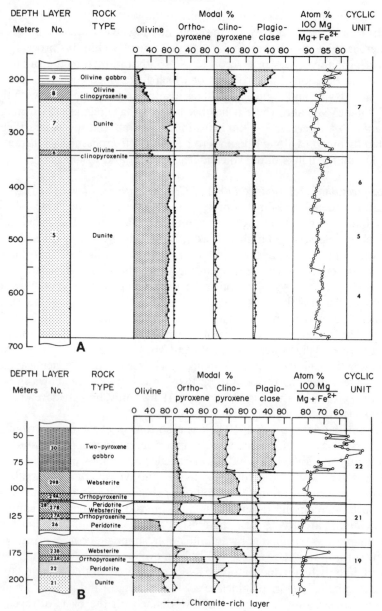

Figure 9-10. Modal and Mg/Fe^{2+} variations in class I (A) and class III (B) cyclic units in the Muskox intrusion. In the mineral columns, the dashed pattern denotes material that is mainly cumulus; the stippled pattern, material that is almost wholly postcumulus.

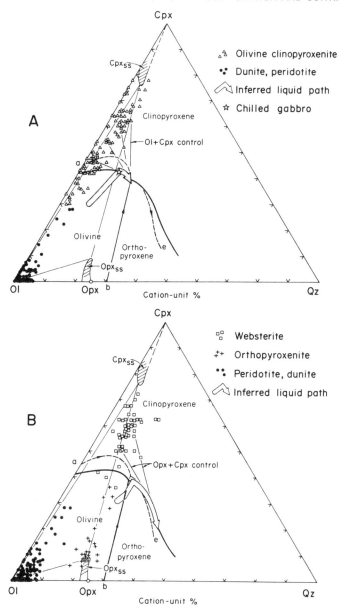

Figure 9-11. Chemical analyses of the rocks in Muskox Class I (A) and Class III (B) cyclic units plotted on the Pl projection from Fig. 9-1.

phase throughout the dunite, but characteristically is absent in the pyroxenite layers because of a chromite → augite reaction relationship (Irvine, 1967).

In the succession of class III units in Fig. 9-10B, the repetition of the sequence peridotite–orthopyroxenite–websterite is a prime feature, but in addition, thin chromite-rich layers occur at the contacts between peridotite and orthopyroxenite in two of the units and so are part of the systematics of repetition. In cyclic unit 22, the ultramafic cumulates are capped by a layer of two-pyroxene gabbro in accord with the full crystallization order listed for the silicate minerals in Table 9-3. Also notable is that where orthopyroxene appears as a cumulus phase, olivine immediately disappears, as would be expected in the products of liquid fractionated across the olivine-orthopyroxene reaction boundary (cf. Fig. 9-6A,B). The pyroxenites typically contain small amounts of interstitial quartz, evidence that they formed from Qz-normative liquid (Irvine and Smith, 1967, Fig. 26).

In the normative plots in Fig. 9-11, the analyses of the cumulates are shown in comparison with approximate liquid fractionation paths. The liquid path for the Class I units is computed from the intrusion chilled margin composition by a method described below. The path for the class III units is only schematic but is consistent with a path computed for the uppermost cyclic unit in the layered series (unit 25) from the compositions of what apparently were the last rocks to crystallize from the Muskox magma. The locations of the cotectic boundaries indicated by these paths are slightly different from those defined in Fig. 9-1 but are well within the range of locations possible, depending on the details of the liquid compositions. The cumulates generally cluster along control lines from the liquid paths—as they should if the fractionation mechanism was efficient. If the clustering is compared with the distribution patterns of the lavas of Kilauea and Mauna Loa in Figs. 9-7A and 9-8A, it is seen that there is considerable similarity between the spread of the olivine clinopyroxenite analyses and that of the Kilauean lavas showing Ol + Cpx control. As was noted previously, the spread of the lavas would appear to reflect accumulation of olivine and augite in proportions that gradually shifted in favor of the pyroxene as the liquid was fractionated along a curved path on the olivine-clinopyroxene cotectic. This same interpretation for the intrusion rocks is supported in that the shift is evident as an upward increase in the ratio of augite to olivine through the olivine clinopyroxenite layers. This variation is illustrated by the modal data for layer no. 8 in cyclic unit 7 in Fig. 9-10A, and has been observed in olivine clinopyroxenite layers in five of the other seven cyclic units in which they occur (Irvine, 1970a, Fig. 8).

In general, the modal and chemical data from the Muskox intrusion indicate that *the fractionated minerals accumulated as they were precipitated with very little sorting.*

Mg-Fe^{2+} Variations There is a general trend of upward decrease in Mg/Fe^{2+} through the Muskox layered series, but as with the effects of crystallization order, the most systematic variations are within cyclic units. Especially well developed trends occur in the class I units 4–6 illustrated in Fig. 9-10A. Through this succession the ratio $100Mg/(Mg + Fe^{2+})$ repeatedly decreases by from 3 to 5 percentage units through dunite layers 100–130 m thick. These trends also show in analyses of the olivine and chromite in the dunite, and in view of the association with modal variations relating to crystallization order in units 6 and 7, the trends seem clearly to be due to fractional crystallization of these minerals, especially the olivine. They may, therefore, be compared with the Mg–Fe^{2+} fractionation curves in Fig. 9-2A (see below).

The trends of Mg/Fe^{2+} in the class III units illustrated in Fig. 9-10B are less systematic than those in the class I units, but the variations are of similar magnitude even though the units are much thinner (note the change in the scales of both Mg/Fe^{2+} and depth between diagrams A and B).

In the dunite of the class I units, the modal abundances of pyroxene and plagioclase are generally so small as to indicate that very little interstitial (intercumulus) liquid was trapped between the cumulus crystals; and from plots of chemical analyses of the dunite on Al–Mg–Fe^{2+} diagrams (not shown) it is apparent that the limits of Mg/Fe^{2+} variation were not appreciably affected by reaction with such liquid. Very different relations are indicated, however, in the class III units represented in Fig. 9-10B. The modal data from these units show that they embody the crystallization products of at least 30% intercumulus liquid, and possibly as much as 40%, so the Mg/(Mg + Fe^{2+}) values of the rocks probably differ considerably from those of the original cumulus minerals. In particular, it may be inferred that this effect is responsible for the sudden general decrease and ensuing erratic variations in Mg/Fe^{2+} that are seen in cyclic unit 22 in passing upward from websterite through the two-pyroxene gabbro. Thus, when plagioclase began to precipitate and formation of the gabbro was initiated, the fraction of the bulk Mg + Fe^{2+} content of the cumulate deriving from the cumulus minerals was reduced to only about 40% of that in the underlying two-pyroxene cumulate (the websterite), and so the contribution of the intercumulus liquid to Mg/(Mg + Fe^{2+}) in the total cumulate was proportionately increased (in accord with the type of relations outlined in the caption of Fig. 9-2).

Liquid Compositions The most magnesian olivine in cyclic units 4–6—$Fo_{89.5}$, occurring at the base of cyclic unit 5 (Fig. 9-10A)—is also the most magnesian that has been found in the whole intrusion. It can be taken, therefore, as indicative of the composition of the most primitive liquid to crystallize in the intrusion. For a K_D value of 0.31, $100Mg/(Mg + Fe^{2+})$ in this liquid would be 72.5, and from the fractionation curve in Fig. 9-2A for a first olivine $Fo_{89.5}$, it can be inferred that by the stage the fractionated olivine had changed in composition by 3 percentage units (as in cyclic unit 4), some 25%–30% of the Mg and Fe^{2+} in the liquid would have been transferred to the crystals, and for a change of 5 percentage units (as in cyclic unit 6), the transfer would amount to about 40%. To make a more complete estimate of the liquid composition, however, and to convert the transfer estimates to percent solidification, a composition is required for some derivative liquid, as was explained in the theory section.

The best available data for the composition of a derivative liquid comes from the chilled gabbro in the marginal zones and feeder dike of the intrusion. In early work on the intrusion, Smith and Kapp (1963) obtained analyses of two samples of this material that seemed most likely to represent liquid compositions on the basis of textures. The analyses of the two samples were very similar, and subsequent work has not uncovered a better sample.[5] Biggar (1974) investigated the 1-atm melting relations of one of the samples at controlled oxygen fugacities (f_{O_2}), and for the geologically reasonable f_{O_2} of 10^{-10} atm his data indicate the crystallization order: olivine + chromite (1225°C); clinopyroxene (1205°C); plagioclase (1200°C). The order is that observed in the class I cyclic units, and since olivine is on the liquidus and clinopyroxene appears at a temperature only 20°C below, the sample would appear to represent the parental liquid of these units at a crystallization stage when dunite was still accumulating, just before olivine clinopyroxenite began to form. This correlation is supported in that a liquid with the composition of the investigated gabbro sample would be in equilibrium with olivine Fo_{83-84}, which is essentially the composition indicated for the olivine in the dunite just below the olivine clinopyroxenite layer in cyclic unit 6 (Fig. 9-10A). A liquid equivalent to the other chilled gabbro sample would be in equilibrium with olivine Fo_{85-86}, which is the composition of the olivine at the top of the dunite layers in cyclic units 4 and 5. The concentrations of Ni and Cr in the chilled gabbro samples also are appropriate

---

[5] All other samples of the gabbro that have been analyzed appear from the results either to include accumulated mafic minerals and plagioclase, or to represent liquids that either were more extensively fractionated or somewhat contaminated with country rock material.

to a liquid in equilibrium with the upper parts of the dunite layers (Irvine, 1977b).

On the basis of the above considerations, a "primitive liquid" composition could be calculated by adding olivine to the chilled gabbro composition in accord with Equations 9-4 and 9-6 in the theory section. What appears to be a better method in the present case, however, is to add the actual dunite from a well-differentiated cyclic unit such as unit 5 or 6 to the chilled gabbro in proportions appropriate to yield a liquid composition that would be in equilibrium with the most magnesian olivine in the dunite. By adding "layer increments" of the dunite in the reverse order in which they were precipitated, a series of compositions can be generated corresponding in theory to the liquid path as it precipitated olivine and chromite. The path can then be extended in the opposite direction (i.e., toward silica) by subtracting increments of the olivine clinopyroxenite and olivine gabbro layers in cyclic unit 7 from the chilled margin composition, weighted in proportion to the previous addition of dunite in accordance with the relative thickness of layers in unit 7. A fractionation path calculated in this way is illustrated in Fig. 9-11A, and a composition for the most primitive liquid is given in Table 9-2.

The composition of the primitive liquid compares closely with those estimated for Kilauea and Mauna Loa and could similarly be identified as picritic tholeiite. Because olivine and chromite are both cumulus phases throughout the dunite, however, even this liquid would presumably be somewhat fractionated, rather than being a primary melt introduced without change from its source area in the upper mantle. Liquidus boundaries like the olivine-chromite cotectic generally shift in composition with pressure, so a primary melt from the mantle is not likely to arrive at crustal depths precipitating both minerals unless it has undergone some crystallization en route (O'Hara, 1965).

ABYSSAL THOLEIITES

Background Information   In recent years, the submarine tholeiitic lavas of the mid-ocean ridges have come to be one of the most extensively studied volcanic rock series. The series has been found to be remarkably constant in petrography and chemistry in sampling that encompasses broad regions of the Atlantic, Pacific, and Indian oceans. A large proportion of the investigated samples contain phenocrysts of both olivine and plagioclase, with augite the next mineral to crystallize. Miyashiro, Shido, and Ewing (1970) pointed out, however, that many samples have only olivine phenocrysts and many, only plagioclase phenocrysts (like the lavas studied by Bowen). They termed these rocks Ol- and Pl-tholeiites and on the basis of chemical differences between them, defined an olivine-plagioclase cotectic for abyssal tholeiite liquid in a normative projection on the join Ol–Pl–Px. Shido, Miyashiro, and Ewing (1971, 1974) additionally recognized Cpx-tholeiites in which augite is the dominant phenocryst (see also Hekinian and Aumento, 1973, and Bougault and Hekinian, 1974), and they gave data on a few lavas with pigeonite phenocrysts. Shibata (1976) examined these different tholeiite types in projections similar to those in Fig. 9-1 and attempted to estimate the pressure of phenocryst formation by comparing apparent phase boundary locations with experimental data on basalt melting relations at various pressures. In a recent deep sea drilling project (DSDP Leg 37), abundant picritic lavas were discovered, and the first data on these and associated rocks have been given by Blanchard et al. (1976).

Crystallization Characteristics and Differentiation   Analyses of abyssal tholeiites are plotted on normative projections in Fig. 9-12 and on the rectangular Al–Mg–Fe^{2+} diagram in Fig. 9-13. Samples classed by previous authors as Ol-, Pl-, and Cpx-tholeiites are distinguished by different symbols, and also distinguished are twenty-nine analyses of abyssal tholeiite glass reported by Mathez (1976). The glass analyses are important because they represent actual liquid compositions. The apparent fractionation trend of the Skaergaard liquid and the trends of oceanic island tholeiite series from Thingmuli volcano, Iceland, and the Galápagos Islands are shown for comparison and contrast.

A major feature of abyssal tholeiites, first recognized by Chayes (1965) and conspicuous in Fig. 9-12, is that they typically have olivine in the norm; very few are Qz-normative. The trend of the lavas richest in Ol cannot be described by a single olivine-control line, but is enclosed by a narrow fan or cone of these lines that projects to the relatively small elliptical area outlined on the plagioclase liquidus in plot B. The other lavas mostly cluster in a zone that trends in the general direction of the olivine-clinopyroxene-plagioclase cotectic from Fig. 9-1, and in preparing the graphs it was noted that many of the individual suites of analyses show well-defined trends in this direction. The different tholeiite types generally segregate into areas appropriate to their mineralogy, but rather than attempt to define liquidus boundaries on this basis, the author has chosen to draw an olivine-clinopyroxene-plagioclase cotectic along the main trend of the glass analyses. According to Mathez (1976), all twenty-nine of the glasses have olivine and plagioclase as "near-liquidus phases"; twelve also have chromite, and five of the others additionally have augite. As it turns out, all but one of the seventeen analyses without chromite

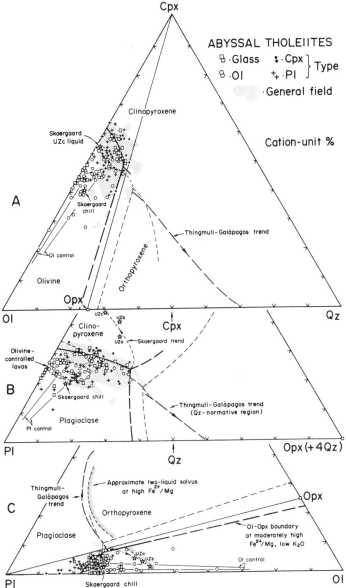

Figure 9-12. Plots of analyses of abyssal tholeiites on the normative projections from Fig. 9-1. Data from Mathez (1976), Blanchard *et al.* (1976), sources listed by Shibata (1976), and many other sources. Shown for comparison are compositions for the Skaergaard chilled margin (from Wager, 1960) and upper zone liquids (from McBirney and Nakamura, 1974), and the Qz-normative trends of oceanic island tholeiite series of Thingmuli volcano, Iceland (Carmichael, 1964) and the Galápagos Islands (McBirney and Williams, 1969). For explanation, see text.

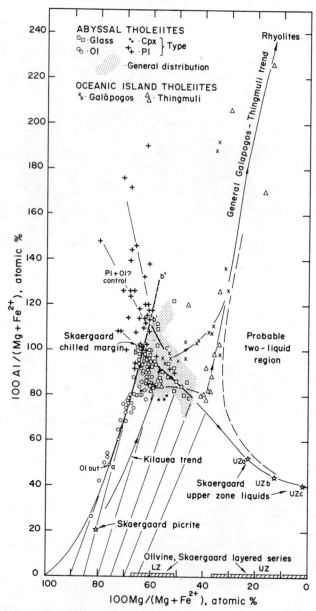

Figure 9-13. Plot of $100Al/(Mg + Fe^{2+})$ vs. $100Mg/(Mg + Fe^{2+})$ in abyssal tholeiites, oceanic island tholeiites series, and Skaergaard liquids and olivines. Data sources as in Fig. 9-12. See text for details.

plot along the suggested cotectic "line," practically within the limits of analytical uncertainty. Moreover, all but five of the other analyses appear to plot on the olivine-plagioclase cotectic "surface," and four of the exceptions plot only a short distance away in the olivine volume. Only one glass analysis clearly falls in the plagioclase volume.

With the boundaries defined in this way, all of the Ol-tholeiites and most of the Cpx-tholeiites plot in the appropriate liquidus volumes. Some of the Pl-tholeiites fall in the plagioclase volume, but more than half plot along the olivine-plagioclase cotectic surface. (Unfortunately, because of the crowding of data points, not all of the Pl-tholeiite analyses could be shown on plot C where this relationship is evident.) On the basis of these relationships, then, it would appear that the olivine-rich lavas represent the most primitive fractionation products of a series that has generally differentiated according to crystallization order 5 in Table 9-1, viz., olivine, plagioclase, clinopyroxene, with orthopyroxene or pigeonite possibly appearing as late phases. Notable in this respect is that the olivine-controlled lavas are very poor in potential Qz—much poorer, for example, than the olivine-controlled lavas of Kilauea and Mauna Loa (cf. Figs. 9-7 and 9-8). This feature would seem fundamentally to underlie the paucity of Qz-normative differentiates in the series.

The relationships revealed in the Al–Mg–Fe^{2+} diagram in Fig. 9-13 are generally consistent with the above analysis. In this diagram, the data points for the olivine-controlled lavas have been fitted with an olivine fractionation curve, and it is seen that most of the glass analyses fall on a trend line of iron enrichment (bc) that stems from this curve. On the basis of correlation with relations in Fig. 9-12, this trend line may be interpreted as representing the olivine-plagioclase cotectic over the first part of its length, and the olivine-clinopyroxene-plagioclase cotectic at lower Mg/Fe^{2+}. With the exception of the Pl-tholeiites, which are given specific attention in the next section, most of the lava analyses plot in the vicinity of this line, and as in the case of the cotectic in the norm projections, many of the individual suites of analyses show the same trend.

A feature that might appear discrepant in the above interpretations is that in plots A and C in Fig. 9-12, the olivine-clinopyroxene-plagioclase cotectic inferred from the glass analyses begins at compositions considerably poorer in Ol than the corresponding reference boundary from Fig. 9-1, and then trends toward the latter boundary rather than in parallel with it. It would appear, however, that these differences can largely be explained on the basis that the olivine liquidus volume decreases in size relative to the liquidi of clinopyroxene and plagioclase as Mg/Fe^{2+} in the liquid decreases (see introduction to Fig. 9-1). The reference boundaries are largely based on the melting relations of Kilauean tholeiites, and as may be seen by comparing Fig. 9-13 with Fig. 9-9A, the cotectic trend for abyssal tholeiites begins at considerably higher Mg/Fe^{2+} than the corresponding trend for Kilauea, and then extends to equally low values even though it does not reach Qz-normative compositions.

Decrease in Mg/Fe^{2+} may also be a contributing factor in the absence of Qz-normative compositions. It appears likely, as indicated by the suggested phase boundary relations in Fig. 9-12, that before most abyssal tholeiite liquids could reach the Qz-normative region, their Mg/Fe^{2+} would have dropped to such an extent that the olivine-orthopyroxene liquidus boundary would have receded into the Ol-normative region, as it does, for example, in the system $MgO-FeO-SiO_2$ (Bowen and Schairer, 1935). Further advance toward Qz would then be blocked, or at least impeded, by the crystallization of Ca-poor pyroxene. This effect would necessarily be of secondary importance because Ca-poor pyroxene is rare in abyssal tholeiites, but nevertheless it is of interest to know whether the olivine → Ca-poor pyroxene reaction is functional. The compilation of data in Fig. 9-12 suggests that if it is, the relationship is probably like that observed in the Skaergaard intrusion (Wager and Deer, 1939; Wager and Brown, 1968) where olivine crystallization ceased only temporarily and then resumed at lower Mg/Fe^{2+}.

Possible Origins for the Pl-tholeiites The Pl-tholeiites warrant special attention inasmuch as they are an anomaly in the differentiation scheme outlined above, and because they are akin to the plagioclase porphyries studied by Bowen. They are generally found in close association with the other tholeiite types, so it is unlikely that they represent a completely different parental liquid. The questions arise, however, as to whether they reflect a crystal sorting process whereby the plagioclase phenocrysts were separated from olivine and augite phenocrysts; or whether they formed through some process that modified the general differentiation trend so that plagioclase became the only liquidus phase, as in the liquid blending process in Fig. 9-6B.

Origin by the liquid blending process appears to be precluded on the basis of two observations. First, Pl-tholeiites generally have relatively high Mg/Fe^{2+} ratios rather than values intermediate to liquids along the cotectics between plagioclase and the mafic minerals (Fig. 9-13); and second, fractionated liquids with Ca-poor pyroxene on the liquidus, analogous to daughter liquid D_2 in Fig. 9-6B, are evidently rare among abyssal tholeiites. (Certainly, they are much less common than the Pl-tholeiites themselves.)

The possibility of an origin by crystal sorting is not precluded but may be questioned on the basis of the liquidus relations inferred in Fig. 9-12, in that more than half of the Pl-tholeiites plot along the olivine-plagioclase cotectic rather than in the plagioclase liquidus volume. It would seem that if the examples on the cotectic contain accumulated plagioclase, then they must also contain accumulated olivine. The relations in the $Al–Mg–Fe^{2+}$ diagram (Fig. 9-13) are compatible with this interpretation in that the Pl-tholeiites generally have high Mg/Fe^{2+}. Another possible origin for the Pl-tholeiites is suggested by the observation by several groups of authors (e.g., Bryan et al., 1976; Clague and Bunch, 1976; Thompson et al., 1976) that the differentiation of abyssal tholeiites commonly appears to have involved fractionation of more clinopyroxene than would be suspected on the basis of its frequency of occurrence as phenocrysts. Similarly, the chemistry of Pl-tholeiites suggests fractionation of more olivine than would be inferred from its reported absence as phenocrysts. The suggested origin is that two stages of crystallization are generally involved. Most of the differentiation occurs in the first stage under conditions of higher pressure and, probably more critically, higher H_2O fugacity such that the cotectics between plagioclase and the mafic minerals are shifted toward plagioclase (e.g., to about $b'c'$ in Fig. 9-13). In the detailed plots from which Fig. 9-13 was compiled, many analyses fall along the olivine fractionation curve between b and b' and would appear to represent liquids that overshot the cotectic bc. The second stage occurs after the magma is moved to shallower depths, or some H_2O is lost, or both. The cotectics then take approximately the positions defined by the glass analyses, and more phenocryst plagioclase is crystallized while phenocrysts of mafic minerals tend to be resorbed. The latter process would increase Mg/Fe^{2+} in all cotectic liquids, but the effect would be most noticeable in the earlier types in which olivine was the only mafic mineral. It may be significant in this regard that a few of the tholeiite glasses analyzed by Mathez (1976) do have relatively high $Al/(Mg + Fe^{2+})$ as might develop in the postulated first stage (Fig. 9-13).

Primitive Liquid Composition   A "primitive" abyssal tholeiite liquid composition is given in Table 9-2. The composition was calculated on the basis of Equations 9-4 and 9-6 in the theory section from the average of a selection of analyses that plot close to the olivine fractionation curve in Fig. 9-13, using $K_D = 0.31$ and a most magnesian olivine $Fo_{90.5}$. The olivine composition is the most magnesian that has been reported to occur in the lavas (Blanchard et al., 1976, p. 4233). The composition of the primitive liquid is similar to those estimated for the Hawaiian tholeiite series and the Muskox intrusion, but it has lower Ca/Al and would crystallize less olivine at low pressure.

Skaergaard Intrusion

General Differentiation   The Skaergaard intrusion stands in marked contrast to the Muskox intrusion in its size and shape, and in features such as the nature of its border units and the character of its layering. The main differences, however, are in composition. At the present erosion surface, the Skaergaard intrusion consists almost entirely of gabbroic and ferrodioritic rocks. The only visible ultramafic rocks within its borders [as defined by Wager and Deer (1939) and Wager and Brown (1968)] are some subrounded blocks of olivine cumulates ranging from picrite to feldspathic peridotite in composition that are included in a zone a few tens of meters thick around the lower edge of the exposed part of the layered series (cf. Wager and Brown, 1968, p. 116). Granophyric differentiates are present but are quantitatively minor as compared with the proportion of granophyre in the Muskox intrusion.

Some of this contrast with the Muskox intrusion arises because the Skaergaard intrusion was apparently emplaced essentially as a single injection of liquid and then effectively crystallized as a closed system, rather than being open to repeated additions of fresh liquid. But the fundamental difference evidently lies in the liquid composition, and it is in this respect that the Skaergaard intrusion shows some correlation with abyssal tholeiites (see below).

The main trend of differentiation in the Skaergaard layered series is one of marked iron-enrichment in the gabbro and diorite, distinguished in particular by the reappearance of olivine at high Fe^{2+}/Mg after it had given way to pigeonite at an earlier stage by virtue of their usual reaction relationship in tholeiitic liquids. Wager (1960) divided the visible layered series into lower, middle, and upper zones (respectively denoted LZ, MZ, and UZ) and appropriate subzones on the basis of the crystallization order of the cumulus minerals. In the lowest subzone, LZa, olivine and plagioclase are both cumulus. They are shortly joined by augite in LZb, and then by magnetite in LZc. Cumulus pigeonite appears sporadically in LZb and LZc and finally takes the place of olivine in MZ. At the top of MZ, however, this succession is reversed—olivine reappears as a cumulus phase, and shortly thereafter pigeonite is no longer found. The appearance of cumulus apatite marks the base of UZb, and finally, in UZc, iron-wollastonite takes the place of clinopyroxene. McBirney and Nakamura (1974) have found evidence experimentally that immiscible granophyric liquid probably also separated from the basic magma during formation of the UZ cumulates. If it is assumed (as Wager did) that the picrite-peridotite blocks are cognate inclusions representative of early differentiates,[6] then the order of appearance of the silicate minerals is order 5 in Table 9-1 (Ol; Pl; Cpx; Opx), the same order as was indicated above for abyssal tholeiites.

Parental Liquid Composition   At first sight, a correlation of the Skaergaard intrusion with abyssal tholeiites might seem inappropriate inasmuch as the intrusion is emplaced in continental rocks. In plate tectonic reconstructions for the time at which the intrusion was formed, however, the east coast of Greenland is situated just off the axis of the mid-Atlantic Ridge, with the Skaergaard area opposite the present site of Iceland; and Brooks (1973) has suggested that the Skaergaard magma was fed from the postulated Iceland plume. Furthermore, among the myriads of basaltic dikes that occur along the Greenland coast in the Skaergaard area, C. K. Brooks and T. F. D. Neilsen (pers. comm., 1976) have found some of approximately the same age as the intrusion that are compositionally equivalent to the abyssal tholeiites, and they note similarities between the compositional variations of these dikes and the trend of the Skaergaard liquid calculated by Wager (1960).

Wager's estimates of the liquid trend were based on the chilled margin of the intrusion. He obtained analyses of three samples of this material selected on petrographic grounds and ultimately chose one of the analyses as most likely to be representative of the liquid. His estimate of the size of the intrusion deriving from this analysis is evidently too large, however (Blank and Gettings, 1973), and the question frequently has been raised as to whether his parental liquid composition is appropriate. McBirney (1975) has obtained more analyses of the chilled margin, and he reports that although the Wager composition is about average, the margin as a whole is highly varied. It is of interest, therefore, to examine the Wager composition further.

As seen in Table 9-2, the composition can be matched almost exactly with analyses of abyssal tholeiite glasses. It is, therefore, a bone fide natural liquid composition. Biggar (1974) has investigated the melting relations of the chilled gabbro at oxygen fugacities from 10^{-5} to 10^{-12} atm, and he found that olivine and plagioclase are both on the liquidus throughout this range, with clinopyroxene the next silicate to crystallize—the same order as the order of appearance of cumulus minerals in the layered

---

[6] The picrite-peridotite has recently been found exposed in place in the gneissic basement beneath intrusion (M. A. Kays and A. R. McBirney, pers. comm., 1975), but there is still a possibility that it is an early Skaergaard differentiate.

series. In the normative projections (Fig. 9-12), the Wager analysis plots in the abyssal tholeiite olivine liquidus volume, but only a short distance from the olivine-plagioclase cotectic; in the Al–Mg–Fe^{2+} diagram (Fig. 9-13) it falls practically on the junction of the abyssal tholeiite olivine fractionation curve and the inferred olivine-plagioclase cotectic (*bc*). The data compilation in Fig. 9-13 shows also that Mg/Fe^{2+} in the sample is appropriate to liquid in equilibrium with the cumulus olivine in the picrite blocks, and that the abyssal tholeiite trend leads to the UZ liquid compositions estimated by McBirney and Nakamura (1974) on the basis of melting experiments.[7]

There is one apparent discrepancy in that the first cumulus olivine in the layered series is about Fo_{67} (Wager and Brown, 1968, Fig. 14), a composition much too rich in iron to be in equilibrium with Wager's preferred liquid composition. A first explanation that comes to mind is that the liquid was already extensively fractionated by the time this olivine was precipitated. This cannot be the complete explanation, however. As seen in Fig. 9-2, for the fractionated olivine to change from Fo_{84} (the composition that would be in equilibrium with the chilled margin) to Fo_{67}, about 60% of the Mg and Fe^{2+} in the initial liquid would have to be transferred to crystals. To convert this transfer to percent solidification, it is noted from Fig. 9-12C that an olivine-plagioclase cotectic precipitate should contain some 20%–25% olivine (in cation equivalents), corresponding to about 15 cation % Mg + Fe^{2+}. The chilled gabbro composition contains 18% Mg + Fe^{2+}; hence, as explained in the caption of Fig. 9-2, the solidification would amount to $60 \times 18/15 = 72\%$. A liquid of the chilled gabbro composition could not undergo such extensive fractionation and still have only olivine and plagioclase on the liquidus.

On the other hand, the discrepancy is essentially explained if the liquid was about 15% fractionated by the time LZa began to accumulate, and if the cumulus olivine and plagioclase equilibrated with some 30%–40% intercumulus liquid during postcumulus solidification. After 15% fractionation, the olivine being fractionated would be about Fo_{81}; 100 Mg/ (Mg + Fe^{2+}) in the intercumulus liquid would be 57; and on the basis of the Mg + Fe^{2+} concentrations given above for the cumulate and initial liquid, Mg + Fe^{2+} in the fractionated liquid would have increased to 18.5%. Accordingly, if 35% intercumulus liquid were trapped, 100 Mg/

---

[7] The relationship of these UZ liquid compositions to the abyssal tholeiite differentiation trend in the normative projections (Fig. 9-12) is unusual but could relate to the development of liquid immiscibility and to the fact that Ca-poor pyroxene is not stable at high Fe^{2+}/Mg (e.g., Bowen and Schairer, 1935). The fact that the trend of the UZ liquids is opposite to the trend of silica enrichment shown by the lavas of Thingmuli and the Galápagos accords with the reappearance of olivine and suggests fractionation of immiscible silica-rich liquid.

$(Mg + Fe^{2+})$ in the whole rock would be about 70—a value that compares with the observed olivine composition given above, or, even better, with the value 69 shown by Wager's (1960) average LZa rock.

It may be questioned whether the LZ cumulates retained as much as 35% intercumulus liquid, and there may be other reasons to disqualify the Wager chilled gabbro composition as a possible Skaergaard liquid composition. The purpose of the present considerations, therefore, is not to defend Wager's choice but to illustrate ways of testing possible parental liquid compositions for differentiated intrusions. Chilled margin compositions offer the possibility of a direct sample, but the pitfalls are many. Some samples may contain accumulated minerals; some may represent fractionated or contaminated liquids; still others may be altered by various metasomatic processes; and it may even happen that the parental liquid was not homogeneous. Thus every sample has to be tested on its own merits. On the other hand, if the Wager chilled gabbro composition is representative, then the intrusion exemplifies extreme fractionation of what is apparently the major basaltic magma series of Cenozoic times.

EFFECTS OF CRYSTAL SORTING: A PARADOX

FIELD EVIDENCE OF CRYSTAL SORTING AND MAGMATIC SEDIMENTATION

An examination of the effects of crystal sorting is in large degree an assessment of the importance of the process itself, and pertinent information comes mainly from layered intrusions. Until recently, the rhythmic layering of the Skaergaard intrusion would have been cited as a prime example, but for reasons explained below, the Skaergaard evidence cannot currently be regarded as unequivocal. To illustrate the problem, relations in the Duke Island ultramafic complex in southeastern Alaska will be used as a primary base, and then the Skaergaard relations will be described in comparison.

In the Duke Island complex (Irvine, 1974), olivine and clinopyroxene have been sorted within layers ranging from a few centimeters to a meter or so in thickness, extending for distances up to several hundred meters. The sorting is on the basis of grain size but also shows as modal sorting in that the olivine is generally finer grained than the pyroxene. Thus, many layers grade from coarse pyroxene-rich rock at the base to finer olivine-rich rock at the top; others comprise alternate units of sharply contrasting grain size and modal composition; and the occasional layer even shows reverse grading. Still other layers resemble conglomerate beds, consisting in large part of fragments of pyroxenite derived from older parts of the complex. The fragments are usually concentrated in the lower parts of layers and may also be graded by size. Evidence of the nature of the sorting process comes from a variety of layering structures indicative of current transport, erosion, and deposition, such as scour-and-fill structures resembling cross beds. Many of these structures are associated with blocks of included pyroxenite ranging from about 1 m to possibly as much as 200 m on a side. The layers on which these blocks have fallen are commonly deformed and faulted; some are sharply truncated; and a few are seen to be partly torn up with the disturbed material redeposited a short distance away. Younger layers typically drape over the blocks and thicken off their flanks, with the effect that the depositional surface is smoothed and streamlined in the same way that the bedding in subaqueous turbidity current sediments is smoothed over irregularities. In a few cases, the blocks have cracked open, presumably when they impacted, and layers of olivine and pyroxene have been deposited in the crevasslike openings (Irvine, 1959, Fig. 89).

Virtually identical structures are seen in the Skaergaard intrusion, except that in this body the modal differentiation in the layering is on the basis of mineral density rather than grain size (Wager and Deer, 1939; Wager and Brown, 1968). Thus, a well graded Skaergaard layer has some combination of olivine, the pyroxenes, and magnetite concentrated at its base, and is enriched in plagioclase toward its top. As at Duke Island, however, there are many layers in which the grading is less systematic, and even some with reverse grading.

Through much of the Skaergaard layered series, but especially in the middle zone, there occur dozens of blocks of anorthositic gabbro, ranging like the Duke Island pyroxenite blocks, from a meter to several hundred meters on a side. These blocks presumably represent a part of the intrusion's upper border group that has otherwise been lost through erosion. They have accumulated with the layering and have had almost all the same effects as the blocks at Duke Island. Some have sliced into the layers on which they have fallen, truncating them at high angles; others have ploughed in, folding and deforming layers (see Wager and Brown, 1968, p. 77); layers above blocks typically are draped and smoothed over their tops. Many layers are strewn with occasional small fragments of the anorthositic gabbro (Fig. 9-14C), and there is at least one "conglomeratic" unit, 10–20 m thick, in which fragments are prominent throughout (see Wager and Brown, 1968, Fig. 39).

Through the parts of the Skaergaard layered series where blocks are abundant, the modally differentiated layers commonly occur in rapid succession, but in the basal part of the upper zone they tend to occur individually, alternating with thicker layers of uniform gabbro (Fig. 9-14C,

Figure 9-14. Banding and layering structures in the Skaergaard intrusion. (A) Colloform growth structures in the marginal border group, Invarmiut. The contact of the intrusion is to the left. (B) Cross-bedded belt, LZa, Uttentals Plateau. (C) Rhythmic layering, MZ, Kraemers Ö. Note the anorthositic xenoliths scattered through the section. (D) Intermittent rhythmic layering, UZa, near Home Bay.

D). Because of the latter arrangement, Wager (1963) proposed that two types of current deposition were involved, the uniform layers having accumulated from relatively slow, steady convection currents, and the modally differentiated layers from rapid density or suspension currents akin to the turbidity currents of subaqueous sedimentary environments (but not as turbulent). It appears notable in this regard that the lower and marginal edges of the layered series are fringed by a remarkable "cross-bedded belt" displaying evidence of countless episodes in which plagioclase and the mafic minerals accumulated in irregularly layered units that were then partly eroded away and covered by younger layers of the same type, usually disposed at some different angle (Fig. 9-14B). The author's impression of this belt is that it represents the staging area of the rapid suspension currents postulated by Wager—a place where crystals derived from convecting magma descending the cooling walls of the intrusion piled up in unstable banks from which they slumped (or were jarred free by anorthositic gabbro blocks) to flow *en masse* down into the intrusion and across the flatter interior floor of accumulating regular layers.

Another outstanding feature of the Skaergaard layered series, possibly unique among layered intrusions, is a group of structures known as the trough bands (Wager and Deer, 1939). These are synformal structures, mainly occurring in the basal part of the upper zone, distinguished by chutelike axial zones of modally differentiated layers. The axial zones typically measure 10–20 m in width, and some can be traced for 300 m along strike and more than 100 m stratigraphically. Wager and Deer showed that the troughs focus on the center of the magma body (as it existed when the troughs were formed), and they described alignment of mineral grains and other features that seemed to indicate flow of currents down the troughs, away from the walls of the intrusion toward its interior. Recent mapping of the troughs by the author and D. B. Stoeser has shown that they are essentially constructional features: the troughs alternate with—and the layers fill in between—levee-like ridges of relatively massive gabbroic cumulates that evidently grew concurrently with the first layers in the troughs. Subsidiary troughs occurring locally within the massive gabbro tend to overlap like linguoid cross beds. Evidence of minor current scouring can be seen in the bottoms of most of the major troughs, and some troughs feature systematic distributions of the mafic minerals and plagioclase that appear to reflect sorting by currents flowing in a double helicoidal mode as commonly is observed in currents flowing along chutes (Langmuir circulation or Görtler vortices; cf. Karcz, 1967; Houbolt, 1968; Folk, 1971; Blatt, Middleton, and Murray, 1972).

CRYSTAL SETTLING NOT A FACTOR?

There is, however, a major paradox associated with the Skaergaard layering. On the basis of the molar volumes of oxide components in silicate melts, Bottinga and Weill (1970) calculated densities for liquids of the compositions estimated by Wager (1960) for the different stages of fractionation of the Skaergaard magma, and they found that the liquids would be denser than the plagioclase that was supposed to have precipitated from them. This relationship has since been confirmed experimentally by Murase and McBirney, 1973; pers. comm., 1976). Thus, as McBirney emphasizes, we are faced with a dilemma: In the intrusion that has for years been synonymous with crystal settling in magmas, the most abundant "settled" mineral should have floated!

It appears, furthermore, that the problem is not unique to Skaergaard. Morse (1973) has calculated that the same relationship existed in iron-enriched liquids in the Kiglapait intrusion in Labrador, even though layering structures typical of sedimentation are common in the intrusion, and it seems likely that this relationship obtained in most intrusions featuring moderately strong iron-enrichment (see also Campbell, Dixon, and Roeder, 1977). Moreover, it is apparent that even in basaltic magmas in which plagioclase might be denser than the liquid, the settling velocity of the typical feldspar crystal would be practically negligible.

If one accepts this view, then it becomes apparent that there have been other indications that crystal settling has not been as important in the origin of layered intrusions (and, hence, in magmatic processes in general) as usually has been inferred. For example, Jackson (1961) found that in the Stillwater complex in Montana, coexisting cumulus olivine and chromite commonly are markedly out of "hydraulic equilibrium"—the olivine is not as dense as the chromite, but its crystals are so much larger that they should have settled 10–100 times faster. This relationship also occurs in the Muskox intrusion (Irvine and Smith, 1969), and in this case the modal ratio of chromite to olivine in certain cyclic units varies in parallel with the fractionation trends of Ni and Mg/Fe. On this basis it is evident that the minerals accumulated as they crystallized with little or no sorting (a point that was emphasized in a previous section for the cumulus silicate minerals in the intrusion). Originally it was thought that this relationship implied accumulation under conditions that were almost unaffected by magmatic currents (Irvine and Smith, 1969; Jackson, 1961, 1971), but it now seems likely that the proper interpretation is that accumulation involved very little crystal settling.

PHYSICAL CONDITIONS OF CRYSTALLIZATION
IN LARGE BODIES OF BASIC MAGMA

The above interpretations are not to say that the fractionated minerals in layered intrusions have not accumulated on the bottom on the magma bodies in which they occur (be it by crystal settling or however) or that plagioclase is not fractionated in this way just as well as the mafic minerals. These points were essentially proved by Wager and Deer (1939) in their study of the Skaergaard intrusion and have since been confirmed in numerous other intrusions. In the Muskox intrusion, for example, the direction of fractionation within cyclic units is invariably upward, practically to the roof of the intrusion; cumulus plagioclase occurs at intervals almost throughout the layered series (it is found in cyclic units 2, 7, 18, and 22–25); and there is no rock along the visible section of the roof contact that could be interpreted as representing floated plagioclase. Relations in the Bushveld complex in South Africa are similar and more dramatic. In this body, anorthositic layers occur in abundance at relatively low stratigraphic levels with layers of chromitite, orthopyroxenite, and norite, and then again at relatively high levels above a thick section of gabbroic cumulates in association with magnetitite layers and iron-rich olivine gabbro. But no accumulations of floated plagioclase are observed at the roof contact, despite the marked iron-enrichment (cf. Wager and Brown, 1968; Willemse, 1969).[8]

How then is this dilemma to be explained? On the one hand the fractionated minerals have undoubtedly accumulated from the bottom up, and the evidence of crystal sedimentation is compelling if not overwhelming. On the other, crystal settling does not appear to have been important. The possible answers seem only to be two. One is that the fractionated crystals grew *in situ* on the floor of the intrusion, with the apparent sedimentation structures having some other explanation. The other is that there is something in the mechanics of magmatic current flow that will, in certain circumstances, cause transported crystals to be deposited downward even when they should tend to float.

In terms of the physics of the problem, two general points seem reasonably well established. First, bodies of basic magma of the sizes common to layered intrusions probably undergo continuous convection while cooling and crystallizing. This point has been demonstrated on theoretical grounds by Shaw (1965, 1969) and Bartlett (1969), and is supported petrologically in that the well-developed differentiation patterns observed in many layered intrusions indicate fractionation of bodies of liquid of definite size that were extremely well mixed while crystallizing, as probably can only be accomplished by convection. The second point is that, under

---

[8] On the other hand, it is not meant to imply that plagioclase might not have accumulated at the top of some intrusions by floating. The large anorthositic complexes appear to be particularly likely places for such an event to have occurred. For example, see Emslie (1970) on the Michikamau intrusion in Labrador, and Weiblen and Morey (1975) on the Duluth complex in Minnesota. In both these bodies, extensive units of anorthosite overlie major divisions of layered troctolitic cumulates, and Emslie has tentatively suggested that the Michikamau anorthosite represents an accumulation of floated plagioclase. The anorthositic blocks of the Skaergaard intrusion may also have had this origin.

Nor is it meant to imply that olivine, the pyroxenes, chromite, or magnetite cannot or do not settle in magmas. They clearly are denser than all common magmatic liquids and, therefore, given the opportunity, should sink. The point being made here is that the process of layer formation generally appears to have been such that there was not much occasion for these minerals to settle on an individual basis.

the condition that the fractionated minerals mainly accumulate on the floor of a magma body (be it by growth *in situ* or current deposition), the heat loss that controls their growth will be mainly through the roof and upper walls of the body. This point was recognized by Wager and Deer (1939) and is confirmed in mathematical heat transfer models (Irvine, 1970b, 1974). The rate of accumulation of crystals due to heat loss to the roof and upper walls is so rapid that the main body of residual liquid is effectively insulated against downward heat loss by the crystals themselves. Most heat loss through the floor results only in the solidification of trapped interstitial liquid and cooling of the crystals.

The above conditions would not seem to be the most appropriate for growth of crystals on the floor of a magma body, but such crystallization can occur (in theory at least) from convecting liquid that has been super-cooled in the process of cycling along the roof and upper walls of the body and descending to its floor (Wager, 1963; Wager and Brown, 1968; also Irvine, 1970b, Fig. 1, path *abcdja*). Rocks formed by growth of crystals on the surfaces of accumulating layers have been called *crescumulates* by Wager and Brown (1968), but the examples that have been convincingly demonstrated differ greatly from typical cumulates, and in no case have they unequivocally been shown to have formed by the mechanism of supercooling just mentioned. Wager and Brown cited the "harrisite" of the Isle of Rhum intrusion as a type example. This rock had been described by Brown (1956) and Wadsworth (1960) as consisting of coral-like, upward branching aggregates of elongate olivine crystals grown from the tops of ordinary layers of cumulus olivine. According to Donaldson (1974, 1975), however, there are also occurrences in which the crystals have grown downward, and others in which they are directed laterally. He suggests that they represent local pockets of liquid in normal cumulates in which the coarse crystals grew because of supercooling due to sudden loss of H_2O from the liquid.

Another example of crescumulates cited by Wager and Brown (1968, p. 510) is the comb layering of the Willow Lake intrusion described by Taubeneck and Poldervaart (1960). As Wager and Brown noted, however, this layering is confined to the periphery of the intrusion and so probably was formed because of supercooling associated with outward (or downward) heat loss through the layers themselves, rather than with losses higher along the intrusion walls or roof.

Perhaps a more convincing example of crescumulates formed from magma supercooled high in an intrusion is the inch-scale layering of the Stillwater complex. This layering was distinguished by Hess (1960) and has been further illustrated by Wager and Brown (1968, p. 310). A well-known and particularly impressive occurrence near the Stillwater River consists of dozens of alternate layers of pyroxene and plagioclase with pyroxene layers regularly spaced in doublets (see Wager and Brown, 1968, Fig. 187). The minerals are not strongly oriented normal to the layering, but they are relatively coarse-grained, and it is difficult to imagine the pyroxene-layer doublets as having formed by a current-deposition mechanism. The locality is several thousand meters stratigraphically above the exposed part of the basal contact of the intrusion, so it would seem that if the minerals did grow *in situ*, it was probably because of upward rather than downward heat loss.

The difference between crescumulates and typical primary cumulates seems particularly well illustrated by a comparison of the Skaergaard marginal border group and layered series. The marginal border group is differentiated in a way that is clearly indicative of solidification from the walls of the intrusion inward (Wager and Deer, 1939), and its textures and structures are such as to indicate that the solidification occurred in large part (although by no means entirely) by crystallization *in situ*. Thus, interspersed through the outer "Tranquil Division" of the group are zones in which the fractionated minerals or clusters of these minerals have grown normal to the walls of the intrusion, forming the "perpendicular-feldspar and wavy-pyroxene rocks" of Wager and Deer (1939) and Wager and Brown (1968). The inner, "Banded Division" is characterized by layers or bands of contrasting modal composition from a few centimeters to several meters in thickness. These bands typically are highly irregular (cf. Wager and Brown, 1968) but in places exhibit beautifully developed colloform growth structures directed into the intrusion (Fig. 9-14A). There are also rocks with typical cumulus textures that would seem to represent accumulations of crystals that were formed in convecting magma and then caught along the walls and frozen in place. Otherwise, however, the marginal border group stands in marked contrast to the layered series, where the stratification is smooth and almost planar, and where perpendicular orientation of minerals is inconspicuous, if not totally absent. It might be argued that the layered series also formed by *in situ* crystallization and that the contrast with the marginal border group reflects a difference in the mechanism of crystal nucleation, depending on whether the liquid was losing heat at the site of crystallization (as through the marginal border group) or had already been supercooled (while convecting along the roof and upper walls of the intrusion). But even if this possibility can be supported, it remains to be demonstrated that *in situ* crystallization could produce the apparent sedimentation structures and the overall relations of the layering described earlier.

The alternative explanation, that magmatic currents can deposit minerals downward even when they are less dense than the liquid, is also unorthodox. Among other things, it would require a fluid dynamic effect opposite to those identified with flow differentiation. On the other hand, the possibility is not without geological support. In the Duke Island ultramafic complex, for example, distributed through a 500-meter section of the most extensively layered part of the complex, there occur more than a dozen xenoliths of milky white, vein-type quartz, measuring from about 10 cm to almost a meter in maximum dimension (Irvine, 1974). With a density of only about 2.55 g/cm^3 at the probable temperature of what was evidently an ultrabasic host liquid, these fragments almost certainly should have floated, yet they have been deposited with some of the best graded and sorted layers in the complex.

It might also be emphasized that, although there have been important recent advances in respect to defining the physical properties of magmas, their actual fluid mechanics is still a very little known subject. It is known, for example, that magmas have substantial yield strengths (Shaw, 1969; Murase and McBirney, 1973), that their effective viscosities increase markedly with the concentration of crystals (Shaw, 1965), and that crystals frequently grow in chains and clusters and otherwise interlock with and adhere to one another (e.g., Kirkpatrick, 1977). But it is not currently possible to predict patterns and rates of convection in igneous intrusions; experimental and theoretical studies of transport of crystals in magmas have so far been limited to models of flow differentiation (Bhattacharji, 1967; Komar, 1972a,b); and no quantitative study has been made of possible mechanisms of crystal deposition. Thus, in view of the efficacy of aqueous currents in transporting and depositing sediment (and given the uncertainties that still exist as to the mechanisms involved despite enormous amounts of research in that area; e.g., see Blatt, Middleton, and Murray, 1972), it obviously would be premature to discard concepts of magmatic sedimentation based on the structural and textural similarities of igneous layering and sedimentary bedding. Given that plagioclase floats in some magmas, the term "crystal settling" must be used with more discretion than has been the practice in the past twenty years, but by the same token, it may prove that *deposition of crystals* by magmatic currents is more important than even Wager suggested.

DEPOSITION OF CRYSTALS

As matters stand, Wager's (1963) concept of two types of magmatic currents still seems the most satisfactory as a general framework for the interpretation of layering in differentiated intrusions. Ordinarily, the crystals accumulate from relatively slow, regular convection currents with almost no crystal sorting. Variations in the relative modal proportions of the fractionated minerals are then largely a function of the crystallization characteristics of the parental liquid as determined by its fractionation trend and by processes such as replenishment and blending with fresh liquid. These types of variation commonly involve large volumes of magma, as is evident, for example, in stratiform layered intrusions where they are manifest in cyclic units tens to hundreds of meters in thickness extending for tens of kilometers. By contrast, the layers in which minerals appear to have been sorted by grain size or density typically are only a few centimeters to a meter or so in thickness and can rarely be traced for more than a few hundred meters. In this respect at least, then, they would seem reasonably attributed to deposition from local, rapid density or suspension currents. It might also be noted that in flume studies of sedimentation from aqueous turbidity currents, the deposition of graded beds is observed to be a highly dynamic process (e.g., Middleton, 1967). The process is much more than just a matter of distributing the suspension and having the sediment settle out; the sorting is largely accomplished through the action of the current itself. Present knowledge suggests that similar sorting may occur in magmatic currents through processes of flow differentiation (e.g., Komar, 1972a, p. 980), even though the flow is mainly laminar rather than turbulent.

As for the mechanisms by which crystals might be deposited, it seems clear that if such deposition occurs, it must be under conditions that differ considerably from those obtaining during flow differentiation. In particular, it would seem that decrease in flow velocity should be an especially important factor. In the studies of flow differentiation that have been made to date, the tendency of suspended solids (= crystals) to move away from fixed surfaces bounding the flow has been demonstrated experimentally for the condition that the total flow is constant (e.g., Bhattacharji and Smith, 1964) or theoretically for conditions of steady-state (equilibrium) flow and segregation (Komar, 1972a,b). Thus, either directly in the experiments or by implication in the theoretical analysis, force is continuously being applied to the suspension to sustain the flow. The differentiation mechanisms are such that this force causes shearing of the suspension along bounding surfaces, and the shearing produces grain dispersive pressures and lift forces that deflect the crystals to regions where the shearing is less intense (see Komar, 1972a,b). Deposition of crystals, on the other hand, would presumably occur from flows that were being decelerated by drag along the floor (or roof or walls) of an intrusion after the driving forces that initiated the flow had been relaxed or removed. It seems possible that, under these conditions, the zone of strong shearing would be displaced upward, leaving the relatively crystal-free magma produced immediately adjacent to the floor by flow differentiation during acceleration of the current, and shifting into the more concentrated suspension above. With this shifting, a sharp increase in grain dispersive pressure would be expected within the basal part of the suspension that was still flowing. Some of the crystals in this zone would deflected upward, but it would seem that others should be displaced downward into the now-stagnant magma beneath, and if the shearing continued gradually to shift upward, an overall downward displacement of the crystals might ultimately result. In effect, then, the crystals would be expelled from the flow and plated onto the floor by the action of the current. Settling of mafic minerals would, of course, augment such an effect, and it would seem that the yield strength of the magma might be a significant factor in the deceleration of the current by drag along the floor.

In slow, continuous convection currents decelerating gradually away from the source area of the crystals, plagioclase should accumulate along with the mafic minerals in a quasi-steady-state process. In the intermittent, rapid currents, it might become separated from the mafic minerals through flow differentiation while the current was accelerating, but then, being entrained in the flow system, would be deposited later, as at the top of modally graded layers. After being lodged in a stagnant zone, any tendency for the plagioclase crystals to float free would be opposed by the yield strength of their host liquid.

Whether or not this mechanism is correct, the geological relations suggest that, if crystal sorting does occur in igneous systems, it is largely restricted to currents of relatively small volume. The process, therefore, is probably not as important as Bowen believed.

CONCLUSIONS

In the preceding sections, it has been shown in theory, and to some extent by example, that a great variety of rock compositions can be developed through crystal accumulation directly as a result of fractional crystallization of tholeiitic magmas, given certain demonstrable differences in primitive liquid compositions and in certain auxiliary processes such as liquid blending, change in pressure, loss of volatiles, and reaction between the crystals and the liquids in which they accumulate. A method is described for defining primitive liquid compositions for basic magmas with olivine on the liquidus, and it is shown that two of the major tholeiitic series of Cenozoic time apparently stem from picritic melts—i.e., from melts rich in normative olivine. No general conclusion is drawn with respect to the existence of liquids rich in normative plagioclase, but in the one considered example of magma with plagioclase as the only apparent liquidus phase (the Pl-type abyssal tholeiites), the liquid appears to have formed from an Ol-rich melt by a crystal fractionation process involving a change in depth (pressure) or H_2O fugacity. All these observations and deductions relate to the kinds of principles advocated by Bowen. Effects attributable to mechanical sorting of crystals are reviewed, but the evidence suggests that this process usually occurs in a special type of magmatic current of limited volume and so is probably not as important in the overall differentiation of rocks formed by crystal accumulation as Bowen thought.

Acknowledgments   The manuscript was improved through critical reviews by N. T. Arndt, A. R. McBirney, H. R. Naslund, and H. S. Yoder, Jr. The author is responsible, however, for any errors of fact or interpretion.

References

Arndt, N. T., A. J. Naldrett, and D. R. Pyke, Komatiitic and iron-rich tholeiitic lavas of Munro Township, Northeast Ontario, *J. Petrol.*, *18*, 319–369, 1977.
Bailey, E. B., C. T. Clough, W. B. Wright, J. E. Richey, and G. V. Wilson, The Tertiary and Post-Tertiary geology of Mull, Loch Aline and Oban, *Mem. Geol. Surv. Scotland*, 1924.
Baragar, W. R. A., Petrology of basaltic rocks in part of the Labrador Trough, *Geol. Soc. Am. Bull.*, *71*, 1589–1644, 1960.
Bartlett, R. W., Magma convection, temperature distribution, and differentiation, *Am. J. Sci.*, *267*, 1067–1082, 1969.
Bhattacharji, S., Mechanics of flow differentiation in ultramafic and mafic sills, *J. Geol.*, *75*, 101–112, 1967.
Bhattacharji, S., and C. H. Smith, Flowage differentiation, *Science*, *145*, 150–153, 1964.
Biggar, G. M., Phase equilibrium studies of chilled margins of some layered intrusions, *Contrib. Mineral. Petrol.*, *46*, 159–167, 1974.
Blanchard, D. P., J. M. Rhodes, M. A. Dungan, K. V. Rodgers, C. H. Donaldson, J. C. Brannon, J. W. Jacobs, and E. K. Gibson, The chemistry and petrology of basalts from Leg 37 of the Deep-Sea Drilling Project, *J. Geophys. Res.*, *81*, 4231–4246, 1976.
Blank, H. R., Jr., and M. E. Gettings, Subsurface form and extent of the Skaergaard intrusion, *EOS, Trans. Am. Geophys. Un.*, *54*, 507 (abstract), 1973.
Blatt, H., G. V. Middleton, and R. Murray, *Origin of Sedimentary Rocks*. Prentice-Hall, N.J., 634 pp., 1972.
Bottinga, Y., and D. F. Weill, Densities of liquid silicate systems calculated from partial molar volumes of oxide components, *Am. J. Sci.*, *269*, 169–182, 1970.

Bougault, H., and Hekinian, R., Rift valley in the Atlantic Ocean near 36°50′; petrology and geochemistry of basaltic rocks, *Earth Planet Sci. Lett.*, *24*, 249–261, 1974.

Bowen, N. L., The crystallization of haplobasaltic, haplodioritic, and related magmas, *Am. J. Sci.*, *40*, 161–185, 1915.

———, Certain singular points on crystallization curves of solid solutions, (*U.S.*) *Natl. Acad. Sci. Proc.*, *27*, 301–209, 1941.

Bowen, N. L., and J. F. Schairer, The system, $MgO–FeO–SiO_2$, *Am. J. Sci.*, *29*, 151–217, 1935.

Brooks, C. K., Rifting and doming in southern East Greenland, *Nature (Phys. Sci.)*, *244*, 23–25, 1973.

Brown, G. M., The layered ultrabasic rocks of Rhum, Inner Hebrides, *Phil. Trans. Roy. Soc. London, Ser. B.*, *240*, 1–53, 1956.

Bryan, W. B., G. Thompson, F. A. Frey, and J. S. Dickey, Inferred geologic settings and differentiation in basalts from the Deep-Sea Drilling Project, *J. Geophys. Res.*, *81*, 4285–4304, 1976.

Campbell, I. H., J. M. Dixon, and P. L. Roeder, Crystal bouyancy in basaltic liquids and other experiments with a centrifuge furnace, *EOS, Trans. Am. Geophys. Un.*, *58*, 527, 1977.

Carmichael, I. S. E., The petrology of Thingmuli, a Tertiary volcano in eastern Iceland, *J. Petrol.*, *5*, 435–460, 1964.

Chayes, Felix, The silica-alkali balance in basalts of the submarine ridges, *Carnegie Institution of Washington Yearbook*, *64*, 155, 1965.

Clague, D. A., and T. E. Bunch, Formation of ferrobasalt at East Pacific Midocean spreading centers, *J. Geophys. Res.*, *81*, 4247–4256, 1976.

Darwin, C., *Geological Observations on the Volcanic Islands, visited during the voyage of H.M.S. Beagle, together with some brief notices on the geology of Australia and the Cape of Good Hope. Being the second part of the geology of the voyage of the Beagle, under the command of Capt. Fitzroy, R. N., during the years 1832 to 1836*, Smith, Elder, London, 175 pp., 1844.

Donaldson, C. H., Olivine crystal types in harrisitic rocks of the Rhum pluton and Archean spinifex rocks, *Geol. Soc. Am. Bull.*, *85*, 1721–1726, 1974.

———, A petrogenetic study of harrisite in the Isle of Rhum pluton, Scotland, Unpubl. Ph.D. thesis, Univ. St. Andrews, 1975.

Emslie, R. F., Crystallization and differentiation of the Michikamau intrusion, *in* Origin of anorthosite and related rocks, *N.Y. State Mus. and Sci. Service, Mem. 18*, 163–173, 1968.

———, The geology of the Michikamau intrusion, *Geol. Surv. Can. Paper 68-57*, 85 pp., 1970.

———, Liquidus relations and subsolidus reactions in some plagioclase-bearing systems, *Carnegie Institution of Washington Yearbook*, *69*, 148–155, 1971.

———, Some chemical characteristics of anorthositic suites and their significance, *Can. J. Earth Sci.*, *10*, 54–71, 1973.

Engel, A. E. J., and C. G. Engel, Composition of basalts from the Mid-Atlantic Ridge. *Science*, *144*, 1330–1333, 1964.

Fisher, D. J., A. J. Frueh, Jr., C. S. Hurlbut, Jr., and C. E. Tilley, *eds.*, International

Mineralogical Association, Papers and Proceedings of the Third General Meeting, *Mineral. Soc. Am. Spec. Paper 1*, 322 pp., 1963.

Folk, R. L., Longitudinal dunes of the northwestern edge of the Simpson Desert, Northern Territory, Australia. pt. 1: Geomorphology and grain size relationships, *Sedimentology*, *16*, 5–54, 1971.

Gary, M., R. McAfee, Jr., and C. L. Wolf, *Glossary of Geology*, Amer. Geol. Inst., 805 pp., 1972.

Gibb, F. G. F., Flow differentiation in the xenolithic ultrabasic dykes of the Cuillins and the Strathaird Peninsula, Isle of Skye, Scotland, *J. Petrol.*, 9, 411–443, 1968.

Green, D. H., and A. E. Ringwood, The genesis of basaltic magmas, *Contrib. Mineral. Petrol.*, *15*, 103–190, 1967.

Hekinian, R., and F. Aumento, Rocks from the Gibbs fracture zone and Minia Seamount near 53°N in the Atlantic Ocean, *Marine Geol.*, *14*, 47–72, 1973.

Hess, H. H., Stillwater igneous complex, Montana, a quantitative mineralogical study, *Geol. Soc. Am. Mem. 80*, 230 pp., 1960.

Houbolt, J. J. H. C., Recent sediments in the Southern Bight of the North Sea, *Geol. en Mijnbouw*, 47, 245–273, 1968.

Irvine, T. N., The ultramafic complex and related rocks of Duke Island, southeastern Alaska, Ph.D. thesis, California Institute of Technology, 320 pp., 1959.

———, Chromian spinel as a petrogenetic indicator; pt. 2. Petrologic applications, *Can. J. Earth Sci.*, *4*, 71–103, 1967.

———, Crystallization sequences in the Muskox intrusion and other layered intrusions, I. Olivine-pyroxene-plagioclase relations, *Spec. Publs. Geol. Soc. S. Africa*, *1*, 441–476, 1970a.

———, Heat transfer during solidification of layered intrusions. I. Sheets and sills, *Can. J. Earth Sci.*, *7*, 1031–1061, 1970b.

———, Petrology of the Duke Island ultramafic complex, southeastern Alaska, *Geol. Soc. Am. Mem. 138*, 240 pp., 1974.

———, The silica immiscibility effect in magmas, *Carnegie Institution of Washington Yearbook*, *74*, 484–492, 1975a.

———, Olivine-pyroxene-plagioclase relations in the system Mg_2SiO_4–$CaAl_2Si_2O_8$–$KAlSi_3O_8$–SiO_2 and their bearing on the differentiation of stratiform intrusions, *Carnegie Institution of Washington Yearbook*, *74*, 492–500, 1975b.

———, Axelgold layered gabbro intrusion, McConnell Creek Map-area, British Columbia, *Geol. Surv. Can. Paper 75-1, Part B*, 81–88, 1975c.

———, Metastable liquid immiscibility and MgO–FeO–SiO_2 fractionation patterns in the system Mg_2SiO_4–Fe_2SiO_4–$CaAl_2Si_2O_8$–SiO_2, *Carnegie Institution of Washington Yearbook*, *75*, 597–611, 1976.

———, Origin of chromitite layers in the Muskox intrusion and other stratiform intrusions: a new interpretation, *Geology*, *5*, 273–277, 1977a.

———, Relative variations of substituting chemical components during igneous fractionation processes, *Carnegie Institution of Washington Yearbook*, *76*, 539–541, 1977b.

Irvine, T. N., and Baragar, W. R. A., The Muskox intrusion and Coppermine River lavas, Northwest Territories, Canada, *Int. Geol. Congr.*, *24th, Montreal, Field Excursion A29 Guidebook*, 70., 1972.

Irvine, T. N., and C. H. Smith, The ultramafic rocks of the Muskox intrusion, in *Ultramafic and Related Rocks*, P. J. Wyllie, ed., John Wiley and Sons, New York, 38–49, 1967.

Irvine, T. N., and C. H. Smith, Primary oxide minerals in the layered series of the Muskox intrusion, *in* H.D.B. Wilson, ed., *Magmatic Ore Deposits*, Econ. Geol. Mon. 4, 76–94, 1969.

Isachsen, Y. W., ed., Origin of anorthosite and related rocks, *N.Y. State Museum and Sci. Service Mem. 18*, 466 pp., 1969.

Ito, K., and G. C. Kennedy, Melting and phase relations of natural peridotite to 40 kilobars, *Am. J. Sci., 265*, 519–538, 1967.

Ito, K., and G. C. Kennedy, Melting and phase relations in the plane tholeiite-lherzolite-nepheline basanite to 40 kilobars with geological implications, *Contrib. Mineral. Petrol., 19*, 177–211, 1968.

Jackson, E. D., Primary textures and mineral associations in the ultramafic zone of the Stillwater Complex, Montana, *U.S. Geol. Surv. Prof. Pap., 358*, 1961.

———, The cyclic unit in layered intrusions—a comparison of the repetitive stratigraphy in the ultramafic parts of the Stillwater, Muskox, Great Dyke and Bushveld Complexes, *Spec. Publs. Geol. Soc. S. Africa, 1*, 391–424, 1970.

———, The origin of ultramafic rocks by cumulus processes, *Fortschr. Miner., 48*, 128–174, 1971.

Jameison, B. G., Phase relations in some tholeiitic lavas illustrated in the system R_2O_3-XO-YO-ZO$_2$, *Mineral. Mag.*, 37, 538–554, 1970.

Karcz, I. Harrow marks, current-aligned sedimentary structures, *J. Geol., 75*, 113–121, 1967.

Kirkpatrick, R. J., Nucleation and growth of plagioclase, Makaopuhi and Alae lava lakes, Kilauea volcano, Hawaii, *Geol. Soc. Am. Bull., 88*, 78–84, 1977.

Komar, P. D. Mechanical interactions of phenocrysts and flow differentiation of igneous dikes and sills, *Geol. Soc. Am. Bull., 83*, 973–988, 1972a.

———, Flow differentiation of igneous dikes and sills: Profiles of velocity and phenocryst concentration, *Geol. Soc. Am. Bull., 83*, 3443–3448, 1972b.

Kuno, H., High-alumina basalt, *J. Petrol., 1*, 121–155, 1960.

Kushiro, I., and Yoder, H. S., Jr., Anorthite-forsterite and anorthite-enstatite reactions and their bearings on the basalt-eclogite transformation, *J. Petrol., 7*, 337–362, 1966.

Longhi, J., and D. Walker, Fe-Mg distribution between olivine and lunar basaltic liquids, *EOS, Trans. Am. Geophys. Un., 56*, 471 (abstract), 1975.

Macdonald, G. A., Petrography of the Island of Hawaii, *U.S. Geol. Surv. Prof. Pap. 214*-D, 51–96, 1949.

MacRae, N. D., Ultramafic intrusions of the Abitibi area, Ontario, *Can. J. Earth Sci., 6*, 281–303, 1969.

Mathez, E. A., Sulfur solubility and magmatic sulfides in submarine basalt glass, *J. Geophys. Res., 81*, 4269–4276, 1976.

McBirney, A. R., Differentiation of the Skaergaard intrusion, *Nature, 253*, 691–694, 1975.

McBirney, A. R. and Y. Nakamura, Immiscibility in late-stage magmas of the

Skaergaard intrusion, *Carnegie Institution of Washington Yearbook*, *74*, 348–352, 1974.

McBirney, A. R., and H. Williams, Geology and petrology of the Galapagos Islands, *Geol. Soc. Am. Mem. 118*, 1969.

Middleton, G. V., Experiments on density and turbidity currents. III. Deposition of sediment, *Can. J. Earth Sci.*, *4*, 475–505, 1967.

Miyashiro, A., F. Shido, and M. Ewing, Crystallization and differentiation in abyssal tholeiites and gabbros from mid-ocean ridges, *Earth Planet. Sci. Lett.*, *7*, 361–365, 1970.

Morse, S. A., The Kiglapait layered intrusion, Labrador, *Geol Soc. Am. Mem. 112*, 204 pp., 1969.

———, The Nain anorthosite project, Labrador: *Field report 1973, Contr. 11*: Geol. dept., Univ. Massachusetts, 156 pp., 1973.

Murase, T., and A. R. McBirney, Properties of some common igneous rocks and their melts at high temperature, *Geol. Soc. Am. Bull.*, *84*, 3563–3592, 1973.

Murata, K. J., and D. H. Richter, Chemistry of the lavas of the 1959–1960 eruption of Kilauea volcano, Hawaii, *U.S. Geol. Surv. Prof. Pap. 537-A*, 1–26, 1966.

Naldrett, A. J., J. G. Bray, E. L. Gasparrini, T. Podolsky, and J. C. Rucklidge, Cryptic variation and the petrology of the Sudbury Nickel Irruptive, *Econ. Geol.*, *65*, 122–155, 1970.

Naldrett, A. J., and G. D. Mason, Contrasting Archean ultramafic bodies in Dundonald and Clergre Townships, Ontario, *Can. J. Earth Sci.*, *5*, 111–143, 1968.

O'Hara, M. J., Primary magmas and the origin of basalts, *Scot. J. Geol.*, *1*, 19–40. 1965.

———, The bearing of phase equilibria studies in synthetic and natural systems on the origin and evolution of basic and ultrabasic rocks, *Earth Sci. Rev.*, *4*, 69–133, 1968.

———, Geochemical evolution during crystallisation of a periodically refilled magma chamber, *Nature*, *266*, 503–507, 1977.

O'Hara, M. J., M. J. Saunders, and E. L. P. Mercy, Garnet peridotite primary ultrabasic magma and eclogite; interpretation of upper mantle processes in kimberlite, *Physics and Chemistry of the Earth*, *9*, 571–604, 1975.

O'Hara, M. J., and H. S. Yoder, Jr., Formation and fractionation of basic magmas at high pressures. *Scot. J. Geol.*, *3*, 67–117, 1967.

Osborn, E. F., and J. F. Schairer, The ternary system pseudowollastonite–akermanite–gehlenite, *Am. J. Sci.*, *239*, 715–763, 1941.

Pearce, T. H., A contribution to the theory of variation diagrams, *Contrib. Mineral. Petrol.*, *19*, 142–157, 1968.

———, Chemical variation in the Palisade Sill, *J. Petrol.*, *11*, 15–32, 1970.

Powers, H. A., Differentiation of Hawaiian lavas, *Am. J. Sci.*, *30*, 57–71, 1935.

Presnall, D. C., The join forsterite–diopside–iron oxide and its bearing on the crystallization of basaltic and ultramafic magmas, *Am. J. Sci.*, *264*, 753–809, 1966.

Presnall, D. C., S. A. Dixon, T. H. O'Donnell, N. L. Brenner, R. L. Schrock, and D. W. Dycus, Liquidus phase relations on the join diopside–forsterite–anorthite from 1 atm to 20 kbar: Their bearing on the generation and crystallization of basaltic magma, *Contrib. Mineral. Petrol.*, *66*, 203–220, 1978.

Pyke, D. R., A. J. Naldrett, and O. R. Eckstrand, Archean ultramafic lava flows in Munro Township, Ontario, *Geol. Soc. Am. Bull.*, *84*, 955–978, 1973.

Roedder, E., and P. W. Weiblen, Petrology of silicate melt inclusions, Apollo 11 and Apollo 12 and terrestrial equivalents, *Proc. Second Lunar Sci. Conf.*, *Geochim. Cosmoch. Acta*, *1*, Suppl. 2, 507–528, 1971.

Roeder, P. L., and R. F. Emslie, Olivine-liquid equilibrium, *Contrib. Mineral. Petrol.*, *29*, 275–289, 1970.

Rossman, D. L., Geology and petrology of two stocks of layered gabbro in the Fairweather Range, Alaska, *U.S. Geol. Surv. Bull. 1121-F*, 50 pp., 1963.

Schairer, J. F., The system K_2O–MgO–Al_2O_3–SiO_2, Results of quenching experiments on four joins in the tetrahedron cordierite–forsterite–leucite-silica and on the join cordierite–mullite–potash feldspar, *J. Am. Ceram. Soc.*, *37*, 501–533, 1954.

Schairer, J. F., and H. S. Yoder, Jr., The nature of residual liquids from crystallization, with data on the system nepheline–diopside–silica, *Am. J. Sci.*, *Bradley Vol. 258A*, 273–283, 1960.

Schairer, J. F., and H. S. Yoder, Jr., Crystallization in the system nepheline–forsterite–silica at one atmosphere pressure, *Carnegie Institution of Washington Yearbook*, *60*, 141–144, 1961.

Schairer, J. F., and H. S. Yoder, Jr. The system albite–anorthite–forsterite at 1 atmosphere, *Carnegie Institution of Washington Yearbook*, *65*, 204–209, 1967.

Shaw, H. R., Comments on viscosity, crystal settling and convection in granitic magmas, *Am. J. Sci.*, *263*, 120–152, 1965.

———, Rheology of basalt in the melting range, *J. Petrol.*, *10*, 510–535, 1969.

Shibata, T., Phenocryst-bulk rock composition relations of abyssal tholeiites and their petrogenetic significance, *Geochim. Cosmochim. Acta*, 40, 1407–1417, 1976.

Shido, F., A. Miyashiro, and M. Ewing, Crystallization of abyssal tholeiites, *Contrib. Mineral. Petrol.*, *31*, 251–266, 1971.

Shido, F., A. Miyashiro, and M. Ewing, Compositional variation in pillow lavas from the Mid-Atlantic Ridge, *Marine Geol.*, *16*, 177–190, 1974.

Simkin, T., Flow differentiation in picritic sills of north Skye, in *Ultramafic and Related Rocks*, P. J. Wyllie, ed., John Wiley and Sons, New York, 64–69, 1967.

Smith, C. H., Notes on the Muskox intrusion, Coppermine River area, District of Mackenzie, *Geol. Surv. Can. Paper 61–25*, 16 pp., 1962.

Smith, C. H., and H. E. Kapp, The Muskox intrusion, a recently discovered layered intrusion in the Coppermine River area, Northwest Territories, Canada, *Mineral. Soc. Am. Spec. Pap. 1*, 30–35, 1963.

Taubeneck, W. H., and A. Poldervaart, Geology of the Elkhorn Mountains, Northwestern Oregon: Part 2. Willow Lake intrusion, *Geol. Soc. Am. Bull.*, *71*, 1295–1322, 1960.

Thompson, G., W. B. Bryan, F. A. Frey, J. S. Dickey, and C. J. Suen, Petrology and geochemistry of basalts from DSDP Leg 34, Nazca plate, *Initial Reports of the Deep Sea Drilling Project*, 215–226, 1976.

Tilley, C. E., Kilauea Magma 1959–1960, *Geol. Mag. 97*, 494–497, 1960.

Tilley, C. E., The occurrence of hypersthene in Hawaiian basalts, *Geol. Mag.*, *98*, 257–260, 1961.

Viljoen, M. J., and R. P. Viljoen, Evidence for the existence of a mobile intrusive peridotite magma from the Komati Formation of the Onverwacht Group, *Upper Mantle Project., Spec. Publs. Geol. Soc. S. Africa*, *2*, 87–112, 1969.

Visser, D. J. L., and G. von Gruenewaldt, eds., Symposium on the Bushveld igneous complex and other layered intrusion, *Spec. Publs. Geol. Soc. S. Africa*, *1*, 763 pp., 1970.

Wadsworth, W. J., The ultrabasic rocks of southwest Rhum, *Roy. Soc. London Phil. Trans., Ser. B.*, *244*, 21–64, 1960.

Wager, L. R., The major element variation of the layered series of the Skaergaard intrusion and a re-estimation of the average composition of the hidden layered series and of the successive residual magmas, *J. Petrol.*, *1*, 364–398, 1960.

——, The mechanism of adcumulus growth in the layered series of the Skaergaard intrusion, *Mineral. Soc. Am. Spec. Paper 1*, 1–19, 1963.

Wager, L. R., and G. M. Brown. *Layered Igneous Rocks*, Oliver and Boyd, Edinburgh, 588 pp., 1968.

Wager, L. R., and W. A. Deer, Geological investigation in East Greenland, Part III, The petrology of the Skaergaard intrusion, Kangerdluqssuaq, East Greenland, *Medd. Groenl.*, *105, No. 4*, 352 pp., 1939 (reissued 1962).

Washington, H. S., Chemical analyses of igneous rocks, *U.S. Geol. Surv. Prof. Pap. 99*, 1201 pp. 1917.

Weiblen, P. W., and G. L. Morey, The Duluth Complex—a petrologic and tectonic summary, *Proc. 48th Ann. Mtg. Minnesota Sect., Amer. Inst. Min. Eng.*, 72–95, 1975.

Willemse, J., The geology of the Buxhveld Complex, the largest repository of magmatic ore deposits in the world, Econ. Geol. Mon. 4, 1–22, 1969.

Wilson, H. D. B., ed., Magmatic ore deposits: a symposium Econ. Geol. Mon. 4, 366 pp., 1969.

Worst, B. G., The Great Dyke of southern Rhodesia, *Bull. Geol. Surv. S. Rhodesia*, *47*, 234 pp. 1960.

Wright, T. L., Chemistry of Kilauea and Mauna Loa lava in space and time, *U.S. Geol. Surv. Prof. Paper 735*, 1–39, 1972.

——, Magma mixing as illustrated by the 1959 eruption, Kilauea volcano, Hawaii, *Geol. Soc. Am. Bull.*, *84*, 849–858, 1973.

Wright, T. L., and R. S. Fiske, Origin of the differentiated and hybrid lavas of Kilauea volcano, Hawaii, *J. Petrol.*, *12*, 1–65, 1971.

Wright, T. L., and P. W. Weiblen, Mineral composition and paragenesis in tholeiitic basalt from Makaopuhi lava lake, Hawaii (abstract), *Geol. Soc. Amer., Spec. Paper 115*, 242–243, 1968.

Wyllie, P. J., *Ultramafic and Related Rocks*, J. Wiley and Sons, New York, 464 pp., 1967.

Yoder, H. S., Jr., Diopside–anorthite–water at five and ten kilobars and its bearing on explosive volcanism, *Carnegie Institution of Washington Yearbook*, *64*, 82–89, 1965.

Yoder, H. S., Jr., and C. E. Tilley, Origin of basalt magmas: An experimental study of natural and synthetic rock systems, *J. Petrol.*, *3*, 342–532, 1962.

Chapter 10

EFFECTS OF ASSIMILATION

A. R. MCBIRNEY

Center for Volcanology, University of Oregon, Eugene, Oregon

At the time Bowen wrote *The Evolution of the Igneous Rocks*, his discussion of assimilation was set in the context of a lively debate. A large segment of petrologic theory hinged on the belief that the compositional variations of igneous rocks resulted mainly from assimilation of crustal material in more primitive magmas. Bowen's long-time adversary, Fenner, was one of the most forceful American proponents of this view, but the importance of assimilation was also advocated by geologists such as Daly (1933) and Kennedy (1933) who, like Fenner, believed that the field relations they observed offered more compelling evidence than laboratory studies of systems that were said to be too simple to represent natural conditions.

Unless the modern reader appreciates the prevailing views of the time, he may find Bowen's long discussion of thermal relations, and particularly the question of superheat, belabored and elementary. It should be recalled, however, that only a few years earlier Fenner (1923) had proposed that the hybrid rocks of Katmai resulted from wholesale melting and assimilation of andesite in a high-temperature rhyolitic liquid. He presented what seemed to be convincing evidence that rocks of calc-alkaline suites are the products of assimilation of crustal rocks in primitive basaltic magmas (Fenner, 1926).

The degree to which such a concept now seems to be a fundamental violation of all that is known about the thermal relations of magmas is a measure of the pervasive impact of Bowen's writing. Superheated magmas, which were endowed by earlier petrologists with dramatic powers of assimilation, were placed in a more realistic perspective when Bowen made the simple calculation, now familiar to every petrology student, of the balance between the quantities of heat required for a liquid to melt or react with xenolithic material. The effect of the wide difference between the heat capacities (0.2 to 0.3 cal/g) and the heats of solution or fusion of silicates (80 to 110 cal/g) was demonstrated by pointing out that temperatures at least 300° above the liquidus would be required to assimilate a given amount of introduced crystalline material in an equal weight of liquid, even if the crystals had already been preheated to their melting temperature. There is no evidence that magmas normally have significant amounts of superheat, and on the special occasions when they do, they cannot retain it for long in the presence of crystalline phases.

Bowen was not the first petrologist to see that superheat is an unlikely attribute of magmas; Becker (1897) and Harker (1904, 1909) had already recognized the thermal limitation of natural magmas. But Bowen put the question in the context of a rigorous interpretation of crystal-liquid equilibria and showed that the large difference between heat capacity and heat of fusion is, by any measure, the dominating feature of the thermal relation of silicates.

Before considering how well Bowen's theoretical principles have been upheld by subsequent petrologic observations, it will be useful to review some of the elementary crystal-liquid relations that govern the effects of assimilation.

EQUILIBRIUM RELATIONS BETWEEN INCLUSIONS AND LIQUIDS

The effects of contamination can best be illustrated by a few simple examples of what happens when a small crystal of some extraneous origin is added to a large volume of crystallizing liquid (Fig. 10-1). Three relationships are possible, depending on whether the crystal is (a) a phase that the liquid is currently precipitating, (b) a phase the liquid could have crystallized earlier but is no longer precipitating, or (c) a phase that the liquid would precipitate only at a later stage in its cooling history.

In the first case, if a crystal of a mineral such as AB in Fig. 10-1 is added to the liquid with the composition and temperature of X and is the same phase with which the liquid is saturated, the crystal will be stable and simply add to the volume of crystalline material without altering the liquid composition in any way. If the added crystal is a small amount of a mineral such as A, which such a liquid could have precipitated earlier if it had evolved from a higher temperature by crystal fractionation, the

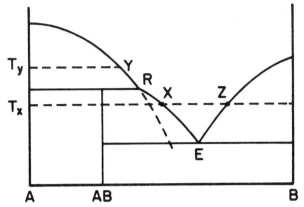

Figure 10-1. Equilibrium relations of crystalline and liquid phases in a two-component system with a eutectic and an intermediate composition that melts incongruently. In the systems forsterite–silica and leucite–silica, which are of this type, forsterite and leucite are phases, such as *A*, that react with the liquid on cooling, and silica has a role like that of component *B*.

liquid cannot melt the crystal, because its liquid is below the melting temperature of pure *A*; it can only react with it. The liquid *X* has a composition that is poorer in *A* than the projected saturation curve or liquidus of *A* at the same temperature. Crystalline *A* will therefore be unstable, and because the liquid cannot become richer in that component without leaving the liquidus of *AB*, crystals of *AB* are precipitated at the same time that the crystals of *A*, combining with components of *B* from the liquid, are converted to *AB*.

If conditions are isothermal, the liquid composition remains unchanged, even though the bulk composition of the system as a whole is changed by the increase in the quantity of crystalline *AB* in equilibrium with the liquid. As more crystals of *A* are added and they react with the liquid to form increasing amounts of *AB*, the bulk composition of the system will eventually become so rich in *A* that it lies between *A* and *AB*, and at that stage the liquid will have been entirely consumed by the reaction.

Under conditions of falling temperature, the composition of the liquid would continue down its course toward *E* and become more depleted in the *A* component, even though the bulk composition of the entire assemblage of crystals and liquid would be richer in that component. If the amount of *A* that is added is enough to place the bulk composition between *A* and *AB*, the liquid may be consumed by reaction before reaching *E*.

If the liquid has the composition and temperature of Y and is in equilibrium with crystalline A, addition of more crystals of A may affect the subsequent course of crystallization. As the temperature falls and the liquid reaches the reaction point, R, the crystals of A react with the liquid to form crystals of AB, and the additional quantity of A increases the amount of liquid consumed in this reaction; if the bulk composition has been changed enough to place it between A and AB, the course of crystallization of the liquid can be changed from one that would reach the eutectic E to one that terminates at R. This effect depends on efficient reaction between the liquid and A; under conditions of strong fractionation there would be little or no visible effect other than an increase in the proportion of A among the final products of crystallization.

If the added crystal is a proportionately small quantity of the mineral B, a phase with which a liquid, such as X, is still undersaturated, the crystal is unstable and will dissolve, even though the temperature of the liquid is below the melting temperature of B. If a large proportion of crystalline AB is present and conditions are isothermal, the composition of the liquid cannot be changed by the addition of a small amount of B without leaving the liquidus where it is saturated with AB. Consequently, the addition of B is balanced by solution of a proportional amount of previously crystallized AB, and the composition of the liquid remains constant. Only after all the crystals of AB are consumed, can the liquid be enriched in component B under isothermal conditions.

Even though these examples have assumed that the added crystals were at the same initial temperature as the liquid, addition of A to a liquid such as X would result in precipitation of a large amount of AB and a liberation of heat, whereas addition of B to the same liquid would require solution of both the added and previously crystallized phases and would result in an absorption of heat. From relations such as these, Bowen formulated the general rule that addition of a phase with which the liquid reacts may result in a liberation of heat, whereas addition of a phase that the liquid has not yet precipitated causes endothermic solution of both the added phase and, in some cases, earlier-formed crystals. In the first instance, the proportion of liquid is reduced; in the second, it is increased.

Under natural conditions, these characteristics of isothermal systems are commonly overshadowed by other effects, particularly when the amount of added material is proportionately large and at an initial temperature below that of the liquid. Heat required to raise the temperature of xenoliths to that of the magma is provided by crystallization of the phase or phases with which the liquid is currently saturated, or the temperature of the liquid is reduced, or both. The final temperature and composition of the liquid will be governed by the liquidus conditions at which equilibrium is restored. Under natural conditions, therefore, the effect of relatively cold material on the liquid is essentially the same, regardless of whether the added material is early or late in the normal crystallization sequence. The heat extracted from the liquid lowers the temperature and results in precipitation of the phase or phases with which the liquid is saturated; the compositional trend of the liquid follows the liquidus just as it would without contamination, and only the proportions of the crystalline end-products differ. An important but commonly overlooked feature of this cooling effect is that, as can be seen in the example in which crystals of A are added to liquid X in Fig. 10-1, the liquid can eventually become more *unlike* the added composition. Under natural conditions, the liquid may follow a course in which it is depleted of the same component that was added.

The same principles hold for systems with two or more eutectics separated by a thermal divide. In Fig. 10-2a, liquids, such as F, that have compositions richer in A than the intermediate compound AB must descend with falling temperature to a eutectic, E_1, between A and AB. If the temperature of the system is below that of the thermal divide, i.e., below the melting temperature of pure AB, addition of a small proportion of crystals of B under conditions of falling temperature cannot affect the subsequent compositional course of the liquid; only the final proportions of phases are changed. The crystals of B react with the liquid and are made over to AB by extracting A from the liquid, and with falling temperature the liquid becomes more *unlike* the composition that was added.

Under certain conditions, however, the amount of contamination may be great enough to consume all of the original liquid, and another liquid may result that could follow a line of descent toward the other eutectic, E_2. Assume, for example, that conditions are isothermal and that the amount of B that is added to liquid F is so great that all of the liquid is consumed in the reaction forming crystals of AB. If crystals of B still remain and the temperature is above that of the eutectic E_2, a new liquid would form with the composition of F' or F'', depending on the proportions of AB and B. Under natural conditions of falling temperature and incomplete reaction or equilibration, a number of liquid compositions may be possible, even in the same body of magma. Differences in the degree of equilibration and rate of cooling will result in localized regimes around xenoliths of various sizes; in each case, the combination of liquid and crystals that develops should approach the equilibrium assemblage for the local conditions and proportions of phases.

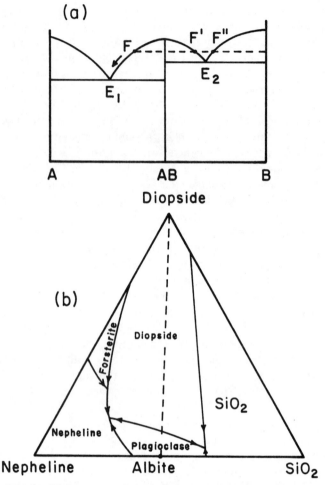

Figure 10-2. (a) Equilibrium relations of crystalline and liquid phases in a two-component system with an intermediate compound that melts congruently to form a thermal high between two eutectics. (b) The three-component system diopside–nepheline–silica (Schairer and Yoder, 1960) is an example of a system in which an intermediate compound (in this case albite) forms a thermal barrier close to its join with a third component, diopside, and divides the system into two subsystems, each with its own eutectic. (Because of solid solution in both albite and diopside, the thermal divide lies near, but not exactly on, the join.) The sub-system nepheline–albite–diopside corresponds to a system in which silica-poor liquids evolve toward an invariant point at which nepheline is precipitated; the sub-system albite–SiO_2–diopside corresponds to a more silica-rich system in which a silica mineral will crystallize. The principles governing the effects of assimilation in such systems are the same as those described in the text for a simple two-component system, such as that shown in (a).

The same principles hold for systems involving one or more solid solution series (Fig. 10-3). Under isothermal conditions, addition of a crystal M that is richer in the high-temperature end member than the composition N in equilibrium with the liquid, results in an exothermic reaction of the added crystal with the liquid. The crystal combines with a proportional amount of liquid and is made over to N, the composition with which the liquid is currently in equilibrium. The volume of liquid is reduced, and if complete reaction takes place the course of crystallization would have a more restricted range and end at a composition and temperature determined by the proportions of components. If, on the other hand, the added crystals have a composition, such as P, that is richer in the low-temperature component, they are superheated in the liquid and will be resorbed at the same time that more crystals of the equilibrium composition are formed. The proportion of liquid is increased and, on cooling, the course of crystallization would be extended.

Minerals that form a solid solution series and fail to attain complete equilibration may be compositionally zoned in ways that record the effects of assimilation. For example, a xenocryst of plagioclase with an anorthite content higher than that being precipitated by the liquid should equilibrate by extracting an albite component from the adjacent liquid; but if cooling proceeds too rapidly for complete equilibration, the result may be a crystal with a more albitic rim and a surrounding groundmass that contains new plagioclase that is more anorthite-rich than that with which the rest of the liquid is in equilibrium. The opposite case, in which

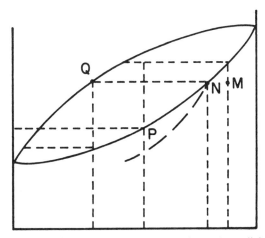

Figure 10-3. Equilibrium relations in a solid-solution series, such as olivine or the plagioclase feldspars. See text for discussion.

the xenocryst is more albitic than the plagioclase with which the liquid is in equilibrium, will result in reversely zoned xenocrysts surrounded by groundmass plagioclase that may be depleted in the anorthite component.

The effects of assimilation can usually be interpreted as an approach to the theoretical relations shown by these simplified examples. In more complex systems, in which the range of possible effects is increased by the presence of more components, there are conditions in which the course of differentiation may be diverted somewhat as a result of an addition of crystalline material. Bowen (1928, pp. 188–194) showed, for example, that in the case of the system diopside–anorthite–albite the rate of change of plagioclase compositions can be altered by addition of crystals of various compositions. But if equilibrium is approached, the course of differentiation followed by the contaminated liquid will normally be one that would be possible for a liquid that had not been contaminated. Bowen emphasized that in most instances, addition of crystalline material, whether it consists of phases that would be early or late in the sequence of crystallization, can take place without greatly altering the course of differentiation of a magma from one that it could follow spontaneously if it were evolving by crystal fractionation; only the proportions of the end products are changed.

If the added material is composed of minerals that the magma would have crystallized earlier in its cooling history, the magma cannot melt the added crystals, but may react with them and make them over to a new equilibrium composition. In the course of this reaction, the magma precipitates more of the crystals which it is currently precipitating. Under conditions of strong fractionation the end result is to reduce the proportion of the differentiated liquid below that which would arise from normal crystallization without the addition of high-temperature minerals.

Addition of crystals that would be precipitated only at a later stage of cooling has the opposite effect. Again the magma gains the heat necessary to dissolve the added crystals by precipitating others with which it is saturated, but the proportion of the end products is shifted in favor of the late crystallizing compositions, and the range of differentiation is usually extended. The conditions under which a distinctly different mineral assemblage can result from contamination are mainly those in which the amount of added material is proportionately large relative to that of the liquid, or in which the rate of cooling is too rapid for equilibrium to be attained.

The geological importance of the principles Bowen demonstrated lies in the fact that, under natural conditions, extensive assimilation could take place. If equilibrium were approached, it would be difficult, if not impossible, to show that the magma had not evolved spontaneously without an addition of material from an outside source.

THE ROLE OF TRACE ELEMENTS AND ISOTOPES

Perhaps the most significant new aspect of assimilation that has emerged since 1928 is the recognition of the important information that can be derived from certain isotopic and trace-element concentrations or ratios. Many of the limitations placed on the effects of assimilation by the phase relations outlined in the preceding section have little or no application to the abundances and distributions of those components that play only a passive role in crystal-liquid relations.

Assimilation, as Bowen showed, normally leads to increased precipitation of minerals that absorb the effect of the added material, and if equilibrium is approached, the resulting variations in abundances of major elements in the remaining liquid may be undetectable. This limitation does not necessarily apply to trace elements, however. If the assimilated rock contains unusual concentrations or ratios of lithophile elements, and if these elements are not preferentially removed by the crystals that are being fractionated, the added element can provide a sensitive measure of assimilation in cases where the effects would otherwise be concealed by equilibration of the major components.

The principles governing the concentrations of trace elements in fractionating magmas have equal application to assimilation, and if the distribution coefficients and relative abundances of these components are known, they can provide a strong constraint on mass-balance relations (Chapter 7). They commonly vary in abundance by much larger factors than the major elements. Moreover, certain of these components are more restricted in the degree to which they may be preferentially removed by crystallizing phases. For example, the abundances of strongly excluded trace elements, such as U, Th, or Zr, and certain isotopic ratios, such as $^{87}Sr/^{86}Sr$, are not likely to be reduced even if large amounts of crystal fractionation take place during the process of assimilation. Several examples are discussed in the sections that follow.

OBSERVED RELATIONS IN NATURAL SYSTEMS

Even in Bowen's time there was no lack of excellent petrographic descriptions of natural examples in which the effects of assimilation could be demonstrated. Lacroix (1893) had provided a comprehensive treatise on the petrographic features of inclusions, and studies of the Tertiary volcanic centers of western Scotland (Harker, 1904; Bailey *et al.*, 1924; Tyrrell, 1928) offered abundant geological evidence of the petrologic effects of assimilation. Daly (1933) later catalogued numerous examples of various types of assimilation, and since that time the number of descriptions in the literature has grown too great to enumerate.

The relations that are commonly shown by quartz xenocrysts in basaltic lavas (Fig. 10-4a, xenocryst at right) illustrate the simplest case of assimilation of a phase with which the liquid is undersaturated. In the earliest stages of reaction between the crystal and liquid, seams of glass penetrate along grain boundaries and fractures, and as the process of solution advances, the grains of quartz are reduced to rounded and embayed remnants surrounded by pale brown glass and a corona of augite. It is less common to see plagioclase in such reaction rims, even when the magma is saturated with this mineral and precipitating it elsewhere. The reason for this relation may lie in the relative diffusion rates of the components of mafic minerals and plagioclase.

The opposite relation, in which the added crystals are a phase that would have crystallized much earlier in the evolutionary course of the liquid, is illustrated by xenocrysts of forsteritic olivine in siliceous magmas (Fig. 10-4a, xenocryst at left). Glass is normally absent from the margins of the olivine; instead, another ferromagnesium mineral, usually pyroxene, jackets the embayed margins of the xenocrysts. The question might be asked why glass is so abundant around quartz xenocrysts that are dissolved endothermically, whereas it is absent from the margins of olivine crystals that react exothermically; consideration of the change imposed on the surrounding liquid in each case explains this apparent inconsistency. In the first case, solution of quartz adds SiO_2 to the ambient liquid and, as a result, its new composition leaves the liquidus and moves into the all-liquid field, whereas reaction of olivine with the same ambient liquid has the opposite effect; it tends to shift the liquid composition below the liquidus (Fig. 10-1). The increased viscosity and lower diffusion rates in the silica-enriched glass may accentuate these differences.

The behavior of feldspar xenocrysts follows similar relations. Figure 10-4b illustrates an example in which basic plagioclase has been added to a rhyolitic liquid saturated with potassium feldspar; Fig. 10-4c shows the opposite case in which crystals of potassium feldspar have been added to a basalt that is precipitating basic plagioclase. In the first example, the plagioclase reacts with its host and is jacketed with alkali feldspar; in the second, the potassium feldspar crystal is partly dissolved.

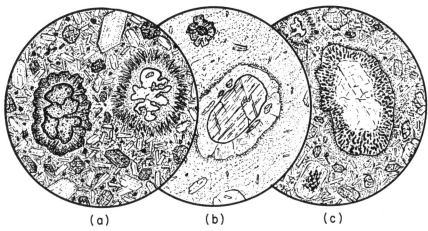

(a)   (b)   (c)

Figure 10-4. Some textures of contaminated volcanic rocks. (a) A tholeiitic basalt from the
Obrajuelo basalt-rhyolite complex near Agua Blanca, Guatemala. The lava
contains xenocrysts of olivine (left) that have reacted with the magma to form a
jacket of pyroxene and quartz crystals (right) that have been partly dissolved
and are surrounded by glass and pyroxene. (b) A rhyolitic lava from the
Gardiner River, Yellowstone Park. Rounded and partly resorbed xenocrysts
of plagioclase have reacted with the liquid and precipitated a rim of potassium
feldspar. A small grain of olivine (near top of field) is rimmed with pyroxene.
(c) An alkali-olivine basalt from the vicinity of Gharian, Libya, contains
contains xenocrysts of potassium feldspar and quartz from the underlying
basement series. The potassium feldspar is rimmed with a vermicular zone of
glass.

Holmes (1936) seems to have been the first to demonstrate that xenocrysts are not simply melted directly to a liquid of their own composition.
By carefully separating and analyzing the glass fraction around quartz
xenocrysts in a contaminated basaltic lava, he showed that it contained
quantities of Al, K, H_2O, and other components that could have come
only from the host magma. The glasses surrounding xenocrysts have been
shown to have regular compositional gradients (Maury and Bizouard,
1974) consistent with the theoretical relations described in the preceding
section. If the host and xenolith are at the same temperature, the glass
immediately adjacent to the phase or phases undergoing dissolution must
approach the composition of a liquid saturated with the added crystals
at the prevailing temperature. At the outer margin of the glassy transition
zone, the glass composition approaches that of the liquid fraction of the
magma in equilibrium with its precipitated phases; between the two
extremes, there is a complete spectrum of concentrations. These same

effects are seen on a larger scale where rocks of nearly monomineralic composition have been partially assimilated in magmas.

Sato (1975) has shown that the compositional gradients around quartz xenocrysts in basalts or andesites can be divided into as many as four distinct zones. Closest to the xenocryst is a zone of Ca-rich pyroxene and glass. If crystallized, this zone may also contain quartz, tridymite, sanidine, biotite, or iron-oxide minerals. The next zone outward consists almost entirely of Ca-rich pyroxene and may be surrounded by a third zone consisting of glass, or, in a crystallized example, some combination of Ca-rich pyroxene, orthopyroxene, quartz, tridymite, plagioclase, sanidine, and iron oxide minerals. The fourth and outermost zone is a leucocratic assemblage in the groundmass of the host rock and may contain one or more pyroxenes, plagioclase, a silica mineral, and alkali feldspar, with or without glass. In some examples, including experimentally produced zones around quartz in basaltic glass, Sato (1975) found alkalies to be preferentially concentrated close to the xenocryst. He attributed this distribution pattern to the non-ideality of alkalies in silicate melts and to a lower activity coefficient of Na and K in silica-rich liquids.

EFFECTS OF ASSIMILATION OF ROCKS OF SEDIMENTARY ORIGIN

Because sedimentary rocks vary between extreme compositional limits, their effects on magmas tend to be very conspicuous. Xenoliths composed mainly of a single mineral and very rich in a component such as silica, lime, or alumina, are assimilated only with difficulty, not simply because they are refractory, but because the process of incorporating such high concentrations of a single component requires dissemination through large volumes of magma. Pelitic sediments composed of silica, alumina, and alkalies in proportions that more nearly correspond to those of felsic magmas tend to be more easily digested than quartzites or carbonate rocks (Pitcher and Berger, 1972).

Quartz sandstones and quartzites have the simplest and most readily predictable behavior of all sedimentary inclusions; their fate can be explained by the rules already noted for individual quartz crystals. Large blocks in basaltic magmas are normally altered to partially fused quartz-rich rocks, known as "buchites," before they begin to dissolve, and blocks in granitic magmas may persist long after other rocks that surrounded them have lost their identity.

Addition of very siliceous rocks to basic magmas has a conspicuous effect, but its influence on the ultimate course of evolution of a magma is limited by the deficiency of the other components that are essential constituents of most strongly differentiated igneous rocks. There is no doubt that siliceous inclusions can change the character of basic magmas. Searle (1962), for example, has shown that assimilation of siliceous sediments in the olivine-alkali basalts of the Auckland district, New Zealand, resulted in olivine becoming unstable and reacting with the liquid to form hypersthene. Few petrologists today, however, would support the once-popular view that large-scale assimilation of siliceous crustal rocks can explain the differences between alkaline and sub-alkaline basalts. As Bowen (1928, pp. 75–77) pointed out, the differences in the concentrations of other components, such as MgO and Al_2O_3, are so great that assimilation of siliceous crustal rocks alone cannot account for the contrasts between alkaline rocks and tholeiites.

Carbonate rocks tend to have an effect opposite to that of quartz-rich rocks in that their addition can result in desilication of the liquid. Bowen defined their reaction with basaltic magma in terms of simple equations, such as

$$CaCO_3 + 3(Mg,Fe)SiO_3 \rightarrow Ca(Mg,Fe)Si_2O_6 + (Mg,Fe)_2SiO_4 + CO_2$$

$$\text{calcite} + \text{hypersthene} \qquad \text{augite} \qquad + \qquad \text{olivine}$$

and
$$CaCO_3 + 2(Ca,Na)(Al,Si)_2Si_2O_8 + \tfrac{1}{2}Al_2O_3$$
$$\text{calcite} + \qquad \text{plagioclase} \qquad + \text{melt}$$

$$\rightarrow 2CaAl_2Si_2O_8 + NaAlSiO_4 + CO_2.$$
$$\text{anorthite} \quad + \text{nepheline}$$

He emphasized that addition of calcium causes increased precipitation of augite and anorthite and depletion of the magma in other components of these minerals not contained in the xenoliths, namely, silica, magnesia, iron, and alumina.

It does not necessarily follow, however, that reactions one can write in terms of simple components are those that proceed in a natural magma. They may be inhibited by the phase relations of the liquid or by the fact that the extreme composition of the xenolith can limit the degree of assimilative effects. It has been shown in an earlier section that an equilibrium reaction producing olivine is impossible if a liquid has evolved to a temperature and composition in which olivine is no longer stable (Fig. 10-1). Natural rocks contaminated with xenoliths of limestone do not contain olivine as an equilibrium phase unless their magmas were precipitating that mineral prior to introduction of the contaminant. Olivine (or more commonly the Ca-bearing analog monticellite) may be found in the undigested xenolith, but not in the liquid. A magma that is not already nepheline-normative but has begun a spontaneous evolution toward differentiation products that eventually precipitate quartz, as in the example in Fig. 10-2b, cannot surmount a thermal barrier and follow a new trend ending in a nepheline-saturated composition, as implied by the second equation above.

Experimental studies by Watkinson and Wyllie (1969) have shown that the simple equations that were thought to illustrate reactions between carbonates and basaltic magmas do not necessarily correspond to the sequences of mineral compositions that appear in nature. Prohibitively large amounts of limestone would have to be incorporated under isothermal conditions before nepheline would become a liquidus phase, and even though nepheline may form by sub-solidus reactions, it would not be precipitated by a felsic magma contaminated with limestone unless the magma would spontaneously crystallize that mineral in its normal course of crystallization.

Aluminous sediments, when assimilated in basaltic magma, should result in an increase in the anorthite and hypersthene content of the final products of crystallization according to Bowen's proposed reaction:

$$2Al_2O_3 + SiO_2 + 2CaMgSi_2O_6 \rightarrow 2CaAl_2Si_2O_8 + 2MgSiO_3.$$
aluminous clay +   diopside   anorthite   + enstatite

Bowen agreed with petrologists, such as Lacroix, Winchell, Read, Tilley, and others (summarized by Daly, 1933, pp. 407–409) who deduced that the group of hypersthene-labradorite gabbros known as norites owed their distinctive mineralogic compositions to assimilation of aluminous sediments. Some large bodies of noritic gabbro, such as those in the Bushveld and Stillwater complexes, are seen to have been intruded into argillaceous sediments and have contact zones and inclusions consisting of pyroxene hornfels with highly aluminous mineral assemblages that include spinel, cordierite, and mullite. Moreover, the fact that the rocks of these intrusions are mainly of cumulate origin (Chapter 9) is consistent with Bowen's interpretation that assimilation of aluminous sediments would have increased the proportions of hypersthene and anorthite that crystallized either as discrete phases or as components of minerals in which they occur in solid solution, and that the nature of the liquid from which they were precipitated may not have been notably altered from a normal trend of basaltic differentiation. Today, however, few petrologists accept such an origin for norites in general; liquids of this composition can be produced by processes that were unrecognized in 1928 when the effects of pressure and volatile components had not yet been explored.

Pelitic rocks contain more of the components of felsic magmatic liquids; for that reason they are more readily assimilated than monomineralic rocks. They also have a greater tendency to extend the evolution of the host magma by increasing the proportions of late differentiated liquids. It is unfortunate that there have been few detailed studies to show precisely how the assimilation process takes place.

Evans (1964) examined the interaction between quartz-rich mica schists and a high-temperature ultramafic intrusion in Ireland, and found that with progressive heating the rocks were converted first from an assemblage of biotite, andesine, quartz, and garnet to one of cordierite, sillimanite, labradorite, magnetite, and minor biotite, and finally to hornfelses consisting of spinel, magnetite, cordierite, and various proportions of corundum and orthopyroxene. The first stages of melting produced a potassium-rich liquid, probably as a result of the initial breakdown of mica, but with increasing temperature the composition of the melt approached that of a granite or granodiorite. Liquids released by xenoliths and absorbed by the enclosing magma were rich, not only in silica and potassium, but also in Mg, Fe, Ti, Mn, and Ca. The trace elements Rb, Ba, Li, and La were also depleted. The magnesium-iron ratio as well as ratios of transition metals in xenoliths were not materially altered, but Ba/K and Rb/K were increased by a preferential loss of K. Alumina, and to a lesser extent Fe, Mg, V, Co, Ni, and Co, were residually enriched, so that the end product was an emery rock almost completely depleted in silica and alkalies. Evans concluded that the differential transfer of components could not have been by diffusion through a pore fluid or by transfer in a hydrous fluid, but rather was by migration of the melt itself.

Strong fractionation of excluded elements and isotopes can occur during rapid melting of xenoliths, especially if mica is one of the first phases to break down. The first liquid may be greatly enriched (relative to the original rock) in potassium, rubidium, and even radiogenic strontium, all of which have high concentration in biotite and muscovite. For example, Pushkar, McBirney, and Kudo (1972) examined the effects of fractional melting of phyllitic schists and gneisses in volcanic rocks and found large contrasts between the Sr isotopic ratios of the first liquid and total rock. The glass fraction of a granitic inclusion in basalt was found to have a strontium isotopic ratio of 0.7231 ± 0.0003, in contrast to a ratio of 0.7060 ± 0.0001 for the total rock. Even without melting, breakdown of micas can result in a transfer of K, H_2O, and other components from xenoliths or wall rocks to the adjacent liquid (Sigurdsson, 1968; Maury and Bizouard, 1974). Although this effect may be less pronounced at depths where biotite is stable (e.g., Ljundgren, 1959), it appears that magmas rising through old crustal rocks may become contaminated with K, radiogenic strontium, and trace elements released by the breakdown of hydrous minerals, even without extensive assimilation or anatexis (Williams and McBirney, 1969). After melting and reaction begin, however, the effects of assimilation of felsic inclusions tend to approach an equilibrium relation in which the concentration of trace elements and isotopes are more nearly proportional to the amount of added contaminants (Al-Rawi and Carmichael, 1967).

EFFECTS OF ASSIMILATION OF IGNEOUS ROCKS

It was recognized during the early studies of the Scottish Tertiary volcanic centers (Harker, 1904; Thomas, 1922) that granitic rocks were readily assimilated in basic magmas and tended to produce increased volumes of granophyric differentiates. The opposite relations had also been reported, however. Harker (1904) believed that the hybrid rocks of Marsco on the Isle of Skye had resulted from partial melting of gabbro by a granophyric liquid, and Tyrrell (1928) described similar relations at Arran. Mention has already been made of Fenner's (1923) account of the hybrid rocks of Novarupta, near Mt. Katmai, where he believed that he saw conclusive evidence for melting of andesite by rhyolite. Somewhat later Fenner (1938) argued for "solution" of solid basalt by a hot, gas-rich rhyolitic lava at Gardiner River in Yellowstone Park. Subsequent studies (Wilcox, 1944; Curtis et al., 1954; Wager et al., 1965; Thompson, 1969) of these and other examples resulted in rejection of the previous interpretation and concluded that the hybrid rocks were the result of mixing of magmas.[1]

Today there remain no undisputed examples of assimilation in which a more refractory igneous rock (i.e., one with a melting temperature higher than that of its host) has been extensively melted by a more evolved felsic magma. There are, however, well-documented instances in which basic rocks or partly crystalline magmas have been partially assimilated in granitic or syenitic magmas. Edwards (1947) provided excellent petrographic descriptions of doleritic inclusions in the syenite of Port Cygnet, Tasmania, and showed that dolerite had been made over to a mineral assemblage close to that of its host. Plagioclase reacted with the liquid and was rimmed by potassium feldspar (as in Fig. 10-4b), and the overall effect on the differentiating syenite was to reduce the volume of late-crystallizing liquid to small pods of shonkinite.

Thomas and Smith (1932) described an example of assimilation of olivine gabbro in a granitic pluton on the northern coast of Brittany. The

---

[1] It is testimony to the marvelous adaptability of geological observations that well exposed and thoroughly studied localities can provide competent petrologists with evidence for diametrically opposed conclusions. Whatever the popularly accepted theory of the moment may be, one can usually find evidence to support it.

original labradorite of the gabbroic inclusions has been changed to sodic oligoclase, and the mafic minerals underwent a sequence of transformations, first from olivine to hypersthene and then to biotite. Augite appears to have been altered first to amphibole and then to biotite.

There are a number of features of the basic inclusions of Brittany that indicate that they were at least partly molten when mixed with the granite (Didier, 1973; Barriere, 1976). They contain clots or ocelli of quartz or alkali feldspar rimmed with clinopyroxene. Apatite, sphene, and titaniferous magnetite are unusually abundant, and large crystals of orthoclase are common, especially in the margins of large blocks. Many of the mafic inclusions have pillow-like forms on their upper sides and load-cast indentations in their base. Some have fine-grained margins with plagioclase aligned parallel to the edge. According to Didier (1973), features such as these are common in mafic inclusions in granitic rocks of other regions as well.

Large crystals of potassium feldspar or other minerals that would not normally be precipitated in abundance from a mafic liquid have been found in marginal parts of xenoliths from a number of localities (Harker, 1904; Thomas and Smith, 1932; Edwards, 1947; Ljundgren, 1959). They appear to result from introduction of unusual amounts of late-crystallizing components during equilibration of the xenolith with its host. The differing conditions and compositions in the xenolith and its host alter the phase relations in such a way that the exchange between the two fractions differs from that which would be expected from simple mixing of their various components.

Consider, for example, the feldspar relations (Fig. 10-5). The compositions of alkali feldspars coexisting with both plagioclase and liquid increase in their potassium content in proportion to the anorthite content of the plagioclase. Hence, it may be possible to have a stable potassium-rich feldspar within a basic plagioclase-bearing xenolith at the same time that a more sodic plagioclase in the host coexists with anorthoclase. Similar relations may affect other components as well. The solubilities of certain elements, such as titanium and phosphorus, are known to differ greatly according to the composition of the liquid, and for this reason, the concentrations of these elements may differ widely in the liquid fractions of the xenolith and its host.

Similar effects may result from conditions that cause crystallization of large amounts of a mineral that includes unusually high proportions of one or more elements. Such a process seems to have been responsible for the complex transfer of components shown by gradients in a basaltic xenolith, enclosed in ferrogabbro, of the Skaergaard intrusion (Figs. 10-6 and 10-7). The most conspicuous feature is the migration of iron from the adjacent

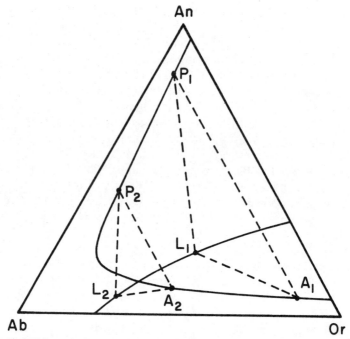

Figure 10-5. Schematic diagram showing the compositions of feldspars in equilibrium with liquids of differing compositions. Note that the alkali feldspar, A_1, in equilibrium with a liquid, L_1, that is precipitating an An-rich plagioclase, P_1, is richer in the potassium end member, Or, than is the intermediate alkali feldspar, A_2, (an anorthoclase), in equilibrium with a liquid, L_2, that precipitates a more albitic plagioclase, P_2.

gabbro and the interior of the block to form a dense layer of magnetite at the contact. Despite the fact that the enclosing liquid had a much higher iron content than the original basalt (about 14 percent vs. 9 percent total iron oxides) there has been a depletion of iron from the margins of the xenolith and an enrichment at the contact with its host. The explanation of this apparent anomaly probably lies in the role of oxygen. At the time the basaltic block entered the magma it must have been highly oxidized by hydrothermal alteration. The anomalously low ^{18}O–^{16}O ratios and their increase from the center of the block outward indicate that meteoric water had affected the basalt at or near the surface and that the oxygen was partly reequilibrated after the block sank to its present position. The oxidizing effect of the block on the adjacent liquid led to increased precipitation of magnetite, which in turn produced chemical potential gradients down which iron, zirconium, and other elements that enter

Figure 10-6. Xenoliths, believed to have come from the early-crystallizing upper part of the Skaergaard intrusion, in the lower part of the layered series. The mineral constituents have been largely re-equilibrated and recrystallized. Large blocks with mafic centers and compositional gradients toward their margins show that the xenoliths have lost some of their mafic components to the host liquid that surrounded them.

magnetite were transferred toward the contact from the block and adjacent liquid. The manner in which this transfer came about is uncertain.

Other types of non-equilibrium effects are found in granitic or felsic metamorphic inclusions in basic magmas (Fig. 10-8). Detailed accounts

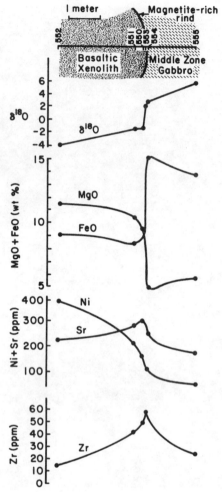

Figure 10-7. Compositional variations in a basaltic xenolith in the middle zone of the Skaergaard intrusion. Oxygen isotopic determinations were made by H. P. Taylor. See text for discussion. (After McBirney, work in progress.)

of the process of alteration, melting, and assimilation have been given by Larsen and Switzer (1939), Al-Rawi and Carmichael (1967), Sigurdsson (1968), and Mehnert *et al.* (1973). During the early stages of thermal alteration, feldspars become less clouded, their optic angles become smaller as the crystal structure is altered to a high-temperature form, and twinning and exsolution lamellae begin to disappear. Heating and oxidation of

Figure 10-8. A xenolith of siliceous gneiss in the margin of the Skaergaard intrusion. The fine-grained central core consists of granophyre and traces of the original minerals of the gneiss. The coarse-grained zone is composed of the same minerals as the host gabbro but appears to have crystallized from a liquid that was enriched in volatile components from the xenolith.

green biotite and hornblende result first in their becoming brown and then in their breaking down and developing rims of magnetite. The first liquid normally appears at three-phase contacts between quartz, plagioclase, and alkali feldspar, but it quickly extends along two-phase boundaries and through cracks in individual crystals. With more advanced melting, feldspar becomes spongy, and quartz tends to be rounded and embayed.

The composition of the liquid is sensitive to a number of factors, such as the stability of hydrous minerals or the rate of heating. If biotite breaks down before the onset of melting, potassium and other components of the biotite may be transferred from the xenolith into the surrounding magma, apparently by vapor transfer, and the ensuing liquid will have a low ratio of potassium to sodium (Maury and Bizouard, 1974). When the temperature rise is rapid or occurs at depths where biotite is stable to high temperatures, there may be little or no fractionation of the alkalies and, as a result, the first liquid is more potassic than it would have been if biotite had been unstable and lost potassium to its host.

CONTAMINATION BY A SECOND LIQUID

Although Bowen did not consider it in his discussions of assimilation, mixing of magmas is another type of contamination that, in some instances, may be as important geologically as assimilation of crystalline material. The strict limitations imposed by the phase relations outlined in preceding sections do not necessarily hold when the added material is a liquid; if the two liquids contain no crystals and are completely miscible, the resulting product is normally a linear combination of the components in the proportions of their abundances in the two liquids and the relative amounts of each liquid.

There are, however, cases in which magma mixing may not lead to a simple linear combination of components. If, for example, one or both liquids contain crystals, and the crystals are not in equilibrium with the composition of the new mixed liquid, they will react or be resorbed, and, depending on the proportions of the various liquid and solid components, the end product may not be simply an intermediate mixture. This type of mixing can be seen from the example in Fig. 10-1. If the two liquids X and Z contain crystals of AB and B respectively and are mixed isothermally in various proportions, the end product could be either liquid X or Z with a different amount of its corresponding crystalline phase; it need not have an intermediate composition.

If a crystalline phase, such as AB in Fig. 10-2a, has a composition intermediate between that of two liquids, such as F and F', and is stable at the prevailing temperature, the two liquids cannot mix without crystallizing, even under isothermal conditions. One or the other of the two liquids could survive unchanged with an amount of the intermediate crystalline phase that would depend on the proportions of the two liquids that have been mixed. Mixing under such conditions would, of course, result in an evolution of heat from crystallization of the intermediate phase, and if this heat raised the temperature of the system as a whole, then an intermediate liquid would be possible. In more complex systems, the composition could be changed somewhat, especially if the intermediate phase were a member of a solid solution series, but as before, the resulting liquid is not a simple proportional mixture but rather a product of the combined effects of mixing and crystallization.

As these simplified examples illustrate, it cannot be assumed that mixing of magmas leads to straight-line variation diagrams (Chapter 7). In natural magmas, which commonly contain large proportions of phenocrysts, non-linear variations could result whenever mixing and extensive reaction take place.

Even if crystalline phases do not affect the composition of mixed liquids, the intermediate compositions may still diverge from a simple pattern of proportional mixtures. Yoder (1973) has shown, for example, that compositional variations between coexisting basaltic and rhyolitic liquids may not be linear, and Sato (1975) has pointed out that the non-ideality of alkalies in silica-rich melts may lead to the same result.

O'Hara (1977) has recently shown that mixing of magmas may explain unusually high abundances of excluded trace elements in liquids that do not otherwise appear to have been affected by large amounts of differentiation. Repeated introduction of fresh primitive magma into a body that is cooling and differentiating can maintain the major elements at concentrations close to their original levels; at the same time, however, strongly excluded elements become progressively concentrated with each successive cycle of mixing and differentiation until they have abundances that would otherwise be found only in strongly differentiated magmas.

THE IMPORTANCE OF ASSIMILATION AS A PETROGENETIC PROCESS

Despite the many small-scale examples of assimilation that are known, the evidence for wholesale interaction between mantle-derived magmas and crustal rocks is at best circumstantial. When one considers the large volumes of continental crust that have been displaced by igneous intrusions, especially in orogenic batholiths, and the prominent role that is sometimes assigned to stoping, zone melting, and wall-rock reaction during the rise of magmas toward the surface, it is remarkable that assimilation is not generally considered to be an important if not dominant feature in the evolution of continental igneous rocks. Few petrologists today would concur with Daly's (1933) view that the importance of assimilation and its effects on igneous differentiation have been greatly underestimated. And yet Daly came to this conclusion after many years of field studies; few geologists had a broader experience and understanding of the geologic aspects of plutonic and volcanic rocks than he did.

It may well be that the reason assimilation is not mineralogically conspicuous in most continental igneous rocks is simply that the effects, as Bowen predicted them, are largely governed by phase relations that tend to produce the same assemblages of crystals and the same major-element compositions that would result from spontaneous fractionation. The disproportionate volumes of different rock types that should result from assimilation are seldom evident, because accurate measurements of the volumes of all components of a magmatic series can rarely be obtained, and even if they are available, it is usually possible to explain them as the result of differing conditions of magma generation or fractionation. In view of these ambiguities, increased importance has been assigned to the evidence provided by trace elements and isotope geochemistry, which have strongly influenced modern interpretations of many rocks that were once thought to owe their compositions to assimilation.

Lee and Van Loenen (1971), however, have carried out a detailed field and mineralogical study of a well-exposed group of hybrid granitic rocks in the Snake Range of Nevada, and have shown that the compositional variations which, in the absence of clear field evidence and data on accessory minerals, would logically be attributed to differentiation by simple crystal fractionation, are in fact the result of assimilation of shales, quartzites, and limestones through which the magmas rose by stoping and assimilating the rocks of their walls and roof.

There are two important groups of rocks that have been widely attributed to magmatic assimilation. One consists of alkaline rocks, which were interpreted by many petrologists as the products of assimilation of carbonates; the other is the calc-alkaline series associated with andesitic volcanoes, which have been thought to result from assimilation of sialic crustal material.

The prolonged debate over the role of limestones in the origins of alkaline rock began soon after the hypothesis was first formulated. Proposed, on the basis of field observations, by Daly in 1910, it was challenged on theoretical grounds by Bowen in 1922. The historical development of these contrasting views has recently been outlined in an excellent review by Wyllie (1974).

Daly (1910, 1933) and Shand (1922, 1930) observed that alkaline igneous rocks are relatively rare compared to sub-alkaline type, and that in 107 of the 155 occurrences known in 1933, there was a clear association with carbonate rocks. In most other places, there was insufficient evidence, but the presence of carbonates could not be ruled out.

Assimilation of carbonates was said to have several effects. According to Daly, the lime combines with silica and other components of the magma to produce pyroxene, plagioclase, and a de-silicated liquid in the manner outlined in an earlier section of this chapter. The limestone would also act as a flux to reduce the viscosity of the magma and facilitate crystal settling and differentiation. The CO_2 released in the same process would form new phases, including alkali-rich compounds, that would rise toward the upper levels of the magma.

Bowen (1922, 1928) argued that the assimilation process would require prohibitively large amounts of heat, but Shand (1930) maintained that this limitation did not apply to natural conditions in which the assimilation process proceeded in the upper levels of a large body of magma. Bowen conceded that if enough carbonate material were dissolved, the path of crystallization of a sub-alkaline magma might be altered to the extent that a new phase, such as melilite could appear, but the degree of contamination required seemed to him unreasonable. Instead, he reasoned that equilibrium relations would result in most magmas precipitating the same minerals that they would crystallize in their normal courses of differentiation. The advocates of carbonate assimilation countered these objections with the argument that the process in nature was not one of equilibration, but rather one of widely varying conditions, and that local concentrations of carbonates could be high enough to produce the relatively small volumes of alkaline rocks found in such associations.

Subsequent studies of field relations appeared to add further support for limestone assimilation. Chayes (1942), for example, re-examined the alkaline rocks of the Bancroft area of Canada, where much of the evidence cited in earlier discussions had been found, and concluded that "the special explanation which places least strain upon the field facts is limestone-syntexis." Tilley (1952) re-examined the associations of limestone with basalts and granites in the Tertiary igneous province of Britain and concluded that reactions had in fact produced residual liquids that precipitated melilite and nepheline. He deduced, however, that the trends of liquid compositions differed from those of normal alkaline igneous rocks and that limestone assimilation was not, therefore, an important petrogenetic process.

Shand (1945), using the same field evidence and even some of the same examples studied by Tilley, had arrived at a diametrically opposed conclusion. He pointed to examples in which there is a complete succession of compositions between granite and highly alkaline rocks in close association in the same district, and argued that, despite the experimental evidence for a thermal barrier that should have stood between the end numbers of the series, the field relations were inconsistent with any interpretation other than a continuous sequence of differentiation.

Credit should go to Holmes (1950) and his student Higazy (1954) for first utilizing trace elements to elucidate the origin of igneous rocks associated with carbonates in East Africa. They showed that the contents of Sr, Ba, Y, Zr, La, and P were much too high in the carbonates for them to have had a sedimentary origin. The isotopic compositions of carbon, oxygen, and strontium were later found to support an igneous origin (Holmes, 1965; Deines, 1968), not only for the carbonatites of East Africa, but also for many of the bodies that Daly (1933) had cited as limestone inclusions.

The reinterpretations that followed this discovery brought an extra-ordinary shift of petrologic reasoning. Holmes (1950), Schuiling (1964a), and others turned the hypothesis of limestone assimilation around and reversed the roles of the original liquid and its contaminants. They proposed that assimilation of granitic crustal rocks in carbonatitic magmas could explain the same alkaline rocks that had previously been attributed to assimilation of limestone in granitic magmas! The motivation for this tenacious defense of the assimilation process was based, among other things, on what was considered to be overwhelming field evidence. As Schuiling (1964b) asserted, the carbonate assimilation theory seemed to have been thoroughly documented by natural examples. Any attempt to refute this evidence with unrealistic laboratory experiments would, in his view, "reduce the only theory which really has been demonstrated in the field with a degree of certainty, scarcely ever achieved by petrological observations, to the common level of guesses unsupported by facts."

In the end, the problem was resolved by geochemical evidence that showed unequivocally that the trace elements and isotopic compositions of alkaline rocks could not be produced by any combination of limestone and sub-alkaline magmas. The same evidence that Holmes used to show that carbonatites have a magmatic origin can be used to demonstrate that limestones are too poor in Sr, Ba, Zr, Ti, P, and rare-earth elements to produce the levels of concentration of these elements found in alkaline rocks.

Nowhere have the interpretations of the role of assimilation been more drastically revised than in the petrologic theories for the origin of calc-alkaline rocks. Until the 1960's, many petrologists reasoned that assimilation of crustal material in basaltic magma, together with the more oxidized conditions resulting from the water and oxidized minerals contained in sediments, could logically account for most of the characteristics of calc-alkaline rocks. The locally high proportions of intermediate compositions, the abundance of phenocrysts, many of them out of equilibrium, and the lack of strong iron enrichment in intermediate stages of differentiation were attributed to assimilation and high degrees of oxidation. These views were the outgrowth of a number of investigations of igneous suites, but two studies were especially influential, namely Kuno's (1950) study of the rocks of Hakone volcano in Japan and Wilcox's (1954) study of the lavas of Parícutin volcano in Mexico.

Kuno (1950) provided what was, and still remains, one of the most thorough documentations of the field relations, petrography, and petro-chemistry of a large volcanic complex. He recognized two suites of rocks that had evolved more or less simultaneously during the development of the volcano and its caldera. The first, which he called the pigeonitic series, was characterized by marked iron enrichment and monoclinic pyroxenes. He conjectured that this series was the product of normal differentiation of a primary tholeiitic basalt. The second group, which he called the hypersthenic series, showed less iron enrichment, and contained both monoclinic and orthorhombic pyroxenes, as well as hornblende and various types of xenoliths; Kuno considered this to be the product of fractionation of basaltic magma that had assimilated granitic crustal rocks. He concluded that the hypersthenic series, and most calc-alkaline rocks in general, resulted from contamination and fractionation of basaltic magma that has risen through continental crustal rocks. Shortly before his death, however, Kuno (1968) reviewed the geological and geochemical evidence bearing on the origin of andesites, and essentially dismissed the contribution of assimilated crustal rocks in favor of simple fractional crystallization of basalt under differing oxidation conditions. The main effect of the crust, he believed, was to contribute water, which affects the course of differentiation, and to provide a thick layer of rocks of low density that may act as a density filter and reduce the proportion of denser magmas reaching the surface.

Wilcox's (1954) study of the lavas of Parícutin was unique in that it dealt with a single batch of magma, 1.4 km^3 in volume, that was erupted from the same vent during a nine-year period. As a result of the close observations that were possible during most of the eruption, the relations and relative order of appearance of the rocks were very well known. Progressive changes seen in the compositions of lava could be explained by a combination of fractionation of phenocrysts observed in the lavas and assimilation of granitic xenoliths found in all stages of melting and incorporation into their host. The silica contents of successively erupted lavas increased from 54 to over 60 weight percent; olivine decreased in abundance and developed reaction rims of pyroxene; and large phenocrysts of plagioclase disappeared while hypersthene became increasingly common. Using a graphical method to relate the compositions of analyzed rocks and the minerals they contained, Wilcox showed that the progressive variations in the major-element compositions could be explained as the result of fractional crystallization of olivine and plagioclase and simultaneous assimilation of granitic basement rocks. The heat liberated by crystallizing minerals was found to be inadequate to melt the xenoliths, but Wilcox postulated that additional heat was contributed by a larger body of convecting magma at greater depth and that no superheat was required for the basalt to assimilate the granitic material.

Wilcox's calculations can be greatly refined by use of computer techniques (A. T. Miesch, in preparation; Wilcox, this volume, Chapter 7) and by use of trace elements that are very sensitive to contamination. Although the basic mechanism proposed by Wilcox still appears to be reasonable for major-element variations, especially if the role of amphibole is considered (Eggler, 1972; Cawthorn and O'Hara, 1976), the possibility of large-scale assimilation has been questioned on geochemical grounds. Tilley, Yoder, and Schairer (1968) reported that there was no significant difference in the isotopic ratios of strontium in the earliest basalt and in an andesite that was erupted late in the sequence (0.7043 and 0.7040, respectively) and inferred that the lack of a marked difference is inconsistent with assimilation on the scale envisaged by Wilcox. The isotopic ratios of the basement rocks have not been reported, however, and one cannot rule out contamination on the basis of these limited data.

Wilcox calculated that approximately 25.4 grams of granitic material were assimilated by each 100 grams of basaltic magma, while 2.9 grams of olivine and 9.6 grams of plagioclase crystallized. If the basalt had a typical strontium content of 600 ppm and that of the granitic xenoliths was only 50 ppm, as it commonly is in rocks of this composition, the strontium ratio of the granite would have to have been at least as high as 0.753 to raise the original ratio of the magma from 0.7040 to 0.7050!

Although it is true that the strontium in the andesites of volcanoes standing on thick continental crust is not systematically more radiogenic than that of volcanoes in island arcs (e.g., Pushkar, 1968), it is probably unwarranted to conclude that this fact in itself rules out the possibility of interaction between the magma and crust. Recent studies, such as that of James, Brooks, and Cuyubamba (1976), have shown that the variation in the isotopic ratios among volcanoes in the same chain, or even among the lavas of the same volcano, can be as great or greater than the differences between one geologic province and another. Isotopic equilibration of the magma with the crust through which it rises can account for most of these variations, even without extensive assimilation of other components.

The problem of the role of assimilation in the origin of andesites has been a much more difficult one to resolve than the comparable problem for limestones and alkaline rocks. Although isotopic and trace-element relations have been conclusive in showing that assimilation of limestone has not been a major factor in the genesis of alkaline magmas, it has been possible to do this only because, in this instance, the differences between the magma and limestone are very great. The same is not true of calc-alkaline rocks and continental crust. The search for the "magic element" that would provide a measure of the contribution of crustal material to andesites, rhyolites, and granites has so far been disappointing, mainly because the trace elements that are depleted or enriched in siliceous differentiates normally have similar patterns of relative abundances in crustal rocks. Although it can be shown that the trace-element contents of andesites are not the result of a simple mixture of basalt and crustal rocks (e.g., Taylor, 1969), it is much more difficult to rule out the possibility of assimilation combined with crystal fractionation. The latter process, as Bowen showed so clearly, is more likely to be the one operating in nature. The role of assimilation in continental igneous rocks cannot, therefore, be understood on the basis of geochemical relations alone, but requires careful consideration of geological relations, petrographic evidence, and the nature of phase relations, both under equilibrium and non-equilibrium conditions.

References

Al-Rawi, Y. and Carmichael, I. S. E., A note on the natural fusion of granite. *Am. Mineral.*, *52*, 1806–1814, 1967.

Bailey, E. B., Clough, C. T., Wright, W. B., Richey, J. E., and Wilson, G. V., Tertiary and post-Tertiary geology of Mull. *Mem. Geol. Surv. Scotland*, 445 pp., 1924.

Barriere, M., Architecture et dynamisme du complexe éruptif centré de Ploumanac'h (Bretagne). *Bull. Bur. Res. Geol. Min.*, *2nd Ser.*, Sec. 1, No. 3, 247–295, 1976.

Becker, G. F., Fractional crystallization of rocks. *Am. J. Sci.*, *4*, No. 4, 257–261, 1897.

Bowen, N. L., The behavior of inclusions in igneous magmas. *J. Geol.*, *30*, 513–570, 1922.

————, The Evolution of the Igneous Rocks. *Princeton University Press*, N.J., 332 pp., 1928.

Cawthorn, R. G. and O'Hara, M. J., Amphibole fractionation in calc-alkaline magma genesis. *Am. J. Sci.*, *276*, 309–329, 1976.

Chayes, F., Alkaline and carbonate intrusives near Bancroft, Ontario. *Geol. Soc. Amer. Bull.*, *53*, 449–512, 1942.

Curtis, G. H., Williams, H., and Juhle, R. W., Evidence against assimilation of andesite by rhyolite in the Valley of Ten Thousand Smokes, Alaska. (abstr.). *Amer. Geophys. Un. Trans.*, *35*, 378, 1954.

Daly, R. A., Origin of the alkaline rocks. *Geol. Soc. Amer. Bull.*, *21*, 87–118, 1910.

————, Igneous Rocks and the Depths of the Earth. McGraw-Hill, New York, 598 pp., 1933.

Deines, P., The carbon and isotopic composition of carbonates from a mica peridotite dike near Dixonville, Pennsylvania. *Geochim. Cosmochim. Acta*, *32*, 613–625, 1968.

Didier, J., Granites and their enclaves; the bearing of enclaves on the origin of granites. Translated by J. T. Renouf. Elsevier Sci. Pub. Co., Amsterdam, 393 pp., 1973.

Edwards, A. B., Alkali hybrid rocks of Port Cygnet, Tasmania. *Proc. Roy. Soc. Victoria*, *58*, 81–115, 1947.

Eggler, D. H., Water-saturated and undersaturated melting relations in a Paricutin andesite and an estimate of water content in the natural magma. *Contr. Mineral. Petrol.*, *34*, 261–271, 1972.

Evans, B. W., Fractionation of elements in the pelitic hornfelses of the Cashel-Lough Wheelann intrusion, Connemora, Eire. *Geochim. Cosmochim. Acta*, *28*, 127–156, 1964.

Fenner, C. N., The origin and mode of emplacement of the great tuff deposit of the Valley of Ten Thousand Smokes. *Nat. Geogr. Soc. Contr. Tech. Paper*, Katmai Series, No. 1, 74 pp., 1923.

———, The Katmai magmatic province. *Geology*, *34*, 673–772, 1926.

———, Contact relations between rhyolite and basalt on Gardiner River, Yellowstone Park. *Geol. Soc. Amer. Bull.*, *49*, 1441–1484, 1938.

Harker, A., The Tertiary igneous rocks of Skye. *Mem. Geol. Surv. Scotland*, 481 pp., 1904.

———, *The Natural History of Igneous Rocks*. London, 384 pp., 1909.

Higazy, R. A., Trace elements of volcanic ultrabasic potassic rocks of southwestern Uganda and adjoining parts of the Belgian Congo. *Geol. Soc. Amer. Bull.*, *66*, 39–70, 1954.

Holmes, A., Transfusion of quartz xenoliths in alkali, basic and ultrabasic lavas, south-west Uganda. *Mineral. Mag.*, *24*, 408–421, 1936.

———, Petrogenesis of katungite and its associates. *Am. Mineral.*, *35*, 772–792, 1950.

———, *Principles of Physical Geology*. Roland Press, New York, 1288 pp., 1965.

James, D. E., Brooks, C., and Cuyubamba, A., Andean Cenozoic volcanism: magma genesis in the light of strontium isotopic composition and trace-element geochemistry. *Geol. Soc. Amer. Bull.*, *87*, 592–600, 1976.

Kennedy, W. Q., Trends of differentiation in basaltic magmas. *Am. J. Sci.*, *25*, 241–256, 1933.

Kuno, H., Petrology of Hakone volcano and the adjacent areas, Japan. *Geol. Soc. Am. Bull.*, *61*, 957–1020, 1950.

———, Origin of andesite and its bearing on the island arc structure. *Bull. Volc.*, *32*, 141–176, 1968.

Lacroix, A., Les enclaves des roches volcaniques. 2v., Masson, Paris, 1893.

Larsen, E. S. and Switzer, G., An obsidian-like rock formed by the melting of a granodiorite. *Am. J. Sci.*, *237*, 562–568, 1939.

Lee, D. E. and Van Loenen, R. E., Hybrid granitoid rocks of the southern Snake Range, Nevada. *U.S. Geol. Survey. Prof. Paper No. 668*, 48 pp., 1971.

Ljundgren, P., Petrogenetic significance of gneiss blocks partly melted by hyperite magma. *Kungl. Fysiogr. Sällsk. Lund Förhe*, *29*, 75–90, 1959.

Maury, R. C. and Bizouard, H., Melting of acid xenoliths in a basanite: an approach to the possible mechanisms of crustal contamination. *Contr. Mineral. Petrol.*, *48*, 275–286, 1974.

Mehnert, K. R., Büsch, W., and Schneider, G., Initial melting at grain boundaries of quartz and feldspar in gneisses and granulites. *Neues Jarb. Mineral. Monatsh.*, 165–183, 1973.

O'Hara, M. J., Geochemical evolution during fractional crystallisation of a periodically refilled magma chamber. *Nature*, *266*, 503–507, 1977.

Pitcher, W. A. and Berger, A. R., *The Geology of Donegal, a Study of Granite Emplacement and Unroofing*. Wiley-Interscience, New York, 435 pp., 1972.

Pushkar, P., Strontium isotope ratios in volcanic rocks of three island arc areas. *J. Geophys. Res.*, *73*, 2701–2714, 1968.

Pushkar, P., McBirney, A. R., and Kudo, A. M., The isotopic composition of strontium in Central American ignimbrites. *Bull. Volc.*, *35*, 265–294, 1972.

Sato, H., Diffusion coronas around quartz xenocrysts in andesite and basalt from Tertiary volcanic region in northeastern Shikoku, Japan. *Contr. Mineral. Petrol.*, *50*, 49–64, 1975.

Schairer, J. F. and Yoder, H. S., Jr., The nature of residual liquids from crystallisation with data on the system nepheline–diopside–silica. *Am. J. Sci.*, *258-A*, 273–283, 1960.

Schuiling, R. D., Dry synthesis of feldspathoids by feldspar-carbonate reactions. *Nature*, *201*, 1115, 1964a.

———, The limestone assimilation hypothesis. *Nature*, *204*, 1054–1055, 1964b.

Searle, E. J., Quartzose xenoliths and pyroxene aggregates in the Auckland basalts. *N.Z. J. Geol. Geophys.*, 130–140, 1962.

Shand, S. J., The problem of the alkaline rocks. *South Africa Geol. Soc. Proc.*, *25*, 19–32, 1922.

———, Limestone and the origin of feldspathoidal rocks: an aftermath of the Geological Congress. *Geol. Mag.*, *67*, 415–427, 1930.

———, The present status of Daly's hypothesis of the alkaline rocks. *Am. J. Sci.*, *243A*, 495–507, 1945.

Sigurdsson, H., Petrology of acid xenoliths from Surtsey. *Geol. Mag.*, *105*, 440–453, 1968.

Taylor, S. R., Trace-element chemistry of andesites and associated calc-alkaline rocks. *Oregon Dept. Geol. Mineral Ind. Bull.*, *65*, 43–63, 1969.

Thomas, H. H., Certain xenolithic Tertiary minor intrusions in the island of Mull (Argyllshire). *Quart. J. Geol. Soc. London*, *78*, 229–259, 1922.

Thomas, H. H. and Smith, W. C., Xenoliths of igneous origin in the Trégastel-Ploumanac'h granite, Côtes du Nord, France. *Quart. J. Geol. Soc. London*, *88*, 274–296, 1932.

Thompson, R. N., Tertiary granites and associated rocks of the Marsco area, Isle of Skye. *Quart. J. Geol. Soc. London*, *124*, 349–385, 1969.

Tilley, C. E., Some trends of basaltic magma in limestone syntaxis. *Am. J. Sci.*, *250-A*, 529–545, 1952.

Tilley, C. E., Yoder, H. S., Jr. and Schairer, J. F., Melting relations of igneous rocks. *Carnegie Institution of Washington Yearbook*, *66*, 450–453, 1968.

Tyrrell, G. W., The geology of Arran, *Mem. Geol. Surv. Scotland*, 292 pp., 1928.

Wager, L. R., Vincent, E. A., Brown, G. M., and Bell, J. D., Marscoite and related rocks of the western Red Hills complex, Isle of Skye. *Phil. Trans. Roy. Soc. London, Ser. A.*, *257*, 273–307, 1965.

Watkinson, D. H. and Wyllie, P. J., Phase equilibrium studies bearing on the limestone-assimilation hypothesis. *Geol. Soc. Amer. Bull.*, *80*, 1565–1576, 1969.

Wilcox, R. E., Rhyolite-basalt complex of Gardiner River, Yellowstone Park, Wyoming. *Geol. Soc. Amer. Bull.*, *55*, 1047–1080, 1944.

————, Petrology of Paricutin volcano, Mexico. *U.S. Geol. Bull.*, *965-C*, 281–353, 1954.

Williams, H. and McBirney, A. R., Volcanic history of Honduras. *Univ. Calif. Publ. Geol. Sci.*, *85*, 1–101, 1969.

Wyllie, P. J., Limestone assimilation, in *The Alkaline Rocks*, ed. H. Sørensen, Wiley & Sons, N.Y., 459–474, 1974.

Yoder, H. S., Jr., Contemporaneous basaltic and rhyolitic magmas. *Am. Mineral.*, *58*, 153–171, 1973.

Chapter 11

THE FORMATION OF SILICEOUS
POTASSIC GLASSY ROCKS

DAVID B. STEWART

U.S. Geological Survey, Reston, Virginia

The extrusion of magma onto the earth's surface can result in such rapid cooling that crystallization is prevented. The cooling rate required to prevent crystallization depends upon the composition of the magma and especially upon its SiO_2 and H_2O contents. Nearly anhydrous silica-rich magma readily quenches to the highly viscous liquid called glass. Obsidian is glass that has rhyolite composition. Although obsidian is found in thicknesses of 10 meters or more where it has been cooled over months or years, its formation is restricted to near-surface environments. Basaltic liquids can be quenched to glass only if cooled in seconds or, at most, hours, and thus can form only beads, shards, or thin selvages.

Natural glasses were very important petrologic evidence to Bowen (1928, p. 125) because he shared a common assumption of petrologists of his day, namely, that glassy rocks "are the only rocks of which we can say with complete confidence that they correspond in composition with a liquid." He was aware that some of the original volatiles in magmatic liquid might be lost when the liquid cooled to glass, and that when devitrification took place, even to very fine grain sizes, differences in composition between otherwise geologically equivalent glassy and crystalline rocks were sufficiently common to cause the petrologist to be constantly on guard. Bowen's consideration of natural glasses led him to believe that the most potassic natural glasses, pitchstones containing as much as 8 wt % K_2O, were the glassy equivalents of the potassic granites, and that both represented magmatic liquids. The petrogenetic problem, of course, was how could a liquid having such a high K_2O content originate?

Bowen proposed a peritectic model for phase relations in the system Or–An in which very potassic liquid formed by fractional crystallization. He considered the mantling of plagioclase by potassic feldspar, common in some volcanic and plutonic rocks, to be evidence against the existence of a eutectic relationship between plagioclase and alkali feldspar, and to constitute support for his hypothesis. He subsequently (Schairer and Bowen, 1947) proved the peritectic model to be incorrect for the anhydrous system, and he repudiated it (Schairer and Bowen, 1947, p. 86): "The suggestion referred to must, therefore, be rejected and other causes must be sought for the development of rock types that were believed to owe their origin to this assumed relation." The cited paragraph stops there, however, and Bowen did not specifically deal with this problem in print again. In the years since 1947, many relevant phase diagrams have been determined for feldspar components in the presence of other silicate and oxide components, as well as the geologically important volatiles H_2O, CO_2, and others (see summary by Luth, 1976). In all these systems, the relationship between alkali feldspar and plagioclase is of eutectic type. The only peritectic relation involves plagioclase and leucite as the potassic phase. The mantling of one feldspar by another that Bowen alluded to as supporting evidence for his hypothesis was considered on a theoretical basis by Stewart and Roseboom (1962). A peritectic relationship between sodic plagioclase and alkali feldspar is theoretically possible in the ternary feldspar system in bulk compositions with Ab > 50 wt % because of rapid movements of phase boundaries with small temperature decreases. The resulting fractionation paths, however, do not yield liquids as enriched in K_2O as required by Bowen's hypothesis.

As peritectic crystallization phenomena are inadequate to yield highly potassic glasses as Bowen had proposed, other processes, such as alkali exchange, liquid immiscibility, gaseous transfer, and partial melting, must be considered, partly "because these processes always do just what one may wish them to do" (Bowen, 1928, p. 227). The following discussion, as did Bowen's, will center on the formation of glassy rocks oversaturated in silica and containing $K_2O > Na_2O$—glassy rhyolite (obsidian) and its successively more hydrated equivalents, perlite and pitchstone. All these varieties may contain small amounts of phenocrysts, but potassic siliceous rocks are notably aphyric, being essentially glass without crystallinity. Obsidian may contain some H_2O that was present in magmatic liquid quenched to glass. Addition of more H_2O to such glass at a later time is commonly referred to as secondary hydration because the H_2O added later usually differs in $^{18}O/^{16}O$ and D/H from that originally present (H.P. Taylor, 1968) and has different effects on the physical properties of the glass, to be discussed below.

The common peralkaline rocks that contain alkalies in molecular excess of alumina will not be specifically considered, because when anhydrous they do not attain the high K_2O contents (~ 8 wt %) that especially concerned Bowen. Furthermore, glassy peralkaline rocks are subject to the same hydration processes that affect obsidian (Noble, 1967; Macdonald and Bailey, 1973). In a recent review Bailey, Barberi, and Macdonald (1974) concluded that peralkaline rocks have been formed by several processes including fractional crystallization of mafic parental magma and by direct derivation within the mantle by vapor-controlled melting. Rare undersaturated mafic potassic lavas have enigmatic, apparently deep-seated, origins that will be discussed in Chapter 14.

Abundant chemical and isotopic evidence exists for significant changes in the compositions of glassy rocks after their initial quenching to glass; this evidence will be reviewed below. In brief, the highly potassic glasses most probably originated by a process where H^+ replaced cations in the glass at different rates, so that hydration and relative potassium enrichment originated simultaneously by solid-state diffusion. This process can be initiated at relatively high temperatures during cooling, but does not cease when ambient temperatures are attained. In fact, most observed changes may result from low-temperature interactions with meteoric water. Although the process is referred to as hydration, the associated chemical changes are always more profound than the simple addition of H_2O.

Hydration causes significant differences in the color and luster of natural glasses, volume changes that lead to the concentric fracturing typical of perlite, strain birefringence, and increases of the refractive index and, to a lesser extent, specific gravity. Ross and Smith (1955) studied obsidian-perlite pairs to show that the effect on the refractive index of the same amount of H_2O was different for obsidian and perlite, suggesting that there were two markedly different ways in which H_2O was combined structurally in these two types of glasses. Obsidians were found to be characterized by less than 1% H_2O, values of a few tenths of 1% being typical, and other presumably magmatic volatiles were also present. They noted that as wt % H_2O increased in certain glasses, wt % Cl_2, F_2, and CO_2 decreased. Perlite typically was found to contain 2 to 5 wt % H_2O. Ross and Smith interpreted the geological and chemical evidence to indicate that such large amounts of H_2O were incorporated in a post-magmatic episode at temperatures as low as ambient temperatures of meteoric water, and that the magmatic volatiles (halogens and CO_2) were simultaneously lost.

Friedman and Smith (1958) studied the deuterium-hydrogen compositions of the suite of coexisting obsidians and perlites that had been studied by Ross and Smith. They found that the H_2O in obsidians was unrelated to local meteoric water. On the other hand, the H_2O in the coexisting perlite was similar to local meteoric water but with isotopic fractionation of deuterium in the perlite about 35 per mil less than in the water. These observations were regarded as evidence that meteoric water was the source of H_2O for the hydration of obsidian to form perlite. H. P. Taylor (1968) confirmed this interpretation by showing that the isotopic ratios of oxygen as well as hydrogen varied together and in a systematic way with local meteoric water (Fig. 11-1). The isotopic composition of local meteoric water is strongly influenced by altitude, latitude, proximity to oceans, and

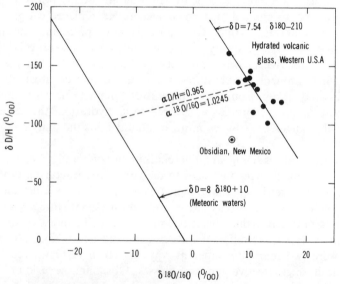

Figure 11-1. Plot of the D/H and $^{18}O/^{16}O$ for perlites and hydrated obsidians from the western U.S. from H. P. Taylor (1968, p. 30). The line through the points for hydrated volcanic glass was drawn parallel to the line for the variation of D/H and $^{18}O/^{16}O$ in meteoric waters determined by Craig (1961), but offset from it by constant fractionation factors for the glass-water exchange process as determined by Friedman and Smith (1958) for D/H (0.965) and by H. P. Taylor (1968) for $^{18}O/^{16}O$ (1.0245). Hydrated obsidians contain local meteoric water exchanged at or close to ambient temperatures.

global climate. Taylor's isotopic data indicate that the process of hydration of obsidian is not simply the addition of meteoric water. The isotopic fractionation of hydrogen suggested by Friedman and Smith (1958) was confirmed, and it was found that oxygen fractionation into the perlite of $+25$ per mil $^{18}O/^{16}O$ accompanies the addition of H_2O. Hydrogen and oxygen from the water must exchange, therefore, for hydrogen and oxygen in the glass. It was suggested, by analogy to the fractionation of $^{18}O/^{16}O$ between alkali feldspar and water, that the fractionation factor observed in perlites would correspond to a temperature of about 50°C for the exchange process. Only obsidians with a few tenths of a wt % H_2O gave $^{18}O/^{16}O$ that can be taken to be representative of the magma from which they formed. Large isotopic differences were caused by even 0.5% of added water, devitrification, or magmatic crystallization. Nearly anhydrous obsidians have oxygen isotopic compositions characteristic of temperatures of the magmatic range, 750°–900°C, and quite different from granitic plutons compositionally equivalent in terms of major chemical components such as SiO_2, Al_2O_3, or alkalies.

The rate of hydration of obsidian has been used as the basis of a method for dating artifacts (Friedman and Smith, 1960). Because of its wide application to archeological problems, this method has led to numerous experimental and empirical evaluations of the kinetics of the hydration process (see summary by R. E. Taylor, 1976). The rate of hydration is known to be particularly sensitive to mean annual temperature and to initial silica content of the glass (Friedman and Long, 1976).

In the early 1960's the technology of cation-selective glass electrodes was applied to natural glasses of a wide range of SiO_2 contents by Garrels and his students (Garrels et al., 1962; Truesdell, 1962, 1966). Truesdell (1966) experimentally determined the ion-exchange constants of obsidians at 25°C to indicate qualitatively what chemical changes may be expected from the interaction of obsidians with meteoric waters. He found the selectivity sequence of glass to be $2H^+ > 2K^+ > 2Na^+ > Ca^{2+} \geq Mg^{2+}$ for a wide range of compositions so that the continued passage of meteoric water through glassy rocks should cause a gain by the glass of K^+ and H_2O and a loss of Na^+. His plot of alkali ratio vs. H_2O content (Fig. 11-2) for natural glasses demonstrated that the relationship predicted for a low-temperature exchange process occurred. A strong positive correlation exists between increasing hydration and increasing K_2O/Na_2O; this relationship had first been noted by R. L. Smith from analyses of hydrated obsidians (Truesdell, 1962, p. 77). Truesdell (1966, p. 121) observed: "The finer the state of division, the wetter the climate, and the older the glass, the more likely it is to have a high K_2O/Na_2O ratio and high percent

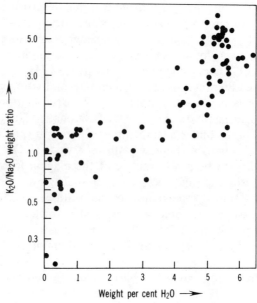

Figure 11-2. The relation between wt % H_2O and K_2O/Na_2O (log scale) for natural glasses found by Truesdell (1966, p. 120). Glasses with high K_2O/Na_2O invariably have high H_2O contents.

H_2O."[1] Although few of the glasses discussed by Truesdell attained the very high K_2O contents of principal concern to Bowen, the very potassic glasses Bowen referred to were all hydrated, and doubtless had been subjected to ion-exchange processes, as recognized earlier by Terzaghi (1935, p. 379).

If water from rain or melting snow reacts with glass, Na^+ will be preferentially leached, and K^+ will be preferentially retained in the glass. Thus, potassium enrichment may occur together with overall mass decrease.

[1] The maximum content of H_2O possible in hydrated glass from the regime of meteoric waters is not known from experimental data. The rate of hydration is known to increase with increasing temperature and to be apparently independent of H_2O pressure (Friedman and Long, 1976, p. 347). The amount of hydration possible increases with decreasing temperature, however, and increases with increasing H_2O pressure (Friedman, Long, and Smith, 1963). H. P. Taylor (1968, p. 21) reported a vitrophyre from Wolf Creek, Montana, with 8.05 wt % H_2O, the highest value known to the present author. Zeolites are found in the rocks at this locality, and possibly their presence in submicroscopic crystals increases the H_2O content.

Lipman (1965) studied analyses of different parts of the same igneous bodies, mostly rhyolitic, to determine the effects of primary crystallization, secondary devitrification, and hydration. Compared to crystalline parts of the same bodies, hydrated glasses contained more H_2O and Al_2O_3, and less Na_2O, SiO_2, and Fe_2O_3. The relative enrichment of Al_2O_3 was interpreted to have been the result of leaching from the glass of Na_2O and SiO_2 components because the meteoric waters flowing from the leached rocks were rich in SiO_2 and Na_2O, and because there was a positive correlation between the cation deficiencies and the greater surface areas (higher porosity or smaller shard size) within the glassy rocks. Lipman's data indicated that as much as several wt % SiO_2 and 0.5 wt % of Na_2O had been removed from many vitrophyres and that in a few, K_2O had also been lost. In most suites, however, total K_2O increased with hydration and sodium loss. No K_2O contents in excess of 6 wt % were formed in any suites. In tuffs, sodium was removed and potassium was added by exchange proportional to the surface area within the tuff. The hydrated glasses found in the margins of rhyolitic ash flow sheets were more variable in composition and hence less representative of the original magma than internal parts of the same sheets that had undergone primary crystallization.

Original and hydrated glasses can be compared by assuming that Al_2O_3 is not added or removed during any leaching, oxidation, or ion exchange that accompanies hydration. Noble (1967) demonstrated the constancy of Al_2O_3 and reported a mean loss of 0.4 wt % Na_2O, similar to that found by Lipman (1965). In Noble's suite, a mean increase of 0.3 wt % K_2O was also observed, but in some well-documented samples up to 10% of the amount of potassium originally present was lost. Noble suggested that in the early stages of reaction, gain of potassium may occur while sodium is lost, but after continued alteration both may be leached. Some leaching of silica and significant oxidation of ferrous iron accompanied hydration. Noble, Smith, and Peck (1967) demonstrated that losses of F and losses or additions of Cl could occur during hydration; significant losses of both accompanied devitrification. Minor and trace elements are also affected by hydration processes. Zielinski, Lipman, and Millard (1976) reported hydrated glasses showed consistent loss of Li, addition of Sr and Ba, and unchanged U contents.

Many of these chemical and geological observations of the hydration process have been confirmed experimentally. White and Claassen (1977) have demonstrated by dissolution experiments conducted with vitric tuff over the ranges pH 4.5–7.5, 12°–50°C, and glass surface to aqueous volume ratios of $35m^2/1$–$300m^2/1$ that the hydration process consists of a solid state counter-ion diffusion between hydrogen ions and cations contained within the glass. The reaction rate depends on pH, valence of the cations, temperature, and the concentrations of cations in the meteoric water and the glass. The authors deduce a kinetic model to describe the evolution of the compositions of ground water and glass from their mutual interaction.

Comparison of glasses (including hydrated glasses) with their devitrified equivalents by Lipman (1965, p. D20) and by Noble (1970, p. 2684) showed higher K_2O and lower Na_2O in the devitrified samples. One devitrified sample of Lipman contained 8.21 wt % K_2O calculated H_2O-free, and K_2O/Na_2O of 7.75, higher than any point on Fig. 11-2. Such high-K_2O enrichment results from alkali exchange in alkali feldspars. The high partitioning of K^+ into feldspar from fluid at magmatic temperatures (Orville, 1963) increases as temperature decreases so that in meteoric waters at 25°C, Na/K up to at least 400 (more than 14 times the Na/K of sea water) would still result in the fixation of K in metastable alkali feldspars (J. J. Hemley, personal communication, April, 1977). High-potassium compositions are more effectively achieved by alkali exchange with alkali feldspars than with hydrated obsidians under conditions of low-temperature saturation with meteoric water.

Pitchstones, defined on the basis of their luster, commonly have greater amounts of H_2O than the 2–5 wt % H_2O commonly found in perlites, greatly increased oxidation of iron, some silicification, high K_2O/Na_2O, and 5 wt % or more K_2O. A few perlite-pitchstone pairs are known (R. L. Smith, personal communication, January, 1977), but none have yet been described in the literature. The cause of the more extensive hydration in pitchstone than in perlite is unknown. It could be simply an extension of the exchange process that forms perlite. It could be that more intense hydration took place at higher H_2O fugacities (or temperatures, or both) such as would be found where magma intruded into wet sediments at shallow depths. Perhaps the glass structure underwent decomposition into a mineraloid (e.g., hyalite or natural hydrocarbons). Pitchstones might also result from several processes operating simultaneously—hydration and transport of alkali by moving gases, for example.

The associated processes of hydration, chemical exchange, and leaching are capable of forming siliceous potassic glassy rocks that are not representative of the magma originally extruded. Very large volumes of near-surface potassic glassy rocks can be readily explained by the interaction of glass and meteoric water. Are any other of the processes that have been suggested capable of producing these rocks?

Roedder has evaluated thoroughly in Chapter 2 the effects of liquid immiscibility, and concluded that potassic siliceous magma can in fact form by liquid immiscibility in laboratory experiments and, to a limited extent, in nature. To accumulate or erupt large volumes of nearly anhydrous potassic siliceous magma is mechanically difficult because of its high viscosity. The isotopic evidence described above clearly does not support an origin by liquid immiscibility for local potassic siliceous glasses in much more extensive volcanic terrains.

It could be argued that the transfer of potassium in or through a gaseous phase or vapor might enrich some part of the magma in K_2O so that, upon extrusion, a highly potassic liquid might result. Bailey (1970, p. 185) hypothesized differences in behavior of mobile elements such as alkalies and H_2O in the lower part of the mantle compared with that in higher levels of the mantle. The vapor phase that coexists with siliceous granitic magma was described by Luth and Tuttle (1969) for the range 2–10 kbar, and Wyllie, et al. (1976) suggest that the same general character of the liquidus relations in the granite system extends to at least 35 kbar. Luth and Tuttle (1969) found that over the range from 2 to 10 kbar the vapor phase at temperatures above the solidus of the system Or–Ab–Qz–H_2O was depleted in K_2O (Or component) relative to the alkali composition of natural granites. At temperatures below the solidus, the vapor phase is greatly enriched in silica and low in K_2O relative to the composition of natural granites; only at the exceedingly high vapor pressure of 10 kbar did the vapor composition approach even the K_2O content of the first liquid to appear on melting. Therefore, the condensation of such a vapor will not produce highly potassic magma. Macdonald, Bailey, and Sutherland (1970, p. 515) hypothesized that partitioning of sodium into the vapor phase could increase the proportion of K_2O to total alkalies in the residual liquid and crystals, but this hypothesis has not been demonstrated experimentally. Martin (1974) performed some reconnaissance gas-transport experiments at 3 and 5 kbar with mafic rock types, but did not obtain potassium-rich liquid as a product of gaseous transport.

Transfer of potassium through the gas phase at the temperature and pressures that are typical of near-surface magma bodies could be a possible mechanism. The solubility of feldspar components (and silica) in such a vapor phase, however, decreases to small values (< 1 wt %) under these conditions, and dissolution may not be congruent. Excess alkalies in the form of alkali silicate may result in an appreciable increase in the rate of alkali transport, whereas excess alkalies in the form of halogen salts may reduce alkali transfer. Considerable fractionation of potassium into feldspars takes place by high temperature ($250°–700°C$) alkali exchange between crystalline alkali feldspars and H_2O-rich gases containing alkali halides (Orville, 1963), and especially at low H_2O pressures (Fournier, 1976). Rhyolite glasses devitrify rapidly under such experimental conditions, so that the pertinence and efficacy of this conjectural mechanism have not been verified for rhyolite liquids at the lower temperature limits of their formation under low (< 1 kbar) H_2O pressure.

Partial melting of granitic igneous rocks, gneisses, and sedimentary rocks will produce liquids that contain amounts of plagioclase and quartz each nearly equal to the amount of alkali feldspar (Tuttle and Bowen, 1958, p. 78). Only if the rock to be melted consists exclusively of alkali feldspar and reaches very high temperatures will liquids as rich in K_2O as 8 wt % be generated. Source rocks made of alkali feldspar are rare, and the probability of generating large volumes of potassic liquid by remelting them is small. Furthermore, the possibility of contaminating a very hot potassic liquid with low-K_2O bearing materials as it moves toward the surface is large.

Highly perfect fractional crystallization of granitic magma in the late stages of its crystallization might yield liquids with high K_2O contents. Aplites are commonly considered to be fractional crystallization products from granites. On the other hand, Steiner, Jahns, and Luth (1975, pp. 95–96) found all analyzed examples of aplite-granite or pegmatite-granite pairs known to them to be consistently richer in Qz, or to plot toward the Ab-Qz sideline of the granite system, compared to the composition of the parental granite. In experimental work on An-free granitic compositions with insufficient H_2O for saturation of the liquid, these authors found a tendency for fractional crystallization paths to yield final liquids near $Ab_{20}Or_{45}Qz_{35}$. Such a composition is richer in K_2O (7.6 wt %) than typical rhyolite (~ 5 wt %), but contains somewhat less than the 8 wt % K_2O that concerned Bowen.

In summary, chemical and isotopic evidence and numerous geological examples indicate that the interaction of glassy siliceous rocks and meteoric water will tend to produce large volumes of highly potassic siliceous glass until devitrification intervenes. Processes such as liquid immiscibility, gaseous transport of alkalies, partial melting, highly perfect fractional crystallization, or peritectic crystallization are incapable of generating the large volumes of potassium-rich rocks observed. Other mechanisms are less well demonstrated in the field and in the laboratory.

References

Bailey, D. K., Volatile flux, heat-focusing and the generation of magma, in *Mechanisms of Igneous Instrusions*, G. Newall and N. Rast, eds., *Geol. J., Spec. Issue 2*, 177–186, 1970.

Bailey, D. K., F. Barberi, and R. Macdonald, eds. *Oversaturated Peralkaline Volcanic Rocks. Bull. Volc.*, *38*, No. 3, 497–860, 1974(5).

Bowen, N. L., *The Evolution of the Igneous Rocks.* Princeton University Press, 334 pp., 1928.

Craig, H., Isotopic variations in meteoric waters. *Science, 133,* 1702–1703, 1961.

Fournier, R. O., Exchange of Na^+ and K^+ between water vapor and feldspar phases at high-temperature and low-vapor pressure. *Geochim. Cosmochim. Acta, 40,* 1553–1561, 1976.

Friedman, I., and R. L. Smith, The deuterium content of water in some volcanic glasses. *Geochim. Cosmochim. Acta, 15,* 218–228, 1958.

Friedman, I., and R. L. Smith, A new dating method using obsidian: Part I, The development of the method. *Am. Antiquity, 25,* 476–522, 1960.

Friedman, I., W. Long, and R. L. Smith, Viscosity and water content of rhyolite glass. *J. Geophys. Research, 68,* 6523–6535, 1963.

Friedman, I., and W. Long, Hydration rate of obsidian. *Science, 191,* 347–352, 1976.

Garrels, R. M., M. Sato, M. E. Thompson, and A. H. Truesdell, Glass electrodes sensitive to divalent cations. *Science, 135,* 1045–1048, 1962.

Lipman, P. W., Chemical comparison of glassy and crystalline volcanic rocks. *U.S. Geological Survey Bull., 1201-D,* pp. D1–24, 1965.

Luth, W. C., Granitic rocks, in *The Evolution of the Crystalline Rocks,* D. K. Bailey and R. Macdonald, eds., Academic Press, London, 335–418, 1976.

Luth, W. C., and O. F. Tuttle, The hydrous vapor phase in equilibrium with granite and granite magmas. *Geol. Soc. Am., Mem. 115,* 513–548, 1969.

Macdonald, R., and D. K. Bailey, Chapter N. Chemistry of Igneous rocks, Part I. The chemistry of the peralkaline oversaturated obsidians. *U.S. Geol. Surv. Prof. Paper 440-N1,* 1973.

Macdonald, R., D. K. Bailey, and D. S. Sutherland, Oversaturated peralkaline glassy trachytes from Kenya. *J. Petrol., 11,* 507–517, 1970.

Martin, R. F., Role of water in pantellerite genesis. *Bull. Volc., 38,* 666–679, 1974.

Noble, D. C., Sodium, potassium, and ferrous iron contents of some secondarily hydrated natural silicic glasses. *Am. Mineral., 52,* 280–286, 1967.

———, Loss of sodium from crystallized comendite welded tuffs of the Miocene Grouse Canyon member of the Belted Range Tuff, Nevada. *Geol. Soc. Am. Bull., 81,* 2677–2688, 1970.

Noble, D. C., V. C. Smith, and L. C. Peck, Loss of halogens from crystallized and glassy silicic volcanic rocks. *Geochim. Cosmochim. Acta, 31,* 215–223, 1967.

Orville, P.M., Alkali ion exchange between vapor and feldspar phases: *Am. J. Sci., 261,* 201–237, 1963.

Ross, C. S., and R. L. Smith, Water and other volatiles in volcanic glasses. *Am. Mineral., 40,* 1071–1089, 1955.

Schairer, J. F., and N. L. Bowen, The system anorthite–leucite–silica. *Bull. Soc. Geol. Finlande, 20,* 67–87, 1947.

Steiner, J. C., R. H. Jahns, and W. C. Luth, Crystallization of alkali feldspar and quartz in the haplogranite system $NaAlSi_3O_8$–SiO_2–H_2O at 4Kb. *Geol. Soc. Am. Bull. 86,* 83–98, 1975.

Stewart, D. B., and E. H. Roseboom, Lower temperature terminations of the three-phase region plagioclase-alkali feldspar-liquid. *J. Petrol., 3,* 280–315, 1962.

Taylor, H. P., The oxygen isotope geochemistry of igneous rocks, *Contrib. Mineral. Petrol.*, *19*, 1–71, 1968.

Taylor, R. E., *Advances in Obsidian Glass Studies*. Noyes Press, Park Ridge, N.J., 360 pp., 1976.

Terzaghi, R. D., The origin of the potash-rich rocks. *Am. J. Sci.*, 5th series, *29*, 369–380, 1935.

Truesdell, A. H., Study of natural glasses through thier behavior as membrane electrodes. *Nature*, *194*, 77–79, 1962.

———, Ion-exchange constants of natural glasses by the electrode method. *Am. Mineral.*, *51*, 110–122, 1966.

Tuttle, O. F., and N. L. Bowen, Origin of granite in the light of experimental studies in the system $NaAlSi_3O_8$–$KAlSi_3O_8$–SiO_2–H_2O, *Geol. Soc. Am., Mem. 74*, 153 pp., 1958.

White, A. F., and H. C. Claassen, Kinetic model for the dissolution of a rhyolitic glass, *Geol. Soc. Am., Abst. with Programs*, *9*, 1223, 1977.

Wyllie, P. J., W. L. Huang, C. R. Stern, and S. Maaløe, Granitic magmas: possible and impossible sources, water contents, and crystallization sequences. *Can. J. Earth Sciences*, *13*, 1007–1019, 1976.

Zielinski, R. A., P. W. Lipman, and H. T. Millard, Jr., Minor element variations in hydrated and crystallized calc-alkalic rhyolites, in *Geological Survey Research 1976, U.S. Geol. Surv. Prof. Paper 1000*, 175, 1976.

Chapter 12

THE FELDSPATHOIDAL ALKALINE ROCKS

J. GITTINS

Department of Geology, University of Toronto, Toronto, Ontario, Canada

DEFINITION AND USAGE

The term alkaline rock has been used in a variety of ways and applied to rocks as diverse as alkali olivine basalt, trachyte, syenite, and feldspathoidal rocks. It implies enrichment in alkalies ($Na_2O + K_2O$), but this, in turn, can be related to either the SiO_2 or the Al_2O_3 content of the rock. Silica-undersaturated rocks contain feldspathoidal minerals, and there is common agreement on calling these rocks alkaline. Silica-oversaturated rocks are not generally called alkaline unless a sodic pyroxene or amphibole (such as acmite or riebeckite) is present. In other words, a high content of alkali feldspar is not sufficient grounds for classifying a rock as alkaline. Thus, a hornblende granite is not alkaline, but a riebeckite granite is, because it contains a sodic mineral deficient in Al_2O_3. Riebeckite granites are an example of rocks in which the peralkaline index $(Na_2O + K_2O)/Al_2O_3 > 1$ (mol. prop.). These are known as the peralkaline rocks, in which alkali content is compared to alumina content. Because the alkali content can be related to either silica or alumina, there are, consequently, silica-undersaturated peralkaline rocks and silica-oversaturated peralkaline rocks.

Silica-undersaturated peralkaline rocks are feldspathoidal but also contain non-aluminous or alumina-deficient minerals such as acmite, riebeckite-arfvedsonite amphiboles, and occasionally, rare minerals such

The Evolution of the Igneous Rocks: Fiftieth Anniversary Perspectives
0-691-08223-5/79/0351-40$02.00/0 (cloth)
0-691-08224-3/79/0351-40$02.00/0 (paperback)

as the Na-Ti silicate ramsayite. A further distinction is sometimes made between silica-undersaturated peralkaline rocks and agpaitic rocks. The latter are peralkaline and silica-undersaturated but are characterized by notable enrichment in Zr, Sr, Nb, Ba, REE, Cl, and F that expresses itself in the presence of such rare minerals as eudialyte, villiaumite, catapleiite, lamprophyllite, britholite, the wöhlerite group, and a host of others.

Only the feldspathoidal alkaline rocks are discussed in this chapter. The commonest are the silica-undersaturated, non-peralkaline group composed of nepheline and alkali feldspar together with one or more of pyroxene, amphibole, phlogopite, or biotite, and more rarely of combinations of leucite, kalsilite, hauyne, nosean, sodalite, and analcite. Less common are the peralkaline varieties, and the agpaitic rocks are extremely rare. It is important to understand that the transition from non-peralkaline to peralkaline is usually gradual.

EXPERIMENTAL STUDIES

In his *The Evolution of the Igneous Rocks*, Bowen (1928) was concerned to present the case for fractional crystallization as a powerful petrological process that was largely responsible for developing a diversity of rock types from a small number of rather simple parental magmas. His approach was, therefore, to combine the results of experiments on synthetic systems with a profound knowledge of the rocks as they occur in the field, to create a unified petrogenetic theory. The concept of fractional crystallization as governing the evolution of magmas and Bowen's approach to its study are just as valid fifty years later. The major difference today is that increasing emphasis is being given to fractional melting, generally within the earth's mantle, as the dominant process in generating the parent magma that is subsequently modified by fractional crystallization. Great importance is also attached nowadays to the composition of and the presence or absence of fluid phases during both the initial melting and the subsequent crystallization, and it is realized that some of the older ideas of liquid immiscibility and gaseous diffusion can play important roles. In some respects there are fewer constraints on the modern petrologist than were experienced by Bowen, who sought to explain the most extreme trends, such as alkaline rock development, through fractional crystallization. Today the petrologist pays great attention to the conditions of partial melting within the mantle, for it is here that the principal characteristics of the magma are acquired. Fractional crystallization is generally thought of now as a process that modifies these characteristics at a subsequent period, rather than as the process responsible for the primary development of the alkaline magma.

Against this background it is useful to examine the development of what has come to be one of the single most important systems in igneous petrology, the system $NaAlSiO_4$–$KAlSiO_4$–SiO_2–H_2O. Its great significance is, of course, that it represents the compositions of all the major felsic rocks both over- and undersaturated and the compositions of the residual magmas that derive from more mafic magmas through fractional crystallization. The condensed system was, therefore, most aptly termed by Bowen (1937) "petrogeny's residua system."

The phase equilibrium diagram for the system at 1 atm is shown in Fig. 12-1. The primary phase fields are nepheline solid solutions, leucite,

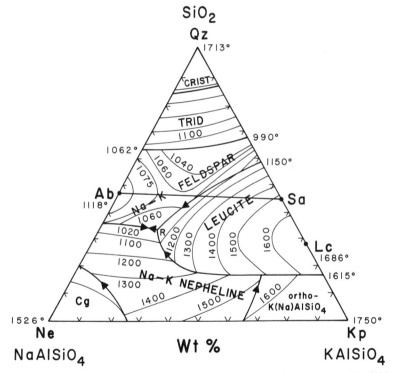

Figure 12-1. Phase equilibrium diagram (1 atm pressure) of the system $NaAlSiO_4$–$KAlSiO_4$–SiO_2 (Schairer and Bowen, 1935, p. 326; Schairer, 1950, p. 514). Abbreviations: Crist, cristobalite; Trid, tridymite, Cg, carnegieite; Ne, $NaAlSiO_4$; Kp, $KAlSiO_4$; Qz, SiO_2; Ab, albite; Sa, potassium feldspar; Lc, leucite.

feldspar solid solutions, and silica minerals. In the study of the feldspathoidal rocks one is concerned with the undersaturated portion of the
diagram. Here there is a ternary minimum that is extremely close to the
reaction point marking the limit of leucite stability. The two most important features of the isotherms in the diagram are a broad thermal
valley extending from the $NaAlSi_3O_8$–$KAlSi_3O_8$ join to both ternary
minima, and a ridge with a depression (a binary minimum) lying along
the alkali feldspar join. This ridge is a thermal barrier that prevents most
undersaturated liquids from developing into oversaturated liquids (or vice
versa) by fractional crystallization. Under water-saturated conditions the
barrier persists to at least 10 kbar (Morse, 1969, 1970).

Unfortunately, this important system has not been studied under
water-deficient conditions, and so it can be discussed only for the condition
$P_{H_2O} = P_{total}$. As P_{H_2O} ($= P_{total}$) increases, several significant changes
occur (Fig. 12-2). Most immediately noticeable is the shrinking of the
leucite field as H_2O enters the liquid.

At approximately 2.6 kbar potassium feldspar ceases to melt incongruently and hence will crystallize directly from a water-saturated melt
of appropriate composition. The leucite field is not completely eliminated,
however, until 8.4 kbar (Scarfe, Luth, and Tuttle, 1966), and so suitably
potassic liquids can still crystallize leucite to quite high pressures. Concurrently with the shrinking of the leucite field, the cotectic between
nepheline and alkali feldspar moves away from the silica apex, and the
separation of the ternary minimum and the reaction point R steadily
increases. The minima on the cotectics develop into eutectics at some
pressure between 2 and 5 kbar, although not necessarily the same pressure
for both eutectics. One eutectic, e, involves nepheline solid solution, sodic
feldspar, potassic feldspar, and aqueous vapor; the other eutectic, d, involves a silica mineral, sodic feldspar, potassic feldspar, and aqueous
vapor.

Thus, at 5 kbar the system involves eutectics rather than minima, in
both the undersaturated and oversaturated portions, and the incongruent
melting of potassium feldspar is eliminated. At P_{H_2O} up to approximately
8.4 kbar, leucite or potassium feldspar can crystallize from melts of appropriate compositions. The thermal barrier along the alkali feldspar
join that was found at 1 atm is still present, and indeed is known to persist
to at least 10 kbar (Morse, 1969). Thus, at crustal pressures, the passage
of granitic to nepheline syenitic liquids and vice versa is prevented in this
system by any process of equilibrium crystallization or fractionation.

As already noted, knowledge of this system at pressures above 1 atm
is limited to conditions of water saturation, a situation that is rarely

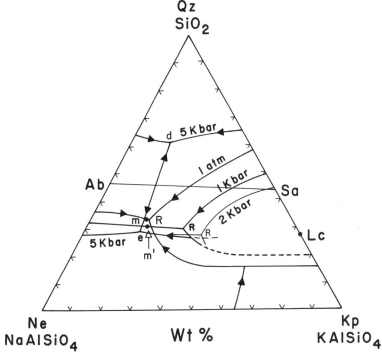

Figure 12-2. Phase equilibrium diagram of the system $NaAlSiO_4$–$KAlSiO_4$–SiO_2–H_2O
for $P = 1$ atm, $P_{H_2O} = 1$ kbar, 2 kbar, 5 kbar (Schairer and Bowen, 1935;
Schairer, 1950; Hamilton and MacKenzie, 1965; Taylor and MacKenzie, 1975;
Morse, 1969; Roux and Hamilton, 1976). R = reaction point, m (black dot) =
ternary minimum at 1 atm, m' (black dot) = ternary minimum at $P_{H_2O} = 1$ kbar,
d = ternary eutectic (oversaturated portion of the system), e = ternary eutectic
(undersaturated portion of the system). The analcite field has been omitted for
simplicity. For a discussion concerning this problem, see Roux and Hamilton,
1976, and Morse, 1970.

achieved in natural magmas except at the very latest stage of crystallization, and not always then. Application to natural magmas requires a
knowledge of the equilibria under conditions of $P_{H_2O} < P_{total}$ and of
$P_{total} = P_{H_2O + CO_2}$. It should be recalled that the solubility of the condensed phases in the aqueous vapor increases with pressure.

Some of the general applications of the system may be appreciated by
following a small selection of crystallization paths in the system. Composition 1 in Fig. 12-3 represents a simplified nepheline syenite magma

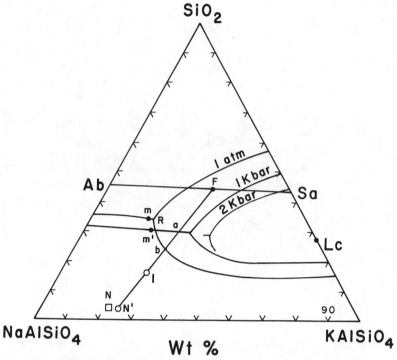

Figure 12-3. Alternative crystallization paths in the system $NaAlSiO_4$–$KAlSiO_4$–SiO_2–H_2O with phase fields shown at $P = 1$ atm, $P_{H_2O} = 1$ kbar, 2 kbar. N = Morozewicz "normal nepheline" composition, m (black dot) = ternary minimum at 1 atm, m' = ternary minimum at $P_{H_2O} = 1$ kbar. At $P_{H_2O} = 1$ kbar composition 1 reaches the nepheline-feldspar cotectic at a and crystallizes nepheline N' and feldspar F simultaneously. At $P = 1$ atm it reaches the leucite field boundary at b and crystallizes nepheline and leucite simultaneously.

without ferromagnesian minerals. Consider its crystallization at $P_{H_2O} =$ 1 kbar, which may be taken to represent plutonic conditions as opposed to extrusive volcanism, which will be examined next. The liquid crystallizes nepheline, N, close to the Morozewicz-Buerger composition[1] and changes composition until it reaches the nepheline-alkali feldspar cotectic at a. Under perfect equilibrium conditions it will follow a slightly curved path, and under fractional crystallization it will follow a more moderately curved path, reaching the cotectic at a temperature below a. Upon reaching the

---

[1] See pp. 360–361 for further discussion.

cotectic, the liquid crystallizes nepheline solid solution and alkali feldspar simultaneously as the liquid composition progresses toward the ternary minimum m', and the completely crystallized rock is a mixture of nepheline solid solution (N') and alkali feldspar, F. At 1 atm pressure, which may be taken to represent volcanic conditions, the liquid composition 1 will still crystallize nepheline solid solution initially but now the leucite field is very much larger, and so nepheline is joined by leucite when the liquid composition reaches the nepheline-leucite field boundary at b. Nepheline and leucite continue to crystallize simultaneously as the liquid composition progresses along the field boundary to the reaction point R. At this temperature leucite begins to react with the liquid to form potassium feldspar (because of the incongruent melting relations of potassium feldspar below $P_{H_2O} = 2.6$ kbar already referred to). Under equilibrium conditions the leucite will be completely resorbed, and the remaining liquid will change composition along the nepheline-alkali feldspar cotectic, crystallizing nepheline and alkali feldspar. At the ternary minimum, m, the liquid will complete its crystallization and, as in the previous example, will end up as a nepheline-alkali feldspar rock. Of course, it should be recalled that there is only a very slight separation at 1 atm between the reaction point and the ternary minimum. Under volcanic conditions, where cooling generally is rapid, it is more likely that fractional crystallization will prevail over equilibrium crystallization and, consequently, leucite will not be completely resorbed at R. Reaction with the liquid may cause a protective mantle of alkali feldspar to be deposited on the leucite, which is then shielded from further reaction. Leucite is effectively removed from the system as a participating phase and the liquid moves from R to the ternary minimum. In this case the final rock is a mixture of nepheline solid solution and alkali feldspar with leucite grains mantled by feldspar. An example of this course of fractional crystallization with the mantling of leucite is the leucite-nepheline dolerite of Meiches in the Vogelsburg region of Germany, which was described by Tilley (1958).

It was noted previously that where $P_{H_2O} = P_{total}$ leucite can crystallize at pressures up to 8.4 kbar, but also that probably very few natural magmas are water saturated. Under dry conditions leucite is stable to at least 20 kbar (Lindsley, 1967), and thus it is reasonable to suppose that in natural magmas where $P_{H_2O} < P_{total}$ leucite can crystallize at total pressures markedly higher than 8.4 kbar, and indeed throughout most of the earth's crust. It is interesting to inquire why leucite appears to be found only in extrusive rocks or very high-level intrusions and not in deep-seated plutonic intrusions. The first important consideration is that the bulk compositions of most leucocratic plutonic alkaline rocks plot on the sodic

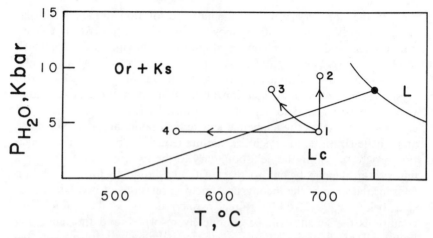

Figure 12-4. The subsolidus breakdown of leucite (Lc) to kalsilite and potassium feldspar (Ks + Or). Breakdown paths are indicated as follows: 1–2 with rising water pressure at constant temperature, 1–4 with cooling at constant water pressure, and 1–3 with cooling accompanied by rising water pressure (after Scarfe, Luth, and Tuttle, 1966). The breakdown curve is intersected by the melting curve at the invariant point 750°C, 8.4 kbar.

side of a line joining nepheline solid solution to $KAlSi_3O_8$, and that at any water pressure slightly over 1 kbar crystallization of nepheline will not result in the liquid reaching the leucite field boundary; hence, at modest water pressures the bulk composition of most natural alkaline rock magmas precludes the crystallization of leucite. The second important consideration is the reaction leucite $\rightleftharpoons$ kalsilite + potassium feldspar that limits the stability of leucite. The curve for this reaction is reported by Scarfe, Luth, and Tuttle (1966) to have a slope of approximately 30°C/kbar. It intersects the temperature axis at about 490°C and passes through an invariant point (8.4 kbar, 750°C) at which potassium feldspar, leucite, kalsilite, liquid, and vapor coexist (Fig. 12-4). Any leucite that crystallized in a plutonic intrusion would be converted to kalsilite + potash feldspar either by an increase in P_{H_2O} as the crystallization of the magma approached completion or by cooling of the crystalline assemblage in the presence of aqueous vapor, or by a combination of the two processes. Patchy intergrowths of kalsilite and potassium feldspar believed to represent the breakdown of leucite in this way are found in the Batbjerg intrusion, East Greenland (Gittins et al., 1977).

The further applicability of the residua system to leucocratic alkaline rocks can be examined in Figs. 12-5 and 12-6. Here are plotted those

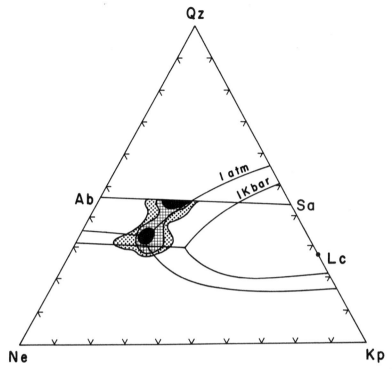

Figure 12-5. Contour diagram illustrating the distribution of analyses of 102 plutonic rocks in Washington's tables (1917) that carry 80% or more of normative Ab + Or + Ne (after Hamilton and MacKenzie, 1965).

analyses in which normative nepheline + albite + orthoclase is at least 80 percent. Figure 12-5 shows 102 analyses of plutonic rocks and Fig. 12-6 represents 122 analyses of extrusive rocks. There is not complete coincidence of the distribution with the thermal troughs in the liquidus surface and the ternary minima, but the agreement is very close. This agreement is not of itself definitive proof of a crystal-liquid equilibrium process in the crystallization of nepheline syenite magmas, but it is a powerful argument in that direction. Particularly interesting is the proximity of the statistical maximum for plutonic nepheline syenites with the ternary minima at $P = 1$ atm and $P_{H_2O} = 1$ kbar, whereas the ternary eutectic at $P_{H_2O} = 5$ kbar is significantly removed from this statistical maximum. The most likely explanation of the behavior is the view, already expressed, that most alkaline rock magmas have crystallized at a water pressure that rarely exceeded 1 kbar whatever P_{total} might have been.

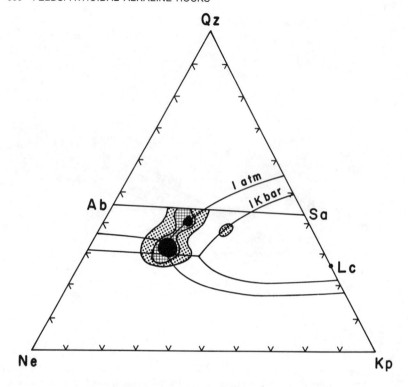

Figure 12-6. Contour diagram illustrating the distribution of analyses of 122 extrusive rocks in Washington's tables (1917) that carry 80% or more of normative Ab + Or + Ne (after Hamilton and MacKenzie, 1965).

NEPHELINE–FELDSPAR TIE LINES IN NATURAL ROCKS

Although nepheline in alkaline rocks might be expected to display a wide range of composition, it does not always do so. Morozewicz (1928) discovered by careful chemical analysis of nephelines from natural rocks that nephelines from plutonic rocks and nepheline-bearing gneisses all have a very restricted range of composition. (Na:K = 3:1 or $Ne_{73}Ks_{27}$ wt %) This discovery was later confirmed by Buerger, Klein, and Donnay (1954) by structural determination, and by other workers who contributed additional chemical data. The 3:1 composition is generally referred to as "normal nepheline" in order to distinguish it from nepheline solid solutions that may have a range of compositions. The compositional limitation is explained by the phase relations in the system $NaAlSiO_4$–$KAlSiO_4$ (Tuttle and Smith, 1958), by work in the system $NaAlSiO_4$–$KAlSiO_4$–SiO_2–H_2O by Hamilton and Mackenzie (1960), and by the crystal structure. Analyses of natural nephelines show that most have a deficiency of Na and Al; Na is replaced by vacant lattice sites, and Al by Si. Thus, there is an apparent excess of SiO_2 in the analyses over the stoichiometric requirements; this is expressed as normative Qz and can be as much as 10 wt %.

Tilley (1952, 1954, 1956) drew particular attention to this behavior of nepheline in different petrological settings and related it to the composition of the coexisting feldspar. In this way he was able to define the limits of different types of nepheline-bearing rocks as plotted in the residua system, and to define tie lines between nepheline and feldspar that characterize plutonic and metamorphosed rocks.

A selection of analyses is plotted in Fig. 12-7. In each case the compositions of the coexisting nepheline and feldspar and of the whole rock are reduced to the parameters Ne, Kp, Qz by calculating the salic normative constituents. For nepheline-feldspar rocks the three points lie on a straight tie line. It is found that most plutonic rocks generate tie lines that lie within the triangular field nepheline (N)–albite–potassium feldspar.

In volcanic rocks the composition of the nepheline may reflect the bulk composition of the rock by being either more sodic or more potassic than the "normal nepheline" composition. Consequently, the tie lines of volcanic rocks may lie within the area that contains the plutonic tie lines or well outside it.

On the basis of his extensive acquaintance with nepheline-bearing rocks, Tilley (1954, 1956) proposed that the compositions of coexisting nepheline and feldspar are not necessarily the compositions that originally crystallized, but that they represent an adjustment toward equilibrium compositions with falling temperature. He surmised that as the igneous assemblage cools, the nepheline and feldspar, aided by a fluid phase, react to give a less siliceous nepheline together with an adjustment in the Na:K ratio of both nepheline and feldspar to bring the nepheline into the convergence field. Experimental confirmation of this reaction has been provided by Hamilton (1961) and Hamilton and MacKenzie (1965) who have shown that two types of exchange occur during the cooling of a nepheline crystal in equilibrium with magma: $Na \rightleftharpoons K$ and $Si \rightleftharpoons Al$. After crystallization is complete, the Si–Al framework ceases to change but $Na \rightleftharpoons K$ continues between nepheline and alkali feldspar resulting in a nepheline that approaches the "normal nepheline" composition (Na:K = 3:1). By establishing experimentally the silica content of

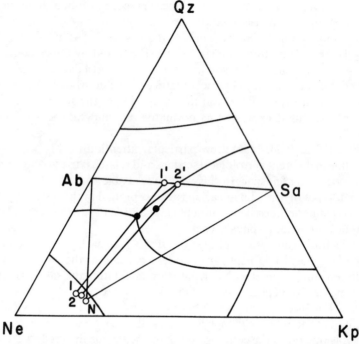

Figure 12-7. Nepheline-alkali feldspar tie lines of plutonic nepheline syenites in the system
$NaAlSiO_4$–$KAlSiO_4$–SiO_2 at 1 atm plotted in terms of the salic normative
constituents. N = Morozewicz "normal nepheline" composition. Most leucocratic nepheline syenites have tie lines within the triangular field N–Ab–Sa.
1–1' and 2–2' are typical tie lines passing through the whole-rock composition
(solid circle) with nephelines slightly more sodic than the ideal composition
(after Tilley, 1958).

nephelines in equilibrium with alkali feldspar and water-saturated liquid at
various temperatures, Hamilton (1961) has been able to show how nepheline can be used as a geothermometer as long as caution is exercised in
the interpretation of the results. In Fig. 12-8 are plotted the compositions
of a selection of natural nephelines from volcanic rocks in which the accompanying feldspar is sanidine or anorthoclase, and plutonic rocks in
which the feldspar is orthoclase microperthite or albite and microcline.
As would be expected, the volcanic nephelines show a higher crystallization temperature than the plutonic nephelines, which are grouped around
the "normal nepheline" composition.

The method is most sensitive for volcanic rocks where relatively rapid
cooling preserves the high-temperature equilibrium composition because

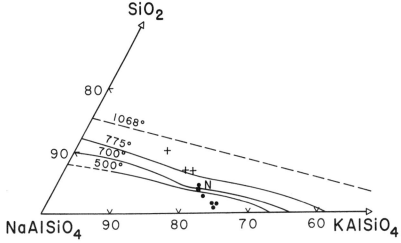

Figure 12-8. Part of the system $NaAlSiO_4$–$KAlSiO_4$–SiO_2 showing the limits of nepheline
solid solution at 500°C (P_{H_2O} = 2 kbar), 700°C, and 775°C (P_{H_2O} = 1 kbar) as
determined experimentally, and at 1068°C as estimated. Plotted on this diagram
are a number of analyzed natural nephelines from nepheline syenites (indicated
by solid circles) and phonolites (indicated by crosses). N (solid square) =
Morozewicz "normal nepheline" composition (after Hamilton, 1961).

cooling time is insufficient for the exchange reactions to take place. In
contrast, plutonic rocks cool very much more slowly and adequate time
is available for the exchange reactions to take place, especially if an
aqueous vapor has developed. As a result, the nepheline may indicate a
temperature markedly lower than that at which it initially crystallized
from the magma.

NATURAL ROCKS AND THE RESIDUA SYSTEM

The applicability of the residua system to natural rocks can now be
examined. An example is provided by the South Qôroq center of the
Igaliko nepheline syenite complex in the Gardar alkaline province of
South Greenland (Stephenson, 1972, 1976). The intrusive center is a series
of concentric stocks and ring intrusions that cut the basalts and sandstones
of the Eriksfjord Formation and are emplaced at a high structural level.
The southern part of this sequence has been obliterated by the intrusion
of the Igdlerfigssalik center. The rocks are a series of early concentric

intrusions of foyaite (augite–nepheline syenite) followed by a ring dike of transitional foyaite–augite syenite and finally by a ring dike of augite syenite.

The foyaites contain about 80 percent nepheline and perthitic alkali feldspar in roughly equal amounts, together with sodalite and mafic minerals that occur interstitially and as clots. These mafic minerals are aegirine augite, arfvedsonitic amphibole, biotite, fayalite, aenigmatite, iron-titanium oxides, and apatite. Some of the rocks are chilled against their margins and show flow lamination.

The augite syenites are composed of schillerised microperthite with interstitial nepheline. Some of the feldspar contains blebs of nepheline. Mafic minerals are titanaugite, arfvedsonite amphibole, olivine, iron titanium oxide, and apatite. Some mafic bands show cumulus development of titanaugite, olivine, iron-titanium oxide and apatite with intercumulus microperthite, nepheline, amphibole, and biotite.

The transitional rock is similar to the augite syenite but its pyroxenes are zoned to rims of aegirine augite.

By employing the method of Hamilton (1961) that was described previously, it can be ascertained that nepheline crystallized from the magma in the temperature range 900° to 700°C. Most of the nepheline, however, is close to the "normal nepheline" composition, which suggests that it has continued to equilibrate with the magma during cooling. It was noted earlier that this continued equilibration is the inherent weakness of this method of geothermometry in plutonic rocks.

In Fig. 12-9 the compositions of the rocks are ploted in the residua system. Most of the analyses show peralkaline indices close to 1.0 but some are slightly peralkaline. The compositions define a curved path that is concave to the Ne–Qz join. At $P_{H_2O} > 0.5$ kbar (approximately) the leucite field has diminished to the extent that all the compositions plot in the alkali feldspar field except for a few in the nepheline solid solution field.

This natural rock behavior can now be compared in Fig. 12-10 with the experimentally determined fractionation paths in the synthetic system. It is immediately apparent that compositions on the potassic side of the thermal valley in the synthetic system are convex to the Ne–Qz join, whereas the natural rocks follow a concave path. It will be seen below that this difference is due to the slightly peralkaline trend of the South Qôroq rocks.

A further example may be taken from the commonly observed field association of trachytes and phonolites. Members of these rock types constitute a trend that has been observed for many years by petrologists who concluded that slightly undersaturated trachytes are parental to phono-

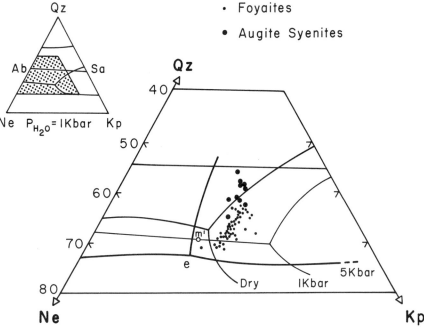

• Foyaites

• Augite Syenites

Figure 12-9. Analyses of the augite syenites and foyaites of the South Qôroq center plotted in the residua system in terms of normative Ne, Kp, and Qz. Phase equilibrium relationships are shown at $P = 1$ atm, $P_{H_2O} = 1$ kbar, 5 kbar, m' = ternary minimum at $P_{H_2O} = 1$ kbar; e = ternary eutectic at $P_{H_2O} = 5$ kbar. The rocks follow a curved fractionation path from the feldspar join to the vicinity of the ternary minimum for $P_{H_2O} = 1$ kbar (after Stephenson, 1976).

lites. Studies in the residua system have shown (Hamilton and MacKenzie, 1965) that removal of feldspar by fractionation from such trachytic liquids does lead to a phonolitic residuum, whereas fractionation of feldspar from a slightly oversaturated trachytic magma will lead to alkali rhyolite, the thermal barrier of the feldspar join separating the two trends.

It is instructive to examine a suite where the trachytic to phonolitic trend is well developed and compare it to the behavior of liquids in the residua system. Such an example is provided by Mount Suswa, Kenya (Nash, Carmichael, and Johnson, 1969), which consists of sodalite trachyte, sodalite phonolite, and phonolite, most of which are mildly peralkaline, and in all of which alkali feldspar is the most abundant phase. The evolution of these rocks can now be examined by reference to the residua system.

It may reasonably be assumed that the first feldspar to crystallize is represented by the cores of phenocrysts, or if the rock is non-porphyritic,

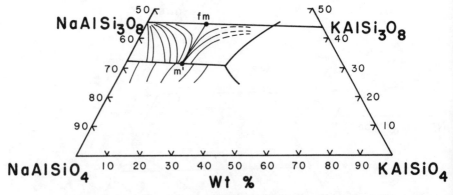

Figure 12-10. Experimentally determined fractionation curves in part of the system NaAlSiO$_4$–KAlSiO$_4$–SiO$_2$–H$_2$O at P_{H_2O} = 1 kbar (projected to the anhydrous base). m' = ternary temperature minimum; fm = temperature minimum on the feldspar join. The straight line joining fm to m' is a unique fractionation curve that does not begin at either NaAlSi$_3$O$_8$ or KAlSi$_3$O$_8$ but extends along the thermal valley from fm (Hamilton and MacKenzie, 1965).

of the most calcic groundmass feldspar. Its composition (expressed as Ab/Or) is joined to the whole rock composition that is taken to represent the composition of the magma. Tie lines are similarly constructed between the composition of the outermost part of groundmass feldspars, and the residual glass. The extension of these tie lines through the liquid compositions gives the initial direction in which the residual liquid moves as feldspar crystallization continues. The array of tie lines is shown in Fig. 12-11.

These tie lines define a set of conjugation lines from which fractionation curves can be constructed (Fig. 12-12) in exactly the same way as was done in the synthetic system by Hamilton and MacKenzie (1965, pp. 220–225). In that system convergent conjugation lines defined a low-temperature valley extending from the minimum on the alkali feldspar join to the ternary minimum on the nepheline-alkali feldspar boundary (Fig. 12-10). The valley separates liquids that initially crystallize feldspars more sodic than the liquid from those that crystallize feldspars more potassic than their parental liquid. An examination of Fig. 12-12 indicates the same behavior in the natural rocks and one may surmise that a similar thermal valley exists in the natural magma.

The significance of the diagram can now be examined. Magmas that lie on the sodic side of the valley will crystallize feldspars that become progressively more potassic, whereas magmas on the potassic side crystallize feldspars that are initially more potassic than the liquid but become

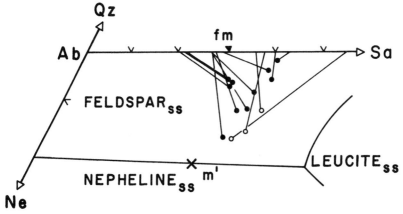

Figure 12-11. A series of analyzed lavas, feldspars, and residual glasses from Mount Suswa, Kenya, plotted in terms of their salic normative constituents in a part of the system $NaAlSiO_4$–$KAlSiO_4$–SiO_2–H_2O at $P_{H_2O} = 1$ kbar. fm = temperature minimum on the alkali feldspar join; m' = ternary minimum. Tie lines connect the alkali feldspar compositions (expressed as Ab:Or) to rock compositions (solid circles) and residual glass compositions (open circles) (after Nash, Carmichael, and Johnson, 1969).

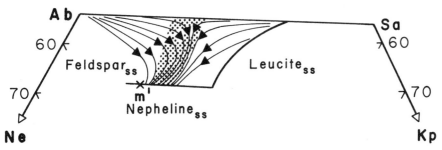

Figure 12-12. Crystallization trends of Mount Suswa lavas based on the experimentally determined trends (Fig. 12-10), but modified after the behavior of the natural tie lines in Fig. 12-11. Stippled zone indicates the limits of the Mount Suswa data (after Nash, Carmichael, and Johnson, 1969).

progressively more sodic as the magma cools. Because of the inflection in the curves, however, there is a limit to this sodic enrichment beyond which the feldspar must become steadily more potassic again. Such behavior is found in some of the lavas and is portrayed in Fig. 12-13. Phenocrysts are zoned from potassic in the core to more sodic compositions, and then

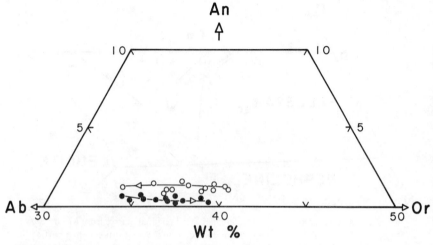

Figure 12-13. Reversal of the zoning in feldspar phenocrysts from a Mount Suswa lava
plotted in terms of anorthite–albite–orthoclase (weight percent). Open
circles are analyses of phenocryst cores, and closed circles are analyses of
phenocryst rims. Zoning direction is indicated by arrows (after Nash,
Carmichael, and Johnson, 1969).

sharply reverse to become more potassic. The groundmass feldspars
parallel this latter trend with zonation from sodic to potassic.

Again it is apparent that the crystallization paths as determined in the
synthetic system are an indication of the evolutionary path of magmas,
but the agreement is not perfect because of the greater complexity of the
magma. In both the South Qôroq and Mt. Suswa rocks the thermal valley
for these mildly peralkaline rocks is displaced toward potassic compositions from its position in the synthetic system. This greater complexity of
the natural magmas may be turned to advantage through a more detailed
examination of the feldspar behavior, both to locate the thermal valley
and to show how feldspar crystallization can develop a peralkaline trend
in the magma.

Bowen (1945) showed that, in general, the feldspar that crystallizes from
any CaO-bearing silicate system is a plagioclase rather than pure albite.
This tendency of plagioclase to capture Ca and Al from the crystallizing
liquid is known as the "plagioclase effect" and is clearly exhibited in the
trends of many trachytes and phonolites that crystallize not a pure alkali
feldspar but an anorthoclase. This incorporation of Ca and Al into feldspar
depletes the residual liquid in both, at the same time enriching it in Na and
Si and generating the peralkaline condition $(Na_2O + K_2O) > Al_2O_3$

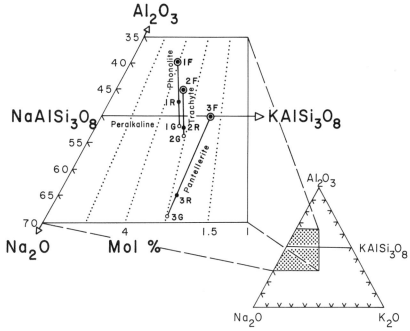

Figure 12-14. Compositions of whole rock, feldspar, and residual glass for a phonolite (1), a nepheline trachyte (2), and a pantellerite (3), plotted in terms of molecular proportions of Na_2O, K_2O, and Al_2O_3. The alkali feldspar join contains all compositions with alkali/alumina of 1:1. Dotted lines show equal Na_2O/K_2O. Circled dots = feldspar phenocrysts, solid dots = whole rocks, and open circles = residual glass (after Carmichael, Turner, and Verhoogen, 1974).

(mol. prop). Figure 12-14 shows, in terms of their molecular proportions of Na_2O, K_2O, and Al_2O_3, the whole rock (i.e., initial liquid), feldspar phenocrysts, and coexisting glass (i.e., liquid composition in equilibrium with feldspar), of a phonolite, a nepheline trachyte, and a pantellerite. In this projection the alkali feldspar join contains all compositions with alkali/alumina of 1:1. Compositions below the join are peralkaline and compositions above it have more alumina; in their words, they contain $CaAl_2Si_2O_8$.

Phonolite no. 1, which is not peralkaline, precipitates an anorthoclase (1F). Its crystallization depletes the initial liquid (1R) in alumina and changes its composition so that it becomes peralkaline (1G). Trachyte no. 2 is already peralkaline so that crystallization of anorthoclase (2F) increases its peralkalinity (2G). The "plagioclase effect" can, therefore,

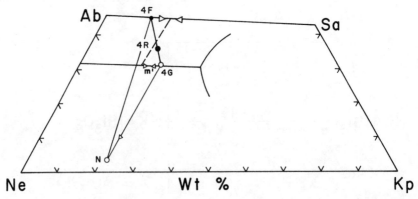

Figure 12-15. Phonolite (large solid circle), feldspar (small solid circle), and residual glass (open circle) plotted in terms of salic normative constituents in a part of the system $NaAlSiO_4$–$KAlSiO_4$–SiO_2–H_2O at $P_{H_2O} = 1$ kbar. Arrows on the Ab–Sa join indicate the temperature minimum; m' = ternary minimum; N = Morozewicz "normal nepheline" composition. The thermal valley, which in the residua system extends from the feldspar minimum to the nepheline-feldspar boundary curve, is shown by a dashed line. The tie lines joining feldspar–whole rock–glass compositions constitute a three-phase triangle, the apex of which lies to the potassic side of the ternary minimum (after Carmichael, 1964).

develop and enhance peralkaline trends, and when a strongly peralkaline composition has already been developed, such as pantellerite no. 3, the crystallization of even an alkali feldspar without an anorthite component (3F) enhances the peralkalinity (3G).[2]

A further result of the plagioclase effect is to shift the ternary minimum in the residua system when rock and mineral compositions are projected into it, as has been seen already. Again this shift can be deduced from the behavior of natural rocks, and Carmichael (1964) has provided a clear illustration in a phonolite that has phenocrysts of anorthoclase within thin rims of more potassic feldspar, ferroaugite, fayalitic olivine, ilmenite, and nepheline, in a very finely crystalline groundmass. After careful separation, both the groundmass and the feldspar phenocrysts were analyzed chemically (Fig. 12-15). Feldspar (4F) contains 19.8 mol % anorthite. A tie line drawn from (4F) through the whole rock composition (4R) to the groundmass composition (4G) on the nepheline-feldspar

---

[2] These examples are discussed fully by Carmichael, Turner, and Verhoogen (1974). A similar set of examples is provided by the lavas of Mt. Suswa (Nash, Carmichael, and Johnson, 1969).

boundary combines with a tie line joining (4G) to the nepheline composition to generate a three-phase triangle whose apex must point in the direction of the ternary minimum. The ternary minimum for rocks containing an anorthite component must, therefore, be at a more potassic composition than the minimum in the lime-free residua system.

THE "PERALKALINE RESIDUA SYSTEM"
$Na_2O-Al_2O_3-Fe_2O_3-SiO_2$

Important as the residua system is, it is limited to those feldspathoidal rocks in which alkali/alumina does not exceed unity. There is, however, the important group of peralkaline rocks. Peralkaline magmas can, of course, be either undersaturated or oversaturated with respect to silica. Undersaturated magmas do not evolve toward the Ne–Ks–Qz minimum but rather toward a eutectic considerably richer in alkalies. Consequently, these peralkaline rocks cannot truly be represented in the residua system. The work of Bailey and Schairer (1966) has partly solved this problem by providing data on several joins and planes within the system Na_2O–Al_2O_3–Fe_2O_3–SiO_2. This system has been called by them the "peralkaline residua system" and is characterized by equilibria involving acmite, albite, nepheline, quartz, and a peralkaline liquid. The f_{O_2} of the experiments was not controlled by a gas-mixing technique, but all experiments were done in air to minimize the reduction of Fe_2O_3. The univariant and divariant relations in this system are dominated by the incongruent melting of acmite, which is not a liquidus phase on the join acmite–nepheline–silica. Instead of acmite there is a broad field of hematite, and acmite is found to crystallize only from liquids that contain potential sodium silicate. In consequence, the oversaturated and undersaturated eutectics, which correspond to peralkaline granitic and nepheline syenitic liquids, are rich in sodium silicate and contrast markedly with the compositions at the granitic and nepheline syenitic minima in the residua system. The temperatures of these eutectics at 1 atm (728° ± 5°C and 715° ± 5°C) are also considerably lower than the residua system ternary minimum (1020° ± 5°C).

It is, of course, an extremely complex system, but from the determined joins and planes it is possible to construct a flow-diagram (Fig. 12-16) showing schematically how all the lines of univariant equilibrium and the invariant points are related within the tetrahedron. This diagram may then be used to deduce the course of crystallization of any liquid in the system. The flow diagram has a completely arbitrary geometrical arrangement, but high-temperature relations are at the top of the diagram passing

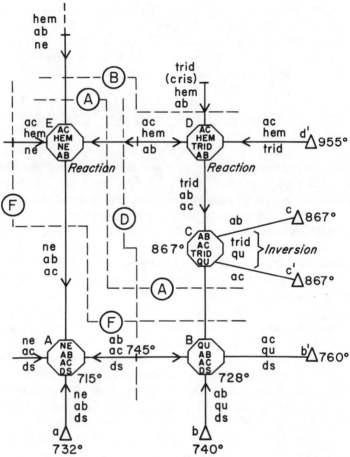

Figure 12-16. Simplified schematic flow diagram showing the univariant and invariant equilibria that involve liquids in the system $Na_2O-Al_2O_3-Fe_2O_3-SiO_2$. Ternary invariant points in the system $Na_2O-Al_2O_3-SiO_2$ are shown by triangles labeled a, b, and c; ternary invariant points in the system $Na_2O-Fe_2O_3-SiO_2$ are shown by triangles labeled b', c', and d'; quaternary invariant points are shown by octagons labeled A, B, C, D and E; ternary joins are shown as dashed lines labeled Ⓐ, Ⓑ, Ⓓ, and Ⓕ, that represent their projection onto the plane $Na_2O-Al_2O_3-SiO_2$ from the acmite composition that is a common apex of all the joins. Temperature maxima on the joins are shown as bars labeled with the appropriate temperature. Univariant lines, with three solid phases plus liquid, link the ternary invariant points of the bounding ternary systems and the quaternary invariant points. The direction of falling temperature along each line is indicated by arrows, and the solid phases crystallizing from the liquid are shown by abbreviations. In this type of schematic diagram the geometric arrangement is arbitrary, and so only the phase relations may be read. Abbreviations are: ab, albite; ac, acmite; cris, cristobalite; ds, sodium disilicate; hem, hematite; ne, nepheline; qu, quartz; trid, tridymite (after Bailey and Schairer, 1966).

to low temperature at the bottom and silica-oversaturated compositions at the right.[3]

Two important quaternary reaction points E and D lead to quaternary eutectics A and B that are respectively undersaturated and oversaturated because they are separated by the silica saturation plane $\textcircled{D}$. Eutectics A and B thus have as their natural analogs the peralkaline nepheline syenites and granites, respectively. Because of the incongruent melting of acmite no plane in the system is truly ternary. Furthermore, although join $\textcircled{D}$ is a silica-saturation plane, it is not a barrier, because liquids in the hematite field leave the plane when they begin to crystallize. They return to it by reaction of hematite with liquid to form acmite, but if the hematite fails to react, perhaps by fractionation, the liquid will reach the quaternary eutectic B where it will crystallize quartz, albite, acmite, and sodium disilicate. Sodium disilicate, of course, is not a mineral. In magmas, which have many more components than this synthetic system, it becomes a component of such minerals as aenigmatite or arfvedsonite, remains in the glass phase of some volcanic rocks, or goes into solution in the alkalic aqueous (and perhaps halide-rich) fluid phase that separates from certain alkaline magmas in their later evolutionary history.

The peralkaline residua system provides a means whereby some silica-undersaturated liquids can yield silica-oversaturated liquids, and alkaline liquids can yield peralkaline liquids. Eutectic A is the peralkaline, under-saturated eutectic in the volume nepheline–albite–acmite–sodium disili-cate. It is instructive to examine the univariant curve acmite–nepheline–albite–liquid that extends from reaction point E at 908°C (nepheline–albite–acmite–hematite–liquid) to eutectic A at 715°C (nepheline–albite–acmite–sodium disilicate–liquid). The curve spans a considerable compositional range, and liquids on it are progressively enriched in sodium disilicate and impoverished in acmite as they crystal-lize acmite, albite, and nepheline. Thus an "ijolite" liquid at E (with about 50 percent acmite) follows a "foyaitic" trend to a "nepheline syenite" liquid at A (with about 10 percent acmite). This trend in the synthetic system is probably analogous to the development of leucocratic nepheline syenites from mafic alkaline magmas and of phonolite residua from nephelinitic magmas.

An additional possibility of considerable interest derives from the extensive solid solution in the system of Fe^{3+} for Al^{3+}, causing the forma-tion of $NaFe^{3+}Si_3O_8$ (iron albite) as well as of $NaFe^{3+}Si_2O_6$ (acmite). A liquid on the slightly oversaturated side of join $\textcircled{D}$ thus crystallizes

---

[3] A full discussion of crystallization in this system is beyond the scope of this chapter. A complete analysis is given in the original discussion of Bailey and Schairer (1966).

iron-bearing albite in considerable amounts. The effect is to remove more silica from the liquid that can then develop a trend toward the undersaturated eutectic. In natural syenitic magmas the effect would be even more striking because substitution of Fe^{3+} for Al^{3+} in the potassium feldspars can be substantial. Carmichael (1967) has, for example, reported feldspars in the ultrapotassic rocks of the Leucite Hills of Wyoming with up to 18 wt percent of the iron-orthoclase component. It can be seen, therefore, that the "peralkaline residua system" offers possibilities whereby the thermal barrier of the "residua system" can be penetrated. To what extent these possible processes operate in natural magmas is largely conjectural at present.

PSEUDOLEUCITES

It has already been observed that in the residua system there is a reaction point at which leucite reacts with liquid to generate nepheline and potassium feldspar. This leucite-eliminating reaction was called by Bowen (1928, p. 253) the pseudoleucite reaction, and in seeking to explain the generation of alkaline rocks in terms of fractional crystallization, he attributed considerable importance to it. Because leucite is a rare mineral even in alkaline rocks, the reaction is no longer considered so important but the problem of pseudoleucites does illustrate some of the difficulties that can arise in interpreting experimental results. Pseudoleucites are complex intergrowths of nepheline and feldspar (usually a sanidine) that frequently have the morphology of leucite crystals. This appearance has led to the assumption that they are in some way derived from leucite crystals, yet the pseudoleucite reaction would be expected to create corroded grains rather than splendidly preserved crystals.

One of the earliest descriptions of pseudoleucite is by Knight (1906), who argued that if a sodium-rich leucite rather than the normal potassium leucite were to crystallize, it might break down under subsolidus conditions to nepheline and orthoclase. The difficulty with this explanation is that sodium-rich leucite is unknown in natural rocks, and this objection was voiced by Bowen and Ellestad (1937) in refuting Knight's interpretation. Their further studies of leucitic lavas led them to suggest that normal potassic leucite crystallizes but eventually undergoes reaction with an increasingly sodic magma, so that the leucite crystal is converted to a nepheline-feldspar pseudomorph.

The problem was not investigated experimentally for nearly fifty years, until Fudali (1963) showed that extensive substitution of Na^+ for K^+ is possible at 1 atm but decreases with increasing water pressure. It was argued by Fudali that such sodic leucites could undergo subsolidus breakdown to intergrowths of nepheline and potassium feldspar. The fact remains that natural sodic leucites are unknown, and one would expect that if such a mineral exists then, somewhere, a lava would have cooled sufficiently rapidly that a sodic leucite would be preserved.

More recently, the role of ion exchange processes has been considered as a possible solution to the dilemma. Experimental studies by Taylor and MacKenzie (1975) and Gupta and Fyfe (1975) have found that leucite solid solutions undergo exchange reactions with sodium-rich glass or sodium-rich hydrous vapor. In natural rocks it seems likely that ion exchange mechanisms can change a potassic leucite into a more sodic type in the subsolidus region and that subsequent cooling causes exsolution of nepheline and alkali feldspar. The leucite structure is destroyed but the resulting pseudomorph preserves the leucite crystal morphology. An element of complexity is thus introduced into experimental petrology that has not generally been recognized. Although equilibrium may exist between crystals and liquid, the distribution of Na and K among crystalline phases, glass, and vapor may be changed by later subsolidus reaction unless the experiment is quenched rapidly and the crystalline charge is removed from the aqueous phase. Exchange processes have also been shown to have considerable importance in the alkali feldspars (Orville, 1963, 1972).

Another type of nepheline-alkali feldspar intergrowth has sometimes been referred to as pseudoleucite, but presents a different set of problems from the pseudomorphs just discussed. It is an intergrowth of vermicular pattern or fingerprint appearance and shows no crystal outline. Examples are known from several alkaline rock intrusions, and one in particular, from Kaminak Lake, N.W.T., Canada, has been discussed by Davidson (1970), who interprets it as the result of cotectic crystallization in the residua system. Another instructive example is found in the Batbjerg intrusion, East Greenland (Gittins *et al.*, 1977), where feather-shapted grains of nepheline and potassium feldspar in fingerprint intergrowth coexist with nepheline and grains that are a patchy intergrowth of kalsilite and potassium feldspar. The kalsilite-potassium feldspar intergrowth is probably a subsolidus breakdown product of leucite (Fig. 12-4). Initially the magma crystallized nepheline, followed by leucite and nepheline when the magma reached the leucite boundary curve. Subsequent increase in P_{H_2O} contracted the leucite stability field, and the magma composition was then on the cotectic between nepheline and alkali feldspar. If this liquid became supercooled, it might have crystallized rapidly as a fingerprint intergrowth of nepheline and potassium feldspar rather than as discrete grains of nepheline and potassium feldspar in fingerprint intergrowth as pseudoleucite because of confusion with the pseudomorphic type that displays crystal morphology.

MELTING STUDIES

Of prime importance to petrologists is a knowledge of the order in which the constituent minerals of a rock have crystallized from the parent magma. Interpretations of crystallization order have usually been made from petrographic observation based on textural relations between individual minerals. The presence of phenocrysts in glassy extrusive rocks will often give the crystallization order of some of the minerals but, in general, unequivocal interpretation is rarely possible, particularly in plutonic rocks. Furthermore, while sometimes indicating the order in which phases began to crystallize, the textural approach is generally of little value in indicating the extent of overlapping crystallization, for many rocks will have had three phases, at least, crystallizing together.

More complete information is obtained from melting natural rocks and observing the order in which the minerals crystallize as the melt is quenched from temperatures lowered a few degrees at a time. This approach, of course, is also fraught with problems because of the difficulty of duplicating the pressure, volatile pressure, fluid composition, and f_{O_2} that prevailed during natural cooling of the magma, even when the petrologist has found a rock that reasonably represents the composition of the magma from which it crystallized. Examples of this experimental method are reported by Tilley and Thompson (1972), Thompson (1973), Piotrowski and Edgar (1970), and Sood and Edgar (1970).

A full understanding of alkaline magma crystallization requires knowledge of the effects of any volatile constituents that might have been present in the magma but are not always preserved in the rocks. Common volatile constituents are H_2O, CO_2, S, Cl, and F. In this regard, classical petrology can generally guide the experimentalist through the mineralogy of the rocks: the former presence of Cl is indicated by sodalite, CO_2 by cancrinite, and F by substitution for OH groups in micas and amphiboles.

The initial approach is generally to study crystallization under the limiting conditions of dry and water-saturated magma, and in a range of intermediate states where $P_{H_2O} < P_{total}$. The investigation can then be extended to take account of volatile constituents other than H_2O.

An illustration of this approach is furnished by experimental work on a nepheline syenite from the Blue Mountain intrusion, Methuen Township, Ontario, Canada (Millhollen, 1971). The crystallization behavior of the nepheline syenite melt was determined under both water-saturated conditions and where $P_{total} = P_{H_2O} + P_{CO_2}$ through the simple expedient of using oxalic acid dihydrate, which decomposes under the conditions of the experiment to 50 mol% H_2O + 50 mol% CO_2. From experimental results with the mole fraction of water $(X_{H_2O}) = 1.0$ and 0.5, the solidus was calculated for X_{H_2O} of 0.2, 0.1, and 0.0 (dry) (Fig. 12-17). Figure 12-18 illustrates the profound effect of incorporating CO_2 into the fluid phase: the solidus is raised almost to the water-saturated liquidus.

The effect of fluid composition on the crystallization temperature of a nepheline syenite magma at relatively low pressure is apparent from that

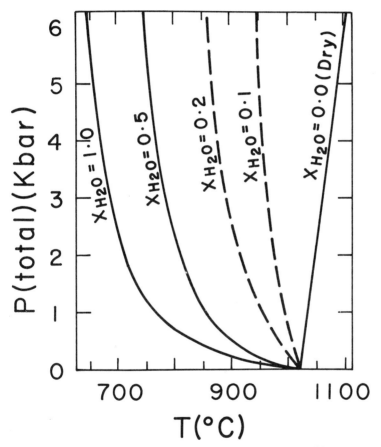

Figure 12-17. $P(total)–T$ projection of experimentally determined solidus curves of a nepheline syenite from Blue Mountain, Ontario, for $X_{H_2O} = 1.0$ and $X_{H_2O} = 0.5$; calculated solidus curves for $X_{H_2O} = 0.2$ and $X_{H_2O} = 0.1$; and calculated dry ($X_{H_2O} = 0.0$) solidus (after Millhollen, 1971).

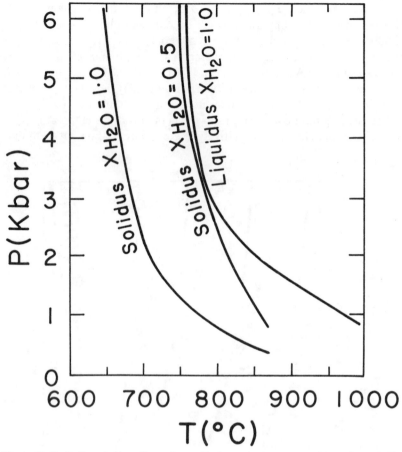

Figure 12-18. *P–T* projection of experimentally determined melting relations of a nepheline
syenite from Blue Mountain, Ontario, showing the solidus and liquidus for
$X_{H_2O} = 1.0$, and the solidus for $X_{H_2O} = X_{CO_2} = 0.5$. Note that the effect of the
CO_2 is to raise the solidus almost to the water-saturated liquidus (after
Millhollen, 1971).

study. The potential role of CO_2 in the evolution of alkaline magmas at
near mantle pressure can now be examined. The solubility of CO_2 in
magmas is very variable. In leucocratic nepheline syenite magmas it is
small at pressures as low as 1 kbar. It is smallest in highly polymerized
siliceous liquids, and increases with basicity, temperature, and pressure
(Eggler, 1973; Mysen, 1975). In addition, CO_2 is found to be more soluble
in the presence of water than under dry conditions, and the effect of CO_2

on melting processes is opposite to that of H_2O, in that melting in the presence of both CO_2 and H_2O produces liquids that are less silica-saturated than those produced by hydrous melting alone. Partial melting studies on natural peridotite using various mixtures of CO_2 and H_2O and pressures between 10 and 15 kbar have demonstrated the profound effect that CO_2 can have on the composition of the melt formed (Boettcher, Mysen, and Modreski, 1975; Mysen and Boettcher, 1975). With pure water the glasses resulting from quenching of melt are quartz-normative, and as the vapor composition is changed from 100 mol % H_2O to 20 mol % H_2O, the composition of the glass changes through olivine-hypersthene

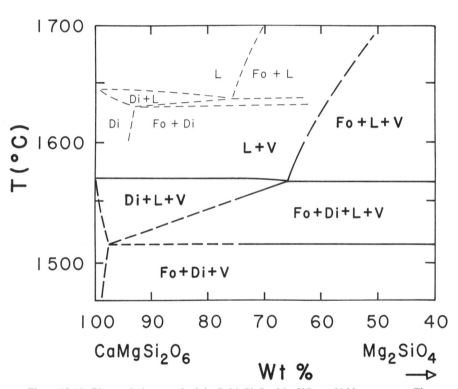

Figure 12-19. Phase relations on the join $CaMgSi_2O_6$–Mg_2SiO_4 at 20 kbar pressure. The join with excess CO_2 (solid and broken lines) is compared with volatile-absent melting relations (light dashed lines). The piercing point at which diopside, forsterite, CO_2-bearing liquid, and CO_2 vapor coexist is at $Di_{65}Fo_{35}$ and about 1570°, compared to the piercing point for diopside, forsterite, and liquid coexisting under volatile-absent conditions at $Di_{75}Fo_{25}$ and 1635°. A further effect of P_{CO_2} is to increase the temperature range between liquidus and solidus (after Eggler, 1974).

normative to olivine-nepheline normative to nepheline-larnite normative. It appears, then, from these experiments, that CO_2 might be the principal factor in determining whether a strongly alkaline magma can be generated by partial melting of peridotite in the mantle of the earth.

Mantle-type peridotites are generally thought to contain orthopyroxene. Thus, an important requirement for the development of an alkaline magma by partial fusion of such a peridotite is that, under the melting conditions, the magma must be in equilibrium with orthopyroxene. Under dry and water-saturated conditions orthopyroxene does not appear to be a liquidus phase at any pressure, but it is found that CO_2 stabilizes orthopyroxene on the liquidus both in synthetic mixtures such as the join $CaMgSi_2O_6(Di)–Mg_2SiO_4(Fo)$ and in natural rocks. Eggler (1974) has shown that at 20 kbar the effect of CO_2 is to lower the liquidus temperatures and to increase the separation of liquidus and solidus, and at 30 kbar to introduce orthopyroxene as a liquidus phase (Figs. 12-19, 12-20). The

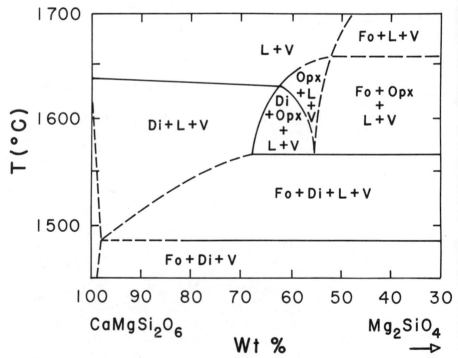

Figure 12-20. Phase relations on the join $CaMgSi_2O_6–Mg_2SiO_4$ at 30 kbar pressure under CO_2-saturated conditions. Orthopyroxene is stable on the liquidus (Eggler, 1974).

same effect is seen in a natural olivine melilitite (Brey and Green, 1975) where, under dry and wet conditions (up to 40 wt % H_2O) and pressures up to 40 kbar, only olivine and clinopyroxene appear as near-liquidus phases, yet at 30 kbar the effect of CO_2 is to suppress the near-liquidus appearance of both minerals and to bring orthopyroxene and garnet on to the liquidus. Because orthopyroxene is not a liquidus phase under dry or wet conditions but is a liquidus phase in the presence of CO_2, it appears that in order for olivine melilitite to be a mantle-melting product, the melting must occur in the presence of CO_2.

A strong case can thus be made for CO_2 playing an essential role in the development of some strongly alkaline magmas. The fact that CO_2 is present in the mantle has long been suspected from CO_2 emissions in volcanic eruptions, but more recently its presence has been established by the recognition of CO_2-bearing inclusions and carbonate inclusions in the minerals of mantle-derived xenoliths (Roedder, 1965). The hypothesis is further supported by the widespread occurrence of carbonate in kimberlites, and by the existence of magmatic igneous carbonate rocks known as carbonatites (Heinrich, 1966; Tuttle and Gittins, 1966).

SCHEMES OF MAGMA GENESIS

In the earlier part of the present century the origin of the alkaline rocks was generally sought within the crust, the emphasis being placed on processes that might suitably modify a relatively simple basaltic or granitic magma. Thus, there were appeals to (1) gas streaming hypotheses whereby alkalies might be concentrated at higher crustal levels, (2) crustal contamination and assimilation schemes such as the desilication arguments involving the reaction of magma with limestone and the removal of ferromagnesian minerals by crystal fractionation, and (3) even more complex processes involving the resorption of earlier crystallized biotite or hornblende. Implicit in all these arguments was that the events occurred within the crust. One of the most popular views for many decades was that alkaline rocks are the result of reaction of magma with limestone to generate ferromagnesian minerals that sink, leaving the magma enriched in alkalies and alumina. At first only basaltic magma was considered (Daly, 1910) but later on Shand (1922) applied the argument to granitic magmas and produced a lengthy catalogue of localities alleged to demonstrate the process in the field. The validity of most of these examples has now been negated by more detailed geological mapping, which is as fundamental to petrogenetical theory as is experiment. A few examples remain where the development of alkaline rocks by limestone assimilation seems to be demonstrable in the field, albeit in very small volumes. Among these are Scawt Hill, County Antrim (Tilley and Harwood, 1931), Camas Mor, Muck (Tilley, 1947), and the Christmas Mountains, Texas (Joesten, 1977), all of which involve the intrusion of gabbroic magma into limestone. In each example the zone of nepheline-bearing rock has a thickness of only a few centimeters to 3 meters, which seems to emphasize the very limited capacity of limestone assimilation to generate alkaline rocks.

A case against the limestone assimilation origin of alkaline rocks was put on a firmer footing by the experimental work of Watkinson and Wyllie (1969) in a study of joins in the system $CaCO_3$–$Ca(OH)_2$–$NaAlSi_3O_8$ in the presence of 25% H_2O at 1 kbar pressure. The experimental results, of course, apply more directly to granitic than to basaltic magmas. They suggest that limestone assimilation consumes heat from the liquid and results in its crystallization. An additional effect is the liberation of CO_2, which has a low solubility in magmas at upper crustal pressures, and tends to remove H_2O from the magma, dehydrating it and inducing crystallization through raising its solidus temperature. The experiments are not definitive, but they have built a persuasive case against limestone assimilation as an important means of generating large volumes of alkaline rocks.

As was noted previously, the early emphasis in theories of alkaline rock genesis was largely on fractional crystallization within the crust. Gradually the emphasis has shifted to the role of fractional melting processes within the mantle in developing the essential alkaline character of the magma. Fractional crystallization is now seen as a later process, within both the mantle and the crust, that accentuates an already developed alkaline character. The magmatic history of the alkaline rocks becomes, then, a two-stage process of partial melting of source rocks followed by crystallization differentiation.

Such a scheme does not rule out the possibility that crustal processes might be important in the development of some alkaline rocks on the continents, and Bailey (1964, 1974) has discussed the possibility that some phonolite, trachyte, and pantellerite, and their plutonic equivalents might have formed by anatexis in the deep crust, aided by influx of volatiles from the mantle and relief of pressure during crustal warping. The association of voluminous eruptions of phonolites and trachytes with domal uplift is remarked by many writers on volcanism in East Africa (Baker *et al.*, 1971; Williams, 1970, 1972).

In general, however, the ultimate origin of the alkaline rocks is sought in the mantle rather than in the crust.

Bowen (1928, pp. 236–240) discussed the development of trachytic rocks both on the continents and in the oceanic islands. He argued that slow differentiation accompanied by magma stirring that allowed the complete resorption of early-formed olivine would prevent the accumulation of SiO_2 in the magma and lead to the formation of trachyte. He held the conviction (p. 239) that this trachytic trend could proceed far enough to develop phonolite, although experimental systems then determined did not provide enough information to enable him to explain how it would develop. He noted the common association of basalt and trachyte in the oceanic islands and suggested that differentiation does not extend beyond trachyte because of the tendency for oceanic basaltic intrusions to be in the form of dikes or small plugs that cool relatively quickly. On the continents, however, the intrusions tend to be of much larger size and differentiation commonly extends to the development of phonolite.

Modern support for Bowen's views is to be found in the volcanic fields of East Africa. In Kenya, widespread occurrence of phonolite, nephelinite, and basalt suggests that magmas of these compositions were of primary origin. Yet at some central volcanoes there is a well-differentiated series of olivine basalt, mugearite, hawaiite, trachybasalt, trachyte, and phonolite, whereas the basalts of fissure eruptions seldom show extensive differentiation (Williams, 1970).

Further field evidence in Kenya suggests, however, that the origin of trachytes and phonolites is more complex than Bowen envisaged. Williams (1970) and Lippard (1973) have both shown that phonolites can be developed by the fractional crystallization of nephelinitic magma. These phonolites occur as small plugs and flows within and on the flanks of very large nephelinitic volcanoes.

The phonolites both of basalt association and of nephelinite association are contrasted with a third type that is not associated with central volcanoes. This is the plateau-type phonolite that occurs in flows of enormous volume and very uniform composition (Williams, 1970, 1972; Lippard, 1973). The plateau-type phonolites exhibit a gradation from sparsely microporphyritic (3–5% phenocrysts) to coarsely porphyritic (15–30% phenocrysts). Phenocrysts are dominantly nepheline and sanidine–anorthoclase in roughly equal proportions; oxides, ferroaugite, apatite, and biotite never exceed 3% of the mode. Most of the phenocrysts have corroded margins and there is a sharp distinction in grain size between phenocrysts and groundmass. The groundmass is a flinty, aphanitic mixture of alkali feldspar and interstitial alkali pyroxene, amphibole, and aenigmatite.

The phonolites deriving from basaltic and nephelinitic parents are quite different. Those associated with basaltic parents range from soda–trachyte to phonolite, are characterized by parallel alignment of groundmass alkali feldspar (60–75% of the mode), and have only sanidine as phenocrysts. Nepheline is a groundmass mineral and alkali mafic minerals occur interstitially. Phonolites derived from nephelinites range from phonolitic nephelinite to phonolite. They are rich in aegirine–augite and sphene and have equal amounts of alkali feldspar and nepheline. They are markedly porphyritic, with phenocrysts of aegirine–augite, sphene, oxides, nepheline, and anorthoclase in a groundmass of alkali feldspar, nepheline, aegirine, oxides, and zeolite (analcite?). The more phonolitic types have 25–30% nepheline and alkali feldspar as phenocrysts. Alkali amphibole and aenigmatite are notably absent from the groundmass, in contrast with the other types.

The major- and minor-element composition of the rocks (high Zr, Nb, La, Ce, Rb; low Ba, Sr) support the idea of their derivation through fractional crystallization, but suggest three distinctly different sources. The plateau-type phonolites differ from the other types because of their huge volume, short eruptive period, uniformity, and sparseness of associated basic to intermediate rocks. These are characteristics that seem at variance with the highly differentiated chemical composition conferred by the minor elements. It is possible that such a composition can be imparted to a magma by partial melting, and that these phonolites may be due to partial fusion beneath domed crust as suggested by Bailey (1964, 1974), or that fractional crystallization may have generated these chemical characteristics in the magma after its formation by partial fusion.

After the eruption of these large volumes of plateau-type phonolites, volcanism in Kenya changed abruptly to trachytes (including ignimbritic trachytes), and this change coincided with a period of major faulting. This type of trachyte seems unrelated to differentiation of basaltic magmas (Williams, 1970).

It appears likely that both phonolite and trachyte can result from fractionation of basaltic magma but that not all originate in this way. Very large volumes of both types seem to be formed as direct partial melts in the lower crust or upper mantle under continents. Those of the ocean basins, however, appear to be restricted to intrusions of basaltic magma.

The effect of crystallization on the differentiation of alkaline magmas is displayed to a spectacular extent by some agpaitic intrusions. One that Bowen used as an example in 1928 is the Ilímaussaq intrusion of southwest Greenland. It is composed of an incomplete ring of augite syenite, chilled against the country rocks, that encloses a layered series of agpaitic nepheline syenites and more extreme differentiates. The bottom of the intrusion is nowhere exposed, but in places the roof of basaltic lavas and quartzite can be seen.

Figure 12-21. Rhythmic layering in the layered series of the Ilímaussaq intrusion, southwest Greenland, exposed in the 400 meter-high wall of Kangerdluarssuq Fiord. The layering consists of light-colored nepheline syenite and dark layers of arfvedsonite-rich rocks and eudialyte-rich rocks. The layering about half way up the section is draped over a block that sank into the lower part of the intrusion from the vicinity of the roof. *Photo by J. Gittins.*

The layered series is exposed in a fiord wall 400 meters high (Fig. 12-21). The layers are sharply bounded by the layers above and below and are of strongly contrasted composition. The general appearance is similar to the well-known layered series of the Skaergaard intrusion but is rendered the more striking by distinct color differences. Thus layers of white nepheline syenite are bounded by black layers of arfvedsonite-rich rock and by red layers of eudialyte-rich rock. The classical interpretation of the intrusion as crystallization in a closed system is developed by Ferguson (1964, 1970) and Engell (1973). The layered series are considered to be bottom accumulates. Sodalite-rich syenites on the other side of the fiord are considered to be flotation accumulates formed by the floating of sodalite crystals to the top of the magma chamber and the subsequent crystallization of alkali feldspar, arfvedsonite, and eudialyte which enclose the sodalite poikilitically.

According to the closed-system view, the parent magma is the augite syenite and the entire sequence of rock types was formed *in situ* by fractional crystallization. In a more recent view Larsen (1976) proposes a crystallizing magma with periodic injections of magmas of different compositions that are themselves derived from differentiation of magma in a source region below the currently exposed intrusion.

Whether the intrusion is the result of closed- or open-system differentia-
.tion, it serves as an illustration of the ability of fractional differentiation accompanied by sinking and floating of crystals to produce rocks of strongly contrasted composition.

The breadth of meaning inherent in the term alkaline rock was mentioned at the beginning of this chapter. Even among the feldspathoidal rocks the range of compositions and mineralogical variations is enormous, and so a very limited group has been discussed here. There has been no consideration of the ultra-potassic kalsilite lavas of Zaire and Uganda; the halide-rich feldspathoidal rocks such as the hauyne- and nosean-bearing varieties, or the sodalite syenites; or the melilite-rich rocks (see Chapter 13), to name but a few of those that have to be omitted from a short chapter.

Bowen's approach in 1928 was to apply the principles of phase equilibrium studies in systems at one atmosphere to selected problems of alkaline rock petrogenesis. He did not offer much insight into the ultimate origin of alkaline magmas, but concentrated on the way in which crystallization differentiation might generate alkaline rocks from basaltic magmas or further modify magmas that were already alkaline.

Fifty years later the petrologist taking essentially the same approach has at his disposal a far greater accumulation of phase equilibrium data at pressures up to 40 kbar and under a variety of fluid compositions and oxygen fugacities. Furthermore, he has data both on fractional and equilibrium *crystallization*, and on fractional and equilibrium *melting*. He is thus able to consider not only the processes that modify existing magmas, but how those magmas might, initially, have been generated within the mantle. (Melting studies are not nearly so advanced as crystallization studies, and therefore it is, as yet, possible to see only a few indications of the way in which a very small part of the spectrum of alkaline rock magmas might have been generated.) In addition to phase equilibrium data, the modern petrologist has a much greater knowledge of the rocks themselves, their distribution and relation to each other in complex intrusions, and their tectonic settings.

Bowen combined his understanding of phase equilibrium principles with an intensive knowledge of the petrography and field relations of the rocks. This approach still holds the greatest promise of unraveling the multitude of alkaline rock problems that remain.

References

Bailey, D. K., Crustal warping—a possible tectonic control of alkali magmatism, *J. Geophys. Res.*, *69*, 1103–1111, 1964.

———, Melting in the deep crust, in Sørensen, H., *The Alkaline Rocks*, Wiley, New York, London, 436–441, 1974.

Bailey, D. K., and J. F. Schairer, The system $Na_2O–Al_2O_3–Fe_2O_3–SiO_2$ at 1 atmosphere, and the petrogenesis of alkaline rocks, *J. Petrol.*, *7*, 114–170, 1966.

Baker, B. H., L. A. J. Williams, J. A. Miller, and F. J. Fitch, Sequence and geochronology of the Kenya rift volcanics, *Tectonophys.*, *11*, 191–215, 1971.

Boettcher, A. L., B. O. Mysen, and P. J. Modreski, Melting in the mantle: phase relationships in natural and synthetic peridotite–H_2O and peridotite–H_2O–CO_2 systems at high pressures, *Phys. Chem. Earth*, *9*, 855–867, 1975.

Bowen, N. L., *The Evolution of the Igneous Rocks*, Princeton University Press, 334 pp., 1928.

———, Recent high-temperature research on silicates and its significance in igneous geology, *Am. J. Sci.*, *33*, 1–21, 1937.

———, Phase equilibria bearing on the origin and differentiation of alkaline rocks, *Am. J. Sci.*, *243A*, 75–89, 1945.

Bowen, N. L., and R. B. Ellestad, Leucite and pseudoleucite, *Am. Mineral.*, *22*, 409–415, 1937.

Brey, G., and D. H. Green, The role of CO_2 in the genesis of olivine melilitite, *Contr. Mineral. Petrol.*, *49*, 93–103, 1975.

Buerger, M. W., G. E. Klein, and G. Donnay, Determination of the crystal structure of nepheline, *Am. Mineral.*, *39*, 805–818, 1954.

Carmichael, I. S. E., Natural liquids and the phonolitic minimum, *Geol. Jour.*, *4*, 55–60, 1964.

———, The mineralogy and petrology of the volcanic rocks from the Leucite Hills, Wyoming, *Contr. Mineral. Petrol.*, *15*, 24–66, 1967.

Carmichael, I. S. E., F. W. Turner, and J. Verhoogen, *Igneous Petrology*, McGraw-Hill, New York, 739 pp., 1974.

Daly, R. A., Origin of the alkaline rocks, *Geol. Soc. Amer. Bull.*, *21*, 87–118, 1910.

Davidson, A., Nepheline–K-Feldspar intergrowth from Kaminak Lake, Northwest Territories, *Canad. Mineral.*, *10*, 191–206, 1970.

Eggler, D. H., Role of CO_2 in melting processes in the mantle, *Ann. Rept. Dir. Geophys. Lab., Washington*, *73*, 457–467, 1973.

———, Effect of CO_2 on the melting of peridotite, *Ann. Rept. Dir. Geophys. Lab., Washington*, *73*, 215–224, 1974.

Engell, J., A closed system crystal-fractionation model for the agpaitic Ilímaussaq intrusion, south Greenland with special reference to the lujavrites, *Bull. Geol. Soc. Denmark*, *22*, 334–362, 1973.

Ferguson, J., Geology of the Ilímaussaq alkaline intrusion, south Greenland. Description of map and structure, *Medd. om Grønland*, *172*, 4, 1–82, 1964.

———, The significance of the kakortokite in the evolution of the Ilímaussaq intrusion, south Greenland, *Medd. om Grønland*, *186*, 5, 1–193, 1970.

Fudali, R. F., Experimental studies bearing on the origin of pseudoleucite and associated problems of alkalic rock systems, *Geol. Soc. Amer. Bull.*, *74*, 1101–1126, 1963.

Gittins, J., J. J. Fawcett, J. C. Rucklidge, and C. K. Brooks, Kalsilite, leucite, nepheline and potash feldspar in the Batbjerg intrusion, East Greenland; natural examples of phase relations in the system $NaAlSiO_4$–$KAlSiO_4$–SiO_2–H_2O, *Geol. Soc. Amer., Abstracts*, *9*, 989–990, 1977.

Gupta, A. K., and W. S. Fyfe, Leucite survival: the alteration to analcime, *Canad. Mineral.*, *13*, 361–363, 1975.

Hamilton, D. L., Nephelines as crystallization temperature indicators, *J. Geol.*, *69*, 321–329, 1961.

Hamilton, D. L., and W. S. MacKenzie, Nepheline solid solution in the system $NaAlSiO_4$–$KAISiO_4$–SiO_2, *J. Petrol.*, *1*, 56–72, 1960.

Hamilton, D. L., and W. S. MacKenzie, Phase equilibrium studies in the system $NaAlSiO_4$–$KAlSiO_4$–SiO_2–H_2O, *Mineral Mag.*, *34*, 214–231, 1965.

Heinrich, E. W., *Geology of Carbonatites*, Rand McNally, Chicago, 555 pp., 1966.

Joesten, R., Mineralogical and chemical evolution of contaminated igneous rocks at a gabbro-limestone contact, Christmas Mountains, Big Bend region, Texas, *Geol. Soc. Amer. Bull.*, *88*, 1515–1529, 1977.

Knight, C. W., A new occurrence of pseudoleucite, *Am. J. Sci.*, *21*, 286–293, 1906.

Larsen, L. M., Clinopyroxenes and coexisting mafic minerals from the alkaline Ilímaussaq Intrusion, South Greenland, *J. Petrol.*, *17*, 258–290, 1976.

Lindsley, D. H., P-T projection for part of the system kalsilite–silica, *Ann. Rept. Dir. Geophys. Lab., Washington*, *65*, 244–247, 1967.

Lippard, S., The petrology of phonolites from the Kenya Rift, *Lithos*, *6*, 217–234, 1973.

Millhollen, G. L., Melting of nepheline syenite with $H_2O + CO_2$, and the effect of dilution of the aqueous phase on the beginning of melting, *Am. J. Sci.*, *270*, 244–254, 1971.

Morozewicz, J., Ueber die chemische zussammensetzung des gesteinsbidelden nephelins, *Fennia*, *22*, 1–16, 1928.

Morse, S. A., Syenites, *Ann. Rept. Dir. Geophys. Lab., Washington*, *67*, 112–120, 1969.

———, Alkali feldspars with water at 5kb pressure, *J. Petrol.*, *11*, 221–251, 1970.

Mysen, B. O., Solubility of volatiles in silicate melts at high pressure and temperature: the role of carbon dioxide and water in feldspar, pyroxene and feldspathoid melts, *Ann. Rept. Dir. Geophys. Lab., Washington*, *74*, 454–468, 1975.

Mysen, B. O., and A. L. Boettcher, Melting of a hydrous mantle, II, Geochemistry of crystals and liquids formed by anatexis of mantle peridotite at high pressures and high temperatures as a function of controlled activities of water, hydrogen and carbon dioxide, *J. Petrol.*, *16*, 549–593, 1975.

Nash, W. P., I. S. E. Carmichael, and R. W. Johnson, Mineralogy and petrology of Mount Suswa, Kenya, *J. Petrol.*, *10*, 409–439, 1969.

Orville, P. M., Alkali ion exchange between vapor and feldspar phases, *Am. J. Sci.*, *261*, 201–237, 1963.

————, Plagioclase cation exchange equilibria with aqueous chloride solution: results at 700°C and 2000 bars in the presence of quartz, *Am. J. Sci., 272*, 234–272, 1972.

Piotrowski, J. M., and A. D. Edgar, Melting relations of undersaturated alkaline rocks from South Greenland, *Medd. om Grønland, 181*, no. 9, 62 pp., 1970.

Roedder, E., Liquid CO_2 inclusions in olivine-bearing nodules and phenocrysts from basalts, *Am. Mineral., 50*, 1746–1782, 1965.

Roux, J., and D. L. Hamilton, Primary igneous analcite: An experimental study, *J. Petrol., 17*, 244–257, 1976.

Scarfe, C. M., W. C. Luth, and O. F. Tuttle, An experimental study bearing on the absence of leucite in plutonic rocks, *Am. Mineral., 51*, 726–735, 1966.

Schairer, J. F., The alkali–feldspar join in the system $NaAlSiO_4$–$KAlSiO_4$–SiO_2, *J. Geol. 58*, 512–517, 1950.

Schairer, J. F. and Bowen, N. L., Preliminary report on equilibrium relations between feldspathoids, alkali-feldspars, and silica, *Trans. Am. Geophys. Union 16th Ann. Meeting*, Nat. Res. Council, Washington, D.C., 1935.

Shand, S. J., The alkaline rocks of the Transport Line, Pretoria District, *Geol. Soc. S. Africa Trans., 25*, 85, 1922.

Sood, M. K., and A. D. Edgar, Melting relations of undersaturated alkaline rocks from the Ilímaussaq intrusion and Grønnedal-Ika complex, South Greenland, *Medd. om Grønland, 181*, no. 12, 41p., 1970.

Stephenson, D., Alkali clinopyroxenes from nepheline syenites of the South Qôroq Centre, South Greenland, *Lithos, 5*, 187–201, 1972.

————, The South Qôroq Centre nepheline syenites, South Greenland, *Grønlands Geol. Undersøg. Bull., 118*, 1976.

Taylor, D., and W. S. MacKenzie, A contribution to the pseudoleucite problem, *Contr. Mineral. Petrol., 49*, 321–333, 1975.

Thompson, R. N., One atmosphere melting behaviour and nomenclature of terrestrial lavas, *Contr. Mineral. Petrol., 41*, 197–204, 1973.

Tilley, C. E., The gabbro-limestone contact of Camas Mor, Muck, Inverness-shire, *Geol. Soc. Finlande Comptes Rendus, 20*, 97–105, 1947.

————, Nepheline parageneses, *Sir Douglas Mawson Anniversary Volume*, University of Adelaide, 167–177, 1952.

————, Nepheline–alkali feldspar parageneses, *Am. J. Sci., 252*, 65–75, 1954.

————, Nepheline associations, *Kon. Ned. Geol. Mijnb., Goel. Ser., Brouwer Vol.*, 403–413, 1956.

————, The leucite nepheline dolerite of Meiches, Vogelsberg, Hessen, *Am. Mineral., 43*, 758–761, 1958.

Tilley, C. E., and H. F. Harwood, The dolerite-chalk contact of Scawt Hill, *Mineral Mag., 22*, 439–468, 1931.

Tilley, C. E., and R. N. Thompson, Melting relations of some ultra alkali volcanics, *Geol. Jour., 8*, 65–70, 1972.

Tuttle, O. F., and J. Gittins, *Carbonatites*, Wiley-Interscience, New York, 591p., 1966.

Tuttle, O. F., and J. V. Smith, The nepheline–kalsilite system II: Phase relations, *Am. J. Sci., 256*, 571–589, 1958.

Washington, H. S., Chemical analyses of igneous rocks, *U.S. Geol. Survey, Prof. Paper 99*, 1201 pp., 1917.

Watkinson, D. H., and P. J. Wyllie, Phase equilibrium studies bearing on the limestone-assimilation hypothesis, *Geol. Soc. Amer. Bull.*, *80*, 1565–1576, 1969.

Williams, L. A. J., The volcanics of the Gregory Rift Valley, East Africa, *Bull. Volcanol.*, *34*, 439–465, 1970.

————, The Kenya Rift volcanics: a note on volumes and chemical composition. In Girdler, R. W. (ed.), East African Rifts, *Tectonophysics*, *15*, 83–96, 1972.

Chapter 13

MELILITE-BEARING ROCKS AND
RELATED LAMPROPHYRES

H. S. YODER, JR.

Geophysical Laboratory, Carnegie Institution of Washington, Washington, D.C.

The lamprophyres were described by Bowen as a broad, ill-defined group of dike rocks, characteristically porphyritic. The phenocrystic crystals are highly femic (olivine, hornblende, mica), whereas the groundmass is alkalic. The richness of ferromagnesian and alkalic constituents is distinctive. Because of his belief that there was a strong tendency for idiomorphic crystals or porphyritic texture, or both, and a lack of well defined tachylitic, spherulitic, or aphanitic borders of the same composition as the main mass of the dike, he did not regard these rocks as ever having existed in an entirely liquid condition. He objected, therefore, to their origin by the remelting of accumulations of early formed crystals as proposed by Beger (1923). Bowen had little more to say in general about lamprophyres in his chapter entitled "Lamprophyres and Related Rocks."

The major mineral assemblages characteristic of the principal mesocratic to melanocratic dike rocks called lamprophyres are as follows:

hornblende + plagioclase	= spessartite
biotite + plagioclase	= kersantite
hornblende + orthoclase	= vogesite
biotite + orthoclase	= minette
barkevikite[1] + augite + plagioclase	= camptonite
barkevikite[1] + augite + analcite + glass	= monchiquite

The Evolution of the Igneous Rocks: Fiftieth Anniversary Perspectives
0-691-08223-5/79/0391-21$01.05/0 (cloth)
0-691-08224-3/79/0391-21$01.05/0 (paperback)
For copying information, see copyright page

There is almost continuous variation between the assemblages. The terminology is highly debated because many lamprophyres are intensely altered and the primary assemblage is difficult to identify. Much of the difficulty stems from the lack of definitive criteria that distinguish the lamprophyres from their counterparts that have the same principal mineralogy and are closely associated with syenitic or dioritic sources.

The first four assemblages, as identified by the names assigned, are almost indistinguishable in chemical composition on the average (Metais and Chayes, 1963, pp. 156–157), but the mineralogical differences suggest that not all the assemblages are isochemical. Vogesite may be related to an augite kersantite for the most part by means of the following simplified reaction:

$$NaCa_2Mg_4Al(Si_6Al_2)O_{22}(OH)_2 + KAlSi_3O_8 + SiO_2 \rightleftharpoons$$
$$\text{pargasite} \qquad\qquad \text{orthoclase quartz}^2$$
VOGESITE

$$KMg_3AlSi_3O_{10}(OH)_2 + (CaAl_2Si_2O_8 + NaAlSi_3O_8)_{ss} + CaMgSi_2O_6.$$
$$\text{phlogopite} \qquad\qquad \text{plagioclase} \qquad\qquad \text{diopside}$$
AUGITE KERSANTITE

The latter assemblage would probably be on the high-temperature side of the reaction. It is noteworthy that the reaction would presumably run under vapor-absent conditions.

Most varieties of lamprophyre contain a feldspar. Olivine, usually pseudomorphed or altered, is alleged to be common to all lamprophyres, although it is not always considered to be an essential phase, except in the monchiquites and perhaps the rare alnoites (biotite + melilite + augite + olivine), and is only rarely found in the matrix. All lamprophyres contain hydrous phases,[3] yet the chemical compositions of the rocks on an anhydrous basis are not very different from those of some alkali-rich lavas (e.g., nepheline basalt, leucite basalt) and plutonic rocks (e.g., shonkinite, teschenite); other lamprophyres are equivalent to granodiorite or even

---

[1] The amphibole in both camptonite and monchiquite is more closely related to kaersutite, according to a recent study by N.M.S Rock (personal communication, 1977).

[2] Natural pargasite is stable with quartz, but the synthetic end member is not stable in laboratory experiments (Boyd, 1956). Quartz may be eliminated from the reaction if tremolite is added to pargasite in the ratio 1:4; only a change in the ratio of the products results.

[3] Knopf (1936) included rocks completely lacking in hydrous phases; however, Métais and Chayes (1964, p. 197) concluded from a thorough review of all published analyses of lamprophyres that a mica, an amphibole, or both are major phases.

granitic compositions. Bowen chose, however, to direct his attention to the rarer olivine- and melilite-bearing varieties, the alnoites, as well as to those alkali-rich rocks containing other lime-rich minerals such as monticellite, garnet, and vesuvianite. A considerable amount of data has been collected on those feldspar-free rocks since 1928, and his interpretations will be examined in the light of the new observations. The lack of discussion of the feldspar-bearing varieties in no way reflects on their importance; however, they are beyond the scope of this commentary on Bowen's chapter.

The principal questions, in Bowen's view, concerning the lamprophyres as a group, and specifically the alnoites with which he was most familiar, were (1) whether there had ever been liquids corresponding approximately in bulk composition to the final products or (2) whether there had been interaction of crystals, accumulated from a complex liquid, with a later, probably different,[4] alkaline, highly mobile liquid that permeated the cumulate and served as the vehicle for intrusion. His observations (1922) on the alnoites from Isle Cadieux, Quebec, and those of Ross (1926) on alnoites from Winnett, Fergus County, Montana, were believed by Bowen to support the latter concept. Ross in fact had come to a different conclusion and stated that "secondary magma introduced subsequent to the initial crystallization had little part in the process." He thought that "hydrothermal solutions carrying mineralizers (essentially water) [H_2O, Cl, F, and SO_3] and alkalies were the controlling factor in the development of the second generation of minerals. . . ." Whereas Bowen considered large amounts of idiomorphic crystals as evidence of accumulation, it is difficult to reconcile the idiomorphic character with a reaction process that would tend to produce corroded crystals. Some observers (Hatch, Wells, and Wells, 1973, p. 421), on the other hand, noted the "strongly corroded and highly altered state of the phenocrysts" that supports Bowen's view of the phenocrysts reacting with their host liquid. Reports of aphanitic borders (e.g., Velde, 1967, p. 215, Fig. 1), spherulites, and glassy varieties (e.g., the lamproites of Niggli, 1923, p. 105) tend to reduce the force of his argument that lamprophyres were never in an entirely liquid condition. Experimental observations on the join diopside–

---

[4] Bowen (1928, p. 259) was not explicit about the relationship of the "liquid with a more complex composition" to the "liquid which permeated and replaced this largely crystalline mass." In his 1922 paper, however, he had referred (p. 12) to "an alkalic liquid (magma) which was, in part at least, its [the nearly consolidated mass of olivine and augite] own interstitial liquid. . . ." He implied from his analysis of the experimental systems presented, and considered it probable in his summary (p. 32), that the olivine and augite were attacked "by their own interstitial liquid" on cooling. In addition, he supposed "that there was a movement of liquid of the same kind through the interstices of the mass. . . ."

nepheline, however, were viewed by Bowen as strong confirmation of the effect he observed in the rocks, of an alkalic liquid, especially a nepheline-rich liquid, on olivine and augite.

THE DIOPSIDE–NEPHELINE JOIN

A revision of the diopside–nepheline join (Bowen, 1922) given by Schairer, Yagi, and Yoder (1962) is reproduced in Fig. 13-1. The system is not binary and should be considered a join in the quinary system Na_2O–CaO–MgO–Al_2O_3–SiO_2. From their physicochemical behavior, electron microprobe analyses, and calculated modes (Yoder and Kushiro, 1972, p. 415), all crystalline phases appear to be solid solutions. The join was found to be

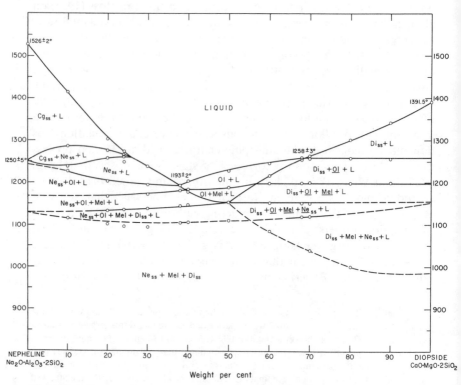

Figure 13-1. Temperature-composition diagram for the nepheline–diopside system at 1 atm. From Schairer, Yagi, and Yoder, 1962, p. 97, Fig. 29.

of such complexity that over 500 runs were made in an effort to determine the stability regions of the various mineral solid solutions. The subsolidus relations were not established by Bowen because of difficulties in the determination of minute crystalline phases under the microscope and without the aid of X-ray techniques. Even with the advances in techniques forty years later, it was not possible to fix the solidus of the system with assurance because of the difficulty of recognizing small amounts of glass in the quench products or, for some compositions, the presence or absence of small amounts of olivine. Crystal growth is sluggish, and equilibrium could not be established with certainty.

Bowen could not have been more perceptive in choosing the most important join in the five-component system to deduce the phase relations. Its critical position in a wide range of nephelinitic and basaltic compositions will become evident. The new results of Schairer, Yagi, and Yoder (1962) reaffirmed Bowen's observation that melilite and olivine separate from liquids whose total composition can be expressed as a mixture of nepheline and diopside. He recognized that the liquid in equilibrium with melilite and olivine was necessarily more siliceous than the compositions in the nepheline–diopside join, and he emphasized the low temperatures at which liquid existed. He did not, however, call the reader's attention to the fact that the principal assemblage of the common nephelinites, nepheline + augite, did not appear by itself, although in another context, (1922, pp. 27–28) he did suggest that rocks consisting of "augitic pyroxene with some nephelite, could not have formed from liquids of their own composition but must have formed by accumulation of their crystals from much more salic liquids (nephelite syenite, etc.)." As will be seen below, some but not all of these observations and suggestions have since been demonstrated in a quantitative way.

The behavior of any particular mixture in the nepheline–diopside system may be deduced from the diagram (Fig. 13-1), which is merely a summary of the many experiments. For example, a liquid of the composition $Ne_{50}Di_{50}$ begins to crystallize at 1228°C with the separation of forsterite. This phase continues to crystallize until a temperature of 1188°C is reached, where melilite also begins to crystallize. The norms of the subsequently analyzed liquids in equilibrium with melilite and olivine are given in Table 13-1. At 1152°C[5] the mixture also begins crystallizing both a diopside solid solution and a nepheline solid solution. The remaining liquid reacts with olivine over a range of approximately 40°C, a consequence of the five-component nature of the system. Both the olivine and

---

[5] A more accurate temperature of 1140° ± 5°C is obtained from more favorable compositions near the O1 + Mel + Di_{ss} + Ne_{ss} "invariant" point (see Figs. 13-3 and 13-4).

Table 13-1. Norms of analyzed glasses from runs on $Ne_{50}Di_{50}$
quenched from various temperatures (Yoder
and Kushiro, 1972, p. 414, Table 28)

Norm	Temperature, °C			
	1185	1180	1160	1125
Ne	50.6	49.0	49.9	40.9
Ab	6.3	5.3	13.8	24.7
Di	38.7	40.6	32.4	27.6
Wo	4.5	4.3	4.6	—
Fo	—	—	—	9.7
Ns*	0.8	1.0	0.1	6.2
	100.9	100.2	100.8	100.1

* Only the value for 1125°C may be significant, because
errors in the determination of Na probably account for the
small amounts of Ns at other temperatures. On the other hand,
the hyperalkaline character of the liquid could be a fundamental
feature of the system.

the liquid are consumed at 1108°C, and the final assemblage is Mel +
Ne_{ss} + Di_{ss}.

The compositions of the four crystalline phases at 1125°C in the
$Ne_{50}Di_{50}$ composition, according to Yoder and Kushiro (1972, p. 415),
are:[6]

$$Mel: \quad Ak_{61}Sm_{38}Geh_1$$
$$Ol: \quad Fo_{97}Mo_3$$
$$Di_{ss}: \quad Di_{92}Jd_4CaTs_4$$
$$Ne_{ss}: \quad Ne_{71}Ab_{29} \text{ (estimated)}[7]$$

The Mel is close in composition to the common melilite in igneous rocks
and to the maximum solid solution observed along the Ak–Sm join
(El Goresy and Yoder, 1974, pp. 359, 367).

The normative composition of the liquid at 1125°C is given in Table
13-1. It is concluded from the increase in Ab in the norms of the glasses

[6] Ab = Albite, $NaAlSi_3O_8$ Geh = Gehlenite, $Ca_2Al_2SiO_7$
Ak = Akermanite, $Ca_2MgSi_2O_7$ Jd = Jadeite, $NaAlSi_2O_6$
CaTs = Ca-Tschermak's molecule, $CaAl_2SiO_6$ Mo = Monticellite, $CaMgSiO_4$
Di = Diopside, $CaMgSi_2O_6$ Ne = Nepheline, $NaAlSiO_4$
Fo = Forsterite, Mg_2SiO_4 Sm = Sodium melilite, $NaCaAlSi_2O_7$

[7] Nepheline composition was derived from the bulk composition after the other analyzed
phases were accounted for. The calculated mode is in accord with visual estimates of the
mode of grain mounts in oil.

with decreasing temperature that the extraction of melilite and olivine from liquids in the nepheline–diopside join does indeed lead to a concentration of salic components in the liquid as suggested by Bowen. The concentration of albite in the liquid was expressed by Yoder and Kushiro (1972, p. 413) by the following equation:

$$3 \text{ diopside} + 2 \text{ nepheline} = (\text{akermanite} + \text{soda melilite})_{ss} +$$
$$\text{forsterite} + \text{albite}.$$

The course of the liquids can be described to a first approximation in the pseudoquaternary system forsterite–larnite–nepheline–quartz.

THE SYSTEM Fo–Ln–Ne–Qz

From the phase relations then known for nepheline–diopside, Bowen deduced the critical relations in the relevant portion of Ne–CaO–MgO–SiO_2, the liquidus relations of the oxide base having been determined by Ferguson and Merwin (1919). Bowen did not attempt to map out the phase volumes within the tetrahedron because that "would be a labor of many years." Much of that labor has now been carried out, mainly by Schairer and Yoder (see Yoder, 1976, Chapter 7, for summary), and the pertinent relations were found to lie in the subtetrahedron Ab–Di–Ne–Wo (Fig. 13-2). The three-phase (two crystalline phases + liquid) surfaces are shown schematically in Fig. 13-3. Within that subtetrahedron lie the "invariant" points pertaining to the olivine- and melilite-bearing rocks of interest.

The relations may be visualized with the aid of projections from Ne to the planes Ab–Di–Wo and Ab–Di–Fo of those surfaces, boundary curves, and "invariant" points *saturated with nepheline solid solution* (Fig. 13-4). The data used in constructing the projections are from the investigated systems listed in Table 13-2. The subtetrahedron is only an approximation of the five-component system, and the projection is not wholly accurate because of the extensive solid solution, for example, of Ab in nepheline. The extensive penetration of the melilite primary phase volume into the Ab–Di–Ne–Wo subtetrahedron is evident.[8] An appreciation is thereby gained of the close association of melilite-bearing rocks and plagioclase-bearing rocks even though the two distinguishing phases are incompatible

[8] There is an analogous penetration of a melilite primary phase volume in the Ne–CaO–Al_2O_3–SiO_2 system where a field of gehlenite solid solution plunges through An–Ne–Wo into Ab–An–Ne–Wo (Bowen, 1945, p. 85, Fig. 8).

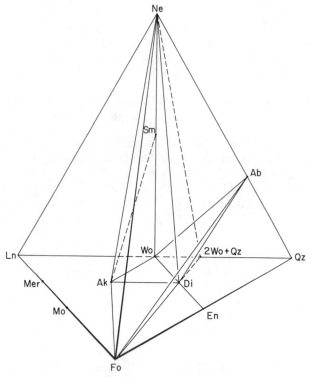

Figure 13-2. A portion of the Ne–CaO–MgO–SiO₂ system of Bowen (1928), Fo–Ln–Ne–
        Qz, illustrating the location of the key subtetrahedron Ab–Di–Ne–Wo and
        the plane Fo–Ne–(2Wo + Qz). Mole percent. Ab, albite; Ak, akermanite; Di,
        diopside; En, enstatite; Fo, forsterite; Ln, larnite; Mer, merwinite; Mo,
        monticellite; Ne, nepheline; Qz, quartz; Sm, sodium melilite; Wo, wollastonite.
        From Yoder and Velde, 1976, p. 581, Fig. 41.

in the same lava.[9] The displacement of the diopside solid solution primary
phase volume toward more salic compositions, as predicted by Bowen,
is also confirmed.

At the corners of the Di_{ss} + Ne + L surfaces are the four "invariant"
points described by Schairer and Yoder (1964, p. 72). These "invariant"
points were displayed in a flow sheet (Schairer, 1942, 1954) without regard
to spatial arrangement. The arrows on the "univariant" curves, indicating

---

[9] Melilite and plagioclase do coexist in some hybrid contact aureoles; however, Schairer
and Yoder (1970, p. 214) explained their incompatibility in igneous rocks as due to the
presence of olivine. See Ak–An–Fo (Yang, Salmon, and Foster, 1972, p. 167) for an illustration of the incompatibility of melilite and plagioclase in the presence of olivine.

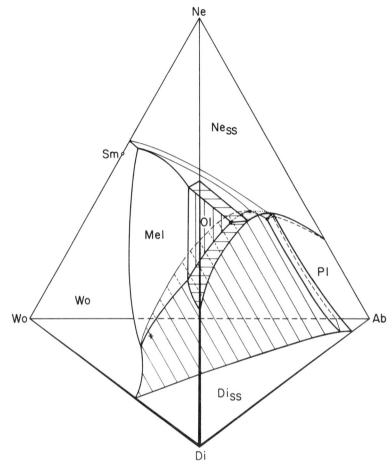

Figure 13-3. A schematic drawing of the liquidus surfaces in the subtetrahedron Ab–Di–Ne–Wo based on the determined equilibria (Table 13-2). The cross marks the penetration of the join Ab–Ak through the Di–Ne–Wo plane; the relations in Ab–Ak–Ne were used in the construction.

the direction of decreasing temperature and change of liquid composition, have not been determined with certainty, especially for $Di_{ss} + Fo + Ne_{ss} + L$, $Fo + Ne_{ss} + Pl + L$, and $Di_{ss} + Ne_{ss} + Pl + L$, The reason for this uncertainty lies in the five-component nature of the system and the lack of knowledge of solid solutions within it. For example, it appears from Fig. 13-4 that C, the "invariant" point $Di_{ss} + Fo + Mel + Ne_{ss} + L$, could be a reaction point of the distributary type and not a simple reaction

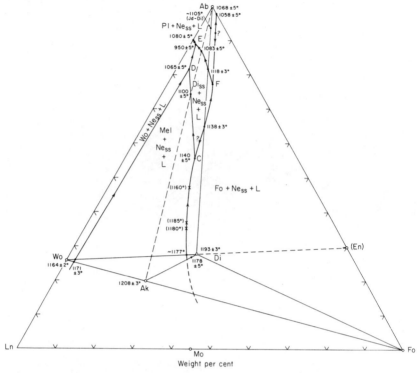

Figure 13-4. Projection of liquidus surfaces saturated with nepheline solid solution from Ne to the planes Ab–Di–Wo, Ab–Di–Fo, Ak–Di–Wo, and Ak–Di–Fo. Spinel field neglected. Crosses are estimates of composition of liquids in equilibrium with with Mel + Fo only (i.e., no Ne_{ss}) based on data in Table 13-1. (The Mel + Fo + L surface is subparallel to the lines of projection, and the determined composition points therein are believed to be suitable for defining the trace of its intersection with the field of Ne_{ss}.) Diagram deduced from phase equilibrium determinations listed in Table 13-2.

point.[10] The "invariant" point F may or may not be a eutectic, depending on the solid solutions in nepheline, diopside, and albite. A temperature maximum probably occurs along the $Pl + Fo + Ne_{ss} + L$ curve mainly because of the formation of plagioclase (see Jd–Di system, Table 13-2).

[10] The "invariant" point C is believed by Schairer and Yoder (1964, p. 69) to be a simple reaction point, on the assumption that a thermal maximum exists on the "univariant" curve $Di_{ss} + Fo_{ss} + Ne_{ss} + L$. The temperature of C, 1140° ± 5°C, and the piercing point temperature of the "invariant" curve through the Ab–Di–Ne plane, 1138° ± 3°C, are within the error of measurement, and the nature of the "invariant" point is therefore open to ques-

Table 13-2. Systems used to construct the projection from
nepheline in Fig. 13-3. Component end-member
minerals* are listed alphabetically.

System	References
Ab–Ak–Ne	Schairer and Yoder (1964, p. 70, Fig. 6)
Ab–Di–Ne	Schairer and Yoder (1960a, p. 278, Fig. 2)
Ab–Fo–Ne	Schairer and Yoder (1961, p. 142, Fig. 35)
Ab–Ne–Wo	Foster (1942, p. 165, Fig. 6)
Ak–Di–Ne	Schairer and Yoder (1964, p. 66, Fig. 2); Onuma and Yagi (1967, p. 238, Fig. 4)
Ak–Fo–Ne	Estimated (two-component joins done)
Ak–Ne–Wo	Schairer and Yoder (1964, p. 67, Fig. 3)
Di–Fo–Ne	Schairer and Yoder (1960b, p. 70, Fig. 18)
Di–Ne–Wo	Schairer and Yoder (1964, p. 71, Fig. 7)
Di–Jd†	Schairer and Yoder (1960a, p. 278, Fig. 4)
Di–Ne	Schairer, Yagi, and Yoder (1962, p. 97, Fig. 29)

*Ab $= NaAlSi_3O_8$ Jd $= NaAlSi_2O_6$
Ak $= Ca_2MgSi_2O_7$ Ne $= NaAlSiO_4$
Di $= CaMgSi_2O_6$ Wo $= CaSiO_3$
Fo $= Mg_2SiO_4$
† Di–Jd cuts the $Ne_{ss} + Pl + L$ curve at about 6 wt % Di.

The discovery of Ns in the norm of the liquid in equilibrium with Di_{ss}, Fo, Mel, and Ne_{ss} at 1125°C (Table 13-1) adds further complications to the simplified representation of the five-component system.

The projection in Fig. 13-4 clearly illustrates the surface $Di_{ss} + Ne_{ss} + L$ from which nephelinites could form, but from salic liquid compositions displaced considerably from Di–Ne, as predicted by Bowen.[11] It is

tion. In the light of the five-component nature of the system and the orientation in the Fo–Ln–Ne–Qz tetrahedron of the Di_{ss}–Fo_{ss}–Ne_{ss} plane, when maximum solid solutions of CaTs in Di, Ab in Ne, and Mo in Fo are taken into consideration, the cutting of $Di_{ss} + Fo_{ss} + Ne_{ss} + L$ by that plane, and hence the existence of a thermal maximum, is debatable.

[11] Bailey (1974; 1976, p. 452) gave the impression that "Olivine-free nephelinites (without melilites) cannot be fitted to the experimental model at all." In addition, the assemblage representing the olivine nephelinites is regarded by Bailey "only as a device to fit the proposed model of crystallization." The assemblage $Di_{ss} + Ne + Fo + L$ is to be found in the plane Ab–Di–Ne, as Bailey noted, and there is no reason known to the writer that requires the assemblage to appear in the Di–Ne–Fo plane itself as Bailey seems to suggest. The formation of the olivine nephelinites would indeed occur with equilibrium crystallization along the four-phase curve connecting the "invariant" points C and F (Figs. 13-4 and 13-5) from compositions greatly displaced from Di–Ne–Fo. During perfect fractionation, most liquids along the four-phase curve C–F would, however, yield a nephelinite and not an olivine nephelinite. The experimental observation of the appropriate assemblage is considered adequate proof that "olivine-nephelinite" is not just a device to fulfill an "experimental model."

Table 13-3. Systems studied containing the join nepheline-diopside*

System	References
Ab–Ne–Di	Schairer and Yoder (1960a, p. 278, Fig. 2)
Ab–Sa–Ne–Di	estimated by Sood, Platt, and Edgar (1970, p. 386, Fig. 5)
Ab–Wo–Ne–Di	Onuma and Yamamoto (1976, p. 351, Fig. 2)[†]
Ac–Ne–Di	Yagi (1963, p. 134, Fig. 50)
Ak–Ne–Di	{ Schairer and Yoder (1964, p. 66, Fig. 2) / Onuma and Yagi (1967, p. 238, Fig. 4)
An–Ne–Di	Schairer, Tilley, and Brown (1968, p. 469, Fig. 69)
Ap–Ne–Di	Kogarko et al. (1977, p. 33, Fig. 5)
$CaTiAl_2O_6$–Ne–Di	Yagi and Onuma (1969, p. 520, Fig. 4)
Fo–Ne–Di	Schairer and Yoder (1960b, p. 70, Fig. 18)
Lc–Ne–Di	Gupta and Lidiak (1973, p. 233, Fig. 1)
Sa–Ne–Di	Platt and Edgar (1972, p. 227, Fig. 2)
Wo–Ne–Di	Schairer and Yoder (1964, p. 71, Fig. 7)

* Abbreviations as in Table 13-2 and $Ac = NaFe^{3+}Si_2O_6$, $An = CaAl_2Si_2O_8$, $Ap = Ca_5(PO_4)_3F$, $Lc = KAlSi_2O_6$, $Sa = KAlSi_3O_8$.

[†] Data presented in their Table 2 are not consistent with the behavior predicted from Fig. 13-4. See also Ab–Ak–Ne of Schairer and Yoder (1964, p. 70, Fig. 6).

important to note that of all the systems studied (Table 13-3) in which Ne and Di are components (Ab–Ne–Di, Ac–Ne–Di, Ak–Ne–Di, An–Ne–Di, Ap–Ne–Di, $CaTiAl_2O_6$–Ne–Di, Fo–Ne–Di, Lc–Ne–Di, Sa–Ne–Di, and Wo–Ne–Di) only Ab–Ne–Di and Sa–Ne–Di exhibit a common boundary curve between Ne_{ss} and Di_{ss} on the liquidus.[12] Further evidence for contiguous fields of Ne_{ss} and Di_{ss} is to be found in the section Ab–Ak–Ne. Nephelinitic lavas, therefore, need not have formed solely by crystal accumulation, and their not uncommon aphanitic character attests to this deduction. The feldspar-free nephelinites and ankaratrites are in fact richer in SiO_2 and normative An than the melilite-bearing rocks, as would be anticipated from Fig. 13-4 (Velde and Yoder, 1976, p. 578). The nephelinites (as well as the urtites, ijolites, melteigites, and fasinites that also consist predominantly of a clinopyroxene and nepheline) appear to be normal differentiation products of alkali-rich magmas.

The relatively close relationship between the "invariant" points D and E (see Figs. 13-4 and 13-5), involving melilite and plagioclase, respectively, is consistent with the rare intermingling of melilite-bearing and plagioclase-bearing lavas from a single volcanic edifice (e.g., Birunga, Zaire; Geurie Hill, New South Wales, Australia; Alban Hills, Italy).

[12] The section through Ak–Di–Lc–Ne given by Gupta et al. (1973, p. 216, Fig. 6d), purporting to show the contiguous fields of Ne_{ss} and Di_{ss}, appears to have been incorrectly drawn. The field of Ne_{ss} appears on the join $Ne_{100-x_4}Ak_{x_4}$–Lc_{100}, not on the join $Ne_{100-x_4}Ak_{x_4}$–Di_{100}. The fields of Lc_{ss} and Ak_{ss} separate the fields of Ne_{ss} and Di_{ss}.

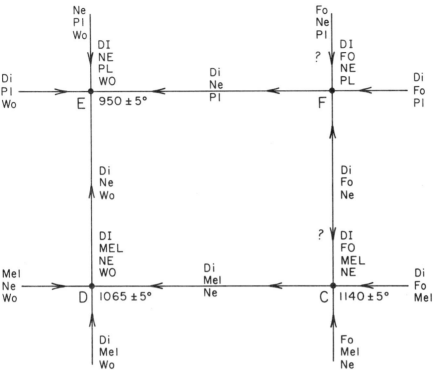

Figure 13-5. Flow sheet for a portion of the system Ab–Fo–Ln–Qz after Schairer and Yoder (1964, p. 72, Fig. 8). The quaternary form of the flow sheet is considered a close approximation of the five-component system. Large dots are "invariant" points. Arrows designate direction of decreasing temperature on "univariant" boundary curves. Solid solutions and high-temperature polymorphs are not indicated. The spatial arrangement is schematic and patterned after Fig. 13-3.

Evidence for reaction relations of melilite with liquid, indicated at "invariant" points D and possibly C, has been searched for in over 300 thin sections of melilite-bearing rocks (Yoder and Velde, 1976, p. 585) but not found. The reason such evidence has not been found in the multicomponent natural rocks is not known.

The appearance of melilite as phenocrysts or groundmass seems to depend in part on the alkali content of the magma (Velde and Yoder, 1976, p. 577). Those melilite-bearing lavas with $(Na_2O + K_2O) > 7.25\%$ have melilite as phenocrysts, whereas those with total alkalies $< 7.25\%$ have melilite only as a groundmass phase. Experimental studies of natural melilite-bearing rocks support these conclusions; however, Yoder (1973, p. 159) noted a concomitant increase in iron enrichment (FeO + Fe_2O_3/

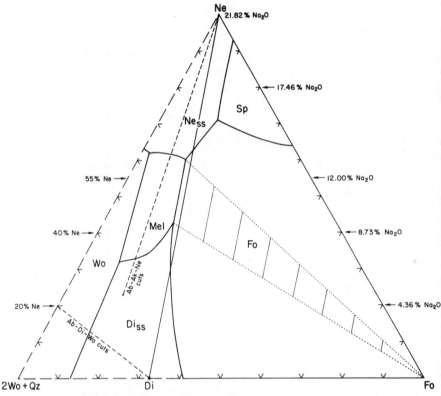

Figure 13-6. Schematic planar extension of the determined system Ne–Di–Fo (Schairer and
Yoder, 1960b, p. 70, Fig. 18) to include more siliceous compositions using the
data in the bounding planes Ab–Ne–Wo and Ab–Di–Wo as well as the
Ab–Ak–Ne section. Weight percent. High-temperature polymorphs are
neglected. The dotted lines outline the maximum region of compositions of
parental materials with relatively high soda content that will yield melilite on
crystallization. From Yoder and Velde, 1976, p. 585, Fig. 44.

$MgO + FeO + Fe_2O_3$) with alkalies. In a study of Ne–Di–Ac, Yagi (1963)
showed that melilite appears on the liquidus surface with the addition of
only about 12 wt % Ac. It would appear that high iron enrichment, resulting mainly from a decrease in MgO (Velde and Yoder, 1976, p. 575),
as well as high alkali content may result in phenocrystic melilite.

An extension of the plane Ne–Di–Fo to include 2Wo + Qz, illustrated
in Fig. 13-2, can be used to indicate the effect of alkalies on the magma.
The deduced phase relations, based on the data listed in Table 13-2, are
presented in Fig. 13-6. Olivine-bearing magmas in the ruled regions

having relatively high alkalies (amounts indicated on right-hand border) would be expected to yield melilite phenocrysts. Somewhat lower alkali content in the magma may yield melilite after augite, perhaps rimming the latter as is not uncommonly observed. Those magmas with still lower alkali content would not be expected, in general, to precipitate melilite, but would more likely yield plagioclase. Again, as Bowen predicted, magma with relatively high alkali content is evidently required to produce melilite-bearing rocks.

OCCURRENCE OF MONTICELLITE IN ALNOITE

The field of monticellite was also believed by Bowen to move toward more salic compositions in the presence of nepheline. He noted the appearance on the Ak–Fo join of a liquidus field for monticellite and presumed that it expanded almost to the Di–Ne–Wo plane. Observations on the Ak–Di–Ne system (Schairer and Yoder, 1964; Onuma and Yagi, 1967) indicate a field for only one olivine, forsterite; therefore, the expansion of the monticellite field, if any, is limited.

The monticellite in the alnoites studied by Bowen may not have resulted from precipitation from liquid, but in fact may be a metamorphic product. Walter (1963) and Yoder (1968) found that akermanite and forsterite react in the solid state at relatively low temperatures (865°C, 1 atm; 1095°C, 10.6 kbar) to form diopside + monticellite. The reaction is clearly displayed at Hendricksplaats, eastern Bushveld, Republic of South Africa, where intergrowths of augite and monticellite formed between olivine and melilite in carbonate blocks (Yoder, 1973, p. 152, Fig. 9). Akermanite itself was found by Yoder (1968, p. 473) to break down to wollastonite + monticellite at somewhat lower temperatures (680°C, 1 atm; 750°C, 6.3 kbar). In the monticellite alnoites of Isle Cadieux, Quebec, the monticellite occurs only in the groundmass as a "replacement mineral" or a "pseudomorph" and then only where augite is present (Bowen, 1922, p. 13). Monticellite includes corroded remnants of forsterite and augite, forms rims about forsterite, and may be interposed between biotite and augite. Similar relations are observed in the monticellite olivine nephelinite of Shannon Tier (Tilley, 1928), where the melilite is presumed to have been consumed in the reaction. On the basis of these experimental and petrographic observations, the monticellite would appear to be a metamorphic product.

In view of the liquidus phase relations in the system Ak–Di–Fo determined by Ferguson and Merwin (1919) and the phase analysis of Tilley and Yoder (1968, p. 458), it would be anticipated, on the other hand, that some monticellite alnoites (Ak + Fo + Mo + Ne) would form by direct crystallization from liquid. The monticellite alnoites (no clinopyroxene) of Southerland Commonage, Cape Province, Republic of South Africa (Taljaard, 1936, p. 295; Tilley and Yoder, 1968, p. 459), and of Haystack Butte, Highwood Mountains, Montana (Buie, 1941), appear to be high-temperature assemblages that crystallized directly from a liquid.

ORIGIN OF MELILITE-BEARING MAGMAS

The presence of a lime-rich mineral such as melilite in a magma was not considered by Bowen to indicate the existence of a magma particularly rich in lime, but rather of an alkali-rich magma reacting with olivine and augite. He opposed Daly's (1910) view that the lime-rich minerals resulted from the contamination of magma with lime-rich rocks such as limestone. Daly (1918, p. 131), however, did not miss Bowen's admission in 1915 (p. 89) "that some melilite rocks are formed by the crystallization of syntectic magma formed by the solution of limestone. . . ." Bowen noted the absence of limestone in some areas (e.g., Winnett, Montana) and quoted Ross's statement "that an alnoitic rock can develop without the assimilation of limestone and in the absence of an excessively large lime content." In the Daly Volume, Bowen (1945) was willing to acknowledge at least one example of the effect of limestone on a doleritic magma as furnishing "a complete and final demonstration of the truth of Daly's hypothesis. . . ."

Where limestone or dolomite beds were absent, Daly (1910, p. 108) appealed to the action of CO_2 itself on subalkaline magma to yield fractions high in alkalies. The presence of CO_2 in simple systems has been shown in general to produce melts with less silica (Eggler, 1973, p. 467), and larnite-normative liquids[13] result from the partial melting of peridotite with CO_2 at high pressures (Eggler, 1974; Boettcher, Mysen, and Modreski, 1975). It is probably one of the most important factors in the generation *at high pressures* of some magmas that would form melilite-bearing rocks at relatively lower pressures. Because of the low solubility of CO_2 in silicate liquids at low pressures, its influence, however, on liquid compositions at low pressures would be negligible.

---

[13] It should be noted that normative larnite is not an infallible indicator of modal melilite. Chayes and Yoder (1971, p. 206) pointed out that in feldspar-free lavas the norm of a clinopyroxene rich in Tschermak's molecule will include Ln. Some 60 out of 214 analyses examined of nonmelilite-bearing, feldspar-free ankaratrites and nephelinites contain normative Ln (Velde and Yoder, 1976, p. 578).

Akermanite is stable in the presence of excess CO_2 at pressures less than about 6 kbar within a limited temperature range (Yoder, 1975, p. 887), reacting to form diopside and calcite at lower temperatures, still within magmatic limits. The field of stability of akermanite would be expected to expand at reduced partial pressure of CO_2 because the diopside + calcite reaction would take place at still lower temperatures. In short, the addition of lime itself to a magma may produce compositions that yield melilite and consequently high melting temperatures, whereas the addition of limestone may desilicate the feldspathic constituents and form feldspathoids. The contained CO_2 may prevent the formation of melilite, except under volcanic or subvolcanic conditions (e.g., Tilley and Harwood, 1931; Sabine, 1975), because of its reaction with CO_2 to form diopside + calcite.

The stability of akermanite is limited to pressures less than 14 kbar, but solid solution with sodium melilite increases the pressure limit of stability considerably (Kushiro, 1964). On the other hand, melilite-bearing rocks were found to transform at relatively low pressures (<10 kbar) into clinopyroxenites holding one or more of the phases olivine, hornblende, biotite, magnetite, and perovskite as accessories (Tilley and Yoder, 1968). In the presence of excess water there is a prominent development of hornblende along with clinopyroxene. It is apparent that the melilite-bearing rocks are confined strictly to volcanic and subvolcanic facies represented at depth by pyroxenites or by the hornblende-bearing or biotite-bearing variants of pyroxenites. These observations led Tilley and Yoder (1968, p. 460) to the view that partial melting of an olivine pyroxenite at depth would yield an alkalic liquid, and the resulting crystal mush, holding relict clinopyroxene from the high-pressure environment, would consolidate as the alnoite assemblage at higher, subvolcanic levels in the crust. The occurrence of xenoliths having high-pressure assemblages in melilite nephelinites (Jackson and Wright, 1970) supports the concept of a deep-seated origin of the host magma where melilite itself is not stable. This view, involving the pressure effect, is in keeping with Bowen's general conclusion, which emphasized the action of an alkalic liquid on femic minerals as the principal cause of the formation of olivine-bearing lamprophyres.

An alternative origin for melilite-bearing rocks relates to the remelting of hornblende-bearing rocks, some of which are larnite-normative, and a discussion of that process is presented in the following chapter.

Acknowledgments The critical reviews of Drs. D. K. Bailey, F. Chayes, D. H. Eggler, B. O. Mysen, D. Presnall, N. M. S. Rock, and D. Velde are greatly appreciated. The writer is especially indebted to Dr. T. Neil Irvine, who suggested the Ne projection of Fig. 13-4 for clarity in presenting the complex relations.

References

Bailey, D. K., Nephelinites and ijolites, in *The Alkaline Rocks*, H. Sørensen, ed., John Wiley and Sons, New York, pp. 53–66, 1974.
———, Applications of experiments to alkaline rocks, in *The Evolution of the Crystalline Rocks*, D. K. Bailey and R. Macdonald, eds., Academic Press, New York, pp. 419–469, 1976.
Beger, P. J., Der Chemismus der Lamprophyre, in *Gesteins und Mineralprovinzen*, by P. Niggli, Vol. 1, Part 6, Gebrüder Borntraeger, Berlin, pp. 217–586, 1923.
Boettcher, A. L., B. O. Mysen, and P. J. Modreski, Melting in the mantle: Phase relationships in natural and synthetic peridotite–H_2O and peridotite–H_2O–CO_2 systems at high pressure, *Phys. Chem. Earth*, 9, 855–867, 1975.
Bowen, N. L., The later stages of the evolution of the igneous rocks, *J. Geol.*, 23, Suppl., 91 pp., 1915.
———, Genetic features of alnoitic rocks at Isle Cadieux, Quebec, *Am. J. Sci.*, 3, 1–34, 1922.
———, *The Evolution of the igneous Rocks*, Princeton University Press, Princeton, New Jersey, 332 pp., 1928.
———, Phase equilibrium bearing on the origin and differentiation of alkaline rocks, *Am. J. Sci., Daly Vol.*, 243A, 75–89, 1945.
Boyd, F. R., Amphiboles, *Carnegie Institution of Washington Yearbook*, 55, 198–200, 1956.
Buie, B. F., Igneous rocks of the Highwood Mountains, Montana, *Geol. Soc. Am. Bull.*, 52, 1753–1808, 1941.
Chayes, F., and H. S. Yoder, Jr., Some anomalies in the norms of extremely undersaturated lavas, *Carnegie Institution of Washington Yearbook*, 70, 205–206, 1971.
Daly, R. A., Origin of the alkaline rocks, *Geol. Soc. Am. Bull.*, 21, 87–118, 1910.
———, Genesis of the alkaline rocks, *J. Geol.*, 26, 97–134, 1918.
Eggler, D. H., Role of CO_2 in melting processes in the mantle, *Carnegie Institution of Washington Yearbook*, 72, 457–467, 1973.
———, Effect of CO_2 on the melting of peridotite, *Carnegie Institution of Washington Yearbook*, 73, 215–224, 1974.
El Goresy, A., and H. S. Yoder, Jr., Natural and synthetic melilite compositions, *Carnegie Institution of Washington Yearbook*, 73, 359–371, 1974.
Ferguson, J. B., and H. E. Merwin, The ternary system $CaO–MgO–SiO_2$, *Am. J. Sci.*, 48, 81–123, 1919.
Foster, W. R., The system $NaAlSi_3O_8–CaSiO_3–NaAlSiO_4$, *J. Geol.*, 50, 152–173, 1942.
Gupta, A. K., and E. G. Lidiak, The system diopside–nepheline–leucite, *Contrib. Mineral. Petrol.*, 41, 231–239, 1973.

Gupta, A. K., G. P. Venkateswaran, E. G. Lidiak, and A. D. Edgar, The system diopside–nepheline–akermanite–leucite and its bearing on the genesis of alkali-rich mafic and ultramafic volcanic rocks, *J. Geol.*, *81*, 209–218, 1973.

Hatch, F. H., A. K. Wells, and M. K. Wells, *Petrology of the Igneous Rocks*, 13th ed., Hafner Press, New York, 551 pp., 1973.

Jackson, E. D., and T. L. Wright, Xenoliths in the Honolulu volcanic series, Hawaii, *J. Petrol.*, *11*, 405–430, 1970.

Knopf, A., 1936, Igneous geology of the Spanish peaks region, Colorado, *Geol. Soc. Am. Bull.*, *47*, 1727–1784, 1936.

Kogarko, L. N., L. D. Krigman, Ye. N. Petrova, and I. D. Solovova, Phase equilibria in the fluorapatite–nepheline–diopside system and the origin of the Khibiny apatite deposits, *Geochem. Int.*, *14*, 27–38, 1977.

Kushiro, I., The join akermanite–soda melilite at 20 kilobars, *Carnegie Institution of Washington Yearbook*, *63*, 90–92, 1964.

Métais, D., and F. Chayes, Varieties of lamprophyre, *Carnegie Institution of Washington Yearbook*, *62*, 156–157, 1963.

Métais, D., and F. Chayes, Kersantites and vogesites; a possible example of group heteromorphism, *Carnegie Institution of Washington Yearbook*, *63*, 196–199, 1964.

Niggli, P., *Gesteins- und Mineralprovinzen*, Vol. 1, Gebrüder Borntraeger, Berlin, 586p., 1923.

Onuma, K., and K. Yagi, The diopside–akermanite–nepheline system, *Am. Mineral.*, *53*, 227–243, 1967.

Onuma, K., and M. Yamamoto, Crystallization in the silica-undersaturated portion of the system diopside–nepheline–akermanite–silica and its bearing on the formation of melilites and nephelinites, *J. Fac. Sci., Hokkaido Univ., Ser. 4*, *17*, 347–355, 1976.

Platt, R. G., and A. D. Edgar, The system nepheline–diopside–sanidine and its significance to the genesis of melilite- and olivine-bearing alkaline rocks, *J. Geol.*, *80*, 224–236, 1972.

Ross, C. S., Nephelinite–hauynite alnoite from Winnett, Montana, *Am. J. Sci.*, *11*, 218–227, 1926.

Sabine, P. A., Metamorphic processes at high temperature and low pressure: The petrogenesis of the metasomatized and assimilated rocks of Carneal, Co. Antrim, *Philos. Trans. R. Soc. London*, *280*, 225–269, 1975.

Schairer, J. F., The system $CaO–FeO–Al_2O_3–SiO_2$: I. Results of quenching experiments on five joins, *J. Am. Ceram. Soc.*, *25*, 241–274, 1942.

———, The system $K_2O–MgO–Al_2O_3–SiO_2$: I. Results of quenching experiments on four joins in the tetrahedron cordierite–forsterite–leucite–silica and on the join cordierite–mullite–potash feldspar, *J. Am. Ceram. Soc.*, *37*, 501–533, 1954.

Schairer, J. F., C. E. Tilley, and G. M. Brown, The join nepheline–diopside–anorthite and its relation to alkali basalt fractionation, *Carnegie Institution of Washington Yearbook*, *66*, 467–471, 1968.

Schairer, J. F., K. Yagi, and H. S. Yoder, Jr., The system nepheline–diopside, *Carnegie Institution of Washington Yearbook*, *61*, 96–98, 1962.

Schairer, J. F., and H. S. Yoder, Jr., The nature of residual liquids from crystallization, with data on the system nepheline–diopside–silica, *Am. J. Sci.*, *Bradley Vol.*, *258A*, 273–283, 1960a.

Schairer, J. F., and H. S. Yoder, Jr., The system forsterite–nepheline–diopside, *Carnegie Institution of Washington Yearbook*, *59*, 70–71, 1960b.

Schairer, J. F., and H. S. Yoder, Jr., Crystallization in the system nepheline–forsterite–silica at one atmosphere pressure, *Carnegie Institution of Washington Yearbook*, *60*, 141–144, 1961.

Schairer, J. F., and H. S. Yoder, Jr., Crystal and liquid trends in simplified alkali basalts, *Carnegie Institution of Washington Yearbook*, *63*, 65–74, 1964.

Schairer, J. F., and H. S. Yoder, Jr., Critical planes and flow sheet for a portion of the system $CaO–MgO–Al_2O_3–SiO_2$ having petrological applications, *Carnegie Institution of Washington Yearbook*, *68*, 202–214, 1970.

Sood, M. K., R. G. Platt, and A. D. Edgar, Phase relations in portions of the system diopside–nepheline–kalsilite–silica and their importance in the genesis of alkaline rocks, *Can. Mineral.*, *10*, 380–394, 1970.

Taljaard, M. S., South African melilite basalts and their relations, *Trans. Geol. Soc. S. Afr.*, *39*, 281–316, 1936.

Tilley, C. E., A monticellite nepheline basalt from Tasmania: A correction to mineral data, *Geol. Mag.*, *65*, 29–30, 1928.

Tilley, C. E., and H. F. Harwood, The dolerite-chalk contact of Scawt Hill, Co. Antrim. The production of basic alkali-rocks by the assimilation of limestone by basaltic magma, *Mineral. Mag.*, *22*, 439–468, 1931.

Tilley, C. E., and H. S. Yoder, Jr., The pyroxenite facies conversion of volcanic and subvolcanic, melilite-bearing and other alkali ultramafic assemblages, *Carnegie Institution of Washington Yearbook.*, *66*, 457–460, 1968.

Velde, D., Sur un lamprophyre hyperalcalin potassique: la minette de Sisco (île de Corse), *Bull. Soc. Fr. Minéral. Cristallogr.*, *90*, 214–223, 1967.

Velde, D., and H. S. Yoder, Jr., The chemical compositions of melilite-bearing eruptive rocks, *Carnegie Institution of Washington Yearbook*, *75*, 574–580, 1976.

Walter, L. S., Experimental study on Bowen's decarbonation series: I. *P–T* univariant equilibria of the "monticellite" and "akermanite" reactions, *Am. J. Sci.*, *261*, 488–500, 1963.

Yagi, K., Liquidus data on the system acmite–nepheline–diopside at 1 atmosphere, *Carnegie Institution of Washington Yearbook*, *62*, 133–134, 1963.

Yagi, K., and K. Onuma, An experimental study on the role of titanium in alkalic basalts in light of the system diopside–akermanite–nepheline–$CaTiAl_2O_6$, *Am. J. Sci.*, *Schairer Vol.*, *267A*, 509–549, 1969.

Yang, H.-Y., J. F. Salmon, and W. R. Foster, Phase equilibria of the join akermanite–anorthite–forsterite in the system $CaO–MgO–Al_2O_3–SiO_2$ at atmospheric pressure, *Am. J. Sci.*, *272*, 161–188, 1972.

Yoder, H. S., Jr., Akermanite and related melilite-bearing assemblages, *Carnegie Institution of Washington Yearbook.*, *66*, 471–477, 1968.

———, Melilite stability and paragenesis, *Fortschr. Mineral.*, *50*, 140–173, 1973.

———, Relationship of melilite-bearing rocks to kimberlite: a preliminary report on the system akermanite–CO_2, *Phys. Chem. Earth*, *9*, 883–894, 1975.

———, *Generation of Basaltic Magma*, National Academy of Sciences, Washington, D.C., 265 pp., 1976.

Yoder, H. S., Jr., and I. Kushiro, Composition of residual liquids in the nepheline–diopside system, *Carnegie Institution of Washington Yearbook*, *71*, 413–416, 1972.

Yoder, H. S., Jr., and D. Velde, Importance of alkali content of magma yielding melilite-bearing rocks, *Carnegie Institution of Washington Yearbook*, *75*, 580–585, 1976.

Chapter 14

THE FRACTIONAL RESORPTION OF COMPLEX MINERALS AND THE FORMATION OF STRONGLY FEMIC ALKALINE ROCKS

DAVID R. WONES[1]

U.S. Geological Survey, Reston, Virginia

The origin of the alkaline mafic rocks remains one of petrology's most intriguing puzzles. Modern explanations include partial melting of ordinary mantle at high pressures, assimilation of crustal material, accumulation of volatiles within the mantle and accompanying metasomatism, or some combination of these processes. Most alkaline mafic rocks, including lamprophyres, contain phenocrysts of olivine, plagioclase, diopside, hornblende, and phlogopite, either singly or in combination. Bowen (1928) concluded from this fact that lamprophyric compositions could not exist as pure liquids and proposed that the fractional resorption of complex minerals such as biotite and hornblende by basaltic magma could produce the lamprophyres and other alkaline mafic rocks that are noted for their relatively high mafic content, and silica undersaturation.

Olivine and plagioclase phenocrysts within the lamprophyres are nearly always present as euhedra, whereas grains of diopside, biotite, and hornblende occur both as euhedra and subhedra. Liquids that will resorb olivine and plagioclase must be at a higher temperature than those liquids with which these minerals are in equilibrium at constant pressure and activity of other components within the melt. Bowen demonstrated clearly that it is unreasonable to expect these minerals, of simple composition, to be resorbed when added to a melt already saturated with them. The same argument holds for most pyroxenes. Alkaline pyroxenes, hornblende, and biotite contain so many components that the resorption of these minerals by a magma may increase the liquidus temperature, lower the solidus temperature, and affect the path of crystallization. The mafic phenocrysts and the alkaline groundmass of the lamprophyric rocks are plausibly explained by the assimilation of the alkaline mafic minerals by a basaltic magma.

Bowen, using carefully selected analyses of hornblendites, given here in Table 14-1A, showed that the assimilation of hornblende can be expressed by a reaction in which femic components appear as phenocrysts of olivine and pyroxene, whereas the feldspathic and feldspathoidal components are found within the melt. Helz (1973, 1976) partially melted three basalts of disparate compositions in the presence of an H_2O-rich phase at 5 kbar total pressure. Her results are given in Table 14-1B, where the starting compositions of the three basalts, the compositions of the hornblendes, and the compositions of the melts coexisting with the hornblendes are given for two contrasting temperatures. Although Bowen's prediction is partly correct, it is important to note that, for all of these compositions, the hornblendes coexist with quartz normative liquids. Helz's data confirm that, with increasing temperature, the compositional increments going into the melt by the reaction of hornblende are feldspathic and feldspathoidal.

Cawthorn (1976) summarized the compositions of amphiboles coexisting with a variety of natural lavas by projecting normative compositions of hornblendes and lavas onto the system Ne–Pl(=Ab + An)–Ol(=Fo + Fa)–Qz–Di. Helz's data are plotted in Fig. 14-1, a variant of Cawthorn's diagram, Ne–Qz–Ol(=Fo + Fa)–An. This choice was made to visualize the Ab–An variations better, and because Cawthorn's summary demonstrated that the diopside content of the amphiboles was almost invariant. In this plot, albite and hypersthene components are recalculated as nepheline + quartz and olivine + quartz, respectively. The diagram shows that, with increasing temperature, the amphibole reacts to add nepheline and plagioclase components to the melt. In Helz's experiments, olivine and clinopyroxene also occur on the liquidus at high temperatures. The An component of the hornblende places limits on the composition of melts formed by resorption of the hornblende, as further resorption will add An to the melt, not diopside or larnite.

Because the hornblendes in Helz's and Cawthorn's studies coexist with SiO_2-rich melts, Bowen's proposed reaction can take place only if alkaline

---

[1] Present address: Department of Geological Sciences, Virgina Polytechnic Institute and State University, Blacksburg, Virgina.

Table 14-1. Normative compositions of hornblendes; basalts used by Helz (1973, 1976) in melting experiments and coexisting hornblendes and melts in those experiments

A. *Normative Composition of Hornblendites** [After Bowen, 1928]*

	Ab	Or	Ne	Di	Le	Cs	An	Hy	Ol	Il
1	19.39	2.78		26.88			21.13	9.30	9.47	
2	19.18	7.23	10.79	26.57	3.92				8.05	
3	3.14	10.01	7.10	14.36			26.41		18.12	
4				8.52	25.52	4.36	6.19	23.91	16.75	
5			6.53		3.49	14.62	30.58		27.29	
6	7.07	6.67	8.09	51.94			7.51		11.56	
7	11.00	1.11	4.26	31.04			15.01		28.76	
8	2.62	7.23		47.92			18.35		9.25	
9	2.62	8.34	3.98	26.60			18.36		14.70	
10	5.24	2.22		14.39			20.29	22.65	25.29	

B. *Helz Experiments***

	Ab	Or	Ne	Di	Le	Qz	An	Hy	Ol	Il
PG	26.74	4.55		15.44		0.73	23.05	20.90		3.23
700Hb	8.44	3.66	4.73	38.93			11.32		28.91	1.96
700 L	31.3	21.9				35.3	6.0	2.1		0.1
1000Hb		0.88	12.47	31.52	3.06		17.41		25.69	7.56
1000 L	36.40	6.5		0.8		12.8	31.7	8.3		2.10
K 1921	16.67	2.90		21.96			24.67	19.77	7.30	4.77
700Hb	12.27	2.78		40.41			11.31	3.93	24.72	2.34
700 L	20.30	20.7				43.8	8.9	1.7		0.2
1000Hb	1.26	2.95	9.86	25.69			24.60		25.01	8.11
1000 L	25.40	5.3				20.6	37.5	7.6		2.2
H 1801	19.65	5.73	3.02	18.42			24.44		21.17	4.58
725Hb	8.77	3.90	7.95	34.79			11.98		26.88	2.79
725 L	22.10	14.2				45.2	12.7	2.0		0.1
1000Hb		0.38	11.73	30.95	3.55		17.87		26.36	6.91
1000 L	31.30	5.8				8.1	32.8	17.7		0.9

* Washington's tables, 1917. Magnetite, ilmenite, apatite, etc., omitted.
**Helz (1973, 1976). Magnetite, apatite, corundum omitted.
PG, Picture Gorge basalts; 700, 700°C; Hb, hornblende; L, melt (glass); K 1921, 1921 basalt of Kilauea; H 1801, 1801 basalt of Hualalai.

hornblendes are separated from their original basaltic system and concentrated where they can be resorbed by another basaltic magma. Bowen inferred that this would occur in a magma chamber in which hornblende crystallizing from a cooler H_2O-rich upper zone would sink through the melt to a hotter, deeper zone, where resorption would take place. This model, if applied to nepheline-rich magmas, requires that some region of the deep crust or mantle is enriched in hornblende. Some more current

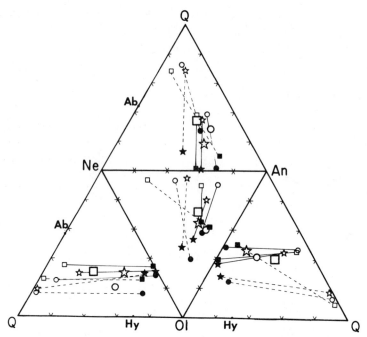

Figure 14-1. Melting experiments on basalts from Picture Gorge (squares), Kilauea 1921 (circles), and Hualalai 1801 (stars) (Helz, 1973, 1976). Large open symbol, rock composition; small open sumbol, melt composition; solid symbol, amphibole composition. Compositions are projected into the $NaAlSiO_4(Ne)$–$CaAl_2Si_2O_8(An)$–$SiO_2(Qz)$–$(Fe,Mg)_2SiO_4(Ol)$ tetrahedron. Dashed connecting lines, 700°C (except for Hualalai 1801, 725°C); solid connecting lines, 1000°C. Pressure = 5 kbar and with the presence of an H_2O-rich gas.

models for the origin of nepheline normative rocks call for small amounts of partial melting at high pressures (Green, 1970). The normative silica content may be indicative of pressure, H_2O activity, or degree of partial melting. It would seem that Bowen's reaction hypothesis, although qualitatively correct, is inadequate for the nepheline normative rocks.

Biotites and amphiboles may directly affect the normative silica values in CIPW normative calculations. In lavas and hypabyssal rocks biotites and amphiboles are commonly oxidized through a process of dehydrogenation (Barnes, 1930; Ernst and Wai, 1970; Wones, 1963) in which the hydrogen leaves the mineral and electrostatic neutrality is maintained by oxidation of Fe^{2+} to Fe^{3+} *in situ*. In rocks where this process has taken place, the Fe component is calculated in the CIPW norm as magnetite or hematite rather than as ilmentite, fayalite, or ferrosilite. The calculated norm yields a higher normative quartz value for the particular rock and gives some bias to the relative silica saturation.

The Bowen model for the potassic mafic rocks remains a possible explanation for these peculiar rocks that are found, to date, only in continental regions. These rocks are rare, and the continental bias may merely reflect the lack of adequate exposure in oceanic regions. Luth (1967), Yoder and Kushiro (1969), and Modreski and Boettcher (1972, 1973) have examined phase equilibria within the system $KAlSiO_4$–Mg_2SiO_4–SiO_2–H_2O at liquidus temperatures appropriate to the earth's crust and mantle. Their studies demonstrate that SiO_2 activity, H_2O activity, pressure, and temperature are the intensive variables that affect the order of crystallization of mafic minerals from an alkaline-rich melt.

In Fig. 14-2 are plotted the compositions of potassium-rich mafic volcanic rocks from New South Wales, Australia (Cundari, 1973), southeastern Spain (Fuster et al., 1967) and Wyoming (Carmichael, 1967). The compositions that lie precisely along the join sanidine–forsterite are an artifact of the CIPW calculation conventions concerning SiO_2 assignments, which make nepheline rather than leucite the feldspathoidal mineral. Also plotted on this diagram are Helz's starting materials from Table 14-1B, representing quartz-normative, hypersthene-normative, and alkali-olivine basalts. Bowen's reaction model requires that these basalts assimilate phlogopite and precipitate forsterite to yield compositions similar to those shown here for the potassium-rich rocks. The compositions of the liquids determined by Modreski and Boettcher (1973) by microprobe analysis, and by Luth (1967) by phase disappearance, all contain less than 15 weight percent Fo. The magmas plotted in Fig. 14-2 are all rich in diopside component, but Modreski and Boettcher showed that this component has little effect on the Fo content of the liquid. Bravo and O'Hara (1975) took exception to Modreski and Boettcher's results and suggested that forsterite precipitated out during the quench of the experiments. The data for the Leucite Hills do plot near Luth's 1-kbar invariant point phlogopite–leucite–forsterite–melt–gas, but the other rocks have compositions whose projections are scattered over the region where forsterite is the primary liquidus phase. Edgar et al. (1976) do not present the composition of the glasses coexisting with olivine and phlogopite in their madupite (phlogopite-diopside ultramafic rock) experiments between 10 and 30 kbar because of ubiquitous quench crystals. They conclude from their experimental results that a source area anomalously rich in K_2O is required for these magmas.

Kalsilite-normative rocks, reported by Combe and Holmes (1944) from Uganda, project in and around phlogopite composition. This projection implies simple melting of that constituent, but those rocks yield normative

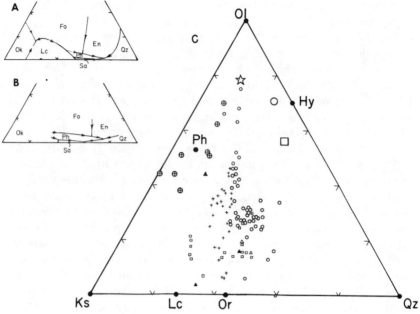

Figure 14-2.
A. Liquidus surface of the system $KAlSiO_4$–Mg_2SiO_4–SiO_2–H_2O projected onto the $KAlSiO_4$–Mg_2SiO_4–SiO_2 plane in the presence of an H_2O-rich gas at 1 kbar (after Luth, 1967).
B. Same as A except at 20 kbar (after Yoder, 1976).
C. Projection of compositions of basalts from Picture Gorge (large open square), Kilauea 1921 (large open circle), and Hualalai 1801 (large open star) onto the $KAlSiO_4$(Ks)–Mg_2SiO_4(Fo)–SiO_2(Qz) plane. Small squares (Carmichael, 1967) and solid triangles (Cross, 1897) are rock compositions from the Leucite Hills, Wyoming. Open triangles are from Smoky Butte, Montana (Velde, 1975). Crossed circles are rocks from Uganda (Combe and Holmes, 1944). Open circles are rocks from southeastern Spain (Fuster *et al.*, 1967). Crosses are rocks from New South Wales, Australia (Cundari, 1973).

compositions rich in diopside, magnetite, and ilmenite, three components not considered in this analysis. Cundari and LeMaitre (1970) have examined the chemical variations within the Roman and Birunga provinces and concluded that fractionation of a melt corresponding to biotite–pyroxenite can account for the variations observed in these localities. The amount of biotite and diopside needed for assimilation to produce such magma requires that these two constituents make up 60 to 80 percent of the magma and is more characteristic of partial melting than of assimilation.

Potassium-rich magmas can be produced by either assimilation or partial melting of potassium-rich mantle or crust by basaltic magmas.

Fractional crystallization or zone refining would yield liquids richer in $Fe/(Fe + Mg)$ and $Na/(Na + K)$ than the observed potassium-rich lavas, but the difference in liquidus and solidus temperatures between the carbonate and silicate melts (Eggler, 1976) makes the assimilation of biotite or hornblende by carbonatite magmas an unlikely process. Sahama (1974, p. 107) after examining the $Fe/(Fe + Mg)$ and $Na/(Na + K)$ values of these rocks as a function of SiO_2 content, concluded that they "can hardly be attributed to . . . one single process. . . ."

Partial melting of biotitic pyroxenite or the assimilation of biotite by basaltic magma require biotite-rich zones within the deep crust or mantle. Studies of inclusions from mantle-derived magmas and carbonatites prompted Aoki (1975), Gittins et al. (1975), and Boyd and Nixon (1975) to suggest that potassium metasomatism created biotite-rich zones within the mantle. Alternatively, magma columns leading to the formation of granitic batholiths would have biotite (Fig. 14-2), hornblende, pyroxene, ilmenite, zircon, and phosphatic minerals occurring on the liquidus. These minerals would sink to the base of the column and form mafic accumulates in the deep crust or upper mantle. Quinn's (1943) study of such accumulations in an 18 m thick dike of Westerly granite demonstrated that this process does take place. Accumulates from early melting events or metasomatism may concentrate K_2O in the mantle source region. Partial melting or assimilation may incorporate that material into a potassium-rich melt.

Isotopic tracers and trace-element patterns are useful aids in the resolution of these possible combinations of origins. Powell and Bell (1974) favor a mixed origin—partial melting of the mantle coupled with assimilation of some continental material—to explain the relatively high values of $^{87}Sr/^{86}Sr$ observed for the potassium-rich mafic rocks. In order to have relatively low values of this ratio, either a young source or one that is low in ^{87}Rb is necessary. Phlogopite concentrates Rb relative to hornblende (Beswick, 1976), and the nepheline-normative and sodium-rich rocks have lower $^{87}Sr/^{86}Sr$ values than their potassium-rich counterparts. Powell and Bell demonstrate that, within a given province, the $^{87}Sr/^{86}Sr$ values are not related to Rb contents in a simple way. Brooks et al. (1976) take the variations in $^{87}Sr/^{86}Sr$ within a province to indicate "mantle" isochrons beneath that province. They interpret the spread in values to indicate an approximate isochron that is the age of the source region. The potassium-rich rocks of Africa yield a Mesozoic "mantle isochron," those in North America one of Tertiary age, and those of Spain and Australia, a Precambrian age. Mantle-related processes of the appropriate age have been ascribed to each of the four areas, and may date the development of the K_2O-enriched region within the mantle.

The highly fractionated REE patterns (Kay and Gast, 1973) of the potassium-rich basalts and nephelinites have been interpreted as the result of small amounts of partial melting of garnet peridotite with chondritic or slightly enriched chondritic REE abundances. These patterns cannot be explained by simple fusion or assimilation of hornblende or biotite because these minerals are not enriched in the light REE. Harris (1974) has argued that zone refining would yield highly fractionated magmas. Such a process could alleviate the mechanical problems of consolidating a widely dispersed melt of low abundance. Zone refining, however, would lead to higher $Fe/(Fe + Mg)$ values than those observed in the potassium-rich mafic rocks, and would not explain the high TiO_2 and K_2O values. Simultaneous assimilation of ilmenite (or sphene), monazite (or apatite), allanite, and zircon along with the biotite or hornblende would explain the observed elemental abundances. These minerals could be present in biotite-enriched regions of the deep crust and mantle.

These alkaline mafic magmas, because of their high liquidus temperatures and inclusions that contain aluminous pyroxene and garnet, must ultimately have their origin within the mantle. Bowen, at least for potassium-rich magmas, may have been close to the correct solution in calling on the assimilation of biotite. The mechanism of concentrating biotite within the deep crust or mantle remains to be explained. There seems to be no need to call on hornblende-rich crust or mantle for the derivation of nephelinites or basanites, but water-rich lamprophyres might be explained by the assimilation of such material by basaltic magma.

Acknowledgments   I am grateful to Dallas Peck for his piercing questions and kind suggestions concerning this problem. Felix Chayes made his collection of rock analyses available to me and provided a listing of pertinent rock analyses. J. G. Arth, R. T. Helz, P. H. Wetlaufer, and H. S. Yoder, Jr., freely provided both education and criticism during reviews of the manuscript. Any errors of judgment or fact are my responsibility, not theirs.

References

Aoki, K., Origin of phlogopite and potassic richterite bearing peridotite xenoliths from South Africa. *Contri. Mineral. Petrol.*, *53*, 145–156, 1975.

Barnes, V. E., Changes in hornblende at about 900°C. *Am. Mineral.*, *15*, 393–417, 1930.

Beswick, A. E., K and Rb relations in basalts and other mantle derived materials. Is phlogopite the key? *Geochim. Cosmochim. Acta*, *40*, 1167–1183, 1976.

Bowen, N. L., *The Evolution of the Igneous Rocks*. Princeton University Press, Princeton, N.J., 332 pp., 1928.

Boyd, F. R., and Nixon P., Origins of the ultramafic nodules from some kimberlites of northern Lesotho and the Monastery mine, South Africa. *Phys. Chem. Earth, 9*, 431–454, 1975.

Bravo, M. S. and O'Hara, M. J., Partial melting of phlogopite-bearing synthetic spinel- and garnet-lherzolites. *Phys. Chem. Earth, 9*, 845–854, 1975.

Brooks, C., James, D. E., and Hart, S. R., Ancient lithosphere: Its role in young continental volcanism. *Science, 193*, 1086–1094, 1976.

Carmichael, I.S.E., The mineralogy and petrology of the volcanic rocks from the Leucite Hills, Wyoming. *Contr. Mineral. Petrol., 15*, 24–66, 1967.

Cawthorn, R. G., Some chemical controls on igneous amphibole compositions. *Geochim. Cosmochim. Acta, 40*, 1319–1328, 1976.

Combe, A. D., and Holmes, A., The kalsilite-bearing lavas of Kabirenge and Lyakauli, south-west Uganda. *Trans. Royal Society Edinburgh, 61*, 359–379, 1944.

Cross, W., Igneous rocks of the Leucite Hills and Pilot Butte, Wyoming. *Am. J. Sci., Ser. 4*, 115–151, 1897.

Cundari, A., Petrology of the leucite-bearing lavas in New South Wales, Australia. *J. Geol. Soc. Australia, 20*, 465–491, 1973.

Cundari, A., and LeMaitre, R. W., On the petrogeny of the leucite-bearing rocks of the Roman and Birunga volcanic regions. *J. Petrol., 11*, 33–48, 1970.

Edgar, A. D., Green, D. H., and Hibberson, W. O., Experimental petrology of a highly potassic magma. *J. Petrol., 17*, 339–356, 1976.

Eggler, D. H., Does CO_2 cause partial melting in the low-velocity layer of the mantle? *Geology, 4*, 69–72, 1976.

Ernst, W. G., and Wai, C. M., Mossbauer, infrared, X-ray and optical study of cation ordering and dehydrogenation in natural and heat-treated sodic amphiboles. *Am. Mineral., 55*, 1226–1258, 1970.

Fuster, J. M., Gastesi, P., Sagredo, J., and Fermoso, M. L., Las rocas lamproiticas del S.E. de Espana. *Estud. Geol., 23*, 35–69, 1967.

Gittins, J., Allen, C. R., and Copper, A. F., Phlogopitization of pyroxenite; its bearing on the composition of carbonatite magmas. *Geol. Mag., 112*, 503–507, 1975.

Green, D. H., The origin of basaltic and nephelintic magmas. *Trans. of Leicester Literary and Philosophical Society, 64*, 28–54, 1970.

Harris, P. G., Anatexis and other processes within the mantle. In *The Alkaline Rocks*, H. Sorensen, ed., John Wiley and Sons, N.Y., 427–436, 1974.

Helz, R. T., Phase relations of basalts in their melting range at $P_{H_2O} = 5$ kb as a function of oxygen fugacity: Part I. Mafic phases. *J. Petrol., 14*, 249–302, 1973.

———, Phase relations of basalts in their melting ranges at $P_{H_2O} = 5$ kb. Part II. Melt compositions. *J. Petrol., 17*, 139–193, 1976.

Kay, R. W., and Gast, P. W., The rare earth content and origin of alkali-rich basalts. *J. Geology, 81*, 653–682, 1973.

Luth, W. C., Studies in the system $KAlSiO_4$–Mg_2SiO_4–SiO_2–H_2O: I, Inferred phase relations and petrologic applications. *J. Petrol., 8*, 372–416, 1967.

Modreski, P. J., and Boettcher, A. L., The stability of phlogopite + enstatite at high pressures: A model for micas in the interior of the earth. *Am. J. Sci.*, *272*, 852–869, 1972.

Modreski, P. J., and Boettcher, A. L., Phase relationships of phlogopite in the system $K_2O–MgO–CaO–Al_2O_3–SiO_2–H_2O$ to 35 kilobars: a better model for micas in the interior of the earth. *Am. J. Sci.*, *273*, 385–515, 1973.

Powell, J. L. and Bell, K., Isotopic composition of strontium in alkalic rocks. In *The Alkaline Rocks*, H. Sorensen, ed., John Wiley and Sons, N.Y., 412–421, 1974.

Quinn, A. W., Settling of heavy minerals in a granodiorite dike at Bradford, Rhode Island. *Am. Mineral.*, *28*, 272–281, 1943.

Sahama, Th. G., Potassium-rich alkaline rocks. In *The Alkaline Rocks*, H. Sorensen, ed., John Wiley and Sons, N.Y., 96–109, 1974.

Velde, D., Armalcolite-Ti-phlogopite-analcite-bearing lamproites from Smoky Butte, Montana. *Am. Mineral.*, *60*, 566–573, 1975.

Washington, H. S., Chemical analyses of igneous rocks. *U.S. Geol. Surv. Prof. Paper*, *99*, 1201p., 1917.

Wones, D. R., Physical properties of synthetic biotites on the join phlogopite–annite. *Am. Mineral.*, *48*, 1300–1321, 1963.

Yoder, H. S., Jr., *Generation of Basaltic Magma*, Natl. Acad. Sci., Washington, D.C., 265p., 1976.

Yoder, H. S., Jr., and Kushiro, I., Melting of a hydrous phase: Phlogopite. *Am. J. Sci.*, *267-A*, 558–582, 1969.

Chapter 15

FURTHER EFFECTS OF FRACTIONAL RESORPTION

E. D. JACKSON*

U.S. Geological Survey, Menlo Park, California

The major role of fractional resorption in the formation of iron-rich alkaline rocks was discussed in the preceding chapter. In this chapter, Bowen turned his attention to the possible effects of this same process on two different, but then topical, igneous phenomena: oscillatory zoning of feldspar crystals, and concentrations of spinel, particularly chromite, in ultramafic rocks. His approach to both problems was the same. In the case of feldspar zoning he examined the effects of changing physical conditions on the binary continuous reaction system albite–anorthite. In the case of spinel formation he considered similar changes in the ternary discontinuous reaction series anorthite–forsterite–silica.

This chapter, like the one before it, reflected developments of the early 1900's, when precise ideas concerning the behavior of crystallizing systems were formulated. The possibility of magmatic resorption of crystals as an important equilibria process in natural magmas was raised by Vogt (1905), Brand (1911), and Shepherd and Rankin (1911). In reporting on the system $CaO–Al_2O_3–SiO_2$, Rankin and Wright (1915) described naturally-occurring mineral compounds that showed magmatic resorption on cooling. Bowen and Andersen (1914) demonstrated that forsterite crystals in the system $MgO–SiO_2$ were always partly, and sometimes completely, re-dissolved during cooling of melts in which forsterite was a primary phase. Work on the system anorthite–forsterite–silica (Andersen, 1915) showed that magmatic resorption of crystals of forsterite, anorthite, and spinel could be explained as simply a consequence of cooling of a complex liquid not dissimilar to natural magma. It was natural, then, that Bowen should extend the concept of equilibrium resorption to that of fractional resorption.

REVERSAL OF NORMAL ORDER OF ZONING

In 1913, Bowen had made use of his experimental data on the binary loop in the system albite–anorthite to confirm earlier suggestions by Day and Allen (1905) that simple normal zoning of feldspar could be caused by incomplete reactions between solid phenocrysts and the liquid in equilibrium with them. In reviewing the geological implications of this system, Bowen noted that in view of the quantitative importance of feldspars in igneous rocks, it was satisfying to find that they conformed to the laws of physical chemistry. He noted that reversals of the order of zoning in feldspars could be explained in the system albite–anorthite if: (1) undercooling of magma and release of latent heat by crystallization were cyclic processes; (2) simultaneous crystallization of other lime-bearing minerals with plagioclase occurred during cooling; or (3) movements of magma took place during crystal growth.

In 1926 Fenner had published a provocative discussion of oscillatory zoning of feldspars in which he pointed out that simple application of the feldspar phase diagram did not explain the complex and commonly oscillatory compositional variations observed in feldspar phenocrysts from Katmai. He proposed three mechanisms that might account for them: (1) changes in magma composition by external causes; (2) cycling of phenocrysts into calcic and sodic portions of magma chambers; and (3) changes in magma composition through periodic escape of volatiles from magma chambers. Bowen's (1913) paper was not cited in Fenner's discussion.

It was in this context that Bowen in 1928 considered the application of fractional resorption specifically to what Homma (1936) would later call normal-oscillatory normal zoning. In this chapter Bowen expanded on his (1913) suggestion that magma movement might cause reversals in feldspar zoning, which had already been amplified as Fenner's second mechanism. He replaced Fenner's terms "calcic" and "salic" with "hotter" and "cooler," and pointed out that during slow cooling of a large body of magma it was not unlikely that normally zoned feldspar crystals grown in the portion cooling from its upper surface, would, through crystal

---

* E. D. Jackson died on July 28, 1978, before his manuscript was edited. His colleague, M. H. Beeson, kindly reviewed the proofs.

settling, sink into hotter portions and be partially resorbed. If the lower, hotter magma were somewhat above the temperature of saturation of the sinking crystal, but saturated with a slightly more calcic phase, then continued growth would result in a reversal of the normal order of zoning. Continued cooling would result again in normal zoning, and thus crystal settling could account for one reversal. If the whole mass were carried up to a higher level by surging, a new period of cooling and sinking could result in a second reversal. Repeated sinking and surging would thus account for repeated reversals, whereas continued cooling and crystallization would result in the overall increasingly sodic trend of the zoned shells.

Fenner (1926) had noted that the ubiquity of oscillatory zoning in plagioclase phenocrysts in lavas, as well as the extreme range of compositions represented in these phenocrysts, argued for long suspension times of feldspars in magmatic liquids. He then observed (1926, p. 703), "The theory of crystallization by differentiation rests upon the ability of a magma to free itself of crystals. These phenomena of the feldspars raise a question as to whether, in the primary magma reservoirs, separation by gravity is always competent against the play of other forces." This statement offended Bowen's total commitment to crystallization-differentiation by crystal settling. He paraphrased Fenner as expressing the opinion that oscillatory zoning (Bowen, 1928, p. 275) "disproves crystallization-differentiation," and vigorously defended processes requiring crystal settling. Indeed, he proposed a second means of producing oscillatory zoning by repeated filtering of hotter liquids through settled crystal mushes, an idea that has received little subsequent notice or support.

It was implicit in Bowen's discussion that oscillatory zoning of igneous plagioclase could be a key to the record of physical and chemical conditions during intrusion, convection, and consolidation of magmas. After fifty years of subsequent research it is still not possible to read these inorganic "tree rings" in detail, but considerable progress has been made (see Maaløe, 1976). Most modern workers would agree that a number of igneous and metamorphic processes result in feldspar zonation, and the problem today is to relate specific patterns of zonation to particular genetic processes.

Most of the processes now thought to result in feldspar zoning had been suggested by 1928. Bowen (1913) demonstrated the basis for normal zoning in incomplete reaction of solid and liquid in the system albite–anorthite. His (1913) mention of magma movement as a cause of reversals of zoning was expanded on not only by Fenner (1926) and Bowen himself (1928) but was used by Homma (1932) and Carr (1954) in attempts to reconstruct convection patterns in magmas through oscillatory zoning patterns in feldspars.

Bowen's (1913) suggestion that some kinds of zoning might be related to simultaneous crystallization of other lime-bearing minerals was elaborated on by Wenk (1945), who considered the effects of co-precipitation of hornblende and biotite. It was later pointed out, however, that oscillatory zoning is common in rocks that contain no phases that compete for calcium or aluminum (Van der Kaaden, 1951; Vance, 1962). The effects of co-precipitation of plagioclase with pyroxene were best assessed by Wyllie (1963), who showed that both liquidus and solididus crystallization paths in the system diopside–anorthite–albite exhibited changes in slope. Projections of the three phase paths on the albite–anorthite binary system revealed both steeper slopes and shelves. The likelihood of oscillatory zoning would be enhanced along the flatter slopes, whereas changes in slope would produce unzoned plagioclase cores surrounded by other zoned rims.

Fenner's (1926) suggestion that reversals of zoning might be caused by periodic dilutions of crystallizing magma from external sources has not attracted much attention. This mechanism would be expected to lead to abrupt and irregular zonal patterns, which might well be looked for in areas of rapid eruption of feldspar-phyric basalts.

Fenner's (1926) third mechanism, which attributed oscillatory zoning to periodic escape of volatiles from magma chambers, received considerable subsequent attention. Phemister (1934) pointed out that zones formed by this process would be coarse and irregular, and would favor a more sodic plagioclase. Hills (1936) agreed that loss of volatiles from a crystallizing magma chamber would favor normal zoning of its plagioclase, but pointed out that increases in volatile pressure would favor reverse zonations. These relations were experimentally verified by Yoder, Stewart, and Smith (1956, 1957). Vance (1962) noted that corroded cores were characteristic of otherwise thin, normally zoned plagioclases of quartz diorites and granodiorites, and specifically suggested that periods of corrosion were caused by falling pressure and volatile saturation. Thus, he suggested, late-stage sodic zoning was due to discharge of a separate volatile phase (resurgent boiling).

By 1927 Harloff had suggested that the regular, closely-spaced type of oscillatory zoning resulted simply from a lack of balance between the rate of crystallization of plagioclase and the rate of diffusion in the liquid adjacent to plagioclase crystals, and that no external chemical or physical changes need be involved. This revolutionary idea, which implied that equilibrium was not maintained in the close vicinity of growing crystals, was not addressed by Bowen. The idea intrigued a number of later students of feldspar zonation, although nearly forty years would pass before direct observation of the phenomena was accomplished. Phemister (1934) suggested that diffusion-reaction mechanisms of this type should produce fine oscillatory zoning of weak amplitude. Hills (1936) modified the mechanism somewhat and proposed a diffusion-supersaturation model in which (1) diffusion of anorthitic liquid brings about saturation directly adjacent to the growing feldspar; (2) supersaturation is reached, a small normal zone is crystallized, and the adjoining melt is impoverished in anorthite content; (3) diffusion of anorthitic component fails to keep pace with growth, and crystallization slows or stops; and (4) the sequence is repeated. Vance (1962) accepted Hills' modification of Harloff's mechanism as most likely to produce delicate and regular oscillatory zoning in feldspar. Bottinga, Kudo, and Weill (1966) tested this mechanism by measuring concentration gradients in basaltic glass adjacent to plagioclase crystals with fine oscillatory zoning. They observed a depletion in aluminum content and an enrichment of the magnesium, iron, and silicon contents in the glass near the crystal-glass interface. Sodium and calcium contents of the glass near the crystal were neither enriched nor depleted, which the authors attributed to relatively greater mobility of these ions. They concluded that these experiments demonstrated that plagioclase zoning is diffusion-controlled, and to the extent, they confirmed Harloff's theory. In addition, they suggested that the common sharp increase in calcium content (Bowen's reversal of zoning) in fine, regular oscillatory zoning is formed in a growth cycle at times when diffusion is more rapid than crystal growth, and when supersaturation in anorthite content of liquid near a growing crystal is building up. After nucleation is triggered, a normal zone crystallizes rapidly, depleting the adjacent zone in plagioclase nutrients, and the process is repeated.

Lofgren (1974) has succeeded in experimentally reproducing discontinuous zoning, reverse zoning, and patchy zoning. In a series of experiments, the temperature of plagioclase melts was dropped in increments of $50°C$ at intervals of time sufficiently long for crystallization to proceed. Each stepwise temperature drop was marked by a sharp discontinuity in plagioclase composition, with each growth plateau successively more albitic as supersaturation increased. Within the growth intervals, however, plagioclase was either homogeneous or reversely zoned, never normal. Reverse zoning occurred in intervals even when temperature was held constant. Whereas the discontinuous steps toward more albitic compositions could be explained as the result of ordinary crystal fractionation and response to a temperature drop, it is more difficult to account for the reverse zoning. Lofgren demonstrated that calcium enrichment resulted from a complex function of the effective distribution coefficients between crystals and melt, the growth rate, the amount of supercooling, the composition of the melt, and the amount of water present. Patchy zoning appeared to be related to restricted diffusion in interior or marginal areas of growing crystals. Although unable to offer an explanation for failure to achieve continuous normal zoning, Lofgren returned full circle and noted that temperature drops producing progressively more albitic zones in natural systems could result from rapid movement of magma up volcanic feeder vents or from subsequent eruption.

LIMITS OF RESORPTION

Inquiring into the natural limits of resorption, Bowen pointed out that, given an intermediate but constant composition of basalt at its source, the maximum magnesium and calcium contents of natural aphanites provided some limits to the possible extent of resorption. He concluded that resorption of olivine could not result in a natural liquid that exceeded 10–15 percent normative olivine, nor could resorption of plagioclase result in a liquid with normative feldspar more calcic than about An_{66}. Whereas one could not necessarily appeal to the process of solution of "sunken" (accumulated) crystals to account for the composition of any particular liquid, the upper limits were thus set by the bounds of natural liquids. The reasoning is cogent, lucid, and leaves nothing further to be said. Bowen's lifelong denial of the existence of natural ultramafic liquids has, however, been challenged by a number of workers who have presented a considerable body of data in support of the occurrence of very mafic liquids (e.g., komatiite) in Archean time (Naldrett and Mason, 1968; Viljoen and Viljoen, 1969; Nesbitt, 1971; Pyke, Naldrett, and Eckstrand, 1973; Brooks and Hart, 1974). If liquids of komatiite composition did exist, then Bowen's limits for olivine resorption would increase to 50 percent normative olivine, although the normative anorthite content of these liquids remains about An_{60} (see Green et al., 1975).

LOCAL RESORPTIVE EFFECTS

Bowen next considered the stronger effects of the resorption process when reacting crystals constitute only a small proportion of a mass of magma. These conditions would result in the maintenance of temperature in the vicinity of settling crystals even though endothermic changes might be occurring in their immediate vicinity. The result of this process would be the growth of an earlier member of the same reaction series to which a settling crystal belonged. Inasmuch as the diffusivity of temperature is much greater than the diffusivity of mass, a corollary is that a chemical gradient should be maintained in the vicinity of the crystal. Bowen noted that this is the same effect called on earlier to explain the reversal of plagioclase zoning, but that more notable marginal changes should be visible if applied to less abundant species.

For reasons not clearly understood, feldspars much more commonly exhibit repeated reversals of zoning than other minerals of continuous reaction series, and to that extent Bowen's specific fractional resorption mechanism has not proved demonstrably important. His intuition about the importance of diffusion gradients in the neighborhood of crystals in liquids, however, was quite accurate. With the advent of the electron microprobe, Bottinga, Kudo, and Weill (1966) demonstrated that compositional concentration gradients exist in glass adjacent to plagioclase crystals in chilled basalt, and Lofgren (1974) reproduced these gradients experimentally. Anderson (1967) reported Mg-depleted compositional gradients near olivine in natural basaltic glass, and Kushiro (1974) observed similar gradients near forsterite crystals in synthetic systems. Donaldson (1975) calculated diffusion coefficients and growth rates from similar data. No doubt the perennial cycle of petrologic observation of natural materials, followed by experimental approximation of natural conditions, will soon lead to a much firmer understanding of the physical conditions of crystal growth.

FORMATION OF SPINEL IN ULTRABASIC ROCKS

Bowen contended that even greater effects of resorption would be evidenced if the crystals concerned were members of discontinuous rather than continuous reaction series. He used Andersen's (1915) study of the system anorthite–forsterite–silica as an example. If a large body of saturated magma on the anorthite-forsterite cotectic cools from its upper surface, the crystallization products will settle into warmer portions of the magma chamber at depth. At the appropriate higher temperature the liquid will react with forsterite and anorthite to form spinel. Of course, the spinel is a transient phase unless it is protected by some kind of corona or unless it is concentrated to such an extent as to exhaust local supplies of liquid prior to complete reversal of the reaction on cooling. Indeed, Bowen noted that such effects are not limited to liquids that lie on a forsterite-anorthite cotectic; any complex magma capable of precipitating magnesian olivine and calcic feldspar would yield early crystals of these minerals that would be transformed to spinel if they sank to hotter regions in the spinel stability field. As an example of this process, Bowen cited Barth's (1927) description of spinels spatially associated at olivine-plagioclase contacts in rocks that appeared to be crystal accumulates. Bowen did not remark on the fact that Barth's spinel resided in spinel-hypersthene symplectites rimming olivine and separating olivine from plagioclase. Nor did he discuss Barth's (1927) interpretation that the symplectites formed as result of the late-stage (metasomatic) crystallization of an aluminous magma in which the reaction

$$Mg_2SiO_4 + Al_2O_3 \leftrightarrows MgSiO_3 + MgAl_2O_4$$

occurred as crystallization progressed.

Shand (1945) doubted that such coronas were formed by magmatic processes, but suggested rather that they were due to later thermal metamorphism of solidified olivine- and plagioclase-bearing igneous rocks. Kushiro and Yoder (1966) demonstrated that in subsolidus assemblages anorthite and forsterite indeed react to form pyroxene and spinel under conditions of falling temperature or rising pressure. That is,

anorthite + forsterite $\rightleftarrows$ clinopyroxene$_{ss}$ + orthopyroxene$_{ss}$ + spinel.

Presnall (1966), on the other hand, contended that olivine and liquid can react to form pyroxene-spinel symplectites under certain magmatic conditions. Gardner and Robins (1974) believed that pyroxene-spinel coronas form in both metamorphic and magmatic environments; some coronas have formed subsequent to plastic deformation of primary olivine and plagioclase, whereas other symplectites have formed at pressures so low as to require solid-liquid reaction at the time of formation.

ORIGIN OF PICOTITE AND CHROMITE

Bowen noted that the spinels of ultramafic rocks generally contain large amounts of iron and chromium substituting for the magnesium and aluminum of pure end-member spinel. He found it difficult to believe, however, that a spinel as rich in chromium as chromite could form as a result of ordinary crystallization processes. He noted that Vogt's (1924) ideas on remelting of sunken crystals to form ultramafic liquids provided a rather simple mechanism for developing chromite through a refractory refining process, but he repeated his objections to the existence of ultramafic magmas: (1) the lack of field evidence for the very high temperatures required; and (2) the absence in nature of aphanites of ultramafic compositions. Bowen proposed instead that spinel is formed by the reaction of magmatic liquid with phenocrysts of olivine and calcic plagioclase in regions of locally elevated temperature. The reverse reaction would take place on falling temperature, but the olivine and plagioclase cannot accomodate Cr_2O_3, and a small amount of chromian spinellid would persist. As the process is repeated, the Cr_2O_3 content would be gradually increased. Chromite so formed would ultimately lose its chromium to clinopyroxene, except where crystal settling had resulted in concentrations of chromite, in which case the remaining liquid supply would be inadequate to effect the reaction. In this context Bowen added that basaltic liquid reacts with spinel to form only olivine plus basic plagioclase. Any Cr_2O_3 residing in spinel (1928, p. 281) "would of necessity be thrown out in some form" and the occurrence of chromite as small euhedral grains would make them likely candidates. Bowen noted similar textures in magnetite in many rocks and perovskite in alnoites, for which similar origins had been proposed.

Henderson and Suddaby (1971) estimated that the chrome spinels of the Rhum intrusion, in the Inner Hebrides, are enriched in Cr_2O_3 by a factor of about 800 with respect to the magma from which they crystallized. The estimated enrichment factor for the Stillwater complex, Montana, is about 1,000 (E. D. Jackson, unpublished data). Bowen's reluctance to accept that simple crystallization of chromite involved enrichments of this magnitude was realistic in 1928. Not until 1954, when Keith investigated the system $MgO-Cr_2O_3-SiO_2$, did it become apparent that a very small amount of Cr_2O_3 saturates silicate liquids with the spinel phase. Later investigations of chromium-bearing systems (Dickey and Yoder, 1972; Arculus, Gillberg, and Osborn, 1974; Muan, 1975; Navrotsky, 1975) confirm the very large primary phase areas of spinel in these systems, and the dominance of the chromite stability field even in melts containing as little as 0.1 percent Cr_2O_3.

Thayer (1946) observed that feldspar-free peridotites contain chromite rich in the magnesiochromite component, that pyroxene-rich layered intrusions contain iron-rich chromites, and that peridotites associated with gabbro contain chromite rich in normative spinel. He described the intrusion of brecciated, high-aluminum chromitites by gabbroic magmas in Cuba, and gave numerous examples of reaction textures in these rocks in which feldspar plated chromite, olivine rimmed feldspar, and enstatite and diopside lay midway between chromite grains. In 1946, Thayer believed these observations substantiated Bowen's hypothesis for the origin of chromite of various compositions. By 1956, however, Thayer minimized the importance of reaction-resorption in the origin of chromite, and emphasized the roles of primary crystallization and concentration by crystal settling. He stated (pp. 41–42) that "The two basic processes involved in the formation of the Bushveld class of chromite deposits probably were crystal settling and magmatic reaction . . . the basic principles of crystallization require that the first formed spinellids be relatively rich in $MgCr_2O_4$, and once formed, they will be the most stable."

Smith (1958) also pointed out the spatial proximity of chromitites and gabbros. He noted that the chromite deposits of the Bay of Islands complex, Newfoundland, occurred in ultramafic rocks, but were localized near gabbro contacts. He concluded that spinel was formed as a result of reaction between gabbroic and ultramafic magmas, and that the spinel was converted to chromite through diffusion of chromium through interstital liquids in the late stages of magmatic history. He implicitly accepted Bowen's (1928) mechanism.

The role of fractional resorption in the origin and composition of chromites, however, could not be formulated or documented in detail until the role of cumulus processes was understood. Wager and Deer (1939) identified "primary precipitate" and "interprecipitate" materials in the layered rocks of the Skaergaard intrusion, east Greenland, and discussed the interactions of deposited crystals and interstitial melt. In the same year Hess (1939) spoke of "settled crystals" and the "interstitial liquid" between these crystals in the layered rocks of the Stillwater complex. In addition, Hess pointed out that curves of the compositional variation of olivine, pyroxene, and plagioclase as a function of stratigraphic height in the Stillwater showed fluctuations that he correlated with rate of accumulation of crystals. Where accumulation had been relatively slow, diffusion between the main magma and the interstitial magma was effective, and overgrowths on the settled crystals were the same composition as the original crystals. Where the rate of accumulation was relatively rapid, however, there was not enough time for adequate diffusion, and the settled minerals reacted with the interstitial liquid to form somewhat more iron-rich ferromagnesian minerals and somewhat more sodic feldspars. This process, although controlled by diffusion rates rather than by rising and falling temperature, involved fractional resorption of the settled olivine, pyroxene, and plagioclase. The term "fractional resorption," but not the concept, was, however, out-of-date by 1939. In 1960 Hess elaborated on the diffusion process in cumulates, and showed that cumulus phases present in small proportions would have their original compositions more seriously affected than would abundant phases.

Jackson (1961) demonstrated that the chromites of the Stillwater complex were secondarily enlarged to about the same extent as the cumulus silicate phases, and, therefore, they had been subject to about the same amount of reaction with the intercumulus melt as had the olivine, pyroxene, and plagioclase. Irvine (1967) and Irvine and Smith (1967) discussed the reaction and resorption of chromites from the Muskox intrusion, Northwest Territories, both during settling and after deposition, and concluded that a reaction relation existed between chromite and pyroxene in that magmatic system. On the other hand, Cameron and Desborough (1969) demonstrated that orthopyroxene and chromite were cotectic crystallization products in the eastern Bushveld complex in South Africa, and concluded that considerably more than half the chromite of chromitite layers grew after deposition of their cumulus nuclei. They therefore found that the final equilibration of chromite and liquid was postdepositional. Cameron (1969) further clarified thinking on the problem of fractional resorption when he pointed out that three separate subsystems need to be considered in the formation of magmatic sediments:

1. The subsystem of nucleation and crystal growth prior to and during settling.
2. The subsystem between settled crystals and supernatant magma.
3. The subsystem between settled crystals and interstitial liquid.

It would appear that processes involving what Bowen would have called fractional resorption could occur, and in some cases have occurred, in all three of Cameron's subsystems, between minerals of both continuous and discontinuous reaction series.

As for Bowen's (1928) illustrative discontinuous reaction,

$$\text{basaltic liquid} + MgAl_2O_4 \rightleftarrows CaAl_2Si_2O_8 + Mg_2SiO_4,$$

it became apparent that the reaction would not be illustrated by the Great Dyke, Stillwater, Bushveld, or Muskox intrusions. In complex systems the chromite field is much expanded compared to the spinel field in the system anorthite–forsterite–silica, and the fields of magnesian olivine and basic plagioclase must have been completely separated during crystallization of chromite in these natural intrusions (Jackson, 1970). The Rhum layered intrusion proved to be another matter. Brown (1956) established the cumulus nature of the layered rocks of Rhum, and noted the close association of settled olivine, plagioclase, and chromite. The base of one of his units (no. 12) was marked by a thin layer of cumulus chromite resting on a layer composed almost entirely of feldspar. Above the chromitite was an accumulation of olivine, and at this locality three nearly monomineralic rocks could be collected in one hand specimen. Brown observed that the translucency of the chromites in the seam was greater than that of isolated chromites beneath the seam. Henderson and Suddaby (1971) investigated the compositions of the chromites in the cumulus layer at the base of unit 12. They observed that chromites included within silicate phases were relatively enriched in iron and chromium, whereas chromites lying between cumulus silicates were enriched in magnesium and aluminum. Based on the assumption that the included chromites retained their initial compositions and the intergrain chromites had been modified by reaction with postcumulus liquid, they proposed the reaction

plagioclase + olivine + chromite + liquid$_1$ $\rightleftarrows$

aluminous chromite + liquid$_2$.

Hamlyn (1977) found similar texturally related compositional variations in the cumulus chromites of the Panton sill, in western Australia, even though plagioclase was not present as a cumulus phase. He concluded that original cumulus chromite relatively enriched in normative magnesio-chromite reacted with postcumulus liquid to form chromite relatively enriched in normative spinel.

Henderson (1975) noted a second reaction involving chromite in the Rhum intrusion. It did not require the discontinuous reaction, but occurred directly between chromite and interstitial melt; it led to iron enrichment of chromite with little change in its Cr/Al ratio, and it appeared to have occurred over a considerable temperature interval.

Thus, although Bowen's reaction illustrating the extraction of $MgAl_2O_4$ from chromite to enrich its magnesiochromite component has been looked for, it has not been found. It is possible, but not likely, that the mechanism is responsible for the chromium concentrations in some chromitites of alpine peridotites. Thayer's (1956, 1963, 1967, 1969) long-held contention that alpine-type chromitites originated as simple cumulates in ophiolitic systems is now receiving considerable support. Careful field work by Greenbaum (1977) and George (1978) has established that the dunites and their contained chromitites in the Troodos massif, Cyprus, are syntectonically deformed magmatic sediments resting on depleted harzburgite basement, and these authors have extended this interpretation to account for the origin of dunite-chromitite bodies of other alpine-type massifs. If their interpretation is correct, then one would expect the podiform chromitites of alpine complexes (Thayer, 1963, 1969) to have been subject to the kinds of reactions summarized by Cameron (1969, 1975). Among these, no doubt, would be the continuous iron-enrichment reaction of Henderson (1975). Discontinuous reactions of the type documented by Henderson and Suddaby (1971) and Hamlyn (1977) may well be common in cumulus chromitites, particularly those of the Bushveld type. It is not apparent that these reactions can account for the formation of high-aluminum, alpine-type, chromitite deposits by enrichment of high-chromium deposists in normative spinel. The large size, massive textures, and internally homogeneous chemical compositions of high-aluminum and high-chromium podiform chromitite deposits, as well as their common occurrence in feldspar-free dunites and harzburgites, still invite explanation.

References

Andersen, O., The system anorthite–forsterite–silica, *Am. J. Sci.*, 4th Ser., *39*, 407–454, 1915.

Anderson, A. T., Possible consequences of composition gradients in basalt glass adjacent to olivine phenocrysts, *Trans. Am. Geophys. Union*, *48*, 227–228, 1967.

Arculus, R. J., Gillberg, M. E., and Osborn, E. F., The system MgO–iron oxide–Cr_2O_3–SiO_2 : phase relations among olivine, pyroxene, silica, and spinel in air at 1 atm., *Ann. Rept. Dir. Geophys. Lab.*, *Carnegie Institution of Washington*, *73*, 317–322, 1974.

Barth, T., Die Pegmatitgänge der kaledonischen Intrusivgesteine im Seiland-Gebiete, *Skrift. d. Norske Videnkaps-Akad.* Oslo, *2*, 1–123, 1927.

Bottinga, Y., Kudo, A., and Weill, D., Some observations on oscillatory zoning and crystallization of magmatic plagioclase, *Am. Mineral.*, *51*, 792–806, 1966.

Bowen, N. L., The melting phenomena of the plagioclase feldspars, *Am. J. Sci.*, 4th Ser., *35*, 577–599, 1913.

———, *The Evolution of the Igneous Rocks*, Princeton University Press, 332p., 1928.

Bowen, N. L., and Andersen, O., The binary system MgO–SiO_2, *Am. J. Sci.*, 4th Ser., *37*, 487–500, 1914.

Brand, H., Das ternäre System Cadmiumchlorid–Kaliumchlorid–Natriumchlorid, *Neus Jahrb. f. Min. Geol. u. Paläo.* Beil. Bd. *32*, 627–700, 1911.

Brooks, C., and Hart, S. R., On the significance of Komatiite, *Geology*, *2*, 107–110, 1974.

Brown, G. M., The layered ultrabasic rocks of Rhum, Inner Hebrides, *Phil. Trans. Roy. Soc. London*, Ser. B, *240*, 1–53, 1956.

Cameron, E. N., Postcumulus changes in the Eastern Bushveld Complex, *Am. Mineral.*, *54*, 754–779, 1969.

———, Postcumulus and subsolidus equilibration of chromite and coexisting silicates in the Eastern Bushveld Complex, *Geochim. Cosmochim. Acta*, *39*, 1021–1033. 1975.

Cameron, E. N., and Desborough, G. A., Occurrence and characteristics of chromite deposits—Eastern Bushveld Complex, in *Magmatic Ore Deposits*, H.D.B. Wilson, ed., Econ. Geol. Mono. 4, 23–40, 1969.

Carr, J. M., Zoned feldspars in layered gabbro of the Skaergaard intrusion, East Greenland, *Min. Mag.*, *30*, 367–375, 1954.

Day, A. L., and Allen, E. T., The isomorphism and thermal properties of the feldspars, *Am. J. Sci.*, 4th Ser., *19*, 93–142, 1905.

Dickey, J. S., Jr., and Yoder, H. S., Jr., Partitioning of chromium and aluminum between clinopyroxene and spinel, *Ann. Rept. Dir. Geophys. Lab.*, *Carnegie Institution of Washington*, *71*, 384–392, 1972.

Donaldson, C. H., Calculated diffusion coefficients and the growth rate of olivine in a basalt magma, *Lithos*, *8*, 163–174, 1975.

Fenner, C. N., The Katmai magmatic province, *J. Geol.*, *34*, 673–772, 1926.

Gardner, P. M., and Robins, B., The olivine-plagioclase reaction: geological evidence from the Sieland petrographic province, northern Norway, *Contr. Min. Petrol.*, *44*, 149–156, 1974.

George, R. P., Jr., Structural petrology of the Olympus ultramafic complex in the Troodos ophiolite, Cyprus, *Geol. Soc. Amer. Bull.*, *89*, 845–865, 1978.

Green, D. H., Nicholls, I. A., Viljoen, M. J., and Viljoen, R. P., Experimental demonstration of the existence of peridotitic liquids in earliest Archean magmatism, *Geology*, *3*, 11–14, 1975.

Greenbaum, D., Textural, structural, and chemical evidence of origin of chromite in the Troodos ophiolite complex, Cyprus, *Econ. Geol.*, *72*, 1175–1194, 1977.

Hamlyn, P. R., Petrology of the Panton and McIntosh layered intrusions, western Australia, with particular reference to the genesis of the Panton chromite deposits, unpub. Ph.D. thesis, University of Melbourne, 390 pp., 1977.

Harloff, C., Zonal structure in plagioclases, *Leidsche Geol. Med.*, *2*, 99–114, 1927.

Henderson, P., Reaction trends shown by chrome-spinels of the Rhum layered intrusion, *Geochim. Cosmochim. Acta*, *39*, 1035–1044, 1975.

Henderson, P., and Suddaby, P., The nature and origin of the chrome-spinel of the Rhum layered intrusion, *Contr. Min. Petrol.*, *33*, 21–31, 1971.

Hess, H. H., Extreme fractional crystallization of a basaltic magma: The Stillwater igneous complex, *Trans. Amer. Geophys. Union*, *20*, 430–432, 1939.

———, Stillwater igneous complex, Montana, a quantitative mineralogical study, *Geol. Soc. Amer. Mem. 80*, 230 pp., 1960.

Hills, E. S., Reverse and oscillatory zoning in plagioclase feldspars, *Geol. Mag.*, *73*, 49–56, 1936.

Homma, F., Über das Ergebnis von Messungen an zonaren Plagioklasen aus Andesiten mit Hilfe des Universaldrehtisches, *Schweiz. Min. Petr. Mitt.*, *12*, 345–352, 1932.

———, Classification of the zonal structure of plagioclase, *Mem. Fac. Sci. Kyoto Imp. Univ.*, Ser. B, *12*, 39–40, 1936.

Irvine, T. N., Chromian spinel as a petrogenetic indicator, Part 2, Petrologic applications, *Can. Jour. Earth Sci.*, *4*, 71–103, 1967.

Irvine, T. N., and Smith, C. H., The ultramafic rocks of the Muskox intrusion, Northwest Territories, Canada, in *Ultramafic and Related Rocks*, P. J. Wyllie, ed., John Wiley and Sons, N.Y., 38–49, 1967.

Jackson, E. D., Primary textures and mineral associations in the ultramafic zone of the Stillwater Complex, *U.S. Geol. Surv. Prof. Paper 358*, 106 pp., 1961.

———, The cyclic unit in layered intrusions—a comparison of repetitive stratigraphy in the ultramafic parts of the Stillwater, Muskox, Great Dyke, and Bushveld complexes, *Geol. Soc. So. Afr. Spec. Publ. 1*, 391–423, 1970.

Keith, M. L., Phase equilibria in the system MgO–Cr_2O_3–SiO_2, *J. Amer. Ceram. Soc.*, *37*, 490–496, 1954.

Kushiro, I., The system forsterite–anorthite–albite–silica–H_2O at 15 Kbar and the genesis of andesitic magmas in the upper mantle, *Ann. Rept. Dir. Geophys. Lab., Carnegie Institution of Washington, 73,* 244–248, 1974.

Kushiro, I., and Yoder, H. S., Jr., Anorthite-forsterite and anorthite-enstatite reactions and their bearing on the basalt-eclogite transformation, *J. Petrol., 7,* 337–362, 1966.

Lofgren, G., Temperature induced zoning in synthetic plagioclase feldspar, in *The Feldspars,* W. S. MacKenzie and J. Zussman, eds., Manchester University Press, 362–375, 1974.

Maaløe, S., The zoned plagioclase of the Skaergaard intrusion, East Greenland, *J. Petrol., 17,* 398–419, 1976.

Muan, A., Phase relations in chromium oxide-containing systems at elevated temperatures, *Geochim. Cosmochim. Acta, 39,* 791–802, 1975.

Naldrett, A. J., and Mason, G. D., Contrasting Archean ultramafic igneous bodies in Dundonald and Clergue Townships, Ontario, *Can. Jour. Earth Sci., 5,* 111–145, 1968.

Navrotsky, A., Thermochemistry of chromium compounds, especially oxides at high temperature, *Geochim. Cosmochim. Acta, 39,* 819–832, 1975.

Nesbitt, R. W., Skeletal crystal forms in the ultramafic rocks of the Yilgarn Block, Western Australia: Evidence for an Archean ultramafic liquid, in *The Archean Rocks, Geol. Soc. Austr. Spec. Publ. 3,* 331–350, 1971.

Phemister, J., Zoning in plagioclase feldspar, *Min. Mag., 23,* 541–555, 1934.

Presnall, D.C., The join forsterite–diopside–iron oxide and its bearing on the crystallization of basaltic and ultramafic magmas, *Am. J. Sci., 264,* 753–809, 1966.

Pyke, D. R., Naldrett, A. J., and Eckstrand, O. R., Archean ultramafic flows in Munro Township, Ontario, *Geol. Soc. Amer. Bull., 84,* 955–978, 1973.

Rankin, G. A., and Wright, F. E., The ternary system $CaO-Al_2O_3-SiO_2$, *Am. J. Sci.,* 4th Ser., *39,* 1–79, 1915.

Shand, S. J., Coronas and coronites, *Geol. Soc. Amer. Bull., 56,* 247–266, 1945.

Shepherd, E. S., and Rankin, G. A., Preliminary report on the ternary system $CaO-Al_2O_3-SiO_2$, a study of the constitution of portland cement clinker, *Jour. Indust. Eng. Chem., 3,* 211–227, 1911.

Smith, C. H., Bay of Islands igneous complex, western Newfoundland, *Geol. Surv. Canada Mem. 290,* 132 pp., 1958.

Thayer, T. P., Preliminary chemical correlation of chromite with the containing rocks, *Econ. Geol., 41,* 202–217, 1946.

———, Mineralogy and geology of chromium, in *Chromium,* M. J. Udy, ed., Amer. Chem. Soc. Mono. 132, *1,* 14–52, 1956.

———, Geologic features of podiform chromite deposits, in *Methods of Prospecting for Chromite,* R. Woodtli, ed., *Org. for Econ. Coop. and Development Bull.,* Paris, 135–148, 1963.

———, Chemical and structural relations of ultramafic and feldspathic rocks in alpine intrusive complexes, in *Ultramafic and Related Rocks,* P. J. Wyllie, ed., John Wiley and Sons, N.Y., 222–239, 1967.

———, Alpine-type sensu-strictu (ophiolitic) peridotites: refractory residues from partial melting or igneous sediments?, *Tectonophysics, 7,* 511–516, 1969.

Vance, J. A., Zoning in igneous plagioclase: normal and oscillatory zoning, *Am. J. Sci.*, *260*, 746–760, 1962.

Van der Kaaden, G., Optical studies of natural plagioclase feldspars with high- and low-temperature optics, Dissertation, University of Utrecht, 105 pp., 1951.

Viljoen, M. J., and Viljoen, R. P., Evidence for the existence of a mobile extrusive peridotitic magma from the Komati Formation of The Onverwacht Group, in *The Upper Mantle Project*, *Geol. Soc. So. Afr. Spec. Publ. 2*, 87–112, 1969.

Vogt, J.H.L., Physicalisch-chemische Gesetze der Kristallisationsfolge in Eruptivgesteinen, *Tschermaks Min. pert. Mitt.*, *24*, 437–542, 1905.

——, The physical chemistry of the magmatic differentiation of igneous rocks, *Vidinsk.-Selsk. Skr. I. Mat.-Naturv. Kl.*, *2*, 1–132, 1924.

Wager, L. R., and Deer, W. A., Geological investigations in East Greenland, pt. 3, The petrology of the Skaergaard intrusion, Kangerdlugssuaq, East Greenland, *Medd. om Grønland*, *105*, 346 pp., 1939.

Wenk, E., Die Koexistenzbedingungen zwichen Hornblende, Biotit, und Feldspaten und die bedeutung, der oszillierenden Zonarstruktur, *Schweiz. Min. Petr. Mitt.*, *25*, 141–164, 1945.

Wyllie, P. J., Effects of the changes in slope occurring on liquidus and solidus paths in the system diopside–anorthite–albite, *Min. Soc. Amer. Spec. Paper 1*, 204–212, 1963.

Yoder, H. S., Jr., Stewart, D. B., and Smith, J. R., Ternary feldspars, *Carnegie Inst. Washington Year Book*, *55*, 190–194, 1956.

Yoder, H. S., Jr., Stewart, D. B., and Smith, J. R., Ternary feldspars, *Carnegie Inst. Washington Year Book*, *56*, 206–214, 1957.

Chapter 16

THE IMPORTANCE OF VOLATILE CONSTITUENTS

C. WAYNE BURNHAM

Department of Geosciences, The Pennsylvania State University, University Park, Pennsylvania

To many petrologists a volatile component is exactly like a Maxwell demon; it does just what one may wish it to do.

(Bowen, 1928)

INTRODUCTION

Most readers of Chapter XVI in *The Evolution of the Igneous Rocks* are left with the distinct impression that Bowen did not regard himself as one of the many petrologists to whom "a volatile component is exactly like a Maxwell demon." In fact, he repeatedly stressed the view that small amounts of a volatile, such as H_2O, would have correspondingly small effects on the liquidus phase relations appropriate to the "dry" melts in silicate systems. Fifty years of experimentation on H_2O-bearing silicate melts of petrologic interest by scores of investigators have provided abundant confirmatory evidence for this view. This same experimentation, however, coupled with other laboratory and field investigations, also has provided abundant evidence that the evolution of many igneous rocks involves much more than the crystallization-differentiation of magmas, a topic with which Bowen was preoccupied in 1928. In the evolution of granitic rocks, for example, small amounts of H_2O are now widely recognized as playing a highly important role, especially in the generation of magmas. It is significant in terms of the present tribute to Bowen's enormous contributions that he was among the first to recognize this fact (Bowen and Tuttle, 1950; Tuttle and Bowen, 1958).

The full importance of volatile constituents in the evolution of igneous rocks cannot yet be quantitatively assessed, because neither their total amounts nor their relative proportions in the regions of magma generation are precisely known. There is little doubt that H_2O is either the most abundant or the second most abundant (next to CO_2) of these volatile constituents. There also is little doubt that were it not for the presence of H_2O in these regions, even in still-undefined "small amounts," the evolution of igneous rocks would have taken a markedly different course. Calcalkaline volcanism (with its associated ore deposits), granitic batholiths, and migmatite terranes, for example, would be far less conspicuous features of global geology than at present. Moreover, the low-velocity zone in the upper mantle, with its attendant plate tectonics implications, probably would be much more limited in extent, because temperatures generally are below the solidus of upper mantle rocks in the absence of aqueous pore fluids or hydrous minerals.

In a large percentage of the papers dealing with the role of volatiles in magmas that have appeared since the pioneer efforts of Goranson (1931, 1932, 1938), emphasis has been placed on the effects of H_2O in lowering melting temperatures of rock-forming minerals and mineral assemblages. This emphasis was a natural consequence of the realization first enunciated by Bowen and Tuttle (1950) and later enlarged upon by Tuttle and Bowen (1958), that felsic rocks begin to melt in the presence of H_2O at temperatures commonly realizable in the deeper parts of the continental crust, regardless of the amount of H_2O present. Moreover, the melts produced, in amounts proportional to the amount of H_2O present, have compositions appropriate to yield certain types of granitic rock upon crystallization. Thus, the old controversy over the origin of granites, in which Bowen was perhaps the "magmatists'" most eloquent spokesman (Bowen, 1948), appears to have been laid to rest in a manner, although not really satisfying to him, certainly more to his liking than to that of the "transformists."

The success achieved by Tuttle and Bowen in explaining many of the features of granite genesis apparently spurred others, notably Yoder and Tilley (1956, 1962), to investigate the role of H_2O in the genesis of more mafic rocks. One of the more significant results of these investigations is that beginning-of-melting temperatures of some basaltic (gabbroic) rocks at high H_2O pressures are depressed even further than are those of granitic rocks, resulting in a tendency for H_2O-saturated solidus temperatures of basaltic rocks to approach those of granitic rocks at pressures in excess of 10 kbar. These early experimental results on "basalt-H_2O" systems also revealed a potential petrogenetic role for small amounts of H_2O considerably greater than that envisioned in the hornblende fractionation model of Bowen (1928), namely, the role of H_2O in partial melting of amphibole-bearing assemblages in the lower crust and upper mantle. Thus, many of the more recent investigations at high pressures have focused on melting relations in amphibole- and mica-bearing systems, especially in the presence of H_2O or mixtures of H_2O and CO_2.

One of the reasons CO_2, mixed with H_2O, has been employed in these investigations is that it generally is second only to H_2O in abundance in the magmatic volatile phase, as judged by analyses of volcanic gases and fluid inclusions in igneous minerals; greater petrologic credibility might, therefore, be attached to the results by its use. A more important reason, perhaps, is that CO_2 is only slightly soluble in felsic melts at relatively low pressures (≤ 3 kbar), as demonstrated by Wyllie and Tuttle (1959); consequently, it became a conveniently "inert" and geologically reasonable substance to employ in experimental investigations of H_2O-under-saturated melting relations. However, recent work at high pressures, especially by Eggler (1973a) and Mysen (1975), has clearly shown that CO_2 is appreciably soluble in mafic silicate melts, and even in felsic melts in the presence of dissolved H_2O. Moreover, Eggler (1974) and others have found that the compositions of melts produced from model peridotite compositions in the presence of CO_2 are significantly different from those produced in the presence of H_2O. The direct significance of this discovery for the generation of silica-saturated versus silica-undersaturated magmas is tremendous; indirectly it also is of great significance in providing clues to the mechanisms by which these chemically different volatiles dissolve in silicate melts.

An understanding of the mechanisms of solution of volatiles in silicate melts is essential to a full understanding of the role of volatiles in igneous petrogenesis, as the mechanisms of solution largely determine the thermodynamic properties of the solutions and they, in turn, govern the phase equilibrium relations. Ideally, perhaps, these mechanisms might be deduced from first principles, but a sound physical-chemical theory of silicate melts that approach magmas in compositional complexity is just now beginning to emerge. Alternatively, the problem can be inverted, and phase relations in volatile-bearing systems, coupled with thermodynamic and other data, can be used to construct internally consistent solution models, provided certain initial assumptions regarding the structure of silicate melts are adopted. This approach has met with considerable recent success in applications to the system $NaAlSi_3O_8$–H_2O (Burnham and Davis, 1974) and to other hydrous aluminosilicate compositions (Burnham, 1974, 1975a, 1975b). It will, therefore, be adopted here in attempts to construct reasonable solution models for volatiles other than H_2O and, especially, to form a conceptual basis for understanding thermodynamic and melting relations in magmatic systems.

A premise underlying the solution model, in keeping with the views of Weyl and Marboe (1959) and others, is that the short-range order of the crystalline state of a given phase (mineral) is largely preserved to temperatures well above its liquidus, and hence the principal change on melting is the loss of long-range order. This loss is supported by numerous x-ray studies of aluminosilicate glasses (Urnes, 1966), as well as by characteristically small entropies of fusion on a gram-atom basis, as compared with more dissociated substances such as NaCl (Robie and Waldbaum, 1968).

Two corollaries of this quasi-crystalline model of aluminosilicate melts are: (1) the structural units in a melt of given composition tend to mimic those of the crystalline phases that appear upon cooling, and (2) electrical neutrality is preserved on a local, structural-unit scale. The first of these corollaries implies that the lower-pressure (<10 kbar) melting of albite, in which the relative atomic positions are approximately as depicted in Fig. 16-1, takes place largely by disorientation of the four-membered units of $(Al,Si)O_4$ tetrahedra with respect to each other; the second implies that the Na and Al atoms in each unit are spatially associated. The first corollary also implies that mixing of this melt with other substances to form a homogeneous solution, as with H_2O, occurs on the scale of a structural unit. Therefore, one gram formula weight of $NaAlSi_3O_8$ (262.2 g) is taken as the thermodynamic mixing unit, or mole of "albite" (ab) component, in melts from which albite or its solid solutions crystallize, irrespective of the bulk melt composition.

It should be emphasized that it is not essential to choose one gram formula weight of $NaAlSi_3O_8$ to represent a mole of "albite" melt, as is done here. Thermodynamic properties must be evaluated on a molal basis, but in the absence of evidence regarding the structural units in silicate melts, it would have been entirely appropriate to choose as a mole of "albite" melt, for example, that mass which contains two moles (or any other whole number) of oxygen, hereinafter referred to as O^{2-}. The choice of a mass that contains fewer than eight moles of O^{2-}, however, will result in fractional moles of sodium and aluminum, and if either of these elements is involved in chemical interaction with H_2O, then the reaction mechanism may be obscured by such a choice. Moreover, certain features of the melting relations in this and other silicate systems, which will be discussed in appropriate succeeding parts of this chapter, can be most readily explained if the thermodynamic components are selected in accordance with the quasi-crystalline model.

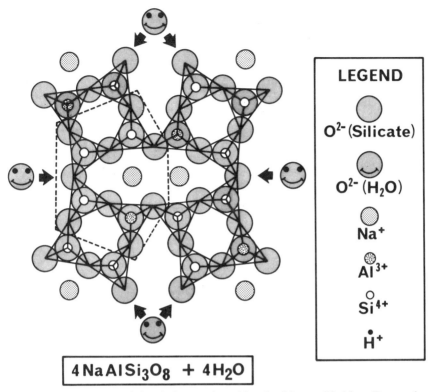

LEGEND

O^{2-} (Silicate)

O^{2-} (H_2O)

Na^+

Al^{3+}

Si^{4+}

H^+

$4\,NaAlSi_3O_8 \; + \; 4H_2O$

Figure 16-1. Projection of the relative atomic positions in albite, modified from Deer *et al.* (1966, Plate 1) and from Burnham (1975a, Fig. 1). Atomic arrangement in $NaAlSi_3O_8$ melt is postulated to mimic that shown in terms of short-range order within the four-membered rings of $(Al,Si)O_4$ tetrahedra.

SILICATE MELT-VOLATILE SOLUTIONS

SOLUTION OF H_2O IN $NaAlSi_3O_8$ MELTS

On the basis of all available experimental and thermodynamic data, as summarized by Burnham (1975a), it now appears to be firmly established that H_2O dissolves in silicate melts principally by reaction (hydrolysis) with oxygen ions (O^{2-}) of the melt to produce hydroxyl ions (OH^-). Prior to the work of Burnham and Davis (1974), it was generally assumed that this reaction was of the form

$$H_2O(v) + O^{2-}(m) \rightleftarrows 2\,OH^-(m), \qquad (16\text{-}1)$$

where (v) and (m) refer to "vapor" ("gas" or aqueous fluid) and melt phases, respectively. This reaction scheme, however, did not adequately explain the thermodynamic relations in the system $NaAlSi_3O_8$–H_2O. Nor could it be reconciled with the fact, as noted by Wasserburg (1957), that the molal solubility of H_2O in $NaAlSi_3O_8$ melts is greater than in silica melts at a given pressure and temperature, even when compared on the basis of equal mole numbers of O^{2-} (Si_4O_8). Therefore, it was proposed (Burnham, 1975a) that, in $NaAlSi_3O_8$ melts where Na^+ provides charge balance on the AlO_4 tetrahedron in each structural unit, dissolved H_2O also undergoes dissociation by exchange of a proton (H^+) for the Na^+. The dissociation resulting from this exchange leads to a linear relationship between the fugacity (f_w, Burnham and Davis, 1974) or activity (a_w, Burnham, 1975b)[1] of H_2O and the square of its mole fraction $[(X_w^m)^2]$, as observed for $X_w^m \leq 0.5$. The exchange also lowers a_w, or the chemical potential of H_2O (μ_w); hence the solubility of H_2O (X_w^m where $a_w = 1.0$) at moderate pressures is greater than in melts that lack exchangeable cations, such as those of Si_4O_8.

In a somewhat idealized fashion, these solution reactions may be visualized as depicted in Fig. 16-2, where only the mole of $NaAlSi_3O_8$ enclosed by the dashed lines in Fig. 16-1 is represented. The purpose of selecting this particular unit for illustration is to focus on the O^{2-} site that is thought to be primarily involved in reaction with the first mole of H_2O, despite the fact that the four-membered ring of $(Al,Si)O_4$ tetrahedra (Fig. 16-1) is regarded as the basic structural entity that maintains its integrity to high values of X_w^m. Thus, as the first mole of H_2O enters the melt (Fig. 16-2, lower left), an Si–O–Si bridge is broken and an H^+ exchanges with an Na^+, the H^+ presumably then forms an OH^- with one of the four O^{2-} coordinating Al^{3+}, and the Na^+ also presumably occupies a new position near the broken-bridge site to maintain local electrical neutrality. This overall reaction, which may be written

$$H_2O(v) + O^{2-}(m) + Na^+(m) \rightleftarrows OH^-(m) + ONa^-(m) + H^+(m), \quad (16\text{-}2)$$

without implication regarding the degree of ionization, depolymerizes the melt to such an extent that, at $N = 1$ ($X_w^m = 0.5$), the original three-dimensional network has been broken into sheets about one structural unit thick (on the average) which are held together only by weak Na–O bonds. As a consequence, the viscosity of the melt, which contains only about 6.4 wt % H_2O ($X_w^m = 0.5$), is a factor of 10^5 to 10^6 lower than that

---

[1] a_w is defined by f_w/f_w°, where f_w° is the fugacity of pure H_2O at the specified pressure and temperature.

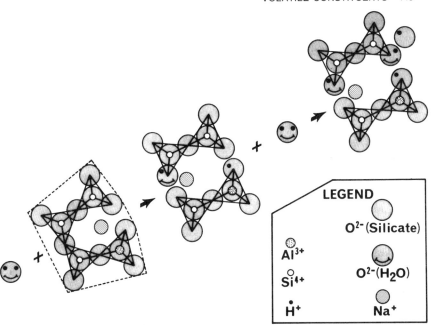

Figure 16-2. Proposed reaction scheme for solution of H_2O in $NaAlSi_3O_8$ melts, as expressed in text Eqs. 16-1 and 16-2 (modified from Burnham, 1975a, Fig. 6). The reaction unit (mole) of $NaAlSi_3O_8$ melt represented is as outlined in Fig. 16-1.

of anhydrous melt at the same temperature (Burnham, 1975a, Fig. 2). Also, the $H^+ \rightleftarrows Na^+$ exchange results in much greater mobility of Na^+, and electrical conductivity is thereby increased by a factor of 10^3 to 10^4 (Burnham, 1975a, Fig. 5) over that of the anhydrous melt.

At higher mole fractions of H_2O ($X_w^m > 0.5$, $N > 1$), the excess H_2O enters the melt primarily by reaction 16-1 (Fig. 16-2, right side), because essentially all of the Na^+ already has been exchanged with H^+ from the first mole of H_2O. This additional H_2O further depolymerizes the melt by breaking the sheets into chains that extend normal to the plane of Figs. 16-1 and 16-2, but the effects on viscosity and electrical conductivity are orders of magnitude smaller. For example, at $N = 2$ ($X_w^m = 0.67$), which is the saturation value ($a_w = 1$) at about 8.3 kbar and 1273K, the viscosity is only about seven times lower, and the conductivity two times higher than at $N = 1$. These small effects of the second mole of H_2O are interpreted to indicate that the chains are not much less resistant to shear stress than are the sheets, and that mobility of the already "liberated"

Na^+ is increased only in proportion to the increased fluidity (depolymerization).

The linear relationship between a_w and $(X_w^m)^2$ alluded to above for values of $X_w^m < 0.5$ leads to the expression (Burnham, 1975b)

$$a_w = k(X_w^m)^2, \tag{16-3}$$

where k is an analogue of a Henry's law activity constant for a dissociated solute. Values for $\ln k$ are presented in Fig. 16-3, from which it can be seen that they are dependent mainly on pressure and, to a lesser extent, on temperature. For values of $X_w^m > 0.5$, where the 2 OH^- produced by reaction 16-1 remain near the former bridging O^{2-} site (far right of Fig. 16-2), a_w increases exponentially with X_w^m; thus

$$a_w = 0.25 \, k \exp (6.52 - 2667/T)(X_w^m - 0.5). \tag{16-4}$$

The activity relations expressed in Eqs. 16-3 and 16-4, coupled with k values from Fig. 16-3 and the thermodynamic properties of pure H_2O (Burnham et al., 1969), suffice to calculate all of the other thermodynamic functions of interest for H_2O dissolved in $NaAlSi_3O_8$ melts. Also, by setting $a_w = 1.0$ in Eqs. 16-3 and 16-4, the maximum molal solubility of H_2O in $NaAlSi_3O_8$ melts can be calculated as a function of pressure and temperature, as illustrated for the 1373K isotherm in Fig. 16-4a. Solubilities calculated in this fashion are maximum values, because setting $a_w = 1.0$ involves an assumption that the "vapor" phase is pure H_2O and this is not strictly valid (see Clark, 1966, p. 436). From a comparison of calculated (solid curve) and experimentally determined (inverted triangles) solubilities in Fig. 16-4, however, it is apparent that this assumption introduces negligible error at pressures as high as 10 kbar. Furthermore, a comparison of calculated (Eq. 16-4) and experimentally determined solubilities (Boettcher and Wyllie, 1969, Fig. 10) indicates that negligible error is introduced even at 15 kbar and 943K. On the basis of these results and the finding that $\ln k$ is essentially linear in $\ln P$ between 2.0 and 10 kbar, $\ln k$ values have been extrapolated to 20 kbar in Fig. 16-3.

SOLUTION OF H_2O IN Si_4O_8 (SILICA) MELTS

It was argued briefly above that solubilities of H_2O in $NaAlSi_3O_8$ melts, where $X_w^m \leq 0.5$ at saturation, are greater than in Si_4O_8 melts, because the $H^+ \rightleftarrows Na^+$ exchange lowers $a_w(\mu_w)$ over that resulting from reaction 16-1 alone. The extent to which a_w is thus lowered can be evaluated at $X_w^m = 0.5$ (where both Eqs. 16-3 and 16-4 are valid) from the relation $[\delta \ln(a_w/a_w')/\delta X_w^m]_{P,T}$, in which a_w and a_w' are the activity expressions from Eqs. 16-3 and 16-4, respectively. At 1420K, for example, it is found

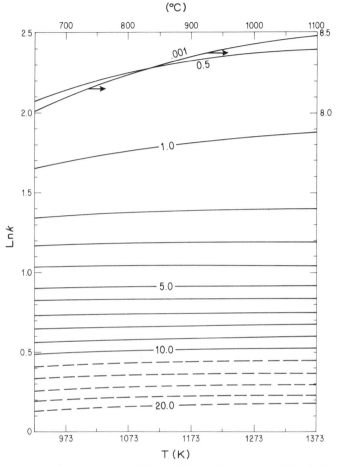

Figure 16-3. Ln k values for text Eqs. 16-3 and 16-4 vs. T at constant P (kbar). The dashed isobars for $P > 10$ kbar were obtained by extrapolation from the data of Burnham and Davis (1974) and Burnham et al. (1969). Values for 0.001 kbar are shown on the right ordinate.

that the $H^+ \rightleftarrows Na^+$ exchange lowers a_w by 0.473, an amount equivalent to reducing X_w^m by 0.14. Thus, were it not for the exchange reaction (Eq. 16-2), the saturation value of X_w^m in $NaAlSi_3O_8$ melt at 1420K and 2.1 kbar presumably would be 0.36, instead of 0.50.

Inasmuch as the mechanism of solution of H_2O in Si_4O_8 melts is presumably the same as in $NaAlSi_3O_8$ melts for $X_w^m > 0.5$ (Eqs. 16-1 and 16-4), the predicted saturation mole fraction of H_2O in Si_4O_8 melt at 1420K

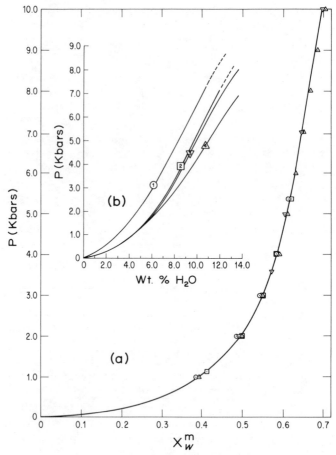

Figure 16-4. Solubility of H_2O in aluminosilicate melts. Circles (1), Columbia River basalt; squares (2), Mt. Hood andesite; inverted triangles (3), albite; and upright triangles (4), Harding pegmatite. (a) Equimolal solubilities at 1373K (1100°C) calculated from experimental weight-percent solubilities in (b), using text Eqs. 16-3, 16-4, 16-5, 16-6, and values of M_e and M'_e from Table 16-1. (b) Experimental weight-percent solubilities at various temperatures, from Burnham and Jahns (1962) and Hamilton et al. (1964).

and 2.1 kbar is 0.36. The experimental solubility data of Kennedy et al. (1962) yield a value of $X^m_w = 0.38$. In view of the potential errors, especially in the above partial derivatives, the agreement is remarkable. Furthermore, solubilities calculated from Eq. 16-4, after substituting 0.38 for 0.50 in the last term, are in agreement with those determined experimentally to pressures of about 7 kbar. At higher pressures, however, experimentally determined solubilities are higher than calculated values, although the highest measured solubility ($X_w^m = 0.67$) at 9.0 kbar and 1385K (Kennedy et al., 1962, Table 4) is not higher than in $NaAlSi_3O_8$ melt at the same pressure and temperature (Fig. 16-4a).

The sharp increase in solubility between 7.0 and 9.0 kbar (from $X_w^m = 0.53$ to $X_w^m = 0.67$) was extrapolated by Kennedy et al. (1962) to a second critical endpoint (complete miscibility) in the system Si_4O_8–H_2O at 9.7 kbar and 1353K. Whether complete miscibility occurs at this, or a somewhat higher pressure (Stewart, 1967), it is apparent that Eq. 16-4, as modified above, is not applicable to this system at high fugacities of $H_2O(f_w)$. This interpretation, consistent with the spectroscopic results of Yin et al. (1971), reflects the differences in the free energies of Si–O–Si and Al–O–Si bridges. As indicated in Figs. 16-1 and 16-2, H_2O is thought to interact first with the Si–O–Si bridges in $NaAlSi_3O_8$ melts, owing to their higher free energies, but at H_2O fugacities of 19 to 25 kbar ($P_w = 10$ to 11 kbar) essentially all of the Si–O–Si bridges already have been broken. At higher values of f_w then, H_2O enters $NaAlSi_3O_8$ melts primarily by reaction with the lower free energy Al–O–Si bridges; hence, the Gibbs free energy of the system is not greatly lowered by reaction, and complete miscibility does not occur even at 20 kbar pressure (Mysen, 1975). In Si_4O_8 melts that obviously lack these lower free energy bridges, however, the Gibbs free energy of the system is lowered practically as much by reaction with one Si–O–Si bridge as with another; hence, complete miscibility occurs abruptly once a threshold f_w has been exceeded (cf. Kennedy et al., 1962, Fig. 14).

Before turning to discussion of the solution of H_2O in igneous-rock melts which, according to the quasi-crystalline melt model, are generally mixtures of several components, it is important to examine further one implication of the differential H_2O solubilities in $NaAlSi_3O_8$ and Si_4O_8 melts. The fact that a_w is lower in $NaAlSi_3O_8$ melt than in Si_4O_8 melt at a given pressure, temperature, and $X_w^m \leq 0.5$ implies that, in a mixture of these two melts, H_2O is selectively concentrated in the aluminosilicate component. For this reason, the melting relations of aluminosilicates (e.g., albite) and non-aluminosilicates (e.g., quartz) are affected differentially in the presence of H_2O, as will be discussed later. Also for this reason, it is essential to choose properly the components of an igneous-rock melt before meaningful comparisons can be made with the system $NaAlSi_3O_8$–H_2O.

SOLUTION OF H_2O IN IGNEOUS-ROCK MELTS

It is apparent from Fig. 16-4b (inset) that, despite general similarities in shape of the curves, H_2O solubilities on a weight percent (mass) basis are markedly different in different rock melt compositions at a given temperature and pressure (cf. Columbia River basalt and Mt. Hood andesite melts). It also is apparent from the foregoing discussions that meaningful comparisons of solubility relations must be made on an equimolal, rather than a mass, basis. The objective of such comparisons, of course, is to extend the H_2O-solution model, hence the thermodynamic relations in the system $NaAlSi_3O_8$–H_2O, to igneous-rock melts.

The crucial test for applicability of the $NaAlSi_3O_8$–H_2O solution model to igneous-rock melts is whether or not the a_w versus X_w^m relationships expressed in Eqs. 16-3 and 16-4 are quantitatively applicable to H_2O dissolved in these melts. It is imperative, therefore, that the mass which constitutes one mole of rock melt (M_e), in terms of its interaction with H_2O, be chosen in such a way as to be consistent with constraints imposed by the model. Inasmuch as the model involves two solution reactions (Eqs. 16-1 and 16-2), there are two sets of constraints, but both sets are imposed by the stoichiometry of $NaAlSi_3O_8$. The set for $X_w^m > 0.5$ ($N > 1$) is readily applied, as the stoichiometry of $NaAlSi_3O_8$ requires only that the equimolal mass of rock melt contain not more than 8.0 moles of O^{2-} and not more than one mole of exchangeable cations. The set of constraints for $X_w^m \leq 0.5$ ($N \leq 1$), on the other hand, must include not only those imposed by the exchange reaction, but that imposed by the fact—demonstrated in the preceding section—that H_2O also interacts with Si–O–Si bridges, which may be formed from Si in excess of the 3 Si/1Na ratio in $NaAlSi_3O_8$. The overriding constraint imposed by the exchange reaction for $X_w^m \leq 0.5$ is that the mole of rock melt contain one mole of exchangeable cations, where exchangeable cations are those not in tetrahedral coordination that can satisfy charge balance on tetrahedrally-coordinated trivalent cations (chiefly Al^{3+}, but also Fe^{3+} in highly alkaline sub-aluminous melts). They include mainly cations of the alkali metals, alkaline earths, and transition metals (with the rare exception of Fe^{3+} just noted); however, their number cannot exceed the number of tetrahedrally coordinated trivalent cations. Thus, in melts that lack normative corundum (CIPW convention), but in which alkali/aluminum is less than one (Columbia River basalt and Mt. Hood andesite in Table 16-1 and Fig. 16-4), the number of exchangeable cations equals the number of Al^{3+}. On the other hand, in corundum-normative melts, such as the Harding pegmatite in Table 16-1, all common silicate-forming cations except Si^{4+}, Al^{3+}, and Ti^{4+} are considered exchangeable.

In Si-rich melts, such as that of the Harding pegmatite, Si in excess of three times the total number of exchangeable cations (as in $NaAlSi_3O_8$) presumably forms Si–O–Si bridges, as in Si_4O_8 melts, which interact with H_2O in accordance with reaction Eq. 16-1. Therefore, the $NaAlSi_3O_8$

Table 16-1. $NaAlSi_3O_8$-Equivalent Masses of Igneous-Rock Melts

Constituent	Columbia River Basalt			Mt. Hood Andesite			Harding Pegmatite		
	1	2	3	1	2	3	1	2	3
SiO_2	50.71	0.844	1.688	58.41	0.972	1.944	75.99	1.265	2.530
Al_2O_3	14.48	0.284	0.426	18.25	0.358	0.537	14.90	0.292	0.438
TiO_2	1.70	0.021	0.042	1.15	0.014	0.028	—		—
Fe_2O_3	4.89	0.061	0.092	1.50	0.019	0.029	0.15	0.002	0.003
FeO	9.29	0.129	0.129	5.06	0.070	0.070	0.58	0.008	0.008
MgO	4.68	0.116	0.116	3.39	0.084	0.084	—		—
CaO	8.83	0.157	0.157	6.70	0.119	0.119	0.24	0.004	0.004
Na_2O	3.16	0.102	0.051	4.35	0.140	0.070	3.97	0.128	0.064
K_2O	0.77	0.016	0.008	0.82	0.017	0.009	2.79	0.059	0.030
Li_2O	—			—			0.61	0.041	0.021
Rb_2O	—			—			0.18	0.002	0.001
P_2O_5	0.36	0.005	0.013	0.26	0.004	0.010	—		—
Total	98.87	1.735	2.722	99.89	1.797	2.900	99.41	1.801	3.099
$M_e(X_w^m \le 0.5)$	98.87(1/0.284) = 348			99.89(1/0.358) = 279			99.41[1/(0.24 + 0.0.1)] = 288		
$M_e(X_w^m > 0.5)$	98.87(8/2.722) = 291			99.89(8/2.900) = 276			99.41(8/3.099) = 258		

Column 1: Chemical analyses in weight percent oxides. Columbia River basalt (Hamilton et al., 1964, p. 23; C. O. Ingamells, analyst) less H_2O and MnO combined with FeO. Mt. Hood andesite (Hamilton et al., 1964, p. 23; V. C. Smith, analyst) less H_2O and MnO combined with FeO. Harding, New Mexico, pegmatite glass (Burnham and Jahns, 1962, p. 726; C. O. Ingamells, analyst) less MgO, Cs_2O, P_2O_5, H_2O and F (0.33%) and MnO combined with FeO.

Column 2: Moles of cationic elements (n_i) in approximately 100g (totals in Column 1). $M_e(X_w^m \le 0.5)$ determined from $1/nAl$. M_e for Harding pegmatite determined from $1/[\sum n_i^\zeta + 0.19(n_{si} - 3\sum n_i^\zeta)]$, where n_i^ζ is number of moles of exchangeable cation (i) as described in text.

Column 3: Moles of O^{2-} in approximately 100g (totals in Column 1). $M_e'(X_w^m > 0.5)$ determined from $8/\sum O^{2-}$

equivalent of this excess Si (Si^{xs}) is given by 0.76 ($Si^{xs}/4$), where the factor 0.76 is the ratio of mole fractions of H_2O in Si_4O_8 and $NaAlSi_3O_8$ melts (0.38/0.50) at the same a_w.

With these constraints in mind, $NaAlSi_3O_8$-equivalent masses of Columbia River basalt, Mt. Hood andesite, and Harding pegmatite were computed for $X_w^m \leq 0.5(M_e)$ and $X_w^m > 0.5(M_e')$ as indicated in Table 16-1. The experimental data from the original sources (see caption to Fig. 16-4) were then recast into equimolal solubilities, using the relations

$$\frac{X_w^m}{1 - X_w^m} = \frac{M_e W_w^m}{18.02(1 - W_w^m)} \tag{16-5}$$

and

$$\frac{X_w^m}{1 - X_w^m} = 1 + \frac{M_e' \Delta W_w^m}{18.02(1 - \Delta W_w^m)}, \tag{16-6}$$

where Eqs. 16-5 and 16-6 are applicable for values of $X_w^m \leq 0.5$ and $X_w^m > 0.5$, respectively, W_w^m is the experimental weight fraction of H_2O, and ΔW_w^m is the weight fraction of H_2O in excess of that required to yield $X_w^m/(1 - X_w^m) = 1$ in Eq. 16-5. From the results of these calculations presented in Fig. 16-4, it is apparent that, within experimental error ($\Delta X_w^m \approx \pm 0.02$), there are no significant differences in the solubilities of H_2O in any of these melts. This result is all the more remarkable when consideration is given to experimental uncertainties that arise from assumptions regarding the oxidation state of Fe in the basalt and andesite melts, and the fact that the experimental data for Harding pegmatite melt had to be adjusted upward in temperature as much as 450K, using Eqs. 16-3 and 16-4.

The fact that H_2O solubilities in the igneous-rock melts of Table 16-1 and Fig. 16-4 are essentially identical to those in $NaAlSi_3O_8$ melts, provided constraints of the solution model are imposed, implies that Eqs. 16-3 and 16-4 are quantitatively applicable to these rock melts. This fact implies, in turn, that other thermodynamic functions for the H_2O component (μ_w, f_w, $\bar{S}_w^m$, $\bar{H}_w^m$, and $\bar{V}_w^m$) have the same dependence upon P, T, and X_w^m, because all of them are derivable from Eqs. 16-3, 16-4, and the corresponding functions for pure H_2O (Burnham et al., 1969).

The full range of igneous rock compositions over which these thermodynamic relations hold has not been established. From discussions in the preceding section and Fig. 16-4a, these relations apparently do not hold rigorously for melt compositions more Si-rich than the Harding pegmatite (Table 16-1) at H_2O pressures (P_w) greater than about 10 kbar, but melts of these compositions with such high H_2O contents ($X_w^m > 0.75$) are of minor petrologic interest. These relations also are not expected to hold for very Al-poor melts of ultramafic compositions even at low H_2O pressures, because they lack both Si–O–Si bridges and significant amounts of exchangeable cations. It is noteworthy, however, that equimolal (8.0 moles of O^{2-}) H_2O solubilities in anorthite, enstatite, diopside, and forsterite melts at 20 kbar are all about the same (Hodges, 1974). Apparently the differences in reactivity (free energy) between Al–O–Si and Mg–O–Si linkages are inconsequential at such high pressures and fugacities of H_2O, where both Mg and Al are predominantly in octahedral coordination.

In calculating the $NaAlSi_3O_8$-equivalent mass (M_e) of Columbia River basalt melt (Table 16-1) for $X_w^m \leq 0.5$, there was an excess of about one mole of $(Mg^{2+} + Fe^{2+} + Fe^{3+})$ per mole of melt (348g), over the allowable number of exchangeable cations that was essentially ignored. The fact that equimolal H_2O solubilities are the same as in $NaAlSi_3O_8$ melt, without considering possible interaction of H_2O with the O^{2-} in melt components containing these excess cations (also Ti^{4+} and P^{5+}), implies that very little interaction occurs with them. It also implies, in turn, that Eq. 16-3 is rigorously applicable only to the sum of aluminosilicate components and, if the quasi-crystalline-melt model is valid, that most of the H_2O (for $X_w^m \leq 0.5$) is concentrated in these latter components. Some consequences of this phenomenon for melting relations will be considered below; first, however, it is appropriate to examine the solution of volatiles other than H_2O in light of the H_2O solution model.

SOLUTION OF OTHER VOLATILES IN SILICATE MELTS

The other volatiles of major petrologic interest can be divided into two major groups on the basis of whether or not the mechanisms of their solution in silicate melts are similar to those of H_2O: that is, whether or not they are hydrolyzed by reaction with the bridging O^{2-} of the melt. The acid volatiles, H_2S, HCl, and HF, which play an important role in ore formation in magmatic environments, are capable of hydrolysis-type reactions such as

$$H_2S(v) + O^{2-}(m) \rightleftarrows SH^-(m) + OH^-(m), \qquad (16\text{-}7)$$

$$HCl(v) + O^{2-}(m) \rightleftarrows Cl^-(m) + OH^-(m), \qquad (16\text{-}8)$$

and

$$HF(v) + O^{2-}(m) \rightleftarrows F^-(m) + OH^-(m). \qquad (16\text{-}9)$$

Because of the dissociated nature of these solutes, Henry's law analogue relations similar to that in Eq. 16-3, are thought to be approximately valid for these substances, with due allowances for variations in k with volatile species. The available experimental evidence (Burnham, 1979) indicates that the Henry's law analogue constant for HF (k_{HF}) is smaller than that for H_2O (k_w), whereas k_{H_2S}/k_w and k_{HCl}/k_w are somewhat greater than unity. The smaller value of k_{HF} implies that HF is more soluble than H_2O in aluminosilicate melts, in keeping with the smaller size of the F^- ion and the stability of the SiF_4 complex relative to $Si(OH)_4$. Conversely, H_2S and HCl are less soluble than H_2O, again in keeping with ionic size and bond strength considerations. Thus, owing to their similarities to H_2O in behavior and to their much lower abundances in most magmas, these other hydrolyzable volatiles may be lumped with H_2O in terms of their roles in the evolution of igneous rocks. This similarity is not meant to imply, however, that they do not play highly important individual roles in certain magmatic processes, such as in the formation of some pegmatites and hydrothermal ore deposits, but only that their roles in silicate melt solutions are basically similar to that of H_2O.

In contrast to these hydrolyzable substances, CO_2, which generally is regarded as the second most abundant magmatic volatile, lacks the capacity to hydrolyze with O^{2-} in melts and, hence, to break the Si–O–Si and Al–O–Si bridges. As a consequence, CO_2 solubility is limited by the extent to which CO_3^{2-} formation $[CO_2(v) + O^{2-}(m) \rightleftharpoons CO_3^{2-}(m)]$ can lower the free energy of the system and this, in turn, is determined largely by the nature of the O^{2-} coordination. In silica melts where all O^{2-} form Si–O–Si bridges, for example, CO_3^{2-} formation would be minimal, because it would involve very close juxtaposition of the small, highly charged C^{4+} and two similarly small and highly charged Si^{4+}. Moreover, a large amount of energy is involved in expanding the three-dimensional network of Si–O–Si bridges to accommodate the large CO_2 molecule. In $NaAlSi_3O_8$ melts, the potential for CO_3^{2-} formation is only slightly greater, as the repulsive forces between C^{4+} and Al^{3+} presumably are only slightly less than between C^{4+} and Si^{4+}; hence, the solubility of CO_2 should be slightly higher. Solubilities of CO_2 in silica melts at high temperatures and pressures have not been measured, but those in $NaAlSi_3O_8$, $Na_{1.33}Al_{1.33}Si_{2.67}O_8$, and $Na_2Al_2Si_2O_8$ melts have, by Eggler (1973a) and Mysen (1975). On an equimolal basis of 8.0 moles of O_2^-, Mysen (1975) has shown that CO_2 solubilities, although small in comparison with those of H_2O even at 30 kbar, increase in the above series roughly in proportion to (Na + Al)/Si, as predicted from the simple CO_3^{2-}-formation model. Moreover, the relative positions of the solubility curves at 1450°C (Mysen, 1975, Fig. 31) suggest that breaking the Al–O–Si bridges by Al entering octahedral coordination (Al^{VI}), where the AlO_6 · octahedra share edges as in jadeite, favors CO_3^{2-} formation and promotes higher CO_2 solubilities. At higher temperatures this effect is not as evident (Mysen, 1975, Figs. 32 and 33); presumably this is due to Al being predominantly in tetrahedral coordination, and hence to reformation of Al–O–Si bridges.

It is possible, of course, that some CO_2 enters these aluminosilicate melts in molecular form, without interaction with the silicate components. This is not likely to be a major mixing mechanism at CO_2 pressures of the order of 10 kbar and temperatures of about 1400°C, because the "holes" in the three-dimensional network of aluminosilicate tetrahedra are not large enough to accommodate the CO_2 molecule without major disruption and, thus, interaction with the network. At higher pressures, however, where Al changes to edge-shared octahedral coordination and the network is thereby disrupted, molecular CO_2 may gain more ready access to the melt. This mode of access might account for the presence of molecular CO_2 in $NaAlSi_3O_8$ glasses quenched from 1450° to 1600°C and 20 kbar, identified spectroscopically by Mysen (1975), but the effects of quenching on speciation, especially as the melt passes through the glass transition, need further assessment.

Experimental studies by Eggler (1973a), Holloway and Lewis (1974), Kadik and Eggler (1975), and Mysen (1975) have shown that addition of H_2O to the system $NaAlSi_3O_8$–CO_2 markedly increases the partitioning ("solubility") of CO_2 in favor of the melt. This increase in "solubility" is readily explained in accordance with the H_2O and CO_2 solution models; thus, as H_2O enters the melt, it preferentially breaks the Si–O–Si bridges (Fig. 16-2), increasing the expansibility of the melt to accommodate more readily molecular CO_2. It also produces non-bridging O^{2-} (as a result of the $Na^+ \rightleftharpoons H^+$ exchange) and OH^-, with which CO_2 can more readily interact to form CO_3^{2-} and HCO_3^-. The extent to which each of these mechanisms contributes to the overall solution of CO_2 in hydrous $NaAlSi_3O_8$ melts cannot be fully assessed at present; here, also, the spectroscopic data of Mysen (1975) suggest that molecular CO_2 is a major C-bearing species at room temperature.

In anhydrous melts more basic than those of feldspar composition, some of the O^{2-} are unshared with neighboring SiO_4 tetrahedra. Because these unshared O^{2-} do not form bridges between two small, highly charged cations in a highly polymerized network, they are more susceptible to CO_3^{2-} formation ("carbonation") and this results in higher CO_2 solubilities, just as in hydrous $NaAlSi_3O_8$ melts. Thus, the equimolal solubilities of CO_2 in $Ca_{1.33}Mg_{1.33}Si_{2.67}O_8$ (diopside) melts are roughly twice those in anhydrous $NaAlSi_3O_8$ melts and comparable to the maximum in hydrous $NaAlSi_3O_8$ melts (Mysen, 1975). Moreover, Eggler (1974) presented evidence that indicates lower CO_2 solubilities in diopside melts than in $Mg_4Si_2O_8$ (forsterite) melts, but higher than $Mg_{2.67}Si_{2.67}O_8$ (enstatite) melts. These higher solubilities in forsterite melt are predictable from the fact that Si–O–Si bridges are lacking and many of the O^{2-} are not linked to Si^{4+} in forsterite, whereas 50% of the O^{2-} in diopside and enstatite form Si–O–Si bridges. The different CO_2 solubilities in the latter two melts, however, must have a different explanation; it is probably related to the fact that the Gibbs free energies of formation of $CaCO_3$ at high temperatures are lower than those of $Ca_{0.5}Mg_{0.5}CO_3$ which, in turn, are lower than those of $MgCO_3$ (Robie and Waldbaum, 1968). Model considerations also lead to the conclusion, confirmed by the experiments of Eggler (1973a), that addition of H_2O to these more basic melts also enhances the solubility of CO_2, through increasing both the expansibility by depolymerization and the proportions of terminal OH^-.

Kushiro (1975) has shown that the effects of adding P_2O_5 to silicate melts that contain non-bridging O^{2-} are similar to those of adding CO_2, in that liquidus field boundaries are shifted markedly toward silica-poor compositions. These effects may be interpreted to indicate a preferential lowering of activities of those melt components that contain the highest proportions of non-briding O^{2-} by formation of the PO_4^{3-} complex anion. Formation of PO_4^{3-}, however, involves more than one O^{2-} $(3O^{2-}:2PO_4^{3-})$; hence, unlike the CO_3^{2-}-forming reaction, local charge balance at each non-bridging O^{2-} site involved is affected by PO_4^{3-} formation. Consequently, the activity of a given component that contains non-bridging O^{2-} should be depressed more by PO_4^{3-} than by an equimolal amount of CO_3^{2-}, a deduction that is consistent with experimental results on the system Mg_2SiO_4–$CaMgSi_2O_6$–SiO_2 by Kushiro (1975, P_2O_5) and Eggler (1974, CO_2). Like CO_2, on the other hand, the solubility of P_2O_5 is expected to be higher in hydrous melts than in anhydrous melts of the same silicate composition.

Sulfur dioxide commonly is a relatively abundant constituent of volcanic gases (Nordlie, 1971), but its abundance probably is due in large part to the hydrolysis of H_2S at magmatic temperatures and very low pressures (low f_w). Combination of the expression for the equilibrium constant of this hydrolysis reaction (K_{SO_2}) with that for dissociation of H_2O (K_{H_2O}) leads to the relation

$$\frac{f_{SO_2}}{f_{H_2S}} = \frac{K_{SO_2}f_{O_2}^{3/2}}{K_{H_2O}^3 f_w} \tag{16-10}$$

from which it can be seen that, with f_{O_2} approximately fixed by Fe-bearing condensed phases, f_{SO_2}/f_{H_2S} decreases with increasing f_w. Thus, at f_w values of 500 bars or greater and f_{O_2} values typical of hydrous magmas

(Carmichael *et al.*, 1974), the principal S-bearing volatile species is H_2S and it dissolves in accordance with reaction Eq. 16-7. At lower values of f_w or higher values of f_{O_2}, however, SO_2 can be an abundant volatile species, but its solubility in melts is very low because it dissolves by mechanisms analogous to those of CO_2.

The last common magmatic volatile to be considered, H_2, is important for the control it exerts on f_{O_2} in silicate melts. In iron-bearing melts, which includes the bulk of magmas, equilibria of the type

$$1/2\ H_2 + Fe^{3+} + O^{2-} \rightleftarrows Fe^{2+} + OH^- \qquad (16\text{-}11)$$

control f_{H_2} and, through the H_2O dissociation equilibrium, also f_{O_2}. From reaction Eq. 16-11, it is apparent that increasing f_{OH^-}, tends to shift equilibrium toward higher ferric-ferrous ratios. At the same time, however, increasing a_w^m also increases f_{H_2} through the $H_2O \rightleftarrows H_2 + 1/2\ O_2$ equilibrium. Hence, whether the ferric-ferrous ratio increases or decreases with increasing a_w^m depends upon the relative magnitudes of the equilibrium constants for the two reactions. On the other hand, in a given magma system where f_{O_2}—hence the ferric-ferrous ratio—is controlled by equilibrium with an Fe-bearing crystalline assemblage, increasing a_w^m causes reduction of the Fe^{3+}-bearing mineral. Therefore, the liquidus temperature of magnetite decreases markedly with increasing a_w, even at constant X_w^m, as observed by Eggler and Burnham (1973, Fig. 2),

MELTING RELATIONS IN SILICATE-VOLATILE SYSTEMS

GENERAL CONDITIONS FOR CRYSTAL-MELT EQUILIBRIUM

The fundamental criterion for heterogeneous equilibrium is that the chemical potential of a given component (μ_i) must be the same in all phases; hence, for equilibrium between a crystalline phase (s) and a melt (m) containing component i, $\mu_i^m = \mu_i^s$. For either phase m or s at equilibrium, $\mu_i = \mu_i^\circ + RT \ln a_i$, where μ_i° is the chemical potential of component i in its standard state at pressure (P), temperature (T), and unit activity ($a_i = 1$). It is convenient for present purposes, however, to choose standard states at temperature T, $P = 1$ bar, and $a_i = 1$ in both pure phases (m) and (s) of component i. With this choice of standard states, the equation of equilibrium at P and T becomes

$$\mu_i^m - \mu_i^s = 0 = \frac{\Delta G_{mi}^\circ + P\Delta V_{mi}}{RT} + \ln\left(\frac{a_i^m}{a_i^s}\right), \qquad (16\text{-}12)$$

where $\Delta G^{\circ}_{mi} = \mu^{\circ m}_i - \mu^{\circ s}_i$ is the Gibbs free energy of fusion at T and 1 bar, and $P\Delta V_{mi}$ is a first approximation to the pressure integral of $\bar{V}^m_i - \bar{V}^s_i$ (the difference in partial molal volumes of component i in melt and solid phases, respectively). The potential error introduced by this approximation is generally less than uncertainties on ΔG°_{mi} in the pressure range of present interest.

MELTING RELATIONS IN THE SYSTEM NaAlSi$_3$O$_8$–H$_2$O

Just as the H$_2$O solution model for the system NaAlSi$_3$O$_8$–H$_2$O can be extended to more complex igneous-rock melts, melting relations in this simple system can serve as a basis for understanding melting relations in H$_2$O-bearing rock systems.

Before Eq. 16-12 can be used to calculate albite-melt equilibrium under a given set of conditions, however, it is necessary to know ΔG°_{mab} as a function of T, a mean value of ΔV_{mab} over the P–T range of interest, and especially a^m_{ab} as a function of T and X^m_{ab} ($a^s_{ab} = 1.0$ in pure albite). Values of $(\Delta G^{\circ}_{mab} + P\Delta V_{mab})/RT$ computed from the data of Robie and Waldbaum (1968) and Burnham and Davis (1971) are graphed versus T in Fig. 16-5; hence, all that remains is to evaluate a^m_{ab} as a function of T and X^m_{ab} (or X^m_w). This evaluation is readily accomplished by application of the Gibbs-Duhem relation, $X^m_w \, d \ln a_w + X^m_{ab} \, d \ln a^m_{ab} = 0$, to Eqs. 16-3 and 16-4; thus, from Eq. 16-3 for $X^m_w \leq 0.5$

$$a^{hm}_{ab} = (X^{hm}_{ab})^2 = (1 - X^m_w)^2 \qquad (16\text{-}13)$$

and from Eq. 16-4 for $X^m_w > 0.5$

$$\ln a^{hm}_{nb} = (6.52 - 2667/T) \left[\ln (1 - X^m_w) + X^m_w \right] - 515/T - 0.127, \qquad (16\text{-}14)$$

where the superscript hm now denotes hydrous melt. It may be noticed that Eq. 16-13 is a statement of an analogue of Raoult's law for the activity of a solvent (NaAlSi$_3$O$_8$ melt) in which the solute (H$_2$O) is dissociated (Eq. 16-2).

It is apparent that, by solving Eq. 16-13 for a given value of X^m_w and substituting the result in Eq. 16-12, the temperature of equilibrium between albite and melt at a given pressure may be obtained from Fig. 16-5. Similar calculations can be carried out for Eq. 16-14, but owing to the T dependence of both $\ln a^{hm}_{ab}$ and $(\Delta G^{\circ}_{mab} + P\Delta V_{mab})/RT$ in Eq. 16-12, iterative procedures must be employed. The results of such calculations are presented in the P–T projection of Fig. 16-6 as constant X^m_w contours (isopleths) on the albite liquidus (S–L) surface.

At some pressure on each of the albite liquidus isopleths in Fig. 16-6 the melt is saturated with H$_2$O. This pressure may be readily obtained by

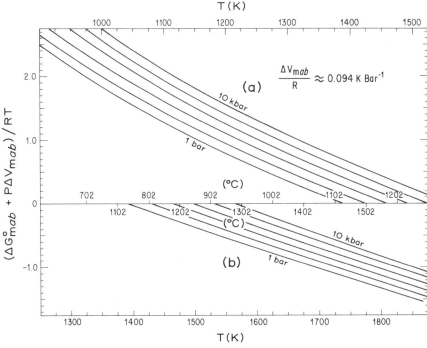

Figure 16-5. Molal Gibbs free energy of melting of albite in the pressure interval from 1 bar to 10 kbar assuming an average $\Delta V_{mab}/R = 0.094$K bar^{-1}. (a) Atmospheric pressure data from Robie and Waldbaum (1968). (b) Atmospheric pressure curve extrapolated on the assumption that $\Delta C^{\circ}{}_{pmab}$ is constant at 1390K value over the extrapolated temperature interval.

solving either Eq. 16-3 or 16-4 (as appropriate) for ln k at $a_w = 1.0$ and finding the corresponding pressure by interpolation in Fig. 16-3. The results of such calculations are also presented in Fig. 16-6 as constant X_w^m contours (isopleths) on the H_2O saturation (L–V) surface.

The intersection of these L–V and S–L surfaces defines a univariant S–L–V curve (H_2O-saturated solidus) where $a_w = 1.0$ and albite is in equilibrium with H_2O-saturated melt and essentially pure H_2O. In reality the aqueous phase is not pure H_2O, but the small discrepancy between the calculated and experimentally determined (dashed curve in Fig. 16-6) curves through 10 kbar[2] indicates that the impurities (dissolved

---

[2] The experimental point at 10 kbar indicated by the cross (+) is from Boettcher and Wyllie (1969). Those of Luth et al. (1964) indicate an equilibrium temperature about 25°C higher, hence a few degrees above the calculated temperature (open circle).

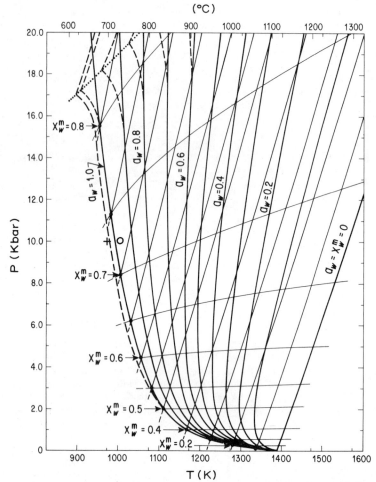

Figure 16-6. Pressure-temperature projection of phase relations in the system $NaAlSi_3O_8-H_2O$. Light solid lines: saturation (solubility isopleths) for H_2O in $NaAlSi_3O_8$ melt, with values indicated at low-temperature terminations. Medium solid lines: albite-melt equilibrium at constant X_w^m, with values of X_w^m indicated at low temperature terminations (intersection with H_2O solubility surface). Heavy solid lines: albite-melt equilibrium at constant a_w; intersection of solubility and liquidus surfaces defines univariant H_2O-saturated solidus ($a_w = 1.0$). Experimentally determined H_2O-saturated solidus indicated by heavy dashed line; data below 10 kbar are from Burnham and Jahns (1962), Luth et al. (1964), Morse (1969), and Tuttle and Bowen (1958), whereas data above 10 kbar are from Boettcher and Wyllie (1969). Dotted line represents the reaction albite ⇌ jadeite + quartz, and the light dashed lines schematically illustrate effects of the $Al^{IV} \rightarrow Al^{VI}$ shift on the constant a_w melting relations.

silicates) have a minor effect on a_w. A more significant aspect of this small discrepancy, however, is the confirmation it provides for the general validity of the thermodynamic data used in the calculations. In fact, the discrepancy at 10 kbar could be accounted for by an error of only 200 calories in $\Delta G^\circ_{mab} + P\Delta V_{mab}$ and this is less than 20% of the total uncertainty on the ΔG°_{mab} function at 298K (Robie and Waldbaum, 1968). Of course, several other causes of the discrepancy, such as an error of only 0.01 in X^m_w at saturation, also are possible.

The $a_w = 1.0$ S–L–V curve in Fig. 16-6 is only one of a family of isoactivity melting curves that define an albite-melt equilibrium surface in P–T–a_w space. This surface, which is of major importance in understanding melting relations in the presence of mixed volatiles, is represented by the heavy isoactivity contours projected onto the P–T plane in Fig. 16-6. Each isoactivity contour was constructed following procedures similar to those employed to locate the S–L–V curve for $a_w = 1.0$.

It is apparent from Fig. 16-6 that the experimental H_2O-saturated solidus curve (Boettcher and Wyllie, 1969) deviates to markedly lower temperatures from the calculated $a_w = 1.0$ equilibrium curve above about 16 kbar. This deviation is caused by the appearance of jadeite as a product of incongruent melting of albite under these conditions. In accordance with the quasi-crystalline melt model discussed previously, the coexistence of jadeite and melt implies the existence in the melt of jadeite-like structural entities in which Al is octahedrally coordinated, instead of tetrahedrally coordinated as in albite. As a consequence of converting part of the $NaAl^{IV}Si_3O_8(ab)$ component to $Na_{1.33}Al^{VI}_{1.33}Si_{2.67}O_8(jd)$ and $Si_4O_8(qz)$ components, a^m_{ab} and, therefore, T are lowered below the values calculated from Eqs. 16-14 and 16-12, respectively. In fact, at 17 kbar and 910K where albite, jadeite, quartz, melt, and "vapor" coexist (Boettcher and Wyllie, 1969), a^{hm}_{ab} (and X^{hm}_{ab}) are about 70% of their corresponding values in a melt that contains only ab component and H_2O.

A further small increase in pressure above the albite $\rightleftarrows$ jadeite + quartz reaction boundary greatly enhances the Al coordination shift ($Al^{IV} \rightarrow Al^{VI}$) and this, in turn, causes a marked further decrease in a^{hm}_{ab}. Of course, a^{hm}_{jd} and a^{hm}_{qz} increase concomitantly, and this causes a sharp rise in the H_2O-saturated solidus temperature of the jadeite-quartz assemblage. After essentially all of the ab component has been converted to jd and qz components (at about 20 kbar), a further increase in pressure, hence increase in X^m_w, causes a^{hm}_{jd} and a^{hm}_{qz} to decrease again, but the very large $P\Delta V_{mjd}$ term in Eq. 16-12 more than compensates for this decrease. Therefore, H_2O-saturated solidus temperatures continue to increase with pressure at a nearly uniform rate (see Boettcher and Wyllie, 1969, Fig. 4).

Thermodynamic formulations other than Eq. 16-12 could have been used to calculate albite-melt equilibrium relations in Fig. 16-6, but none of them is simpler or makes more obvious the fundamental controls of melting relations in this, or any other system. Clearly, one of the controls is $(\Delta G^\circ_{mab} + P\Delta V_{mab})/RT$—a fundamental property of albite substance—and the other is a^m_{ab}—a function of X^m_{ab} and the thermodynamic mixing behavior of $NaAlSi_3O_8$ component in the melt phase. Therefore, mixing any substance with $NaAlSi_3O_8$ melt, in whatever proportions are required to yield a singular value of a^m_{ab}, will lower the albite-melt equilibrium temperature the same amount, provided that substance does not form a solid solution with albite. In this respect, the effects of H_2O on melting relations are not different from those of other substances, such as Si_4O_8. Because of the very large differences in gram formula weight between H_2O (18.02g) and $NaAlSi_3O_8$ (262.2g), however, a small weight percent H_2O in the melt represents a much larger mole fraction and, especially when squared as in Eq. 16-13, has a profound effect on a^m_{ab}. For example, 1.0 wt % H_2O ($X^m_w = 0.128$) in $NaAlSi_3O_8$ melt lowers a^m_{ab} by 0.24 (Eq. 16-13) and the melting temperature of albite by 70K (Fig. 16-5), whereas addition of 1.0 % Si_4O_8 (240.3g) lowers a^m_{ab} by only 0.01 and the melting temperature of albite only one or two degrees. To depress the melting temperature of albite 70K by adding Si_4O_8, it would be necessary to add about 22 wt %, and for other silicate components of larger molal mass the amount would be even greater.

MELTING RELATIONS IN THE SYSTEM $NaAlSi_3O_8$–$CaAl_2Si_2O_8$–H_2O

The equation analogous to Eq. 16-12 for crystal-melt equilibrium in this multicomponent system may be written

$$\ln\left(\frac{a^{hm}_{ab}a^s_{an}}{a^s_{ab}a^{hm}_{an}}\right) = \frac{\Delta G^\circ_{man} - \Delta G^\circ_{mab} + P(\Delta V_{man} - \Delta V_{mab})}{RT}, \qquad (16\text{-}15)$$

from which it is apparent that, for general application, it is first necessary to know not only $\Delta G^\circ_{man}/RT$, ΔV_{man}, a^s_{ab}, and a^s_{an}, but also how the ab and an components in the melt mix with H_2O and with each other. The successful extension of the $NaAlSi_3O_8$–H_2O solution model (Eqs. 16-3 and 16-4) to igneous-rock melts demonstrated above implies that all of the aluminosilicate components, properly defined, mix with H_2O in precisely the same way. Therefore, Eqs. 16-13 and 16-14 also are applicable to the an component in this or any other aluminosilicate melt of petrologic interest. Furthermore, the fact that Eqs. 16-3 and 16-4 are equally applicable to all aluminosilicate components, again appropriately defined, implies that these aluminosilicate components mix with each other in accordance with Raoult's law ($\gamma_a^{am} = 1$).[3] Therefore,

$$a_a^{am} = X_a^{am}, \tag{16-16}$$

where subscript a refers to an aluminosilicate component, such as ab or an in the present case, and superscript am refers to anhydrous melt.

In theory, Eq. 16-16 is rigorously applicable only to those melt compositions where the validity of Eq. 16-3 can be demonstrated. From published phase equilibrium data, however, it can be shown that Eq. 16-16 is valid for mixing aluminosilicate components with Si_4O_8 (qz) in all proportions and with the other common mineral-forming silicate components in the range of melt compositions where a crystalline aluminosilicate is the stable liquidus phase. Therefore, Eq. 16-16 can be combined with Eqs. 16-13 and 16-14 to yield expressions for the activity of a given aluminosilicate component in a hydrous melt where a feldspar or quartz is the stable liquidus phase; thus, for $X_w^m \leq 0.5$

$$a_a^{hm} = X_a^{am}(1 - X_w^m)^2, \tag{16-17}$$

and for $X_w^m > 0.5$

$$\ln a_a^{hm} = \ln X_a^{am} + \left(6.52 - \frac{2667}{T}\right)\left[\ln (1 - X_w^m) + X_w^m\right] - \frac{515}{T} - 0.127.$$

$$\tag{16-18}$$

A detailed discussion of the Raoult's law relation in Eq. 16-16 and all of its petrological ramifications is beyond the scope of this chapter. Its importance to understanding phase equilibrium relations in igneous rock systems, however, is so great that certain general consequences of it deserve brief attention here. First, by application of the Gibbs-Duhem equation to the hypothetical ternary system 1–2–3 ($X_1 d \ln \gamma_1 + X_2 d \ln \gamma_2 + X_3 d \ln \gamma_3 = 0$), it is readily apparent that, if components 1 and 2 obey Raoult's law ($\gamma_1 = \gamma_2 = 1$), then component 3 must obey Henry's law ($\gamma_3 = k$). Therefore, within the composition range where an aluminosilicate mineral is the stable liquidus phase, other silicate (plus Si_4O_8) components—properly chosen on the basis of eight moles of O^{2-}—obey Henry's law. A second consequence, which follows from the first, is that throughout this same composition range, equilibrium separation of two immiscible silicate melts is precluded. Stable silicate melt immiscibility is not precluded outside this composition range, however.

---

[3] The proof of this statement is too lengthy to be presented here. It can be carried out following the treatment of Darken (1950), as extended by Wagner (1952) and Burnham *et al.* (1978), regarding application of the Gibbs-Duhem relation to multicomponent systems.

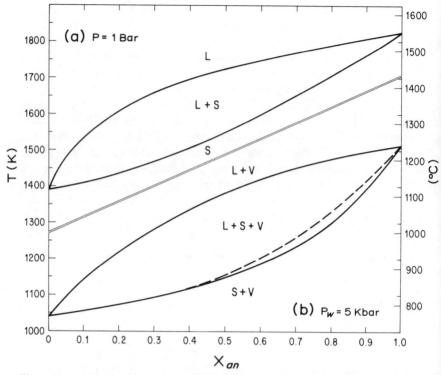

Figure 16-7. Melting relations of plagioclase: L–melt, S–plagioclase, and V–aqueous ("vapor") phase. (a) Experimental data of Bowen (1913) at atmospheric pressure, as modified by Schairer (1957); the experimentally determined solidus is essentially coincident with that calculated from Eqs. 16-12 and 16-16 and the data for ab in Fig. 16-5. (b) Experimental liquidus and solidus (dashed curve) data of Yoder et al. (1957) at $P_w = 5.0$ kbar; the solid solidus curve was calculated as described in the text.

Nor is stable immiscibility precluded between aluminosilicate melt and another melt rich in some non-silicate component such as P_2O_5 or CO_2.

With the expressions for a_{ab}^{hm} and a_{an}^{hm} in Eq. 16-15 established, it is now possible to evaluate a_{ab}^s, a_{an}^s, ΔG_{man}°, and ΔV_{man} from published experimental data. Thus, from the one-bar plagioclase liquidus data of Bowen (1913) and Schairer (1957) shown in Fig. 16-7a, coupled with Eq. 16-12 ($i = ab$ and $P\Delta V_{mab} = 0$), Eq. 16-16, and Fig. 16-5b,[4] calculated values of a_{ab}^s

---

[4] The $\Delta G_{mab}^\circ/RT$ curve from the data of Robie and Waldbaum (1968) was extrapolated to higher temperatures on the assumption that the heat capacity of fusion (ΔC_{pmab}°) remains constant at its 1390K value.

were found to be within ± 0.01 of X_{ab}^s for all plagioclase solidus temperatures in Fig. 16-7a. Hence, plagioclase solid solutions at atmospheric
pressure and solidus temperatures are Raoultian with respect to the ab
component ($\gamma_{ab}^s = 1$). On the assumption that γ_{an}^s also is unity under these
conditions, as contended by Bowen (1913), experimental mole fractions
from Fig. 16-7a were next substituted for activities in Eq. 16-15 and the
equation solved for $\Delta G_{man}^\circ / RT$. The results of these calculations, which
are plotted in Fig. 16-8a, yield an essentially straight line whose slope
gives a heat of fusion (ΔH_{man}°) at 1825K of about 33 kcal mole^{-1}. Although
this ΔH_{man}° is somewhat larger than that obtained by Bowen (29 kcal
mole^{-1}), it is important to note that it would have been even larger if
$\gamma_{an}^s > 1.0$ had been assumed. Of course, values of $\gamma_{an}^s < 1.0$ (negative
deviations from Raoult's law) are not to be expected in this continuous
solid-solution series.

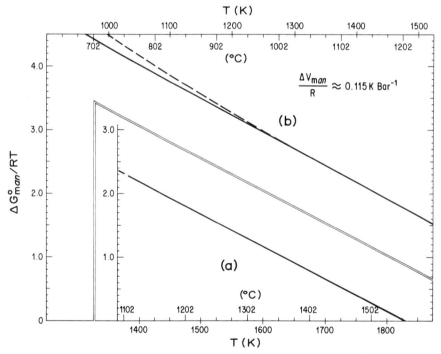

Figure 16-8. Molal Gibbs free energy of melting of anorthite at atmosphere pressure.
(a) Calculated from Eq. 16-15 and the experimental data of Bowen (1913) and
Schairer (1957) in Fig. 16-7a. (b) Linear extrapolation from (a) below 1390K
(solid line), and an extrapolation based on the assumption of constant ΔC_{pman}°
below 1300K (dashed line). Value of $\Delta V_{man}/R \approx 0.115$K bar^{-1} calculated as
described in the text.

The lone remaining parameter in Eq. 16-15, ΔV_{man}, was evaluated from Fig. 16-8a and the experimentally determined melting temperatures of anorthite at 2.0 kbar (Stewart, 1956) and 5.0 kbar (Yoder, 1965) H_2O pressure by appropriate substitutions of Eqs. 16-17 and 16-18 and solving Eq. 16-12 ($i = an$) for $\Delta V_{man}/R$. The average value obtained for this latter parameter ($\Delta V_{man}/R = 0.115$K bar^{-1}) is slightly higher than the corresponding value for $\Delta V_{mab}/R = 0.094$, but reasonable nonetheless. Inasmuch as this calculation involved subtraction of $\Delta G^{\circ}_{man}/RT$ values in Fig. 16-8a from $-(T/P)\ln a^{hm}_{an}$, it is clear that any smaller values of $\Delta G^{\circ}_{man}/RT$, hence smaller values of ΔH°_{man}, would have yielded unreasonably large values of $\Delta V_{man}/R$. Conversely, values of γ^s_{an} significantly greater than 1.0 (larger $\Delta G^{\circ}_{man}/RT$) would have resulted in unreasonably small values for $\Delta V_{man}/R$.

Although albite and anorthite appear to mix ideally ($\gamma^s_{ab} = \gamma^s_{an} = 1.0$) at solidus temperatures and atmospheric pressure, the experimental data of Orville (1972) at 973K and 2.0 kbar and of Windom and Boettcher (1976) at 1373K and 15 to 20 kbar clearly indicate $\gamma^s_{an} > 1.0$ over much of the plagioclase composition range. Hence, at high H_2O pressures, where plagioclase solidus temperatures are greatly depressed, it is possible that $\gamma^s_{an} \neq 1.0$. To test this possibility the H_2O-saturated liquidus data of Yoder et al. (1957) at 5.0 kbar were combined with those in Fig. 16-5 to calculate a^s_{ab} via Eqs. 16-18 and 16-12. The results, assuming $X^s_{ab} = a^s_{ab}$, are shown in Fig. 16-7b as the solid solidus curve. Agreement with the experimentally determined solidus (dashed curve) is precise at higher values of X^s_{ab}, but deviates noticeably at lower values; however, the calculated curve lies everywhere within the experimental brackets (uncertainty). Thus, γ^s_{ab} apparently is essentially unity at temperatures as low as 1050K.

On the other hand, if these calculations are repeated for the an component, using a linear extrapolation of $\Delta G^{\circ}_{man}/RT$ to lower temperatures (Fig. 16-8b), it is found that $a^s_{an} = X^s_{an}$ only at temperatures above about 1375K and $X^s_{an} > 0.88$. At lower temperatures, hence lower values of X^s_{an}, γ^s_{an} appears to increase abruptly at first, to a value of about 1.2, then to increase gradually to about 1.8 at 1075K. The abrupt increase is thought to be due to a change in structural state of plagioclase over a narrow composition interval (Windom and Boettcher, 1976), but the apparent gradual increase at lower temperatures must have another explanation. It may be due in small part to minor experimental errors in location of the plagioclase liquidus, but much of it probably is caused by a change in coordination of Al ($Al^{IV} \rightarrow Al^{VI}$ shift) involving only the an component. Thus, in hydrous melts of intermediate plagioclase composition at high pressures the $Al^{IV} \rightarrow Al^{VI}$ shift is postulated to result in formation of zoisite-like $(Ca_{1.23}Al_{1.85}Si_{1.85}O_{7.38}OH_{0.62})$ structural units, in which AlO_6 octahedra share corners with SiO_4 tetrahedra. As a consequence, a_{an}^{hm} is lowered relative to that calculated from Eq. 16-18 and this has the same effect on lowering liquidus temperatures (Eq. 16-15) as increasing $a_{an}^s = \gamma_{an}^s X_{an}^s$ by increasing γ_{an}^s. Evidence in support of this interpretation comes from the discovery by Furst (1978) that plagioclase of intermediate composition ($X_{an}^s = 0.3$ to 0.75) melts incongruently to zoisite plus an aluminum silicate-enriched plagioclase melt in the presence of H_2O at 8.0 kbar, the lowest pressure investigated.

The $Al^{IV} \rightarrow Al^{VI}$ shift involving only *an* component is not restricted to hydrous melts at high pressures. It can be shown from all available experimental data to occur, even at atmospheric pressure, in all silica-saturated melts that contain octahedrally coordinated cations and is especially pronounced in those that contain Mg. Moreover, it also can be shown that the $Al^{IV} \rightarrow Al^{VI}$ shift involves the *ab* component in silica undersaturated (olivine normative) melts at atmospheric pressure. The overall effect of this shift on plagioclase-melt equilibrium, then, is to lower equilibrium temperatures and, in silica-saturated melts, to increase the albite content of the equilibrium plagioclase.

Increasing pressure on aluminosilicate melts, either hydrous or anhydrous, is well known to result in expansion of the liquidus fields of those phases in which at least part of the Al is octahedrally coordinated, such as corundum, garnets, spinel, aluminous pyroxenes, and zoisite just noted. In accordance with the quasi-crystalline melt model, this implies that pressure enhances the $Al^{IV} \rightarrow Al^{VI}$ shift, which in turn implies that activities of those melt components in which Al is tetrahedrally coordinated decrease with pressure. Hence, despite the positive nature of the $P\Delta V_{mi}$ term in Eq. 16-12, which tends to increase equilibrium temperatures (Fig. 16-6), decreasing a_{ab}^m and a_{an}^m (especially a_{an}^m) through the $Al^{IV} \rightarrow Al^{VI}$ shift tends to decrease equilibrium temperatures in plagioclase-bearing systems. This effect, which is enhanced in hydrous melts by formation of zoisite-like structural units, is clearly evident in the experimental results of Eggler and Burnham (1973) on Mt. Hood andesite (Table 16-1) up to 10 kbar, and of Whitney (1975) on synthetic "haplogranodiorite" compositions up to 8.0 kbar. It is reflected in a negative slope of the H_2O-undersaturated plagioclase liquidi in pressure-temperature projection.

The experimental results of Eggler and Burnham (1973) also show that, as the slope of the plagioclase liquidus at constant $X_w^m = 0.43$ decreases with increasing pressure, the slope of the clinopyroxene liquidus increases. The immediate cause of this increase in slope presumably is increased activities of aluminum pyroxene (Ca- and Mg-Tschermak's) components in the clinopyroxene solid solution in which half the Al are octahedrally coordinated. Of course, these increased activities in the solid solution merely reflect corresponding increases in the activities of these Al^{VI}-bearing components in the melt resulting from a pressure-enhanced $Al^{IV} \rightarrow Al^{VI}$ shift.

There are numerous petrologically important consequences of the $Al^{IV} \rightarrow Al^{VI}$ shift for melting relations in igneous-rock systems that cannot be discussed here. One of them, however, that already has been touched upon above and may have an important bearing on the controversial problem of anorthosite genesis, deserves additional brief comment. Owing to the complementary relationships between slopes of the plagioclase and pyroxene liquidi in dioritic melts, decompression (upward intrusion) of a melt initially in equilibrium with these phases at a pressure greater than about five kilobars will result in crystallization of plagioclase and resorption of pyroxene. This decompression need not be adiabatic, as the heat of crystallization of plagioclase is sufficient to sustain considerable heat loss to the wallrocks without a significant drop in temperature. Because the dioritic melts from which this plagioclase crystallizes are less dense, gravitational settling of crystals is a viable process for forming bodies of cumulate plagioclase.

Thus, despite the numerous complications the $Al^{IV} \rightarrow Al^{VI}$ shift introduces into quantitative assessments of crystal-melt equilibria in plagioclase-bearing systems, recognition of its existence and Raoultian behavior of the aluminosilicate melt components has provided a thermodynamically tested basis for explaining many previously enigmatic melting relations in igneous rock-forming systems. Perhaps in no other apparently simple multicomponent system is this better illustrated than in the system $CaAl_2Si_2O_8(an)–Ca_{1.33}Mg_{1.33}Si_{2.67}O_8(di)–H_2O$, especially as regards the combined effects of H_2O and the $Al^{IV} \rightarrow Al^{VI}$ shift on crystal-melt equilibria.

MELTING RELATIONS IN THE SYSTEM
$CaAl_2Si_2O_8–Ca_{1.33}Mg_{1.33}Si_{2.67}O_8–H_2O$

Crystal-melt equilibria in the systems $CaAl_2Si_2O_8–Ca_{1.33}Mg_{1.33}Si_{2.67}O_8$ and $CaAl_2Si_2O_8–Ca_{1.33}Mg_{1.33}Si_{2.67}O_8–H_2O$ at various pressures are shown, modified from Yoder (1965, Fig. 11), in Fig. 16-9. At atmospheric pressure ($P = 1$ bar, $X_w^m = 0$), anorthite and melt are in equilibrium with a diopside–Ca-Tschermak's solid solution at 1547K and $X_{an}^{am} = 0.41$. The occurrence of Ca-Tschermak's component in diopside implies the existence of cts ($Ca_{1.33}Al_{1.33}^{VI}Al_{1.33}^{IV}Si_{1.33}O_8$) in the melt, having formed from an ($CaAl_2Si_2O_8 \rightleftarrows 0.75$ $Ca_{1.33}Al_{1.33}^{VI}Al_{1.33}^{IV}Si_{1.33}O_8 + 0.25$ Si_4O_8) in re-

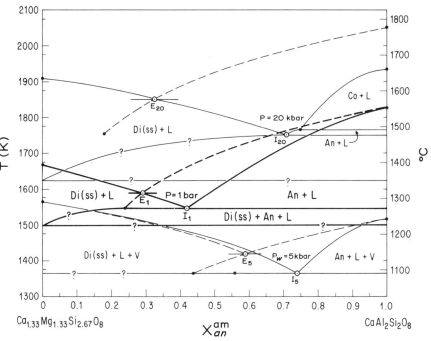

Figure 16-9. Projected melting relations in the system $CaAl_2Si_2O_8–Ca_{1.33}Mg_{1.33}Si_{2.67}O_8–$ H_2O modified from Yoder (1965). The $P = 1$ bar and $P = 20$ kbar projections are for fluid absent ("dry") conditions, whereas the $P_w = 5.0$ kbar projection is for H_2O-saturated conditions. The dashed anorthite liquidus curves were calculated from text Eqs. 16-12 and 16-16, neglecting effects of the $Al^{IV} \rightarrow Al^{VI}$ shift. The points I_1, I_5, and I_{20} are experimentally determined or estimated (I_{20}) isobaric invariant points, and E_1, E_5, and E_{20} are hypothetical eutectics between anorthite and pure diopside (no $Al^{IV} \rightarrow Al^{VI}$ shift), as discussed in the text.

sponse to the $Al^{IV} \rightarrow Al^{VI}$ shift. In consequence, a_{an}^{am} is lowered with respect to a given experimental value of X_{an}^{am} and $X_{an}^{am} - a_{an}^{am}$ is, in accordance with already established Raoultian behavior (Eq. 16-16), the mole fraction of *an* converted to *cts* and *qz* (Si_4O_8). Thus, were it not for the $Al^{IV} \rightarrow Al^{VI}$ shift and consequent *cts* formation, a true binary eutectic (E_1) between pure diopside and anorthite presumably would occur at about 1580K and $X_{an}^{am} = 0.29$, as indicated in Fig. 16-9. Furthermore, the mole fraction of *an* converted ($X_{an}^{am} - a_{an}^{am}$), when divided by the mole fraction of the component that contains the octahedral sites (X_{di}^{am}), yields a ratio $R_{di}^{an} = 0.29$ which is precisely the ratio of octahedral sites to octahedral plus tetrahedral sites that Al can occupy [potential $Al^{VI}/(Al^{VI} + Al^{IV})$ sites] in a 1:1 mole mixture of di and an $[1.33/(2.67 + 2) = 0.29]$.[5]

At 20 kbar and 1753K, on the other hand, $X_{an}^{am} = 0.71$ at the invariant point I_{20}, but $a_{an}^{am} = 0.18$. The difference, divided by X_{di}^{am}, yields a value of $R_{di}^{an} = 1.83$, which is only 0.01 greater than the sum of potential $Al^{VI}/$ $(Al^{VI} + Al^{IV})$ sites in 1:1 mole mixtures of di, co $(Al_{5.33}^{VI}O_8)$, gr $(Ca_2Al_{1.33}^{VI}Si_2O_8)$, and py $(Mg_2Al_{1.33}^{VI}Si_2O_8)$ with an. Hence R_{di}^{an}, which is a measure of the $Al^{IV} \rightarrow Al^{VI}$ shift, increases on the average of 0.077 kbar^{-1}. Furthermore, were it not for this $Al^{IV} \rightarrow Al^{VI}$ shift, a binary eutectic between anorthite and pure diopside (E_{20}) presumably would occur at about 1850K and $X_{an}^{am} = 0.32$. This composition is only about three mole percent richer in an than E_1, an illustration of the fact that ΔH_{mdi} and ΔV_{mdi} are very close to ΔH_{man} and ΔV_{man}, respectively.

In strong contrast to this very minor shift in the anhydrous anorthite-diopside eutectic composition with pressure, the projected H_2O-saturated eutectic at $P_w = 5.0$ kbar$(E_5$, Fig. 16-9) is shifted by 30 mol $\%$ $(X_{an}^{am} = 0.59)$, relative to the one-bar value. This dramatic shift is due, of course, to the differential lowering of a_{an}^{hm} by selective concentration of H_2O in the aluminosilicate structural units (components) of the melt, as mentioned in earlier discussions of H_2O solubilities in Columbia River basalt melts. Thus, the very profound effects of even modest H_2O pressures (5.0 kbar) on the composition of first-formed melts in the system diopside–anorthite–H_2O find ready explanation in the thermodynamic properties of the melts, but only if the quasi-crystalline melt model is adopted in the selection of components.

Melting Relations in Silicate-Mixed Volatile Systems

Space limitations and lack of data on mixing properties of volatiles other than H_2O in silicate melts preclude a quantitative treatment of the thermodynamics of melting in silicate-mixed volatile systems. Therefore, only a few brief generalizations will be made here concerning the differential effects of H_2O and CO_2 on crystal-melt equilibria. The reader interested in the effects of mixed H_2O-CO_2 volatiles on melting relations is referred especially to the experimental results of Eggler (1973a), Eggler and Burnham (1973), Holloway and Burnham (1972), Kadik and Eggler (1975), Mysen and Boettcher (1975a,b), and Rosenhauer and Eggler (1975).

---

[5] A more comprehensive discussion of this important relationship is beyond the scope of this chapter. It is appropriate to note, however, that the relationship appears to hold for all melts in which Mg^{2+} is the octahedrally coordinated cation at atmospheric pressure. Furthermore, in $Mg_4Si_2O_8(fo)$-bearing melts, Al from ab component also participates in the $Al^{IV} \rightarrow Al^{VI}$ shift and in $Fe_4Si_2O_8(fa)$-bearing melts $R_{fa}^a \approx 0.38\ R_{fo}^a$.

Bearing in mind that equations of the type of Eq. 16-12 for each of the liquidus phases in a multicomponent system at equilibrium can be combined as in Eq. 16-15, it is readily apparent from the last section that addition of another substance (H_2O) which differentially lowers the activity of one melt component (a_{an}^{hm}) tends to enrich the equilibrium melt composition in the component thus affected. Hence, owing to the selectivity of H_2O toward reaction with Si–O–Si bridges, especially if the bridges are in an aluminosilicate component where the proton-cation exchange can operate, H_2O-bearing melts at equilibrium are enriched in *ab*, *or*, *an*, and *qz* components relative to their anhydrous counterparts. In contrast, the selectivity of CO_2 toward CO_3^{2-} formation with non-bridging O^{2-} differentially lowers the activities of components such as *fo* ($Mg_4Si_2O_8$), *en* ($Mg_{2.67}Si_{2.67}O_8$), and *di* ($Ca_{1.33}Mg_{1.33}Si_{2.67}O_8$), and hence CO_2-bearing melts at equilibrium tend to be enriched in those components that contain non-bridging 0^{2-} relative to their CO_2-free counterparts.

Consideration of the H_2O solution model leads to the following predicted order in which the activities of melt components (a_i^m) are depressed at $P \leq 10$ kbar: $ab = or = an > qz \gg di = en = fs(Fe_{2.67}Si_{2.67}O_8) > fo = fa(Fe_4Si_2O_8)$. On the other hand, similar consideration of the CO_2 solution model leads to the following order in the same pressure range: $fo = fa > di = en = fs > ne > lc > an > ab = or > qz$. At pressures much greater than 10 kbar, the extensive $Al^{IV} \rightarrow Al^{VI}$ shift introduces complications among the aluminosilicates components, but the general order between then plus *qz*, as a group, and the other components listed (those that contain nonbridging O^{2-}) remains valid. Thus, at pressures up to perhaps 30 kbar, H_2O depresses the activities of aluminosilicates plus *qz* the most, the *olivine* components the least, and the *pyroxene* components to an intermediate extent, whereas CO_2 has just the opposite effect.

It should be emphasized that the order in which individual activities are depressed is not necessarily the order in which liquidus temperatures are depressed, as ΔT_{mi} also is inversely dependent upon the temperature slope of $\Delta G_{mi}^\circ/RT$ in Eq. 16-12 $[d(\Delta G_{mi}^\circ/RT)/dT = -\Delta H_{mi}^\circ/RT^2]$. However, except for the fact that quartz liquidus temperatures in some bulk compositions are depressed more than those of the feldspars at a given pressure and f_w, owing to very small values of ΔH_{mqz}°, the orders of T_{mi} and a_i^m depression are approximately the same. A mafic mineral assemblage, therefore, tends to yield silica-saturated partial melts at high values of f_w and silica-undersaturated partial melts at high values of f_{CO_2}.

These contrasting effects of H_2O and CO_2 have been observed experimentally by several workers. In the volatile-free system $Mg_4Si_2O_8$–$Ca_{1.33}Mg_{1.33}Si_{2.67}O_8$–$Si_4O_8$ at 20 kbar, Kushiro (1969) showed that melt in equilibrium with forsterite, diopside, and orthopyroxene contains only these normative constituents, but under H_2O-saturated conditions the melt in equilibrium with the same phase assemblage is quartz normative. Under CO_2-saturated conditions at 30 kbar, on the other hand, Eggler (1974) found that the melt in equilibrium with this same assemblage is larnite (Ca_2SiO_4) normative, and thus it is grossly undersaturated with respect to silica. On the basis of these results, Eggler (1974) proposed that quartz-normative tholeiitic melts may be produced by partial melting of peridotitic upper mantle in the presence of H_2O-rich volatiles, whereas olivine- and nepheline-normative basalt melts may be produced in the presence of CO_2-rich volatiles.

This proposal finds strong support in the experimental results of Mysen and Boettcher (1975b), who investigated melting relations of natural peridotites in the presence of a mixed H_2O-CO_2 volatile phase. At pressures up to at least 25 kbar, partial melts formed in the presence of a volatile phase where $X_w^v \geq 0.5$ were found to be andesitic in composition, whereas melts formed where $X_w^v \leq 0.4$ were both olivine and nepheline normative; furthermore, the so-called alkalinity of the melts increased with decreasing X_w^v below 0.5. These authors further suggested that high values of f_{CO_2}, coupled with low values of f_w, may be the conditions under which kimberlitic magmas are generated in the mantle; such conditions also may be requisite for generation of carbonatite magmas (Wyllie and Huang, 1975).

In this latter connection, it will be recalled that P_2O_5, like CO_2, has a strong affinity for reaction with the non-bridging O^{2-} in silicate melts to form PO_4^{3-}. Owing to this affinity and the fact that one mole of P_2O_5 yields two formula units of PO_4^{3-} and involves three moles of O^{2-}, the activities of components containing non-bridging O^{2-} are depressed more by P_2O_5 than by equimolal amounts of CO_2. Hence, P_2O_5-bearing partial melts formed from peridotitic compositions tend to be markedly undersaturated with respect to silica, as forcefully demonstrated by Kushiro (1975).

MELTING OF HYDROUS MINERAL ASSEMBLAGES

Since the experimental work of Yoder and Tilley (1962) on basalt-H_2O systems and the realization that hydrous mineral assemblages may play a key role in magma generation, especially in regions of lithospheric plate convergence, the number of publications concerned with partial melting of hydrous mineral assemblages at high pressures has increased dramatically. Most of these publications have dealt with melting of amphibole- or mica-bearing assemblages in the presence of H_2O or a mixture of H_2O and CO_2, and few have dealt with fluid-absent melting of these assemblages—probably the most prevalent melting process in subduction zones and the deep continental crust. The following paragraphs, therefore, will be devoted to an extension of the discussion by Eggler (1973b), on the factors that control hydrous mineral-melt equilibria, with special emphasis on a system of basaltic composition.

In the absence of a fluid phase, a non-porous mafic amphibolite remains stable (no melt present) over a large pressure-temperature region above the H_2O-saturated solidus. This region is bounded at low pressures by the intersection of the sub-solidus hornblende decomposition reaction, which is generally not strictly univariant, and the H_2O-saturated solidus for basalt at about 925°C and 0.5 kbar (Burnham, 1979). At this intersection, which is analogous to the invariant point I, of Eggler (1973b, Fig. 44), $a_w^m = a_w^{hb} = 1.0$, hence $X_w^m = 0.3$ (Eq. 16-3). This X_w^m is equivalent to about 3.0 wt% H_2O and is the lowest possible H_2O content for stable coexistence of melt and hornblende in this particular system, regardless of pressure; therefore, the $X_w^m = 0.3$ solidus isopleth (see discussion of Fig. 16-6) defines the high-temperature boundary for stable existence of amphibolite in the absence of both fluid and melt phases at pressures below about 18 kbar. Also, owing to the generally small positive ΔV of the incongruent melting reaction, equilibrium temperatures along this boundary increase slightly with pressure, reaching values as high as 1000° to 1050°C.

At pressures greater than 18 to 20 kbar, on the other hand, the refractory residuum from incongruent melting changes from olivine + pyroxene + plagioclase to a denser garnet-bearing assemblage, and at pressures above about 23 kbar (Allen et al., 1972) hornblende is not stable at any temperature for this bulk composition. The formation of dense pyrope-rich garnet results in a change in sign of the overall ΔV of reaction from positive to negative, which in turn results in a departure of the fluid-absent beginning-of-melting boundary from the $X_w^m = 0.3$ solidus isopleth toward lower temperatures and, hence, higher values of X_w^m. X_w^m continues to increase with falling temperature to the H_2O-saturated solidus at about 670°C and 21.5 kbar (Allen et al., 1972).

It should be emphasized that everywhere on the boundary just described a_w is controlled by the hornblende-melt equilibrium, hence the amount of melt formed at any given point on the boundary is determined by the bulk H_2O content of the amphibolite. For an amphibolite with a bulk H_2O content of 0.6 wt% (roughly 30% hornblende), approximately 20% melt is produced at pressures up to 18 kbar; but only about 2% melt is produced just above the H_2O-saturated solidus at 21.5 kbar and 670°C.

It also should be emphasized that if a_w is controlled by some other equilibrium, such as that between hornblende and an H_2O-CO_2 fluid mixture, then the melting relations described above have no special significance. On the $X_w^m = 0.3$ solidus isopleth at 10 kbar, for example, $a_w = 0.15$ and hornblende-melt equilibrium is established at about 1000°C; but if a_w is greater than 0.15, melting will begin at a lower temperature and hornblende will persist in equilibrium with more H_2O-rich melt ($X_w^m > 0.3$) to a temperature above 1000°C. Thus, as a_w increases, beginning-of-melting temperatures approach the H_2O-saturated ($a_w \approx$ 1.0) solidus at about 640°C (Hill and Boettcher, 1970), but the upper thermal stability of hornblende passes through a maximum at about 1070°C (Holloway and Burnham, 1972; Burnham, 1979) and $X_w^m \approx 0.5$. This value of X_w^m at the thermal maximum, which was calculated from the experimental data of Holloway (1973), was found to be essentially independent of pressure down to about 2.1 kbar (H_2O saturation).

This phenomenon of a thermal maximum in the stability of a hydrous mineral with increasing a_w is not unique to hornblende; it has been shown by Yoder and Kushiro (1969) to occur in the system phlogopite–H_2O (also at $X_w^m \approx 0.5$) and is to be expected of all hydrous minerals that melt incongruently (Eggler, 1973b). Because hornblende melts incongruently, constraints of the quasi-crystalline melt model require that it be regarded as a limited crystalline solution of several components, including H_2O, whose activities must exceed certain specific values for hornblende of a given composition to coexist stably with melt at a given pressure and temperature. Hence, increasing a_w above 0.15 at 10 kbar increases the thermal stability of hornblende, because H_2O is an essential component. An increase in a_w, however, also increases X_w^m, which in turn decreases the activities of all other components in the melt, especially the alumino-silicates (Eqs. 16-13 and 16-14). Thus, at some value of X_w^m, which is near 0.5 for both hornblende and phlogopite, the activity of one or more of the hornblende-essential components in the melt is depressed below the critical value; further increases in X_w^m, therefore, result in depression of hornblende-melt equilibrium temperatures.

Although this isobaric thermal maximum for coexistence of hornblende and melt at $X_w^m \approx 0.5$ has important petrologic implications, the isobaric thermal minimum at $X_w^m \approx 0.3$ has greater implications regarding the H_2O content of magmas. Inasmuch as the subsolidus breakdown of the common hornblende of most igneous rocks intersects the H_2O-saturated solidus near 0.5 kbar (Burnham, 1979) $X_w^m = 0.3$ (~ 3 wt% H_2O) is essentially the minimum H_2O content of the melt required for crystallization of hornblende from magmas, whether the magmas be of gabbroic or granitic composition. Hence, the occurrence of primary hornblende interstitial to other anhydrous minerals of an igneous rock implies that X_w^m did not reach 0.3 until an advanced stage of crystallization, whereas the occurrence of euhedral hornblende phenocrysts suggests that $X_w^m > 0.3$

at a relatively early stage. Furthermore, inasmuch as the subsolidus breakdown of hydroxyl-rich biotite also intersects the H_2O-saturated solidus of granitic rocks in the vicinity of 0.5 kbar, similar constraints apply to magmatic crystallization of biotite. In using biotite as an indicator of X_w^m, however, care must be taken to ascertain its fluorine content, as appreciable substitution of F^- for OH^- markedly increases the thermal stability of biotite at a given f_w. For the same reason, caution must be exercised in the use of primary muscovite as an indicator of X_w^m in granitic magmas. In contrast to hornblende and biotite however, the minimum H_2O pressure for coexistence of hydroxy-muscovite and granitic melt is about 3.5 kbar, hence the minimum X_w^m for crystallization of fluorine-poor magmatic muscovite is about 0.58 (~ 8 wt% H_2O).

Lest such high H_2O contents be regarded as excessive, it should be remembered that these hydrous minerals melt incongruently to anhydrous crystalline phases and hydrous melt. Conversely, crystallization of these hydrous minerals from a magma generally involves reaction between already crystallized anhydrous minerals and hydrous melt that may constitute only a small percentage of the total mass of magma. Therefore, except in those cases where a hydrous mineral appears at temperatures well above the H_2O-saturated solidus, as in some porphyries (hornblende and biotite) and granite pegmatites (muscovite), the H_2O content of many magmatic liquids initially may have been less than one weight percent.

SECOND BOILING AND EXPLOSIVE VOLCANISM

The phenomenon of second or resurgent boiling, as discussed in detail by Bowen (1928), is a natural consequence of crystallization upon cooling in an H_2O-bearing magma whose bulk composition is such that (1) the mole ratio $(Al + Fe^{3+})$/alkalies is unity or greater and (2) the H_2O content of the initial melt is in excess of the quantity that can be accommodated in whatever hydrous minerals may crystallize from the magma. The bulk of igneous-rock melts, especially those of the calc-alkaline suite broadly defined, satisfy condition (1) generally and condition (2) at pressures below about 0.5 kbar where, as noted above, hydrous minerals are not stable at, or above, the H_2O-saturated solidus. Whether condition (2) is satisfied at higher pressures depends upon the bulk composition of the magma. For example, if a melt of Nockolds' (1954) average hornblende-biotite granodiorite initially contained more than about 0.7 wt% H_2O, it would undergo second boiling before complete solidification, regardless of pressure (assuming the H_2O content of hornblende and biotite remains stoichiometric). On the other hand, if the initial H_2O content was less than 0.7 wt%, all H_2O would be consumed in formation of hornblende and biotite, and melt would disappear at a temperature above the H_2O-saturated solidus without separation of an aqueous phase (boiling). It may be noted that only about half this amount of H_2O is required to produce second boiling in the average granitic magma.

The importance of second boiling in pegmatite formation has been stressed by Jahns and Burnham (1969); however, its importance in explosive volcanism, although long appreciated in a qualitative way, has not been adequately assessed quantitatively. Its importance lies in the fact that the second-boiling reaction, melt → crystals + H_2O "vapor," takes place with an increase in volume (ΔV_r) which, under the low confining pressures of a subvolcanic environment, can theoretically reach enormous values. From the experimental data of Burnham and Davis (1971, 1974), a very close approximation to this ΔV_r for a total pressure (P_t) less than 2.0 kbar is given by

$$\Delta V_r \approx (1 - 2.3 \times 10^{-4} P_t) \left(\frac{RT\, X_w^m}{P_t} \right) - \Delta V_m (1 - X_w^m) \text{ cal bar}^{-1} \text{ mole}^{-1}$$

$$(16\text{-}19)$$

and for $P_t \geq 2.0$ kbar, by

$$\Delta V_r \approx 0.54 \left(\frac{RT\, X_w^m}{P_t} \right) - \Delta V_m (1 - X_w^m) \text{ cal bar}^{-1} \text{ mole}^{-1}, \quad (16\text{-}20)$$

where P_t is in bars, ΔV_m is the weighted average ΔV of melting of the crystalline phases (~ 0.21 cal bar^{-1} for the Mt. Hood andesite and Harding pegmatite in Table 16-1), and the mole is a mole of mixture defined in accordance with the H_2O solution model. The maximum increase in volume, per unit volume of melt, calculated from these equations for complete reaction is about 60% at 0.5 kbar, 30% at 1.0 kbar, and 10% at 2.0 kbar.

At the shallow crustal levels that correspond to these pressures, the wallrocks of a magma chamber or volcanic conduit—owing to their rigidity generally cannot yield plastically to accommodate such large volume changes. Furthermore, because cooling and crystallization proceed inward from the walls, saturation of the interstitial melt in H_2O occurs first near the margins and this tends to form an impermeable barrier to movement of H_2O out of, or into, the magma body. Consequently, as cooling and second boiling proceed inside this impermeable barrier, total pressure (P_t) in the magma must increase. The extent of this increase depends upon how much of the ΔV_r can be accommodated by expansion of the magma body, but it can be seen by rearrangement of Eq. 16-20,

$$P_t \approx \frac{0.54 \, RT \, X_w^m}{\Delta V_r + \Delta V_m (1 - X_w^m)}, \tag{16-21}$$

that, if no expansion can occur, $\Delta V_r = 0$ and P_t increases in direct proportion to $X_w^m/(1 - X_w^m)$. Thus, as crystallization proceeds toward the H_2O-saturated solidus under constant volume conditions, enormous pressures (tens of kilobars) theoretically can be generated, provided the initial H_2O content of the magma was in excess of the amount that could be accommodated in hydrous minerals.

The condition of constant magma volume in a volcanic conduit commonly is maintained during cooling prior to the onset of second boiling, because the ΔV of crystallization ($-\Delta V_m$) is compensated by more magma rising buoyantly into the system. Therefore, only a vanishingly small amount of H_2O can separate from the magma at the onset of second boiling before P_t rises above the confining pressure of the wallrocks. This rise in pressure causes expansion of the magma body through quasi-elastic deformation of the wallrocks, hence $\Delta V_r > 0$ and P_t is less than the theoretical maximum value. Precisely how much less is dependent primarily upon the confining load pressure (P_l), the percentage of melt present at the onset of second boiling, and the extent to which the magma body expands.

To illustrate the relationships among these variables, let it first be assumed that an andesitic magma has been emplaced at a depth of about 2 km ($P_l = 0.5$ kbar) and that second boiling has increased P_t to 1.0 kbar at the stage when 50% of the original mass remains molten ($\sim 900°C$). Under these conditions, $X_w^m = 0.4$ and $\Delta V_r \approx 0.6$ cal bar^{-1} mole^{-1} (Eq. 16-19); hence, the entire mass of magma (crystals + melt) has increased in volume by about 0.3 cal bar^{-1} mole^{-1}, or about 15%, since the onset of second boiling. Such a large increase in volume is unlikely to be accommodated by deformation of the wallrocks without brittle failure. Furthermore, the assumed differential pressure ($P_t - P_l$) of 0.5 kbar exceeds by about 25% the tensile stress required to rupture some of the strongest igneous rocks (Ryan, 1979) under optimum conditions (homogeneous specimens free of macro-cracks) that are rarely, if ever, encountered in a subvolcanic environment. As discussed briefly by Burnham (1979), the escape of H_2O from the magma into the fractures thus produced tends to extend the fractures upward by a mechanism akin to hydrofracting.

From the relationships between P_t and ΔV_r just discussed, it is apparent that large amounts of mechanical energy ($P_t \Delta V_r$) are released in the second-boiling reaction; that is, much work is done on the wallrocks of the magma chamber. The total mechanical energy released from an andesitic magma upon complete reaction at a given P_t and corresponding X_w^m is found, from Eqs. 16-19 and 16-20, to pass through a maximum of about 620 cal mole^{-1} of H_2O-saturated melt at $P_t = 0.7$ kbar and 800°C. Although this amount of energy is small in comparison with the total thermal energy (ΔH_r) released in the second-boiling reaction, it is equivalent to about 3×10^{23} ergs km^{-3} of melt and is released almost instantaneously upon failure of the wallrocks. For comparison, this energy is approximately three times the estimated average kinetic energy released, per cubic kilometer of material erupted, in explosive volcanic eruptions (Burnham, 1972). Moreover, it is reduced only about 25% if P_t is as low as 0.15 kbar at the time of eruption. Hence, even neglecting the additional mechanical energy released in adiabatic expansion of the separated steam to atmospheric pressure, the energy released in second boiling alone is fully adequate to account for the explosivity of volcanic eruptions.

The thermal energy released during second boiling is essentially the latent heat of crystallization $(-\Delta H_m)$ of the melt, as the latent heat of "vaporization" of H_2O is negligible (Burnham and Davis, 1974) and the heats of mixing of the principal components in the melt are either zero (aluminosilicates and silica) or very small. Therefore, the thermal energy released can be closely approximated from the heats of fusion of the crystallizing minerals, weighted by their respective mole or mass fractions. Complete crystallization of anhydrous Mt. Hood andesite melt (Table 16-1) at $P_t = 1.0$ kbar and a hypothetical temperature of 927°C (1200K), for example, would release about 50 cal gm^{-1} of melt. On the other hand, crystallization of the 50% H_2O-saturated melt in the above example ($P_t = 1.0$ kbar, $X_w^m = 0.4$) would release only about 24 cal gm^{-1} of magma (melt + 50% crystals). Under static conditions of conductive heat loss to the surroundings (wallrocks), the thermal energy thus produced merely retards the rate of cooling. Under dynamic conditions, however, as in a slowly rising magma column where the rates of boiling and consequent crystallization are greatly accentuated by decreasing P_t, the rate of heat production may far outstrip the rate of conductive heat loss. Hence, much of the 24 cal gm^{-1} may be consumed in actually raising the temperature of magma ($\sim 80°C$) as it approaches the surface.

CONCLUDING STATEMENT

In closing this chapter it is appropriate to recall Bowen's admonition to the many petrologists for whom "a volatile component is exactly like a Maxwell demon." Perhaps the volatile substance H_2O does not warrant such a status, but no other substance of comparable mass proportions in the source regions of magmas has so profoundly influenced the evolution of the igneous rocks or of the earth's crust.

Acknowledgments   This presentation has benefited greatly from the critical comments of J. G. Blencoe, G. A. Merkel, and H. S. Yoder, Jr., as well as from many helpful discussions with the late L. S. Darken on thermodynamics. The research on which the H_2O solution model is based was supported in large part by the National Science Foundation, Division of Earth Sciences, especially through grants GA-1110, GA-3952, and DES72-00247.

References

Allen, J. C., P. J., Modreski, C. Haygood, and A. L. Boettcher, The role of water in the mantle of the earth; The stability of amphiboles and micas, *24th Intl. Geol. Congress, Sec. 2*, 231–240, 1972.

Boettcher, A. L., and P. J. Wyllie, Phase relationships in the system $NaAlSiO_4$–SiO_2–H_2O to 35 kilobars pressure, *Am. J. Sci., 267*, 875–909, 1969.

Bowen, N. L., The melting phenomena of the plagioclase feldspars, *Am. J. Sci., 35*, 577–599, 1913.

———, *The Evolution of the Igneous Rocks*, Princeton University Press, Princeton, N.J., 332p., 1928.

———, The granite problem and the method of multiple prejudices, in *Origin of Granite*, J. Gilluly, Chm., *Geol. Soc. Am. Mem. 28*, 79–90, 1948.

Bowen, N. L., and O. F. Tuttle, The system $NaAlSi_3O_8$–H_2O, *J. Geol., 58*, 489–511, 1950.

Burnham, C. W., The energy of explosive volcanic erruptions, *Earth Min. Sci., 41* (The Pennsylvania State University), 69–70, 1972.

———, $NaAlSi_3O_8$–H_2O solutions: A thermodynamic model for hydrous magmas, *Bull. Soc. fr. Min. Crist., 97*, 223–230, 1974.

———, Water and magmas; a mixing model, *Geochim. Cosmochim. Acta., 39*, 1077–1084, 1975a.

, Thermodynamics of melting in experimental silicate-volatile systems, *Fortschr. Miner., 52*, 101–118, 1975b.

———, Magmas and hydrothermal fluids, in *Geochemistry of Hydrothermal Ore Deposits*, 2nd ed., H. L. Barnes, ed., John Wiley and Sons, New York, Chapter 3, 1979.

Burnham, C. W., L. S. Darken, and A. C. Lasaga, Water and Magmas: Application of the Gibbs-Duhem equation: a response, *Geochim. Cosmochim. Acta. 42*, 277–280, 1978.

Burnham, C. W., and N. F. Davis, The role of H_2O in silicate melts: I. P–V–T relations in the system $NaAlSi_3O_8$–H_2O to 10 kilobars and 1000°C, *Am. J. Sci., 270*, 54–79, 1971.

Burnham, C. W., and N. F. Davis, The role of H_2O in silicate melts: II. Thermodynamic and phase relations in the system $NaAlSi_3O_8–H_2O$ to 10 kilobars, 700° to 1100°C, *Am. J. Sci., 274*, 902–940, 1974.

Burnham, C. W., J. R. Holloway, and N. F. Davis, Thermodynamic properties of water to 1000°C and 10,000 bars, *Geol. Soc. Am. Sp. Paper, 132*, 96 pp., 1969.

Burnham, C. W., and R. H. Jahns, A method for determining the solubility of water in silicate melts, *Am. J. Sci., 260*, 721–745, 1962.

Carmichael, I.S.E., F. J. Turner, and J. Verhoogen, *Igneous Petrology*, McGraw-Hill Book Co., New York, 739 pp., 1974.

Clark, S. P., Jr., Solubility, in *Handbook of Physical Constants*, S. P. Clark, Jr., ed., *Geol. Soc. Am. Mem., 97*, 415–436, 1966.

Darken, L. S., Application of the Gibbs-Duhem equation to ternary and multicomponent systems, *J. Am. Chem. Soc., 72*, 2909–2914, 1950.

Deer, W. A., Howie, R. A., and Zussman, J., *An Introduction to the Rock-forming Minerals*, Wiley & Sons, New York, 528 pp., 1966.

Eggler, D. H., Role of CO_2 in melting processes in the mantle, *Carnegie Institution of Washington Yearbook, 72*, 457–467, 1973a.

———, Principles of melting of hydrous phases in silicate melt, *Carnegie Institution of Washington Yearbook, 72*, 491–495, 1973b.

———, Effect of CO_2 on the melting of peridotite, *Carnegie Institution of Washington Yearbook, 73*, 215–224, 1974.

Eggler, D. H., and C. W. Burnham, Crystallization and fractionation trends in the system andesite–H_2O–CO_2–O_2 at pressures to 10 kb, *Geol. Soc. Am. Bull., 84*, 2517–2532, 1973.

Furst, G. A., The melting of plagioclase in the system Na_2O–CaO–Al_2O_3–SiO_2–H_2O at high pressure and temperature, unpublished Ph.D. thesis, The Pennsylvania State University, University Park, Pa., 1978.

Goranson, R. W., The solubility of water in granite magmas, *Am. J. Sci., 22*, 481–502, 1931.

———, Some notes on the melting of granite, *Am. J. Sci., 23*, 227–236, 1932.

———, Silicate-water systems: Phase equilibria in the $NaAlSi_3O_8–H_2O$ and $KAlSi_3O_8–H_2O$ systems at high temperatures and pressures, *Am. J.Sci., 35-A*, 71–91, 1938.

Hamilton, D. L., C. W. Burnham, and E. F. Osborn, The solubility of water and effects of oxygen fugacity and water content on crystallization of mafic magmas, *J. Petrol., 4*, 21–39, 1964.

Hill, R.E.T., and A. L. Boettcher, Water in the earth's mantle: Melting curves of basalt–water and basalt–water–carbon dioxide, *Science, 167*, 980–981, 1970.

Hodges, F. N., The solubility of H_2O in silicate melts, *Carnegie Institution of Washington Yearbook, 73*, 251–254, 1974.

Holloway, J. R., The system pargasite–H_2O–CO_2: A model for melting of a hydrous mineral with a mixed-volatile fluid—I. Experimental results to 8 kbar, *Geochim. Cosmochim. Acta, 37*, 651–666, 1973.

Holloway, J. R., and C. W. Burnham, Melting relations of basalt with equilibrium-water pressure less than total pressure, *J. Petrol., 13*, 1–29, 1972.

Holloway, J. R., and C. F. Lewis, CO_2 solubility in hydrous albite liquid at 5 kbar, *EOS, Trans. Am. Geophys. Union*, *55*, 483, 1974.

Jahns, R. H., and C. W. Burnham, Experimental studies of pegmatite genesis: pt. I, a model for the derivation and crystallization of granitic pegmatites, *Econ. Geol.*, *64*, 843–864, 1969.

Kadik, A. A., and D. H. Eggler, Melt vapor relations on the join $NaAlSi_3O_8$–H_2O–CO_2, *Carnegie Institution of Washington Yearbook*, *74*, 479–484, 1975.

Kennedy, G. C., G. J. Wasserburg, H. C. Heard, and R. C. Newton, The upper three-phase region in the system SiO_2–H_2O, *Am. J. Sci.*, *260*, 501–521, 1962.

Kushiro, I., The system forsterite–diopside–silica with and without water at high pressures, *Am. J. Sci., Schairer Vol.*, *267-A*, 269–294, 1969.

———, On the nature of silicate melt and its significance in magma genesis: Regularities in the shift of the liquidus boundaries involving olivine, pyroxene, and silica minerals, *Am. J. Sci.*, *275*, 411–431, 1975.

Luth, W. C., R. H. Jahns, and O. F. Tuttle, The granite system at pressures of 4 to 10 kilobars, *J. Geophys. Res.*, *69*, 759–773, 1964.

Morse, S. A., Feldspars, *Carnegie Institution of Washington Yearbook*, *67*, 120–126, 1969.

Mysen, B. O., Solubility of volatiles in silicate melts at high pressure and temperature: The role of carbon dioxide and water in feldspar, pyroxene, and feldspathoid melts, *Carnegie Institution of Washington Yearbook*, *74*, 454–468, 1975.

Mysen, B. O., and A. L. Boettcher, Melting in a hydrous mantle: I, Phase relations of natural peridotite at high pressures and temperatures with controlled activities of water, carbon dioxide and hydrogen, *J. Petrol.*, *16*, 520–548, 1975a.

Mysen, B. O., and A. L. Boettcher, Melting of a hydrous mantle: II, Geochemistry of crystals and liquids formed by anatexis of mantle peridotite at high pressures and high temperatures as a function of controlled activities of water, hydrogen and carbon dioxide, *J. Petrol.*, *16*, 549–593, 1975b.

Nockolds, S. R., Average chemical compositions of some igneous rocks, *Geol. Soc. Am. Bull.*, *65*, 1007–1032, 1954.

Nordlie, B. E., The composition of the magmatic gas of Kilauea and its behavior in the near surface environment, *Am. J. Sci.*, *271*, 417–463, 1971.

Orville, P. M., Plagioclase cation exchange equilibria with aqueous chloride solution: Results at 700°C and 2000 bars in the presence of quartz, *Am. J. Sci.*, *272*, 234–272, 1972.

Robie, R. A., and D. R. Waldbaum, Thermodynamic properties of minerals and related substances at 298.15°K and one atmosphere (1.013 bars) pressure and at higher temperatures, *U.S. Geol. Surv. Bull.*, *1259*, 256p., 1968.

Rosenhauer, M., and D. H. Eggler, Solution of H_2O and CO_2 in diopside melt, *Carnegie Institution of Washington Yearbook*, *74*, 474–479, 1975.

Ryan, M. P., High temperature mechanical properties of basalt, unpublished Ph.D. thesis, The Pennsylvania State University, University Park, Pa., 1979.

Schairer, J. F., Melting relations of the common rock-forming oxides, *J. Amer. Ceram. Soc.*, *40*, 215–235, 1957.

Stewart, D. B., The system $CaAl_2Si_2O_8-SiO_2-H_2O$, *Carnegie Institution of Washington Yearbook*, *55*, 214–216, 1956.

———, Four-phase curve in the system $CaAl_2Si_2O_8-SiO_2-H_2O$ between 1 and 10 kilobars, *Schweiz. Mineral. Petrogr. Mitt.*, *47*, 35–39, 1967.

Tuttle, O. F., and N. L. Bowen, Origin of granite in the light of experimental studies in the system $NaAlSi_3O_8-KAlSi_3O_8-SiO_2-H_2O$, *Geol. Soc. Am. Mem.*, *74*, 153 pp., 1958.

Urnes, S., X-ray diffraction of glasses and methods of interpretations, in *Selected Topics in High Temperature Chemistry*, T. Förland *et al.*, eds., Universitets-forlaget, Oslo, 97–124, 1966.

Wagner, C., *Thermodynamics of Alloys*, Addison-Wesley Press, Cambridge, Mass., 161p., 1952.

Wasserburg, G. J., The effects of H_2O in silicate systems, *J. Geol.*, *65*, 15–23, 1957.

Weyl, W. A., and E. C. Marboe, Structural changes during the melting of crystals and glasses, *Trans. Soc. Gl. Tech.*, *43*, 417–437, 1959.

Whitney, J. A., The effects of pressure, temperature and X_{H_2O} on phase assemblage in four synthetic rock compositions, *J. Geol.*, *83*, 1–27, 1975.

Windom, K. E., and A. L. Boettcher, The effect of reduced activity of anorthite on the reaction grossular + quartz = anorthite + wollastonite: A model for plagioclase in the earth's lower crust and upper mantle, *Am. Mineral.*, *61*, 889–896, 1976.

Wyllie, P. J., and W. L. Huang, Peridotite, kimberlite, and carbonatite explained in the system $CaO-MgO-SiO_2-CO_2$, *Geology*, *3*, 621–624, 1975.

Wyllie, P. J., and O. F. Tuttle, Effect of carbon dioxide on melting of granite and feldspars, *Am. J. Sci.*, *257*, 648–655, 1959.

Yin, L. I., S. Ghose, and I. Adler, core binding energy difference between bridging and nonbridging oxygen atoms in a silicate chain, *Science*, *173*, 633–634, 1971.

Yoder, H. S., Jr., Diopside–anorthite–water at five and ten kilobars and its bearing on explosive volcanism, *Carnegie Institution of Washington Yearbook*, *64*, 82–89, 1965.

Yoder, H. S., Jr., and I. Kushiro, Melting of a hydrous phase: phlogopite, *Am. J. Sci.*, *Schairer Vol.*, *267-A*, 558–582, 1969.

Yoder, H. S., Jr., D. B. Stewart, and J. R. Smith, Ternary feldspars, *Carnegie Institution of Washington Yearbook*, *55*, 206–214, 1957.

Yoder, H. S., Jr., and C. E. Tilley, Natural tholeiite basalt–water system, *Carnegie Institution of Washington Yearbook*, *55*, 169–171, 1956.

Yoder, H. S., Jr., and C. E. Tilley, Origin of basalt magmas: An experimental study of natural and synthetic rock systems, *J. Petrol.*, *3*, 342–532, 1962.

Chapter 17

PETROGENESIS AND THE PHYSICS
OF THE EARTH

P. J. WYLLIE

Department of Geophysical Sciences, University of Chicago, Chicago, Illinois

It is in many ways desirable to establish the connection of igneous activity with ascertained facts regarding the nature of the earth as a whole and if possible with the early history and the ultimate origin of the earth. Any system of petrogeny must, of course, be reconcilable with geophysical facts, in so far as these are facts, but it is a different matter to suppose that petrology must be based upon some chosen system of cosmogony. From the very nature of its subject-matter cosmogony must be ever less capable than petrology of reaching demonstrable conclusions. This is, perhaps, true of geophysics also, but in less degree. A brief survey of the data and of some present-day conclusions in geophysical matters may be desirable, together with some suggestion as to their connection with the advocated system of petrogenesis (Bowen, 1928, p. 303).

This statement was Bowen's opening paragraph for Chapter 17, written with his customary style and clarity. Many parts of the chapter remain valid today, but knowledge of the physics of the earth has increased enormously since 1928. *A brief survey . . . of some present-day conclusions in geophysical matters* was *desirable* fifty years ago. Today, a survey is essential, because geophysics has become a starting point for many aspects of petrogenesis. Similarly, geophysicists need the data of petrology for characterization of the earth materials whose properties they measure.

Cosmogony, invigorated by the space program, has now become a fruitful source of geochemical data with relevance to petrogenesis, and the intensive study of the rock samples returned from the Moon has provided many insights into magmatic processes. New apparatus for laboratory measurements ranging from the physical properties of minerals, rocks, and melts to the phase equilibrium relationships of minerals and rocks at high pressures has added much useful data for the interpretation of geophysical measurements and magmatic processes. The theoretical approaches of thermodynamics and geophysical fluid dynamics have been applied successfully to the earth's interior.

The theories of sea-floor spreading and plate tectonics, defining large-scale tectonic environments on a global scale, and involving the movement of large assemblages of rock across and within the earth in horizontal and vertical directions, have added dynamism to petrology since the mid-1960s. As a result of movements from physical processes, pressure and temperature change, and rocks undergo phase transitions, which include melting, the first stage of magmatism.

Petrogenesis and the physics of the earth are now simply two overlapping parts of the interdisciplinary earth sciences.

PETROGENESIS, 1928

Although today much more is known about the physics of the earth and the chemistry and phase relationships of rocks and magmas, Bowen's petrogenetic conclusions to a remarkable extent continue to outline the framework of current thought. The following summary of his chapter illustrates the reasoning he followed to reach these conclusions.

After a brief account of the density of the earth and implications for the core, Bowen listed the *Observations throwing light on the physical condition of earth shells*. His review of earthquake-wave propagation focused on velocities above and below a discontinuity variously placed at 37 km or 60 km, and on whether or not the deeper rocks could be equated with dunite, peridotite, or eclogite. Several pages were devoted to a discussion of *The geothermal gradient and the radioactive content of the rocks*, leading only to the conclusion that temperatures at some unknown depth are high enough to generate basaltic magmas. A section on *Tidal deformation and distortional seismic waves* affirmed the high rigidity of the earth. Several pages dealing with *The source of magmas* emphasized *the necessity of deriving all magmas ultimately from matter much more basic than basalt and probably from peridotitic substance as represented in stony meteorites* (p. 311). Bowen argued cogently against the then-popular thesis that basaltic magma was derived from a layer of basaltic glass that became fluid upon release of pressure, or by the melting of crystalline basalt. In the final section on *Production of basaltic magma by selective fusion of peridotite*, he outlined evidence that a suitable source corresponds to material with the composition of achondritic meteorites, leaving open the question of whether or not a layer of eclogite could yield basaltic magma, and the question of whether fusion at depth takes place as a result of release of pressure or as a result of reheating. He noted the increasing probability that all magmas *could be derived from the basaltic* [magma] *by crystallization-differentiation*, and added that *even were this a proven fact it would not follow that magmas are so derived. The decision on this point would still depend principally upon geologic considerations* (p. 319).

Bowen was concerned primarily with the source of basaltic magmas, and the probability that these could be the parents of all other magmas. Note his insistence that decisions must be based on geological considerations. The generation of granitic magmas by crustal anatexis received passing mention. Bowen had no concept of plate tectonics and subducted lithosphere. The prospect that andesites and tonalites could be primary magmas from the "subcrustal regions" is an idea that did not arrive until much later.

PETROGENESIS, 1978

For comprehension of magmatic origins and processes, today as in 1928, one needs to know the composition of the mantle and crust, the structure of the upper mantle and crust and the distribution of material within the various components of these structures, and the temperature distribution at depth as a function of place and time. Today, however, these topics appear more complex than they did in 1928, in part because much more is known about them, and in part because plate tectonics has added new dimensions to the former static models of the world. Directly relevant to the items listed above is the generation of lithosphere at oceanic ridges, its lateral movement away from ridges and its subduction, as well as the vertical and lateral motions of material within the earth (Wyllie, 1973, 1976).

Detailed accounts of recent petrogenetic developments are given in volumes dealing with basalts (Hess and Poldervaart, 1967, 1968; Yoder, 1976), oceanic basalts (Papike, 1976), lunar basalts (Lunar Science Institute, 1975), ultramafic and related rocks (Wyllie, 1967), alkaline rocks (Sørensen, 1974), carbonatites (Tuttle and Gittins, 1966; Heinrich, 1966), and aspects of igneous petrology in general (Hyndman, 1972; Carmichael, Turner, and Verhoogen, 1974).

Petrogenesis can now be related much more successfully to tectonic environment than it could before plate tectonics provided the guidelines.

Basaltic magmas are assumed to be derived from mantle peridotite, but distinctive chemical characteristics for basalts erupted in different environments (ocean ridges, ocean plates, compressive plate boundaries, continental rift zones, continental platforms) suggest differences in source material or petrogenetic processes. The calc-alkaline rock series associated with compressive plate boundaries may involve petrogenetic processes independent of basaltic magmas. The alkalic plutonic complexes of continental platforms, with rare rocks far removed from basalt in composition, and the kimberlites, carbonatites, and highly alkalic lavas, may be derived by crystallization-differentiation of basaltic magmas following paths similar to those that Bowen (1928) tracked through known and schematic phase diagrams, but, if so, the process must have been followed to extreme limits. Most petrologists today invoke sources and processes other than crystallization-differentiation of basaltic magma to account for these rock associations.

MAGMA SOURCES

Bowen (1928) reached the following conclusions: *We find the most probable source of basaltic magma in the selective fusion of a portion of the peridotite layer* (p. 315). *A source of plateau basalts in the selective fusion of peridotite substance analogous to meteoritic material is, therefore, quite within the bounds of possibility and even of probability* (p. 316). . . . *it is not likely that crystalline basaltic substance would usually give basaltic liquid in any process of remelting* (p. 314).

In 1978, most petrologists would accept these statements, but they would extend them by considering three types of source material for magma generation in three distinct environments (Wyllie, 1973; Carmichael, Turner and Verhoogen, 1974, Chapters 7, 11, 12, and 13). (1) Following Bowen, one finds the source of basaltic magma in the selective fusion of mantle peridotite. (2) Subduction of oceanic crust transports basaltic rocks to depths where they undergo partial fusion, yielding magmas of intermediate SiO_2 content. (3) The partial fusion of continental crust produces granitic magmas and rhyolites.

There are many estimates for the composition of mantle peridotite. The different approaches were reviewed by Wyllie (1971, Chapter 6; see also Ringwood, 1975, Chapter 3, 5, 6, and 16; Yoder, 1976, Chapter 2). The first approach involves the use of extraterrestrial chemical data and the formulation of physical and geochemical models for the origin of the solar system and the earth. The second approach involves the study of ultramafic rocks derived from the mantle, specifically, the nodules enclosed in kimberlites and some alkalic basalts. The third approach is petrological, and the mantle composition is calculated by assuming that the primitive mantle peridotite would yield a basaltic magma and leave behind a residual refractory dunite-peridotite (but see Yoder, 1976, Chapter 8).

Despite the different approaches followed, there is general agreement that more than 90% by weight of the mantle is represented by the system $FeO-MgO-SiO_2$. The components Na_2O, CaO, and Al_2O_3 lie within the range 5–10%. More than 98% of the mantle is composed of these six components (with Fe_2O_3 calculated as FeO). No other oxide reaches a concentration of 0.6%. There are large uncertainties about the concentrations and distributions of trace elements and volatile components. Estimated K_2O contents vary by a factor of ten (0.015% to 0.15%). Water and CO_2 are present in small amounts, probably distributed in an irregular fashion. Upper mantle peridotite is certainly heterogeneous, ranging from the original, undifferentiated peridotite to modifications produced by the partial or complete removal or addition of the more fusible fractions. There is evidence from some ultramafic nodules in kimberlites and alkalic basalts that metasomatism by fluids has occurred within the mantle (Lloyd and Bailey, 1975; Harte, Cox, and Gurney, 1975).

The oceanic crust is composed dominantly of low-potassium tholeiite, together with bodies of serpentinite (Carmichael, Turner, and Verhoogen, 1974, Chapter 8). Part of the crust has been altered by penetrative convection of sea-water, at spreading centers, with temperatures ranging high enough to generate greenschists and amphibolites. The subducted ocean crust consists, therefore, largely of basalt and gabbro, amphibolite, and serpentinite (Papike, 1976). This partly hydrated basaltic material, transported to regions of high pressures and temperatures by subduction, is converted into quartz eclogite, and water released by dehydration may contribute to magma generation. Water may also migrate into the mantle peridotite overlying the subducted slab, lowering the temperature of beginning of melting, and influencing also the compositions of magmas generated from the peridotite. The participation of subducted oceanic sediments in magmatic processes is probably trivial.

The composition of the continental crust has been reviewed by Wyllie (1971, Chapter 7), Ringwood (1975, Chapter 2) and Condie (1976, Chapter 4). The average composition is very similar to the average composition of the andesites. Metamorphism of this material would yield a tonalitic gneiss. The main source material for magma generation in the deep continental crust is probably tonalitic gneiss, together with some amphibolite, with both rock types attaining the eclogite facies.

OBSERVATIONS THROWING LIGHT ON
THE PHYSICAL CONDITION OF
EARTH SHELLS

Bowen (1928, p. 304) wrote: *The principal observations which have a bearing on the question of the character and condition of the materials of the outer portion of the earth may be listed under the following heads: Earthquake-wave propagation, The geothermal gradient, The radioactive content of rocks, Tidal deformation of the earth, Isostasy.* After review of some of these, he concluded (p. 310): *Two lines of evidence prove that, not only the earth as a whole, but also the individual shells of which it is composed have a very high degree of rigidity. The information comes from the manner of propogation of earthquake waves and from the magnitude of the tidal deformation of the earth's body.*

It would require a whole book to review the 1978 physical observations. Textbooks such as those of Garland (1971), Bott (1971), Le Pichon, Francheteau and Bonnin (1973), Jacobs, Russell, and Wilson (1974), and Stacey (1977) provide additional information about surface waves, free oscillations of the earth, gravity, the electrical conductivity of rocks within the earth, heat-flow measurements, creep, and anelasticity of the earth's interior. Reviews of recent research are given for more than 125 topics covering all aspects of geophysical sciences in the U.S. National Report edited by Bell (1975).

EARTHQUAKES AND EARTHQUAKE-WAVE PROPAGATION

In 1928, Bowen wrote (p. 304): *Indications as to the nature of materials at various depths is obtained from a study of the velocity of earthquake waves,* and this remains the most important source of information. He concluded (p. 310): *The significance of these general considerations, from the petrologic point of view, is that no definite knowledge is inconsistent with the view that there is a granitic shell as much as 25 km thick and a passage through inter-mediate to basic rocks of normal radioactivity content without attaining the ultrabasic rocks until a depth of some 60 km is reached.* This general picture remains valid today for the continents, although the oceanic section is very different (Wyllie, 1971, p. 148).

Seismic waves passing through the earth provide a measure of its internal properties, and hence of its concentric structure, illustrated in Fig. 17-1. The seismic-wave velocity profiles show the fine structure of the upper mantle. The velocities increase with depth in a series of steps, with high velocity-gradients occurring near 150–200 km, 350 km, and 650 km.

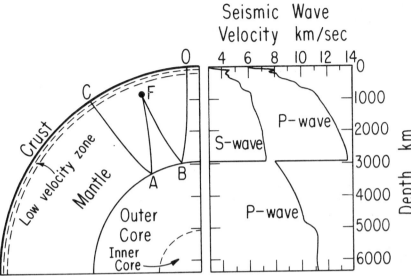

Figure 17-1. Velocities of P-waves and S-waves within the earth, compared with the internal structure of the earth. The paths of ScS waves from deep-focus earthquake F sample the mantle beneath oceans (FBO) and continents (FAC). On this scale, the differences in thickness of crust and lithosphere beneath continents and oceans cannot be distinguished.

Velocities appear to increase more uniformly from 700 km to the mantle-core boundary. There are variations in profile from one tectonic environment to another (see review by Solomon, 1976). These details have been revealed only since the 1960s. Bowen had no knowledge of a low-velocity zone, for example.

The thickness of the crust is variable, and so is that of the lithosphere. The lithosphere is commonly equated with the crust and upper mantle above the low-velocity layer (but see Goetze, 1977). Solomon (1976) reviewed the variations in seismic determinations of oceanic lithosphere thickness, and values for continental lithosphere thickness in tectonic, platform, and shield regions. The low-velocity layer extends between about 100 km and about 200 km in depth. The depth and thickness of this layer vary from one tectonic environment to another. Beneath continental shields, the layer may be much reduced, or even absent.

Petrogenetic considerations relate to the low-velocity layer, which is commonly believed to be the normal site of magma generation, and to depths of 700 km, where subducted lithosphere slabs become indistinguishable from surrounding mantle. According to the hypothesis of thermal plumes and hot-spots, however, some form of energy exchange at the core-mantle boundary may be responsible for the uprise of deep mantle plumes (Morgan, 1972), with magma generation resulting at high levels in the mantle.

Jordan (1975) reviewed evidence indicating that between ocean basins and continental shields, thermal or compositional differences, or both, persist to depths of at least 400 km. This evidence implies that the lithosphere plates that transport drifting continents may be 400 km thick beneath the continental shields. Figure 17-1 illustrates the paths of earthquake waves that contributed to this conclusion. Travel times of S-waves from deep-focus earthquakes to oceanic island stations (FBO) are about 5 seconds later than the times to continental stations (FAC), which could indicate that systematic differences in S-wave velocities between oceans and continents extend to at least 400 km depth. Solomon (1976) reviewed the arguments, and concluded that systematic continent-ocean differences do not appear to extend below 200 km, and that the asthenosphere below 200 km depth is indistinguishable beneath oceans and continents.

The seismic body-wave velocities in Fig. 17-1 are consistent with the conclusion that the mantle portion of the lithosphere is composed of peridotite. Press (1968) used a Monte Carlo statistical method to test randomly generated earth models against available geophysical data. His results suggested that the lower lithosphere may include a substantial proportion of eclogite. Ringwood (1975, Chapter 3) concluded that several independent lines of physical evidence supported the occurrence of peridotite rather than eclogite for this part of the upper mantle.

Experimental data obtained at high pressures and temperatures for possible earth materials place additional constraints on the structure and composition of the upper mantle and crust. Chung (1976) presented laboratory measurements of seismic wave velocities and densities in peridotites and eclogites, and discussed their application to the elasticity and composition of the oceanic upper mantle. He concluded that the elastic properties of olivine eclogite described the seismic structure of the lower lithosphere. D. H. Green and Liebermann (1976) considered the structure and composition of the oceanic upper mantle by combining geophysical data, laboratory measurements of physical properties, the phase relationships of possible mantle rocks, and general petrological considerations.

Seismic results provide information about the dynamic earth, as well as the static picture of Fig. 17-1. The distribution of earthquake foci is fundamental evidence for the theory of plate tectonics, and for the existence of subducted lithosphere slabs. Study of the first-motions of individual earthquakes along plate boundaries gives information about directions of relative movement.

Seismic methods have also been successful in locating bodies of magma, although data are not yet precise enough to define shapes and percentage of melting. Yoder (1976, Chapter 3) reviewed some of the evidence. Eaton *et al.* (1975) combined geological and geophysical data in a study of the Yellowstone National Park. The distribution of earthquake epicenters, the attenuation of *P*-waves, the absence of *S*-waves on some records, and the large delays in teleseismic *P*-waves constitute an important part of the data that are interpreted in terms of a magma body 85 km long and 55 km wide, probably a few kilometers below the surface and a few kilometers thick. This magma body appears to lie above a much larger volume of thermally disturbed crustal and mantle rocks extending to a depth of almost 100 km. Seismic data from the ocean ridges have also been interpreted in terms of magma chambers (Papike, 1976).

TEMPERATURE DISTRIBUTION

Bowen (1928, p. 306) summarized the situation succinctly: *The temperatures that may exist at various depths within the earth present a very difficult problem.* He went on to outline approaches to solution of the problem (p. 306):

In present attacks on the problem, the age of the earth is taken as known from the uranium-lead ratio in the oldest intrusive rocks. From this and from the observed thermal gradient the thickness of the radioactive shell is calculated . . . The conclusion reached is that a thickness of 10–20 km of average granite, which type of rock has more radioactive substance than any other common type, would account for all the radioactive substance of the earth.

The assumption involved in the calculation is that all heat loss is by conduction, yet we know that there is loss from other causes (p. 308).

Upon crustal temperatures definite knowledge is thus reduced to the fact of rising temperature which somewhere within striking distance of the surface is such as to render possible the production of the most refractory magma we know—basaltic magma (p. 310).

There is still great uncertainty about the temperature distribution within the earth, as shown by Fig. 17-2, which illustrates temperature profiles deduced by various approaches.

The geotherm calculated by Lubimova (1967) is based on conduction and the earth's thermal history. Clark and Ringwood (1964) developed a petrological model involving the formation of continents by vertical differentiation of the upper mantle. [Tozer (1967) extended their curves from 400 km to 1400 km]. The uncertainty in the calculations is demonstrated by the fact that Ringwood (1966) changed the contribution of radiative

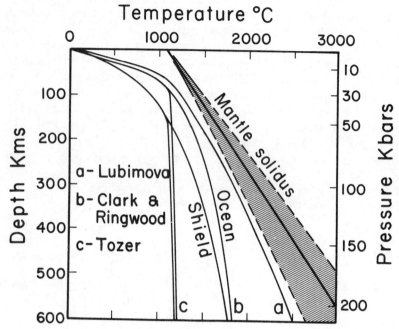

Figure 17-2. Calculated temperature distributions within the earth according to (a) Lubimova (1967), (b) Clark and Ringwood (1964), and (c) Tozer (1967), illustrating the effects of conduction, radiative transfer, and convection, respectively (see text). The shaded region illustrates linear extrapolation of a range of experimentally measured solidus curves for peridotites (compare Figs. 17-5 and 17-6).

transfer to the extent of raising the geotherms by about 250°C at 250 km. Convection in the mantle influences temperature distribution significantly. Figure 17-2 shows Tozer's (1967) evaluation of the effect of convection superimposed on the conduction-radiation model of Clark and Ringwood. There are now many other published estimates of temperature distribution filling in the spaces between the selected examples illustrated. In the depth range 100–200 km, which is probably the most important interval for magma generation, the estimated temperatures differ less than those for greater depths.

Bowen (1928, p. 314) concluded that: *The elastic properties of the earth appear to reduce us to the necessity of considering the remelting of crystalline material.* Melting temperatures of crystalline peridotite within the earth, extrapolated from limited experimental data as shown in Fig. 17-2, are higher than temperatures indicated by the calculated geotherms. Figure

17-14 shows, similarly, that the normal continental geotherm is not high enough to melt crustal rocks, even in the presence of aqueous pore fluids.

The temperature distribution in a cooling oceanic lithosphere plate being transported away from a spreading center has been calculated repeatedly, and solutions agree satisfactorily with seismic data (Le Pichon, Francheteau, and Bonnin, 1973, Chapter 6; Solomon, 1976; Bottinga and Allègre, 1976). Values ranging from 1000°C to 1200°C are commonly suggested for suboceanic depths of 100 km (e.g., D. H. Green and Liebermann, 1976).

The mineralogy of the peridotite nodules enclosed by kimberlites and alkalic lavas provides estimates of mantle temperatures independent of geophysics and geophysical fluid dynamics (Boyd, 1973). Field geologists, mineralogists, experimental geochemists, and geophysicists have together established that the minerals in these nodules provide information about fossil geotherms (temperature distributions at the time of eruption), information that is of vital significance to theoretical geophysicists who seek constraints for thermal regimes and convection models. The basic data consist of calibrations of mineralogy in terms of pressure and temperature on the phase diagram of Fig. 17-6. Values ranging from 800°C to 1000°C are commonly suggested for subcontinental nodules from depths of 100 km (Boyd, 1973; Solomon, 1976), with somewhat higher temperatures of 1000°C to 1100°C for suboceanic nodules from alkalic basalts (MacGregor and Basu, 1974; Mercier and Carter, 1975). There remain some problems of interpretation, as reviewed in detail in the proceedings of a 1975 symposium on "geothermometry and geobarometry" (Boettcher, 1976).

The mineral assemblages in rocks of the continental crust also provide values for temperatures at various depths in regions undergoing metamorphism. In 1940, Bowen proposed that the facies of regional metamorphism could be calibrated by experimental studies of the metamorphic reactions that form the boundaries of the facies. Apparatus suitable for the experiments became available in the late 1940s, and this was followed by many studies aimed at defining what Bowen had termed the petrogenetic grid. About thirty years later, Turner (1968) presented an appraisal of experimentally determined metamorphic reactions and concluded (p. 167) that Bowen's abstract concept had become an instrument "increasingly capable of effective use in calibrating temperature and pressure gradients of crystallization in metamorphic terranes."

Several deduced depth-temperature paths followed by continental rocks during regional metamorphism are plotted in Fig. 17-14. They all involve temperatures that are significantly higher than indicated by the oceanic geotherm of Fig. 17-2. For temperatures to reach these high levels in the continental crust, there must have been large thermal anomalies in the underlying mantle.

One of the main problems in explaining magma generation is to account for the source of heat for raising mantle or crustal rocks to melting temperatures, and for the heat of fusion. Significant perturbations must occur in the temperature distributions shown in Fig. 17-2. Many mechanisms have been proposed to explain how magmas are generated (Yoder, 1976, Chapter 4). Two of them, convective uprise of material and the conversion of mechanical energy to heat, involve divergent and compressive plate boundaries, which are sites of major magma generation.

Bottinga and Allègre (1976) and Oxburgh and Turcotte (1976) reviewed some of the models that have been proposed for the flow field and associated thermal structure in the mantle beneath the axial region of the ocean ridges. The temperature distributions obtained in theoretical models are very sensitive to the parameters adopted in the calculations, as shown by Fig. 17-3.

Figure 17-3A shows the predicted steady-state distribution of isotherms calculated by Oxburgh and Turcotte (1968) for the region near the ridge. The dashed lines are flow lines for the convecting material. They applied boundary-layer theory for the structure of two-dimensional convection cells, assuming a constant, Newtonian viscosity independent of temperature, and neglecting the heat release due to radioactivity in the mantle rocks.

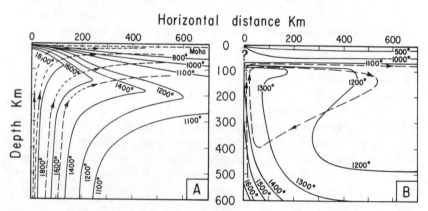

Figure 17-3. Calculated temperature distributions and flow lines for mantle convecting beneath an ocean ridge. These diagrams illustrate portions of large convection cells. A. After Oxburgh and Turcotte (1968), constant-viscosity model. B. After Torrance and Turcotte (1971), temperature-dependent viscosity model. Conduction-cooled lithosphere above 100 km after Forsyth and Press (1971).

Figure 17-3B shows a numerical solution for an improved model obtained by Torrance and Turcotte (1971), using a viscosity appropriate for diffusion creep, which varies with temperature and depth. Calculations were not extended through the 75 km-thick lithosphere because increasing viscosity with decreasing temperature in the lithosphere led to calculations involving unreasonably long computer times. For completeness and for comparison with Fig. 17-3A, lithosphere isotherms have been added following Forsyth and Press (1971); these are quite independent of the calculated deeper isotherms, and the question mark beneath the ridge shows where the results are incomplete. Note that temperatures beneath the ridge are decreased significantly by use of a temperature-dependent viscosity, and the flow pattern is also changed.

The geotherm 700 km from the ridge crest in Fig. 17-3B is similar to the adiabatic convection geotherm of Tozer in Fig. 17-2, and the geotherm beneath the ridge in Fig. 17-3B is similar to the Clark and Ringwood conduction-radiation geotherm in Fig. 17-2. Somewhat higher temperatures are required to generate basaltic magma beneath the ridges, but these are readily achieved by minor changes in the properties of materials used in the calculations.

Solomon (1976) emphasized that no single "oceanic" or "continental" geotherm is adequate for either lithosphere or asthenosphere. Richter (1973) and McKenzie and Weiss (1975) proposed that heat is transported through the asthenosphere by a relatively small convection pattern secondary to the pattern of plate motions. This process could produce temperature differences of 100°C or more compared with the average geotherm at a depth of 100 km, depending on position with respect to upwelling or downwelling portion of the secondary scale convection.

Subduction of the oceanic lithosphere slab at compressive plate boundaries provides a heat sink. The factors influencing the temperature distribution in a downgoing slab are the rate of descent and its thickness, radiative conductivity, heating produced by its adiabatic compression, radioactive heating, latent heat of phase transitions in gabbro and peridotite (solid-solid, dehydration, and melting reactions), stress or frictional heating at the boundaries between lithosphere and asthenosphere, and movement of fluids and magma. The effects of these parameters, and the uncertainties in temperature distributions in subduction zones, are indicated by selections from published estimates illustrated in Fig. 17-4.

Minear and Toksöz (1970) used a finite difference solution of the conservation of energy equation to determine the effects of the factors listed above, with the complexity being increased in a series of steps. Figure 17-4A shows the effect of a cold slab as a heat sink. Figure 17-4B incorporates all of the factors that they believed to affect the thermal regime.

Horizontal distance Km

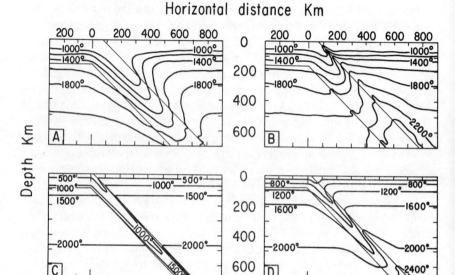

Figure 17-4. Calculated temperature distributions for mantle and subducted lithosphere plates, illustrating the effects of the main factors influencing the thermal structure. A and B. After Minear and Toksöz (1970, Figs. 6 and 10). C. After Griggs (1972). D. After Toksöz, Minear, and Julian (1971). These examples have not taken into account the cooling effect of endothermic dehydration reactions at depths down to 100 km, nor the effects of magma generation, and the migration of solutions and magmas (cf. Oxburgh and Turcotte, 1976).

Toksöz, Minear, and Julian (1971) presented what they considered to be more realistic computations, with one result shown in Fig. 17-4D.

The magnitude of the shear strain heating along the narrow zones of relative movement between slab and mantle is debated. Griggs (1972) concluded that it is not of great importance, and his model in Fig. 17-4B shows the heat-sink effect dominant. Oxburgh and Turcotte (1970) illustrated a slab with similar internal temperature distribution, but with low-temperature isotherms extending even deeper. They concluded that frictional heating along the upper boundary must be sufficient to cause melting, and they set temperatures at this boundary according to their estimates of melting temperatures, based on experimental data and petrological conclusions.

Oxburgh and Turcotte (1976) reviewed the effect of endothermic metamorphic reactions in the subducting oceanic crust, and the migration of dehydration water upwards or downwards. The process of magma generation in subduction zones is strongly dependent on the behavior of water. The latent heat of fusion for magma generation and the subsequent migration of magmas in turn influences the temperature distribution.

PHASE RELATIONSHIPS OF MANTLE PERIDOTITE

For the detailed interpretation of geophysical properties of the earth's interior and correlation with phase diagrams, it is necessary to know not only the composition of the rock material and the temperature at various depths, and the equilibrium phase diagram for the material, but also the properties of the phases in the different phase fields, and the kinetics of the transitions between the fields. Furthermore, there is no guarantee that measured geophysical properties relate to equilibrium conditions within the earth, because non-equilibrium states may persist in a dynamic earth (Wyllie and Schreyer, 1976).

A schematic phase diagram for mantle peridotite is shown in Fig. 17-5. This diagram is based on experimental data from many sources, with much extrapolation to higher pressures. The mineralogical zoning of the mantle can be estimated by following the geotherms reproduced from Fig. 17-2 through the peridotite mineral facies of Fig. 17-5. For a given geotherm, the successive mineral facies and phase transition intervals provide a cross-section through the mantle (see Ringwood, 1975, Chapters 3, 5, 6, 9-14). The narrow intervals of large changes in seismic velocities occurring at depths of approximately 350–450 km and 650–700 km (Fig. 17-1) correspond well with the phase transitions in the peridotite. Furthermore, estimated density changes within the mantle and in the peridotite show satisfactory correspondence (Wyllie, 1971, p. 132 and 135; Ringwood, 1975, p. 495).

The deep phase transitions must influence mantle dynamics, and mantle convection in turn influences magmatic processes. The most significant part of Fig. 17-5 for magma generation is nearer to the surface. Details are shown in Fig. 17-6. Peridotite at depths down to at least 300 km consists predominantly of olivine and two pyroxenes, together with a fourth aluminous phase. Figure 17-6 shows the change with increasing depth from plagioclase-peridotite through spinel-peridotite and into garnet-peridotite, via two narrow phase-transition intervals (plotted as lines). The plagioclase, clinopyroxene, and garnet pass into the liquid within 50°C above the solidus, producing a wide field for the coexistence of liquid with olivine and orthopyroxene. The compositions of near-solidus liquids change from basaltic at lower pressures to picritic at pressures corresponding to depths of 100 km or so, according to O'Hara's (1968) estimated normative olivine contours.

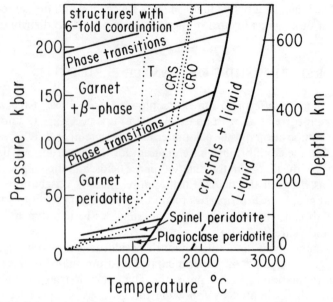

Figure 17-5. Schematic phase diagram for peridotite based on data from various sources, with much extrapolation to higher pressures (Wyllie, 1971, Chapter 6; Wyllie, 1973, Fig. 2; Ringwood, 1975, Chapters 5 and 6; Yoder, 1976, Chapter 2). Compare Fig. 17-2 for uncertainties in position of the solidus, and Fig. 17-6 for details in the depth interval 0–100 km. Dotted lines are geotherms from Fig. 17-2; T = Tozer, CR = Clark and Ringwood, S = shield, O = ocean.

It is now widely believed that most of the geophysical properties in the seismic low-velocity zone shown in Fig. 17-1 can be explained by the presence of a small proportion of interstitial magma. There is nothing in Figs. 17-5 and 17-6 to account for incipient melting in this zone. In shield regions, where the zone is absent or less clearly defined, the smaller decrease in S- and P-wave velocities could be explained by critical temperature gradients combined with the presence of mineralogical and chemical heterogeneity (Wyllie, 1971, Chapters 3 and 6; Ringwood, 1975, Chapter 6). Given a trace of H_2O or CO_2 in the mantle, however, it is difficult to escape the conclusion that a trace of magma is developed at a depth of 100 km or so in suboceanic mantle.

Goetze (1977) reviewed the effect of interstitial melt on the mechanical properties of the upper mantle, and concluded that the rheology of the upper mantle is compatible with experimental data on unmelted olivine. There is no experimental evidence on rocks to demonstrate that their rheology is influenced significantly by a small degree of partial melting.

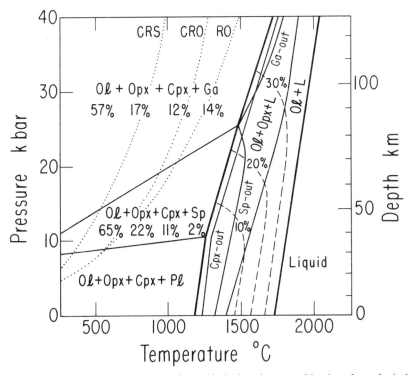

Figure 17-6. Schematic phase diagram for peridotite based on a combination of petrological experimental data from various sources (see Wyllie, 1971, pp. 118, 120, 130, and 194). Details of the phase relationships, and the composition and percentage of liquid through the melting interval, are not well established by direct experiment (Wyllie, 1971, pp. 193–205; Ringwood, 1975, pp. 143–150; Yoder, 1976, pp. 24–29, 110–115). With increasing pressure, the percentage of dissolved normative olivine increases, as shown by the dashed lines estimated by O'Hara (1968). Dotted lines are geotherms: CR = Clark and Ringwood (1964), R = Ringwood (1966), S = shield, O = ocean. Other abbreviations: Ol = olivine, Opx = orthopyroxene, Cpx = clinopyroxene, Pl = plagioclase, Sp = spinel, Ga = garnet, L = liquid. Mineral percentages are estimates from different sources (Wyllie, 1971, p. 120).

Therefore, Goetze argued, the common equation of the zone of decoupling (the asthenosphere) directly with the low-velocity layer and a partially melted layer is probably not justified. There are many inadequately known factors.

Selected results for the influence of H_2O on the phase relationships of peridotite are illustrated in Fig. 17-7. The solidus curve shows that melting begins in the presence of aqueous vapor at temperatures considerably

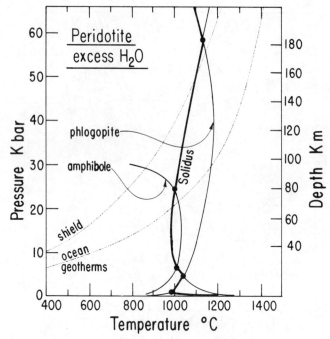

Figure 17-7. Selected phase boundaries for peridotite with excess H_2O, in part schematic. Compare with Fig. 17-6. Solidus with excess H_2O (heavy curves, after Kushiro, Syono, and Akimoto, 1968; see Fig. 17-8A for comparison of four sets of experimental data). Amphibole breakdown curve (Millhollen, Irving, and Wyllie, 1974; see Fig. 17-8B for comparison of four sets of experimental data). Phlogopite breakdown curve, extrapolated from Modreski and Boettcher (1973), see Wyllie (1973, Fig. 3). The solidus is assumed to be almost coincident with that for peridotite without phlogopite, except for very low pressures and very high pressures, where the phlogopite melting reaction (in association with peridotite minerals) causes melting at temperatures lower than the solidus for phlogopite-free peridotite (see Yoder and Kushiro, 1969). The dotted lines are geotherms of Clark and Ringwood (1964), compare Fig. 17-6.

lower than for the dry peridotite (Fig. 17-5). According to the two geotherms plotted, melting would begin at a depth of about 70 km beneath oceanic lithosphere, and about 150 km beneath the shields, if there was sufficient H_2O present to produce an aqueous pore fluid at these depths.

The presence of H_2O produces amphibole and phlogopite (given sufficient K_2O and Al_2O_3) in peridotite. Their breakdown curves in the presence of excess H_2O are shown in Fig. 17-7. Amphibole is limited to depths shallower than 100 km, whereas phlogopite can exist to depths ranging from 100 to 180 km. If there is only enough H_2O present to stabilize amphibole or phlogopite, with no free vapor left over, then the curves for the beginning of melting of the peridotite are situated near the stability boundaries of amphibole and phlogopite shown in Fig. 17-7, where these minerals break down in melting reactions, yielding H_2O for the generation of a trace of H_2O-undersaturated liquid.

The upper boundary of the seismic low-velocity zone in various tectonic environments could be explained by incipient melting of peridotite in the presence of aqueous pore fluid, or by incipient melting of peridotite caused by the melting of amphibole or phlogopite without excess H_2O (Kushiro, Syono, and Akimoto, 1968; Lambert and Wyllie, 1968, 1970; Modreski and Boettcher, 1973). Note the variations possible by the intersections of the geotherms with the three phase boundaries. D. H. Green and Liebermann (1976) and Solomon (1976) have reviewed details of geophysical and petrological aspects of this interpretation of the low-velocity zone.

It is important for geophysicists and other users of phase diagrams to realize that the experimental determination of equilibrium conditions for some parts of complex rock systems is very difficult. There is no doubt that many published phase diagrams for rock compositions are reaction diagrams, or synthesis diagrams, or partly deduced diagrams, that do *not* represent stable equilibrium. For applications of peridotite phase relationships to mantle physics, there is the additional problem of deciding which peridotite composition is applicable. Figures 17-8 and 17-9 compare several sets of results and interpretations for melting conditions and amphibole stability in different peridotites, determined in different laboratories. The differences in Fig. 17-9 are greatest in the pressure interval 20 kbar to 30 kbar, the region most critical for interpretation of the upper boundary of the oceanic low-velocity zone in terms of incipient melting.

Despite the difficulty of selecting the most appropriate peridotite composition for a heterogeneous mantle, and the problem of determining equilibrium phase diagrams for whole-rock compositions, the geotherms in Fig. 17-9 indicate that vapor-absent melting of amphibole-peridotite is a reasonable interpretation for the upper boundary of the oceanic low-velocity zone. But users of such diagrams should be careful not to place too much reliance on the precise details.

It has been established recently that CO_2 is also influential in causing incipient melting of mantle peridotite at depths greater than about 75 km (Huang and Wyllie, 1974; Wyllie and Huang, 1975; Eggler, 1976a). The phase fields intersected by a peridotite composition containing a small

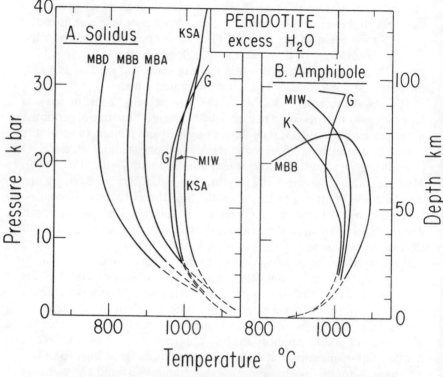

Figure 17-8. Comparison of experimental studies on different peridotite compositions with excess H_2O. Results of G = D. H. Green (1973a), K = Kushiro (1970), KSA = Kushiro, Syono, and Akimoto (1968), MB = Mysen and Boettcher (1975a, rocks A, B, and D), MIW = Millhollen, Irving, and Wyllie (1974). For comparison of experimental brackets, see Millhollen, Irving, and Wyllie (1974, Figs. 2 and 3). A. Solidus curves, compare Fig. 17-7. B. Curves for dissociation or melting reactions of amphibole in peridotite, compare Figs. 17-7 and 17-9. For comparison of kaersutite stability in mantle-derived rocks, see Merrill and Wyllie (1975).

percentage of CO_2 are illustrated in Fig. 17-10. The subsolidus region is divided into two parts by the carbonation reaction TQ:

olivine + clinopyroxene + CO_2 = orthopyroxene + dolomite (calcic)

Free vapor exists only on the low-pressure, high-temperature side of this reaction. On the high-pressure side, CO_2 reacts with the peridotite producing calcic dolomite. No free vapor remains, unless there is enough CO_2 present to react completely with the 11% clinopyroxene in the rock

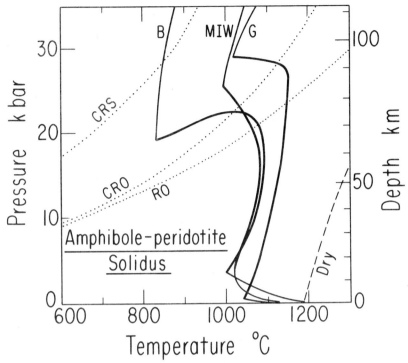

Figure 17-9. Three experimentally-based versions of the solidus for peridotite with a few tenths of a percent of H_2O, enough to generate amphibole with no free vapor left over. The heavy lines show the solidus where vapor-absent amphibole-peridotite begins to melt. The light lines correspond to the solidus with vapor, where amphibole is unstable (compare Figs. 17-7 and 17-8). B = Boettcher (1973), based on excess-H_2O results of Mysen and Boettcher (1975a); G = D. H. Green (1973a), an experimental reaction boundary; MIW = Millhollen, Irving, and Wyllie (1974), based on excess-H_2O results, and experiments and theory for amphibole–H_2O–CO_2 melting by Holloway (1973). Dotted lines are geotherms, see Fig. 17-6.

(Fig. 17-6); complete reaction is most unlikely for normal mantle conditions.

Notice the phase fields intersected by the geotherms, assuming a mantle composed of peridotite containing 0.1% CO_2. According to the Clark and Ringwood (1964) geotherms, CO_2 exists as carbonate, and incipient melting does not begin before the depth exceeds 130 km. According to the Ringwood (1966) oceanic geotherm, free CO_2 may exist for a depth interval of about 20 km before incipient melting begins at a depth near 80 km.

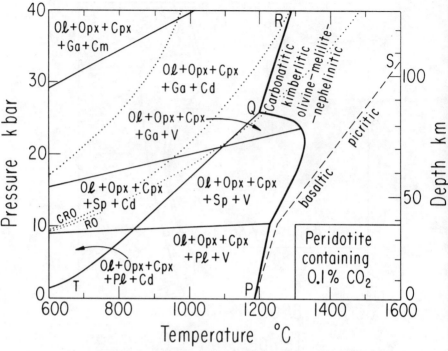

Figure 17-10. Schematic phase diagram for peridotite–CO_2, with insufficient CO_2 to carbonate the peridotite completely in reaction TQ:

olivine(Ol) + clinopyroxene(Cpx) + CO_2(V)
$\qquad$ = orthopyroxene(Opx) + calcic dolomite(Cd).

The solidus is divided into two portions, PQ showing the effect of CO_2 vapor on melting (contrast Figs. 17-7 and 17-8A), and QR showing the effect of calcic dolomite as a mineral additional to the volatile-free peridotite (after Wyllie, 1977a, Figs. 2 and 8A). For graphical convenience, it is assumed that reaction TQ remains univariant even with FeO as a component, and with Al_2O_3 dissolved in the pyroxenes (Wyllie and Huang, 1975; Wyllie, 1977a). The subsolidus phase boundaries between plagioclase-, spinel-, and garnet-peridotite are transferred from Fig. 17-6, assuming that they are not affected significantly by the carbonation reaction TQ and the presence of 0.1% CO_2. The vapor-absent melting of partially carbonated peridotite generates a liquid just above QR that is essentially carbonatitic, with Ca/Mg > 1, and CO_2 solubility that may approach 40% (Wyllie and Huang, 1975) or 27% (Eggler, 1976b). Dotted lines are geotherms. See Fig. 17-6 for abbreviations; additional abbreviations: Cd = calcic dolomite, Cm = magnesite solid solution. Dashed line is the solidus for peridotite, from Fig. 17-6.

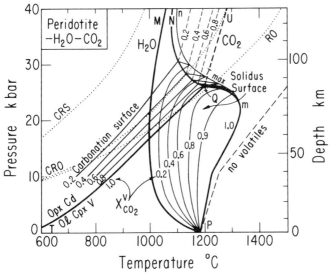

Figure 17-11. Schematic phase diagram for peridotite–CO_2–H_2O, modified after Wyllie (1977a, Fig. 8A), taking into account recent data for the divariant reaction surface for the carbonation reaction TQ in the presence of CO_2-H_2O vapor (Eggler, Kushiro and Holloway, 1976), and the phase relations for the solidus in the system MgO–SiO_2–CO_2–H_2O (D. E. Ellis and Wyllie, unpublished). The divariant carbonation surface is contoured by lines for constant vapor-phase composition. Similar contours are estimated for the solidus surface, based on experimental data of Mysen and Boettcher (1975a), and fitted between solidus curves for excess H_2O (Fig. 17-7) and excess CO_2 (QU differs from QR in Fig. 17-10), as described by Wyllie (1977a, Figs. 1, 2, and 7B). An additional feature is the temperature maximum on the solidus surface along the line mn. The magnitude of the separation between mn and mN, which is a direct function of the magnitude of the temperature maximum, is not known. The contours on each surface intersect the corresponding contours on the other surface, and the curve QN is the projected line of intersection of the two surfaces. For a system with insufficient CO_2 present to carbonate completely all of the clinopyroxene in the peridotite, the beginning of melting is limited to the area $PQNM$ of the solidus (which includes the solidus maximum passing through line mn). Contours for the solidus in the area NQU are dashed; these conditions are not reached in normal mantle. For the solidus surface $PMmn$, on the H_2O side of the temperature maximum mn, H_2O/CO_2 is greater in liquid than in vapor (standard situation for the low-pressure solidus). For the solidus surface mnN on the CO_2 side of the temperature maximum mn, H_2O/CO_2 is greater in vapor than in liquid (standard situation for high-pressure, partly carbonated peridotite). For the beginning of melting of partially carbonated peridotite in the presence of CO_2-H_2O, the vapor composition is buffered along the line QN. If the peridotite composition is suitable for the stabilization of phlogopite, the vapor-phase composition for the beginning of melting of a partially carbonated phlogopite–peridotite is buffered by a line near QN, at slightly lower temperatures, and slightly higher H_2O/CO_2. Dotted lines are geotherms, see Fig. 17-6.

For a more realistic mantle, containing traces of both H_2O and CO_2, the deduced Fig. 17-11 illustrates the processes involved. The divariant surfaces for the solidus and the carbonation reaction are contoured in terms of vapors with constant CO_2/H_2O. The subsolidus stability fields for amphibole and phlogopite (Fig. 17-7) are similarly represented by divariant surfaces extending to lower temperatures (Wyllie, 1977a, Figs. 12 and 13), but these are omitted in order to illustrate the effect of the carbonate alone.

The two divariant surfaces intersect along the line QN, which is the locus for the beginning of melting of carbonated peridotite in the presence of CO_2-H_2O vapors. The vapor composition is buffered along the line QN. At pressures above about 25 kbar, a very small proportion of CO_2 is sufficient to produce calcic dolomite in peridotite, and the coexisting vapor is strongly enriched in H_2O.

For mantle peridotite containing CO_2, H_2O, and no phlogopite, the depth of incipient melting is given by the intersection of a geotherm with the line QN. Oceanic geotherms pass through a partly carbonated lower lithosphere with CO_2-H_2O pore fluid. It is readily shown that amphibole can become stable on the solidus only for high H_2O/CO_2, and for a very limited area on the solidus surface (Wyllie, 1977a). Incipient melting of carbonated peridotite begins for a wide range of fluid compositions at a depth near 80 km. For continental geotherms, partially carbonated lithosphere with H_2O-enriched pore fluids persists to much deeper levels before melting can begin (compare Figs. 17-11 and 17-7).

For peridotite with accessory phlogopite, the vapor-phase composition would be buffered by reactions between phlogopite and carbonate, and a buffering curve with higher CO_2/H_2O would exist on the solidus surface in the area QNU. Alternatively, all of the CO_2 and H_2O could be stored in minerals, with the vapor-absent melting of a phlogopite–carbonate–peridotite beginning along a solidus curve at a temperature somewhat lower than QR (Fig. 17-10). This range of alternatives should be considered in the interpretation of geophysical measurements.

PRODUCTION OF BASALTIC MAGMA BY SELECTIVE FUSION OF PERIDOTITE

There is, at present, a very distinct trend of thought towards identifying the problem of the source of magmas with the problem of the source of basaltic magmas (Bowen, 1928, p. 311). It is, perhaps, worth while to examine somewhat more closely the selective fusion of a peridotite layer (p. 315). On the whole the production of basaltic magma by selective fusion of peridotite at a depth probably as great as 75–100 km seems to be the preferable method in spite of the great difficulties involved (p. 319).

There have been many schemes developed for basalt petrogenesis and, despite Bowen's conclusion in 1928 that partial fusion of peridotite was the preferable method, schemes involving fusion of other materials were promulgated during the next thirty years or so (Turner and Verhoogen, 1960, pp. 226–234, 432; Wyllie, 1971, pp. 168–173; Yoder, 1976, Chapters 1 and 2). By 1960, according to Turner and Verhoogen (1960, Chapter 15), the evidence was compelling for generation of basaltic magma by partial fusion of mantle peridotite.

Since 1960, petrogenetic schemes have been based largely on experimental studies of natural peridotite and basalt compositions, and the phase relationships in simple systems of the principal end-member minerals that make up peridotite and basalts. In their pioneering experimental work, Yoder and Tilley (1962) concluded that most basalts are eutectic-like. The composition of the magma generated in the mantle has apparently adjusted successively on the way to the surface, or equilibrated near the surface in magma reservoirs. The common basalts are the orderly end products of polybaric fractional crystallization.

Yoder and Tilley (1962) developed the concept of a basalt tetrahedron. Yoder (1976, Chapter 7) reviewed this concept and subsequent expansions, and traced the flow sheet of univariant liquid paths connecting invariant points at 1 atmosphere. Whereas Bowen (1928) regarded plateau basalts as the parental basalt from which most other magmas could be derived by crystallization differentiation, Yoder (1976, p. 128) concluded that three "parental" magmas, derived from a single parental material, are required to derive the magma sequences observed: olivine tholeiite, basanite, and olivine–melilitite–nephelinite.

Two comprehensive experimentally-based petrogenetic schemes were compared by Wyllie (1971, pp. 196–208) using Figs. 17-12 and 17-13. These schemes were developed by O'Hara (1965, 1968) and D. H. Green and Ringwood (1967), respectively. Note the mantle mineralogy at various depths in the two figures, and compare with the near-solidus mineralogy shown in Fig. 17-6. The mantle mineralogy adopted for Fig. 17-13 differs because the spinel dissolves in the aluminous pyroxenes before the solidus temperature is reached, and the reaction to produce garnet-peridotite does not occur until about 31 kbar at the solidus. Space limitations prohibit a detailed review, but these figures merit careful study.

O'Hara (1965, 1968) developed the theme of Yoder and Tilley (1962) and O'Hara and Yoder (1967) that the major commitment of basaltic type is made by the melting process at various depths, and that the extrusive basalts are residual liquids of advanced crystal fractionation. He

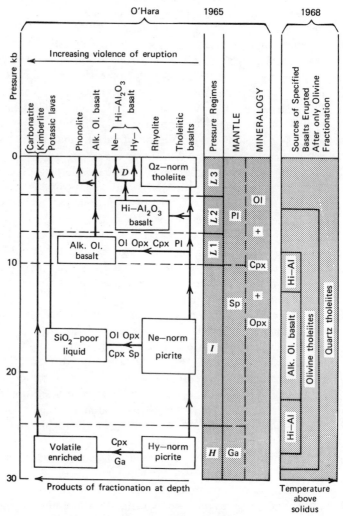

Figure 17-12. Wyllie's (1971, Fig. 8-15) summary of suggested basalt fractionation scheme
dependent upon depth of origin and rate of migration to the surface according to O'Hara (1965, 1968), incorporating the experimentally-based
concepts of Yoder and Tilley (1962) and O'Hara and Yoder (1967). The main
portion is based on O'Hara's (1965) Table 1, and the right-hand portion is a
revised scheme (O'Hara, 1968) showing effect of olivine fractionation only.
Minerals fractionating are listed near the arrows; note eclogite fractionation
in deepest levels. For mantle mineralogy compare Fig. 17-6, which also lists
abbreviations. Additional abbreviations: Hy = hypersthene, Ne = nepheline,
D = thermal divide, Hi-Al = Hi-Al$_2$O$_3$ = high-alumina (basalt).

Green & Ringwood 1967.   Green 1969

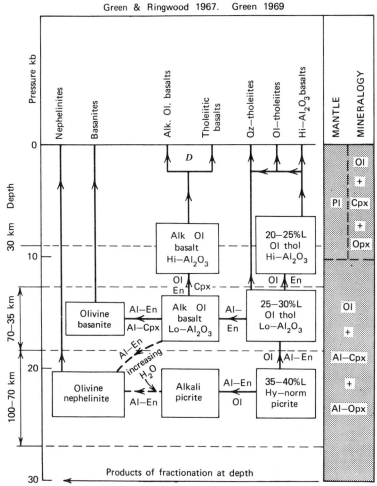

Figure 17-13. Simplified representation of the petrogenetic scheme of D. H. Green and
Ringwood (1967) and D. H. Green (1969) after Wyllie (1971, Fig. 8-16).
Contrast the mantle mineralogy between 10 and 30 kbar in this chart with that
in Figs. 17-12 and 17-6; garnet does not appear on the solidus until about
31 kbar in the scheme of D. H. Green and Ringwood. The right-hand boxes
show products of equilibrium partial melting of peridotite in various depth
intervals. The deduced relationships among various basaltic magmas at
moderate to high pressures are related to "closed system" fractionation.
Minerals fractionating are listed near the arrows; note the absence of eclogite
fractionation, and the role assigned to aluminous enstatite and H_2O in
generation of olivine-nephelinites. Abbreviations, see Figs. 17-12 and 17-6.

suggested that picritic magmas were generated by partial fusion of garnet peridotite and segregated from the mantle source at depths normally greater than 100 km. In 1965, O'Hara concluded that an original picritic magma remaining just saturated with olivine, orthopyroxene, and clinopyroxene during uprise would produce an inevitable eruptive product of quartz tholeiite. This course would be changed if fractional crystallization occurred at depth. In 1968, he considered the composition of erupted surface liquids when a magma batch was transported from a particular depth-temperature area at such a speed that it fractionated only olivine during ascent. He concluded that all volumetrically important basalt magmas had experienced olivine fractionation during ascent.

D. H. Green and Ringwood (1967) determined the phase relationships of a series of synthetic basalt compositions, compared the near-liquidus minerals at various pressures with near-solidus minerals in peridotite, and developed a scheme for generation of the major basaltic types by equilibrium partial melting of peridotite at various depths, normally at depths where garnet peridotite was not stable. The percentage of partial melting and the depth of magma segregation are decisive in determining the composition of a particular batch of magma. Fractional crystallization at various depths would change the eruptive products. Details of the scheme were extended and modified in subsequent papers, and the effect of H_2O was included, as reviewed by Ringwood (1975, Chapter 4).

O'Hara (1968) and Kushiro (1969) criticized several aspects of the petrogenetic scheme of D. H. Green and Ringwood, including the dominant role assigned to aluminous orthopyroxene at intermediate and high pressures. Ringwood (1975, pp. 166–171) replied, ending with the statement: "It must be concluded in the light of the preceding discussion that O'Hara's scheme of basalt petrogenesis is inadequate." Yoder (1976, Chapters 7 and 8) reaffirmed the principles involved in formulation of O'Hara's scheme, stating that (pp. 146–147): "The derivative nature of basalts (and eclogites) has already been demonstrated by Yoder and Tilley (1962) and emphasized by O'Hara (1968)." Yoder took issue with several major aspects of the results of Green and Ringwood (e.g., p. 148), but he did not review the scheme as a whole.

There is no adequate explanation for the generation of the SiO_2-poor olivine–melilitite–nephelinites from volatile-free mantle peridotite or parent basaltic magmas. D. H. Green (1969) suggested, on the basis of reconnaissance experiments, that the field for the primary crystallization of aluminous orthopyroxene was increased in the presence of H_2O at depths greater than about 70 km. He concluded that a small percentage of partial melting with H_2O present would produce these liquids, and later published detailed experiments supporting this thesis (D. H. Green, 1973a). These conclusions and some of the experiments were criticized by O'Hara (1968), considered "erroneous" by Yoder (1976, p. 148), and acclaimed as an "important discovery" by Ringwood (1975, pp. 138–140, 150–153). It now appears that CO_2 rather than H_2O is the most influential component in expanding the field of orthopyroxene (Eggler, 1974).

Eggler (1974) studied the effect of CO_2 on the compositions of liquids coexisting with the model mantle peridotite assemblage forsterite + enstatite + diopside, and discovered that at some pressure between 15 kbar and 30 kbar the liquid became larnite-normative. Huang and Wyllie (1974) studied carbonate-silicate joins, and concluded that there was a liquidus path from CO_2-bearing basaltic liquids through kimberlitic (or melilitic) compositions to carbonate-rich liquids. Combination of these and subsequent results provides the evidence for the deduced phase diagram for periodotite–CO_2 given in Fig. 17-10 (Wyllie and Huang, 1975, 1976a; Eggler, 1976a; Kushiro, Satake, and Akimoto, 1975). At depths greater than about 80 km, partial fusion of mantle peridotite containing CO_2 (vapor-absent melting along QR) produces first a trace of carbonatitic magma, which changes through compositions corresponding to kimberlites or olivine–melilite–nephelinites with a small increase in degree of partial melting.

Liquid compositions must change if H_2O is present in addition to CO_2 (Mysen and Boettcher, 1975b), but Fig. 17-11 shows that at pressures above about 25 kbar, carbonate exists on the solidus even in the presence of vapors enriched in H_2O/CO_2. The liquid compositions are probably influenced more strongly by the carbonate than by dissolved H_2O. The near-solidus liquid may be carbonatitic even with high H_2O/CO_2 in the rock. Brey and D. H. Green (1975, 1976) determined the primary minerals on the liquidus of an olivine melilitite composition with dissolved CO_2 and H_2O. They inferred that the melilitite may be derived by equilibrium partial melting of peridotite containing small amounts of CO_2 and H_2O, with high CO_2/H_2O, at 30 kbar, 1150°–1200°C.

PRODUCTION OF GRANITIC MAGMA BY SELECTIVE FUSION OF THE CRUST

Bowen (1928, p. 319) wrote: *Many granitic magmas may have their immediate origin in the remelting, say by deep burial, of a granite derived in more remote times from basic material.*

The bitterness of the international granite controversy between magmatists and transformists subsided into intermittent local arguments after Bowen and Tuttle (1950) and Tuttle and Bowen (1958) published their high-pressure results on the *Origin of granite in the light of experimental studies in the system $NaAlSi_3O_8$–$KAlSi_3O_8$–SiO_2–H_2O.* They refined the anatectic model, using the experimental results to show how the partial melting of crustal rocks in the presence of small amounts of aqueous pore fluid would produce H_2O-undersaturated granitic magmas.

It is now evident from experimental studies of synthetic systems (Winkler, Boese, and Marcopoulos, 1975; Huang and Wyllie, 1975; Luth, 1976) and of many rocks in the presence of H_2O (for reviews see Wyllie et al., 1976; Wyllie, 1977b), that generation of H_2O-undersaturated granitic liquid is the normal consequence of regional metamorphism. The curves for the beginning of melting of crustal rocks in the presence of aqueous pore fluid are plotted in Fig. 17-14A, together with the curves for the beginning of melting of crustal rocks containing no free pore fluid, but with the hydrous minerals muscovite, biotite, or amphibole. These minerals react to yield H_2O for H_2O-undersaturated granitic liquid (Brown and Fyfe, 1970; Robertson and Wyllie, 1971). The solidus curves are compared with $P-T$ paths of regional metamorphism summarized by Turner (1968, p.359), and two independent estimates of conditions for the formation of granulite-facies rocks.

Figure 17-14B shows the compositions of liquids that could be developed from deep crustal rocks, containing a generous allowance of 2% H_2O. The maximum depth-temperature range for the generation of H_2O-saturated granitic liquids is narrow for the andesitic or tonalitic gneiss that probably constitutes the greater part of the source rocks. At lower pressures, H_2O-undersaturated liquid of granite composition coexists with crystals through a wide range of temperatures, reaching granodiorite composition only at temperatures above 900°C. With increasing temperature and pressure, liquid compositions trend toward tonalite and diorite, at least in terms of feldspar-quartz components. But note that with 2% H_2O, the liquidus temperature for tonalite is about 1100°C, indicating that tonalite and diorite liquids cannot be generated by anatexis of the crust under conditions of normal regional metamorphism, and that these magmas must represent crystal mushes. The only escape from this conclusion is if there has been a significant contribution of heat and material from less siliceous magmas generated in subducted ocean crust and mantle peridotite.

PRODUCTION OF MAGMA OF INTERMEDIATE COMPOSITION BY SELECTIVE FUSION OF SUBDUCTED OCEANIC CRUST

The partial fusion of subducted oceanic crust (amphibolite, quartz eclogite, perhaps minor sediments) produces magma of intermediate

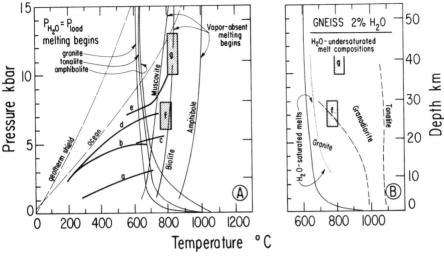

Figure 17-14. Conditions for fusion of crustal rocks in the series granite–tonalite–gabbro with H_2O compared with estimated pressures and temperatures attained during regional metamorphism (Wyllie, 1977b).

A. Calculated geotherms for shield and oceanic regions of Clark and Ringwood (1964) compared with depth-temperature paths for regional metamorphism in different metamorphic terranes: estimated from rock mineralogy, collated by Turner (1968, Fig. 8-4, paths a to e); granulite facies temperatures and pressures estimated by Touret (1971, f) and Wood (1975, g). Curves for the beginning of melting with excess H_2O and for the beginning of melting in the absence of vapor, corresponding to the initial breakdown of hydrous minerals muscovite, biotite, and amphibole, are from various sources reviewed by Wyllie (1977b). Note the wide variety of temperatures attained during regional metamorphism, and the extent to which they exceed the calculated "normal" geotherms. Compare Figs. 17-5 and 17-6.

B. For deep crustal rocks of a wide range of compositions (gabbro–tonalite–granite-metamorphosed greywacke and shale) in the presence of 2% H_2O, melting begins within the narrow band labeled "H_2O-saturated melts." For a rock of fixed composition, this interval of H_2O-saturated liquids is even narrower. Liquids produced by anatexis in amphibolite-granulite facies rocks normally are H_2O-undersaturated. The precise compositions of liquids generated at high pressures are not well established, but they are limited to the granite-granodiorite range. Note that a temperature of 1100°C is required to reach the liquidus for tonalite composition with 2% dissolved H_2O.

SiO_2 content. This idea was not conceived when Bowen wrote his book in 1928, and the topic cannot be treated in detail here.

According to T. H. Green and Ringwood (1968), Oxburgh and Turcotte (1970), and Marsh and Carmichael (1974), among others, these magmas may rise from depths of 100–150 km directly to the surface, as primary andesites and tonalites. Any fractionation at depth is governed by amphibole, pyroxenes, and garnet. According to reviews by Wyllie *et al.* (1976) and Wyllie (1977c), it is more likely that andesites and tonalites represent the end products of polybaric crystallization and other differentiation processes acting on the initial magmas. Ringwood (1975, Chapter 7) developed more complex models, suggesting that the hydrous intermediate-acid magmas generated at depth reacted with overlying mantle peridotite, producing olivine pyroxenite, which rises diapirically and melts at higher levels to yield basaltic andesites and andesites.

At least part of the magma rising from subduction zones must remain trapped in the crust and incorporated into batholiths. Batholiths may thus include magmas derived by partial fusion of mantle peridotite, partial fusion of subducted ocean crust, and partial fusion of deep-seated continental crust, together with differentiates of all these materials.

CONCLUDING REMARKS

There are many factors to be evaluated in considering the identity and origin of magmas, and there are many lines of evidence. These lines include the study of isotopes, rare earth elements, and other trace elements. Nicolaysen (1976) concluded that the study of isotopes and trace elements in the petrology of oceanic basalts "constitutes a true frontier of earth sciences," with applications not only to the petrogenesis of basalts, but also to deep-seated processes and mantle dynamics. He pointed out, however, that at present there are sharply contrasting interpretations of these data. Yoder (1976, pp. 150–161) reviewed the significance of rare earth element data, and concluded that "experimental evaluation of the principal variables is needed before unambiguous conclusions can be reached."

It is important to remember that conclusions reached from the study of isotopes and trace elements, and other lines of evidence, cannot be valid unless they also satisfy the constraints imposed by phase equilibrium experiments for the distribution of major elements. Laboratory experiments conducted under controlled conditions could be expected to provide unambiguous data relevant to the origin of rocks. The review in the preceding pages shows that this is not so. There are many practical factors leading to the publication of different results and interpretations for similar studies conducted in different laboratories (see, for example, Yoder and Tilley, 1962, pp. 372–381, 446–448; Nehru and Wyllie, 1975; D. H. Green, 1976).

Bowen (1928, p. 320) concluded his chapter with the following statement: *The main purpose of this chapter is not, however, to express advocacy of any particular mode of derivation of magma but rather to indicate how inadequate are the data now available to permit a definite decision on this point. The early stages of organic evolution—the early stages of the development of the human race—are shrouded in mystery. So it is with the early stages of the development of the magmatic cycle.* The statement is still appropriate for termination of this 1978 chapter.

Acknowledgments   Thanks are due to the National Science Foundation for Grants EAR 76-20410 and EAR 76-20413, which covered manuscript preparation costs and supported research contributions to this chapter, and also for its general support of the Materials Research Laboratory, University of Chicago.

References

Bell, P. M., ed., *U.S. National Report 1971–1974, Rev. Geophys. and Space Phys.*, *13*, 1–1106, 1975.

Boettcher, A. L., Volcanism and orogenic belts—the origin of andesites, *Tectonophysics*, *17*, 223–240, 1973.

———, ed., *Geothermometry-Geobarometry*, *Am. Mineral.*, *61*, 549–816, 1976.

Bott, M. H. P., *The Interior of the Earth*, Edward Arnold, London, 316 pp., 1971.

Bottinga, Y., and C. Allègre, Geophysical, petrological and geochemical models of the oceanic lithosphere, *Tectonophysics*, *32*, 9–59, 1976.

Bowen, N. L., *The Evolution of the Igneous Rocks*, Princeton University Press, Princeton, 332 pp., 1928.

———, Progressive metamorphism of siliceous limestone and dolomite, *J. Geol.*, *48*, 225–274, 1940.

Bowen, N. L., and O. F. Tuttle, The system $NaAlSi_3O_8$–$KAlSi_3O_8$–H_2O, *J. Geol.*, *58*, 489–511, 1950.

Boyd, F. R., A pyroxene geotherm, *Geochim. Cosmochim. Acta*, *37*, 2533–2546, 1973.

Brey, G. P., and D. H. Green, The role of CO_2 in the genesis of olivine melilitite, *Contrib. Mineral. Petrol.*, *49*, 93–103, 1975.

Brey, G. P., and D. H. Green, Solubility of CO_2 in olivine melilitite at high pressures and role of CO_2 in the earth's upper mantle, *Contrib. Mineral. Petrol.*, *55*, 217–230, 1976.

Brown, G. C., and W. S. Fyfe, The production of granitic melts during ultrametamorphism, *Contrib. Mineral. Petrol.*, *28*, 310–318, 1970.

Carmichael, I. S. E., F. J. Turner, and J. Verhoogen, *Igneous Petrology*, McGraw-Hill, New York, 739 pp., 1974.

Chung, D. H., On the composition of the oceanic lithosphere, *J. Geophys. Res.*, *81*, 4129–4134, 1976.

Clark, S. P., and A. E. Ringwood, Density distribution and constitution of the mantle, *Review Geophysics*, *2*, 35–88, 1964.

Condie, K. C., *Plate Tectonics and Crustal Evolution*, Pergamon Press, New York, 288 pp., 1976.

Eaton, G. P., R. L. Christiansen, H. M. Iyer, A. M. Pitt, D. R. Mabey, H. R. Blank, I. Zietz, and M. E. Gettings, Magma beneath Yellowstone National Park, *Science*, *188*, 787–796, 1975.

Eggler, D. H., Effect of CO_2 on the melting of peridotite, *Carnegie Institution of Washington Yearbook*, *73*, 215–224, 1974.

———, Does CO_2 cause partial melting in the low-velocity layer of the mantle? *Geology*, *4*, 69–72, 1976a.

———, Does CO_2 cause partial melting in the low-velocity layer of the mantle? Reply, *Geology*, *4*, 787 and 788–789, 1976b.

Eggler, D. H., I. Kushiro, and J. R. Holloway, Stability of carbonate minerals in a hydrous mantle, *Carnegie Institution of Washington Yearbook*, *75*, 631–636, 1976.

Forsyth, D. W., and F. Press, Geophysical tests of petrological models of the spreading lithosphere, *J. Geophys. Res.*, *76*, 7963–7979, 1971.

Garland, G. D., *Introduction to Geophysics*, W. B. Saunders, Toronto, 420 pp., 1971.

Geotze, C., A brief summary of our present day understanding of the effect of volatiles and partial melt on the mechanical properties of the upper mantle, in *High Pressure Research: Applications to Geophysics*, ed. M. H. Manghnani and S. Akimoto, Academic Press, New York, pp. 3–23, 1977.

Green, D. H., The origin of basaltic and nephelinitic magmas in the earth's mantle, *Tectonophysics*, *7*, 409–422, 1969.

———, Experimental melting studies on a model upper mantle composition at high pressure under water-saturated and water-undersaturated conditions, *Earth and Planet. Sci. Letters*, *19*, 37–53, 1973a.

———, Conditions of melting of basanite magma from garnet peridotite, *Earth and Planet. Sci. Letters*, *17*, 456–465, 1973b.

———, Experimental testing of "equilibrium" partial melting of peridotite under water-saturated, high-pressure conditions, *Can. Miner.*, *14*, 255–268, 1976.

Green, D. H., and R. C. Liebermann, Phase equilibria and elastic properties of a pyrolite model for the oceanic upper mantle, *Tectonophysics*, *32*, 61–92, 1976.

Green, D. H., and A. E. Ringwood, the genesis of basaltic magmas, *Contrib. Mineral. Petrol.*, *15*, 103–190, 1967.

Green, T. H., and A. E. Ringwood, Genesis of the calc-alkaline igneous rock suite, *Contrib. Mineral. Petrol.*, *18*, 105–162, 1968.

Griggs, D. T., The sinking lithosphere and the focal mechanism of deep earthquakes, in *The Nature of the Solid Earth*, ed. E. C. Robertson, McGraw-Hill, New York, 677 pp., 1972.

Harte, B., K. G. Cox, and J. J. Gurney, Petrography and geological history of upper mantle xenoliths from the Matsoku kimberlite pipe, in *Physics and Chemistry of the Earth*, *9*, Pergamon Press, New York, 477–506, 1975.

Heinrich, E. W., *The Geology of Carbonatites*, Rand McNally, Chicago, 555 pp., 1966.

Hess, H. H., and A. Poldervaart, *Basalts: The Poldervaart Treatise on Rocks of Basaltic Composition, Volume 1*, Interscience Publishers, New York, 482 pp., 1967.

Hess, H. H., and A. Poldervaart, *Basalts: The Poldervaart Treatise on Rocks of Basaltic Composition, Volume 2*, Interscience Publishers, New York, 380 pp., 1968.

Holloway, J. R., The system pargasite–H_2O–CO_2: a model for melting of a hydrous mineral with a mixed-volatile fluid. I. Experimental results to 8 kb, *Geochim. Cosmochim. Acta, 37*, 651–666, 1973.

Huang, W. L., and P. J. Wyllie, Eutectic between wollastonite II and calcite contrasted with thermal barrier in MgO–SiO_2–CO_2 at 30 kilobars, with applications to kimberlite–carbonatite petrogenesis, *Earth and Planet. Sci. Letters, 24*, 305–310, 1974.

Huang, W. L., and P. J. Wyllie, Melting reactions in the system $NaAlSi_3O_8$–$KAlSi_3O_8$–SiO_2 to 35 kilobars, dry and with excess water, *J. Geol., 83*, 737–748, 1975.

Hyndman, D. W., *Petrology of Igneous and Metamorphic Rocks*, McGraw-Hill, New York, 533 pp., 1972.

Jacobs, J. A., R. D. Russell, and J. T. Wilson, *Physics and Geology*, 2nd edition, McGraw-Hill, New York, 622 pp., 1974.

Jordan, T. H., The continental lithosphere, *Rev. Geophys. and Space Phys., 13*, 1–12, 1975.

Kushiro, I., Discussion of the paper "The origin of basaltic and nephelinitic magmas in the earth's mantle" by D. H. Green, *Tectonophysics, 7*, 427–436, 1969.

———, Stability of amphibole and phlogopite in the upper mantle, *Carnegie Institution of Washington Yearbook, 68*, 245–247, 1970.

Kushiro, I., H. Satake, and S. Akimoto, Carbonate-silicate reactions at high pressures and possible presence of dolomite and magnesite in the upper mantle, *Earth and Planet. Sci. Letters, 28*, 116–120, 1975.

Kushiro, I., Y. Syono, and S. Akimoto, Melting of a peridotite nodule at high pressures and high water pressures, *J. Geophys. Res., 73*, 6023–6029, 1968.

Lambert, I. B., and P. J. Wyllie, Stability of hornblende and a model for the low velocity zone, *Nature, 219*, 1240–1241, 1968.

Lambert, I. B., and P. J. Wyllie, Low-velocity zone of the Earth's mantle: incipient melting caused by water, *Science, 169*, 764–766, 1970.

Le Pichon, X., J. Francheteau, and J. Bonnin, *Plate Tectonics*, Elsevier Scientific, Amsterdam, 300 pp., 1973.

Lloyd, F. E., and D. K. Bailey, Light element metasomatism of the continental mantle: the evidence and the consequences, in *Physics and Chemistry of the Earth, 9*, Pergamon Press, New York, 389–416, 1975.

Lubimova, E. A., Theory of thermal state of the earth's mantle, in *The Earth's Mantle*, ed. T. F. Gaskell, Academic Press, New York, 231–323, 1967.

Lunar Science Institute, compilers, *Origins of Mare Basalts and Their Implications for Lunar Evolution*, Papers presented at conference Nov. 17–19, 1975. U.S. Government printing office, 205 pp., 1975.

Luth, W. C., Granitic rocks, in *The Evolution of the Crystalline Rocks*, ed. D. K. Bailey and R. Macdonald, Academic Press, London, 335–417, 1976.

MacGregor, I. D., and A. R. Basu, Thermal structure of the lithosphere: a petrologic model, *Science, 185,* 1007–1011, 1974.

Marsh, B. D., and I.S.E. Carmichael, Benioff zone magmatism, *J. Geophys. Res., 79,* 1196–1206, 1974.

McKenzie, D. P., and N. O. Weiss, Speculations on the thermal and tectonic history of the earth, *Geophys. Jour. Roy. Astron. Soc., 42,* 131–174, 1975.

Mercier, J-C. C., and N. L. Carter, Pyroxene geotherms, *J. Geophys. Res., 80,* 3349–3362, 1975.

Merrill, R. B., and P. J. Wyllie, Kaersutite and kaersutite eclogite from Kakanui, New Zealand—Water-excess and water-deficient melting at 30 kilobars, *Geol. Soc. Amer. Bull., 86,* 555–570, 1975.

Millhollen, G. L., A. J. Irving, and P. J. Wyllie, Melting interval of periodotite with 5.7 per cent water to 30 kilobars, *J. Geol., 82,* 575–587, 1974.

Minear, J. W., and M. N. Toksöz, Thermal regime of a downgoing slab and new global tectonics, *J. Geophys. Res., 75,* 1397–1419, 1970.

Modreski, P. J., and A. L. Boettcher, Phase relationships of phlogopite in the system $K_2O–MgO–CaO–Al_2O_3–SiO_2–H_2O$ to 35 kilobars: a better model for micas in the interior of the earth, *Am. J. Sci., 273,* 385–414, 1973.

Morgan, W. J., Deep mantle convection plumes and plant motions, *Bull. Amer. Assoc. Petroleum Geol., 56,* 203–213, 1972.

Mysen, B. O., and A. L. Boettcher, Melding of a hydrous mantle: I. Phase relations of natural peridotite at high pressures and temperatures with controlled activities of water, carbon dioxide, and hydrogen, *J. Petrol. 16,* 520–548, 1975a.

Mysen, B. O., and A. L. Boettcher, Melting of a hydrous mantle: II. Geochemistry of crystals and liquids formed by anatexis of mantle peridotite at high pressures and high temperatures as a function of controlled activities of water, hydrogen, and carbon dioxide, *J. Petrol., 16,* 549–593, 1975b.

Nehru, C. E., and P. J. Wyllie, Compositions of glasses from St. Paul's peridotite partially melted at 20 kilobars, *J. Geol., 83,* 455–471, 1975.

Nicolaysen, L. B., Geochemical constraints on models of mantle dynamics, in *Geodynamics: Progress and Prospects,* ed. C. L. Drake, Amer. Geophys. Union, Washington, D.C., 109–114, 1976.

O'Hara, M. J., Primary magmas and the origin of basalts, *Scott. J. Geol., 1* 19–40, 1965.

———, The bearing of phase equilibria studies in synthetic and natural systems on the origin and evolution of basic and ultrabasic rocks, *Earth Sci. Rev., 4,* 69–133, 1968.

O'Hara, M. J., and H. S. Yoder Jr., Formation and fractionation of basic magmas at high pressures, *Scott. J. Geol., 3,* 67–117, 1967.

Oxburgh, E. R., and D. L. Turcotte, Mid-ocean ridges and geotherm distribution during mantle convection, *J. Geophys. Res., 73,* 2643–2661, 1968.

Oxburgh, E. R., and D. L. Turcotte, Thermal structure of island arcs, *Geol. Soc. Am. Bull., 81,* 1665–1688, 1970.

Oxburgh, E. R., and D. L. Turcotte, The physico-chemical behaviour of the descending lithosphere, *Tectonophysics, 32,* 107–128, 1976.

Papike, J. J., ed., *The Nature of the Oceanic Crust*, *J. Geophys. Res.*, *81*, 4041–4380, 1976.

Press, F., Earth models obtained by Monte Carlo inversion, *J. Geophys. Res.*, *73*, 5223–5234, 1968.

Ritcher, F. M., Convection and large-scale circulation of the mantle, *J. Geophys. Res.*, *78*, 8735–8745, 1973.

Ringwood, A. E., Mineralogy of the mantle, in *Advances in Earth Sciences*, ed. P. M. Hurley, M.I.T. Press, Cambridge, Mass., 357–399, 1966.

———, *Composition and Petrology of the Earth's Mantle*, McGraw-Hill, New York, 618 pp., 1975.

Robertson, J. K., and P. J. Wyllie, Rock-water systems, with special reference to the water-deficient region, *Am. J. Sci.*, *271*, 252–277, 1971.

Solomon, S. C., Geophysical constraints on radial and lateral temperature variations in the upper mantle, *Am. Mineral.*, *61*, 788–803, 1976.

Sørensen, H., *Alkaline Rocks*, Wiley & Sons, New York, 622 pp., 1974.

Stacey, F. D., *Physics of the Earth*, 2nd edition, Wiley & Sons, New York, 414 pp., 1977.

Toksöz, M. N., J. W. Minear, and B. R. Julian, Temperature field and geophysical effects of a downgoing slab, *J. Geophys. Res.*, *76*, 1113–1138, 1971.

Torrance, K. E., and D. L. Turcotte, Structure of convection cells in the mantle, *J. Geophys. Res.*, *76*, 1154–1161, 1971.

Touret, J., Le facies granulite en Norvège Meridionale. I. Les associations minéralogiques, *Lithos*, *4*, 239–249, 1971.

Tozer, D. C., Towards a theory of thermal convection, in *The Earth's Mantle*, ed. T. F. Gaskell, Academic Press, New York, 325–353, 1967.

Turner, F. J., *Metamorphic Petrology*, McGraw-Hill, New York, 403 pp., 1968.

Turner, F. J., and J. Verhoogen, *Igneous and Metamorphic Petrology*, 2nd edition, McGraw-Hill, New York, 694 pp., 1960.

Tuttle, O. F., and N. L. Bowen, Origin of granite in the light of experimental studies in the system $NaAlSi_3O_8$–$KAlSi_3O_8$–SiO_2–H_2O, *Geol. Soc. Amer. Memoir*, *74*, 153 pp., 1958.

Tuttle, O. F., and J. Gittins, *Carbonatites*, Interscience, New York, 591 p., 1966.

Winkler, H. G. F., M. Boese, and T. Marcopoulos, Low temperature granitic melts, *N. Jb. Miner. Mh.*, 245–268, 1975.

Wood, B. J., The influence of pressure, temperature and bulk composition on the appearance of garnet in orthogneiss—an example from South Harris, Scotland, *Earth and Planet. Sci. Letters*, *26*, 299–311, 1975.

Wyllie, P. J., ed., *Ultramafic and Related Rocks*, Wiley, New York, 464 pp., 1967.

———, *The Dynamic Earth*, John Wiley & Sons, New York, 416 pp., 1971.

———, ed., *Experimental Petrology and Global Tectonics*, *Tectonophysics*, *17*, 187–297, 1973.

———, ed., *Geophysical Measurements and Experimental Petrology*, *Tectonophysics*, *32*, 1–143, 1976.

———, Mantle fluid compositions buffered by carbonates in peridotite–CO_2–H_2O, *J. Geol.*, *85*, 187–207, 1977a.

———, Crustal anatexis: an experimental review, *Tectonophysics*, *43*, 41–71, 1977b.

————, From crucible through subduction to batholiths, in *Energetics of Geological Processes*, ed. S. K. Saxena and S. Bhattacharji, Springer-Verlag, 389–433, 1977c.

Wyllie, P. J., and W. L. Huang, Peridotite, kimberlite, and carbonatite explained in the system $CaO-MgO-SiO_2-CO_2$, *Geology*, *3*, 621–624, 1975.

Wyllie, P. J., and W. L. Huang, Carbonation and melting reactions in the system $CaO-MgO-SiO_2-CO_2$ at mantle pressures with geophysical and petrological applications, *Contrib. Mineral. Petrol.*, *54*, 79–107, 1976a.

Wyllie, P. J., and W. L. Huang, Does CO_2 cause partial melting in the low-velocity layer of the mantle? Comment, *Geology*, *4*, 712 and 787, 1976b.

Wyllie, P. J., W. L. Huang, C. R. Stern, and S. Maaløe, Granitic magmas: possible and impossible sources, water contents, and crystallization sequences, *Can. Jour. Earth Sci.*, *13*, 1007–1019, 1976.

Wyllie, P. J., and W. Schreyer, Geophysical measurements and experimental petrology, *Tectonophysics*, *32*, 1–6, 1976.

Yoder, H. S., Jr., *Generation of Basaltic Magma*, Nat. Acad. Sci., Washington, D.C., 265 pp., 1976.

Yoder, H. S., Jr., and I. Kushiro, Melting of a hydrous phase: phlogopite, *Am. J. Sci.*, *267-A*, 558–582, 1969.

Yoder, H. S., Jr., and C. E. Tilley, Origin of basalt magmas: an experimental study of natural and synthetic rock systems, *J. Petrol.*, *3*, 342–532, 1962.

Chapter 18

PARTITIONING BY DISCRIMINANT ANALYSIS: A MEASURE OF CONSISTENCY IN THE NOMENCLATURE AND CLASSIFICATION OF VOLCANIC ROCKS

FELIX CHAYES

Geophysical Laboratory, Carnegie Institution of Washington, Washington, D.C.

After a long period of neglect the taxonomy of igneous rocks appears to be the object of a modest resurgence of concern.[1] Much of the current activity in this field consists of attempts to redefine and rearrange existing names, and inevitably suffers from the lack of widely accepted, reasonably objective criteria for deciding whether proposed modifications of terms already in the public domain are actually in the public interest. This chapter proposes such a criterion—the partition effected by a discriminant function—for the case in which chemical composition is of critical importance. The underlying method, however, is of broader interest; it should be applicable in any situation in which measurements on objects subject to an existing classification provide a substantively meaningful or widely accepted basis for constructing a new classification. The discriminant function based on these measurements is the linear combination that maximizes the ratio of between-to within-group variance, so it is most unlikely that another classification based on them will be more consistent[2] with the original classification.

RATIONALE

Sound nomenclature performs, at the special or varietal level, about the same function performed by classification at the familial or generic level.

The major objective of each is to group like objects together and separate unlike objects from each other. Distinctions between the various levels of generality seem much less clear in petrology than in most branches of natural science, however, and there is an almost inescapable tendency to use about the same set of names at all levels. Even if the nomenclature were entirely free of ambiguity at the specimen level, the practice of incorporating the names of specimens in the names of broader taxonomic groups would introduce an element of ambiguity into any discussion in which the level of grouping was not at all times clearly in view. In practice, the level of grouping is by no means always clear and rock nomenclature at the specimen level is far from unambiguous.

Indeed, throughout the history of petrology the need to reduce or eliminate inconsistency and ambiguity has underlain recurrent attempts to improve existing classification schemes or introduce new ones (see, e.g., Shand, 1927, Preface). Usually the new classifications either redefine existing group names or continue the practice of borrowing taxonomic terminology from specimen nomenclature, so it is unrealistic to expect them to be free of ambiguity or inconsistency. How then is one to decide whether, in these respects, a new classification scheme is or is not preferable to existing practice?

Clearly, a new scheme of classification will be preferable to existing practice if on the whole, and barring other major defects, it materially reduces the incidence of inconsistencies and ambiguities in the naming and grouping of the objects with which it deals. Probably the *caveat* should be added that, with regard to any subset of the objects to which it is to be applied with more than trivial frequency, it must not be markedly less consistent than existing practice; consistency in the denotation of one set of names ought not to be purchased at the cost of introducing inconsistency in the denotation of another set for which there is more than occasional use.

PROCEDURE

This seems a reasonable enough prospectus, but how is one to evaluate the consistency of existing practice in such fashion as to permit systematic

---

[1] See for example, Chayes (1957, 1966, 1969), Irvine and Baragar (1971), Jung and Brousse (1959), Rittmann (1952, 1973), and Streckeisen (1967, 1973), to cite only some of the longer published reports. Since 1968 the subject has been under continuous review by the International Union of Geological Sciences Subcommission on Systematics in Petrology and its affiliated working groups.

[2] Throughout this chapter the term "consistence" is used in the general sense of accordance, compatibility, or agreement with an alternative classification, not in the statistical sense of convergence to an ideal or objectively true value with increase in sample size.

comparison with that attainable in some new classification? Development of a widely acceptable general measure of the *actual* consistency of existing practice would be a formidable task, and might well prove impossible. For a very important class of rocks, however, it is not difficult to specify, by discriminant functions based on chemical composition, the level of consistency that *might* be attained if it were possible to codify existing usage with a minimum of ambiguity.

It could be argued that the coefficients of the discriminant functions themselves codify the usage, but this codification is probably not one that will appeal to many petrologists. For the purpose of the present discussion, however, it does not matter whether such a codification either has been achieved by more conventional means or is simply impossible. The only requirements are that (a) chemical composition is of major importance in the naming of the objects in question and (b) paired names and chemical analyses are available for a large collection of the objects.

The commoner Cenozoic volcanic rocks satisfy both requirements. Bulk chemical composition is generally considered of critical importance in the naming and classification of these rocks, and many petrographic descriptions of them, accompanied by chemical analyses, occur in the journal, reference, and text literature of petrology. A discriminant function calculated from chemical analyses of sets of specimens denoted by any two names can be used as a descriptive classifying device either on the data from which it is computed or on a new set. The results of such a classification operation may be cast in a 2×2 table in which the elements of the leading diagonal are the frequencies of rocks to each of which the discriminant function assigns the name applied in the source description, and the off-diagonal elements are the frequencies of rocks to which the discriminant function assigns one name and the source reference the other. The arrangement is shown schematically in Table 18-1. Entries in such a table provide a basis for evaluating three kinds of agreement

Table 18-1. Symbolic representation of reclassification of rocks named A or B by a discriminant function based on chemical composition*

Name Used in Source Reference	Name Assigned by Discriminant Function		
	A	B	
A	N_{AA}	N_{AB}	$N_{A.}$
B	N_{BA}	N_{BB}	$N_{B.}$
	$N_{.A}$	$N_{.B}$	N

* First character of subscript of a symbolic frequency denotes name used in source description; second denotes name assigned by discriminant function.

between source-name and discriminant function partition, each of which is certainly involved in the diffuse, largely unspecified, and perhaps not completely specifiable, concept of consistent petrographic usage.

The statistic of most obvious interest in this connection describes what might perhaps be termed self-consistency. In the notation of Table 18-1, $N_{A.}$ is the number of analyses of specimens in the sample identified as A's,[3] N_{AA} is the number of such analyses also classified as A's by the discriminant function calculated from all, or some subset, of the available chemical analyses of A's and B's and $Q_s = N_{AA}/N_{A.}$ is the estimated proportion of A's consistently classified by the function. The second intuitively appealing consistency statistic to be drawn from Table 18-1 is the ratio of N_{AA} to its column rather than row sum, $Q_d = N_{AA}/N_{.A}$, or, in words, the proportion of specimens classified as A's by the discriminant function that were also identified as A's in the source references. Finally, the sum of the leading diagonal of Table 18-1 divided by the total number of analyses, $Q_{AB} = (N_{AA} + N_{BB})/N$, is the estimated proportion of specimens of A and B to *each* of which the discriminant assigns the name attached to it in the source reference from which the analysis was drawn.

If the variables used in a discriminant function are actually indifferent to the discrimination, "classification" by means of the function reduces to the tossing of an unbiased coin, and in the long run one half of the A's (or B's) in the sample will be placed in discriminant class A (or B); the "indifference" or "null" values against which observed values of Q_s and Q_{AB} are to be compared are thus

$$\hat{Q}_s = \hat{Q}_{AB} = 1/2.$$

But if there is a 50% probability that any item in the sample will be classified as an A by the discriminant function, the indifference value for Q_d will be a function of N_A and N_B; specifically, the probability that a specimen classified as an A by an indifferent discriminant function was in fact identified as an A in the source reference is

$$\hat{Q} = N_{A.}/N,$$

the proportion of A's in the sample.

Visual (or other) comparison of Q_d values for a particular A against all other varieties will be much simpler if $N_{A.} = N_{B.}$ throughout. In this

---

[3] Source reference names are used throughout except for alkaline and subalkaline basalt, for which see Chayes (1975, pp. 546–547). In most published descriptions of individual specimens—as opposed to group characterizations—the name basalt is used without modification. The commonest modifiers are mineral names, such adjectives as alkali, alkaline, alkali-olivine, subalkaline, and tholeiitic being in fact quite uncommon.

case, of course, $\hat{Q}_d$ is also equal to $1/2$, so that only one of the sample statistics, presumably Q_{AB}, need be examined. By means of a random number generator yielding uniformly distributed numbers in the range $0 < i < 1$ it is easy to adjust sample sizes so that in any particular comparison $N_{A.} \simeq N_{B.} \simeq K$. If K_A and K_B are, respectively, the numbers of A's and B's available for comparison, this result will be achieved if $i_A \leq K/K_A$ and $i_B \leq K/K_B$ are used as acceptance criteria for A's and B's to be included in a particular comparison. Even in rather small samplings the difference between $N_{A.}$ and $N_{B.}$ is usually small enough so that for all practical purposes $\hat{Q}_d = 1/2$. (Of course, occasional samplings in which $N_{A.}$ and $N_{B.}$ differ materially can be discarded, but this tactic has not been used in the work described here.) The sampling scheme has the further advantage that the discriminant function may be tested on analyses many of which have not been used in its calculation. In the data used in this study $228 < K_A < 2336$ so that $K \leq 229$, and it is evident that the overlap between calculation- and test-sets must vary considerably among varieties.

Every Q_{AB} value in Table 18-2 is an average of three trials, each consisting of the calculation of a discriminant function from one randomly selected sample in which $N_{A.} \simeq N_{B.} \simeq 150$, and its test on another such sample. In all experiments "essential" oxides[4] were included in the discriminant. The coefficients of the calculated discriminant functions are exceedingly variable, but as noted in detail below, the consistency of the partitions they effect is remarkably uniform. In the data so far examined, this evidently holds whatever the power of a particular discriminant. In the form used here, discriminant function analysis thus provides an excellent means of establishing consistency levels against which to test rock classifications, but is not a practical method of classifying rocks.[5]

CONSISTENCY OF GENERAL USAGE OF COMMON NAMES OF CENOZOIC VOLCANIC ROCKS

The only way to obtain sound information about this troublesome subject is to examine a very large number of published petrographic descriptions. For the bibliographical problem raised by presentation of

---

[4] SiO_2, Al_2O_3, Fe_2O_3, FeO, MgO, CaO, Na_2O, K_2O, TiO_2.

[5] A further weakness of the binary discriminant as a taxonomic device, as LeMaitre (1976) points out, is that it will classify an object as a member of one of two groups even though it actually belongs to neither. The same difficulty is often encountered in classifications based upon nonstatistical redefinitions of existing rock names or types.

generalizations based on such a survey there is no entirely satisfactory solution. Clearly, the purveyor of the generalizations ought to be held accountable in the usual scholarly fashion, yet experience shows that most readers will have little or no interest in the necessary documentation, and the cost of publishing it—in the present case a bibliography of well over 600 titles—may well be forbidding if not actually prohibitive. All data used to reach results announced in this section have been drawn from the 1975 edition of the data base of the rock information system RKNFSYS/NTRM and reduced by routine operation of system programs; copies of the User's Manual of the system are available on request to readers who may be interested in checking or extending any of the results given here.

Although 196 nouns have been proposed as names for specimens of Cenozoic volcanic rocks analyses of which are included in the 1975 version of the data base, most are rarely used and many appear only in the papers in which they were originally announced. The data base contains more than 10,000 analyses of named volcanic rocks, so the average incidence of principal nominal nouns is about 50. Actually, only 23 are used that often, and ten occur more than 200 times each. The latter alone are attached to 8,275 specimens, and it is *solely with them that this discussion is concerned.*[6] For all rocks identified in the source references by each pair of these names, discriminant functions were computed and tested as described in the preceding section, with $K = 150$ throughout. The resulting Q_{AB} values, multiplied by 100 to convert them to percentages, are shown in Table 18-2. Each entry in the table is the mean of three independent replications. The range of replicates contributing to a single entry very rarely exceeds 3% and is in most instances considerably less than 1%, despite the fact that the coefficients of the three discriminant functions involved usually differ materially (Chayes, 1976). The argument of Table 18-2 is thus the average chemical consistency of current usage of the appropriate row and column captions, i.e., the percentage of specimens to which binary discriminant functions based on all essential oxides assign the same names used in the source references.

If this proportion is large, as in, for example, rhyolite vs. andesite or phonolite, there would appear to be little need for a new method for distinguishing the rocks in question from each other, and any such novel method would be open to severe criticism if it yielded consistency values

---

[6] The average incidence of each of the remaining 186 is less than 10. The consistency with which such rarely used names are applied cannot be effectively characterized by the method described here.

ıble 18-2. Consistency of discrimination (Q_{AB} × 100) for all pairs of commoner
Cenozoic volcanic rocks*

	Rhyolite	Dacite	Andesite	Subalkaline basalt	Alkaline basalt	Basanite	Trachybasalt	Trachyandesite	Phonolite	Trachyte
ıyolite (547)										
acite (494)	89									
ıdesite (2335)	97	83								
ıbalkaline basalt (1358)	99	98	85							
kaline basalt (1698)	>99	99	95	90						
ısanite (338)	>99	>99	98	98	71					
·achybasalt (229)	>99	>99	96	92	75	89				
·achyandesite (231)	89	85	84	92	97	96	84			
ıonolite (447)	99	99	98	98	98	97	96	94		
·achyte (596)	92	96	95	99	99	98	95	85	80	

* Each entry is the average of three independent determinations with $K = 150$. Number of
ıalyses of each type available in base shown in parentheses.

materially lower than those shown. On the other hand, a small value of
Q_{AB} indicates either that a chemically unrealistic discrimination is being
attempted[7] or that current usage is markedly erratic. In either case, it is
clearly such instances that merit close attention in attempts to improve
petrographic nomenclature and taxonomy.

They are brought into focus more clearly by an examination of
Table 18-3, in which all Q_{AB} values less than 95 are listed in ascending
order. It is perhaps as well to recall at this point that, like the score
on a true-false quiz, Q_{AB} measures performance from a base value of
50%, not zero. The last column in Table 18-3, computed as $2(Q_{AB} - 50)$,
is the percentage of improvement the discriminant function provides
over random selection. Assigning letter grades of the type once used in
the American school system to these percentages, the assemblage of
45 binary comparisons in Table 18-2 includes two F's, four D's, six C's,
five B's, and 28 A's, and all of grade less than A are shown in Table 18-3.

Now although it is possible to grade binary discriminants in much
the way students used to be graded, the working interpretation of the

---

[7] As, for instance, between alkaline basalt and basanite, or andesite and dacite, for both
of which various secondary modal criteria have been proposed but neither widely nor consistently applied.

Table 18-3. Binary discriminations of less than 95% consistency, listed in ascending order of I*

Rhyolite	Dacite	Andesite	Subalkaline basalt	Alkaline basalt	Basanite	Trachybasalt	Trachyandesite	Phonolite	Trachyte	$1000 Q_{AB}$	I
				X	X					71	42
				X						75	50
							X			80	60
	X	X								83	66
	X					X				84	68
					X	X				84	68
X						X				85	70
		X	X							85	70
						X			X	85	70
X	X									89	78
X						X				89	78
					X	X				89	78
			X	X						90	80
X									X	92	84
			X			X				92	84
			X				X			92	84
						X	X			94	88

* $I = 2(Q_{AB} - 50)$ is the percentage of improvement over random selection attributable to the discriminant function.

grades must be quite different. In any large class to which genuinely new material is being presented it is to be expected that A students will be comparatively rare; in an optimum classification, on the other hand, binary discriminants of grade *other* than A should be rare. Of the 45 binary discriminants shown in Table 18-2, no fewer than 17, or three-eighths, are of this type! A general awareness of this situation perhaps accounts for much of the current concern about nomenclature and taxonomy, and it is to be hoped the concern persists until the situation is remedied or at least materially improved.

In some respects, however, improvement is probably not to be anticipated except by enforcement of arbitrary dicta that will have the effect of eliminating *future* inconsistency without lowering the level of inconsistency encountered in data already published. Consider, for instance, the discrimination between andesite and subalkaline basalt, currently of grade C, in which about 3 of every 20 specimens will be misidentified by the discriminant function. If the function itself were now pressed into service as a classifying device there would be little or no future inconsistency, but inconsistency in existing data would survive.[8]

The more closely one examines Table 18-3 the more it seems that most reductions of inconsistency obtained by redefinition of individual names will be of this sort. The relative abundances of alkali basalt, basanite, and trachybasalt, for instance, suggest that the discriminants for alkali basalt vs. the latter two are of grade F because the rocks some petrologists consider worth distinguishing by special names are simply called basalt by other petrologists. A similar explanation almost certainly holds for the andesite-dacite discrimination, which is of grade D. In these, as in many other comparisons included in Table 18-3, agreement about arbitrary compositional boundaries might largely or entirely eliminate inconsistency in future usage, but would leave unchanged that contained in data already published. Despite the tone of some commentary (e.g., Rittmann, 1952, concerning andesite-dacite), in most instances neither choice of name is intrinsically preferable or demonstrably more correct or "useful." What is at issue is more a matter of taste and attitude than of propriety, and probably no schedule of calculation, however elegant or intricate, will wholly resolve it.

Sometimes geological considerations may be useful in this respect— for example, do specimens rich in nepheline occur as intrinsic parts of petrographic units that are elsewhere trachytic in composition, or is it just that the trachytes of some writers are the phonolites of others? In the first instance some taxonomic inconsistency is probably unavoidable; in the second perhaps it is not.

If the potential modifications so far discussed are not merely matters of nomenclature, they are barely past the borderline between nomenclature and taxonomy. Of more general interest are attempts to provide a broad, uniform descriptive basis for the classification of igneous rocks. Current

---

[8] The new literature would of course also be in conflict with that already published. If this meant merely invalidating or discarding old work already obsolete in other respects, it would perhaps be tolerable. Most published information about the chemical composition and petrology of igneous rocks, however, is very recent. Data to be reviewed elsewhere indicate, for instance, that of the present stock of complete analyses of Cenozoic volcanic rocks, less than a fifth were in print in 1950 and only a third by 1960. Drastic redefinition of common rock names and classifications would make virtually the entire *corpus* of the subject obsolete with regard to the most important single property of any rock, its identification. The step may ultimately prove unavoidable, but is not to be lightly undertaken.

usage has been characterized here without reference to such classifying principles, but of course it did not develop without them. The difficulty is that most of the common names of Cenozoic volcanic rocks have been in use since long before the classifying principles in vogue at various times in the last seventy years were proposed. They have been continued in use during the enormous and accelerating growth of the petrographic *corpus*, being applied from time to time in the not entirely compatible ways proposed by Zirkel, Rosenbusch, Osann, Iddings, Johannsen, Lacroix, Niggli, or Shand, to name only some of the more influential systematists. Nowadays they are usually used without close regard to how they are defined in any major petrographic system, and a leading advanced text (Turner and Verhoogen, 1960, p. 3) even asserts that such classifications "do not furnish a satisfactory medium for petrological discussion." But surely it is obvious that people who cannot agree about relations between the names and the properties of the objects they discuss cannot possibly know what they are talking about! The present situation is tolerable only for as long as such knowledge is unnecessary. It is to be hoped that attempts to develop a set of generally acceptable classifying principles for igneous petrology will continue.

The taxonomic parameters developed for that purpose will almost certainly be dominantly chemical, since no other kind will be generally applicable. In this connection it is important to realize that the set of elements used here yields discriminant partitions characterized by a rather considerable measure of consistency with the way in which the commoner names of volcanic rocks have been and are used. Specifically, although it is true that 17 of the comparisons in Table 18-2 are of less than A grade—i.e., yield more than 5% of inconsistent classification in the test procedure—no fewer than 11 of these involve either trachybasalt or trachyandesite, names applied in the source references to only 460, or 5.6%, of the specimens. It clearly does not make much sense to adopt broad, general redefinitions of all 10 common names because these two are inconsistently or ambiguously used.

It might nevertheless be considered desirable for other reasons to reclassify on the basis either of some other set of elements or of differently defined parameters—that is to say, different linear combinations—of the same set of elements used here. When a new classifying procedure uses any two of the names that appear in Table 18-2, the following statistics will be of immediate interest:

I. Of all rocks called A or B in the source references,
   (a) what proportion receive the same names when classified by the new procedure?

(b) what proportion fail to qualify as either A or B under the new procedure?

And, similarly,

II. Of all rocks classified as A or B by the new procedure,
    (a) what proportion were denoted by the same names in the source references?
    (b) what proportion were denoted by neither name in the source references?

Unless I.b and II.b are both small and I.a and II.a are as large as the analogous Q_{AB} entry in Table 18-2, the new scheme will not be, in this respect, an improvement over current practice. If a new classification employs different linear combinations of the same elements used here, the partitions it effects will nearly always be less consistent than those shown in Table 18-2. A small loss may sometimes be tolerable for the sake of other advantages, but it is difficult to imagine other properties so advantageous as to compensate for a major loss of consistency.

References

Chayes, F., A provisional reclassification of granite, *Geol. Mag.*, *94*, 58–68, 1957.
———, Alkaline and subalkaline basalts, *Am. J. Sci.*, *264*, 128–145, 1966.
———, The chemical composition of Cenozoic andesite, in "Proceedings of the Andesite Conference," *Oreg. Dep. Geol. Miner. Ind. Bull.*, *65*, 1–11, 1969.
———, On distinguishing alkaline from other basalts, *Carnegie Institution of Washington Yearbook*, *74*, 546–547, 1975.
———, Characterizing the consistency of current usage of rock names by means of discriminant functions, *Carnegie Institution of Washington Yearbook*, *75*, 782–784, 1976.
Irvine, T. N., and W. R. A. Baragar, A. Guide to the chemical classification of the common volcanic rocks, *Can. J. Earth Sci.*, *8*, 523–548, 1971.
Jung, J., and R. Brousse, *Classification Modale de Roches Eruptives*, Masson et Cie, Paris, 120 pp., 1959.
LeMaitre, R., A new approach to the classification of igneous rocks using the basalt–andesite–dacite–rhyolite suite as an example, *Contrib. Mineral. Petrol.*, *56*, 191–203, 1976.
Rittmann, A., Nomenclature of volcanic rocks, *Bull. Volcanol.*, *12*, 95–102, 1952.
———, *Stable Mineral Assemblages of Igneous Rocks*, Springer Verlag, Berlin, 262 pp., 1973.
Shand, S. J., *Eruptive Rocks*, T. Murby & Co., London, 360 pp., 1927.
Streckeisen, A. L., Classification and nomenclature of igneous rocks, *Neues Jahrb. Mineral. Abh.*, *107*, 144–240, 1967.

————, Classification and nomenclature of plutonic rocks: recommendations, *Neues Jahrb. Mineral. Monatsh.*, 149–164, 1973.

Turner, F. J., and J. Verhoogen, *Igneous and Metamorphic Petrology*, 2d ed., McGraw-Hill Book Company, New York, 694 pp., 1960.

AUTHOR INDEX

The indexes were compiled by the editor with the kind help of Dr. L. W. Finger and Miss Dolores M. Thomas, and prepared on a computer system by key-2-disk entries from page proofs. Several programs were used to alphabetize and reformat the individual paged entries. The printed version was obtained by photo reproduction of the output of a letter-quality printer, courtesy of RDA, Inc., Beltsville, Maryland.

406-409,420,521,522,524,
526,531
Chinner, G. A. 203
Christian, J. W. 238,243
Christiansen, R. L. 491,516
Chung, D. H. 490,515
Claassen, H. C. 345,350
Clague, D. A. 285,301
Clark, L. D. 62,73
Clark, S. P., Jr. 446,480,
491,492,495,498-500,503,
513,516
Clough, C. D. 216,230
Clough, C. T. 5,8,11,171,
172,200,245,300,316,335
Coats, R. R. 231
Cohen, L. H. 161,166
Combe, A. D. 417,418,421
Condie, K. C. 487,516
Coombs, D. S. 45,55
Copper, A. F. 419,421
Cox, J. D. 21,50,51
Cox, K. G. 487,516
Craig, H. 342,349
Cross, W. 418,421
Cundari, A. 417,418,421
Currie, K. L. 16,21,24,32,
37,39,41,42,46,51
Curtis, G. H. 322,335
Cuyubamba, A. 334,336

Daly, R. A. 4,307,316,320,
329-31,335,381,387,406,
408
Daniel, G. H. 20,41,55
Darken, L. S. 87,131,463,
479,480
Darwin, C. 246,301
Davidson, A. 375,387
Davies, R. D. 64,74
Davis, B. T. C. 175,200
Davis, N. F. 441,443,444,
447,452,458,476,478-80
Day, A. L. 424,435
De, A. 16,24,51
Deer, W. A. 10,13,34,56,64,
75,145,168,178,186,203,
247,284,286,290,292,293,
295,296,306,432,438,443,
480
Deines, P. 331,335
Delitsyn, L. M. 35,45,51
Delitsyna, L. V. 35,45,51
Dence, M. R. 24,51
Desborough, G. A. 433,435

DeVries, R. C. 138,166
Dickey, J. S., Jr. 200,201,
285,301,305,431,435
Dickinson, W. R. 60,74
Didier, J. 323,335
Dimroth, E. 33,45,51
Dixon, J. M. 293,301
Dixon, S. A. 250,304
Doherty, P. C. 225,232
Donaldson, C. H. 255,280,
281,285,295,300,301,429,
436
Donnay, G. 360,387
Dons, J. A. 43,50
Doremus, R. H. 22,51
Drake, C. L. 518
Drever, H. I. 24,38,51
Drory, A. 224,232
Duhem, P. 458,463
Duncan, A. R. 155,168,182,
183,202,203
Dungan, M. A. 255,280,281,
285,300
Dycus, D. W. 250,304

Eaton, G. P. 491,516
Eaton, J. P. 66,73
Eckstrand, O. R. 247,305,
428,437
Edgar, A. D. 376,389,402,
409,410,417,421
Edwards, A. B. 322,323,335
Eggler, D. H. 154,160,161,
166,176,177,182,196,197,
201,334,336,378-80,387,
406-408,419,421,441,454-
57,467,470,472-74,480-81,
501,504,505,511,516
Einstein, A. 238
El Goresy, A. 396,408
Ellestad, R. B. 374,387
Ellis, D. E. 505
Elwell, R. W. D. 45,50
Emslie, R. F. 193,247,248,
252,294,301,305
Engel, A. E. J. 5,12,247,
301
Engel, C. G. 5,12,247,301
Engell, J. 385,387
England, J. L. 174,175,200
Erlank, A. J. 64,74
Ernst, W. G. 416,421
Evans, B. W. 321,336
Ewart, E. 155,168,182,183,
202,203

SUBJECT INDEX

SUBJECT INDEX 575

Phanerite 160
Phase boundary
  change with pressure 158,
  160
Phase incompatibility
  plagioclase vs. olivine
  161,162
  pressure dependence 161
Phase rule 40,114
Phase subtraction
  relation to liquidus 160
Phase transition 484
  kinetics of 497
Phase volume 120
  change with alkalies 248,
  250
  change with iron
    enrichment 248,250
  change with pressure 248
  change with water 248,250
Phase-diagram interpreta-
  tion
  new developments 129
Phenocryst 160,163,188,205-
  207,214,221,222,224-26,
  245,246,269,280,284,285
  332,333,340,369,370,376
  383,384,413,414,424,425
  alteration of 393
  correlated with magma
    composition 265
  formation:pressure of 280
  importance of core
    composition 365
  of hornblende 474
  of iron oxide 186
  of melilite 404
  relationship to globule
    or spherule 38
  relict 11
  zoning in 367
Phlogopite
  breakdown curves 500
  formation reaction 191
  in peridotite 500
  melting of 500,501
  subsolidus stability
    field 506
Phlogopite-peridotite
  partially carbonated 505
Phonolite 32,364,368-70,382
  383,529
  basalt association 383
  from nephelinitic magma
  373

nephelinite association
  383,384
  origin of 384
  plateau-type 383,384
  volume of 384
Physicochemical processes 3
  5
Physics
  of earth 483,484
Picotite
  origin of 430
Picrite 196,271,286-88
Picritic lava 280
Picritic liquid 258
Picritic melt 299,497
Picritic tholeiite 270,279
Piercing point 379,400
Pigeonitic rock series 178,
  179,186,192,332
Pitchstone 340
  definition of 346
  potash content 339
  source of water in 241
  water content 241,346
Plagioclase
  changes in structural
    state of 466
  incongruent melting of
    467
  melting relations of 464
Plagioclase cumulate 468
Plagioclase effect 369,370
  definition of 368
Plagioclase flotation 293,
  294,297,299
  Skaergaard paradox 292
Plagioclase liquidus
  effect of water on 160
  water-undersaturated 467
Plagioclase peridotite 196,
  497
  subsolidus phase
    boundaries 504
Plagioclase solidus
  calculation of 466
Plagioclase tholeiite
  association 284
  crystallization of 285
  iron enrichment 284,285
  origin of 284
  role of high pressure 285
  role of water 285
  two-stage process of
    formation 285,300
Plate boundary 486

SYSTEMS INDEX

Page references are to synthetic mineral end-member systems, alphabetically arranged, to synthetic oxide systems, and to natural rock-type and mineral-type systems discussed in the text, figures, or both.

SYNTHETIC MINERAL END-MEMBER SYSTEMS

SYNTHETIC OXIDE SYSTEMS

ROCK-TYPE AND MINERAL-TYPE SYSTEMS

LIBRARY OF CONGRESS CATALOGING IN PUBLICATION DATA

Main entry under title:

The Evolution of the igneous rocks.

Includes indexes.
1. Rocks, Igneous.   2. Bowen, Norman Levi,'
1887–1956.   The evolution of the igneous rocks.
I. . Yoder, Jr., Hatten Schuyler, 1921–QE461.E87   552'.1   79-84023
ISBN 0-691-08223-5
ISBN 0-691-08224-3 pbk.

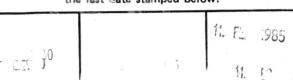

11. FE 1985

11. [?

THE EVOLUTION
OF THE IGNEOUS ROCKS

Fiftieth Anniversary Perspectives

Norman L. Bowen
21 June 1887–11 September 1956